Neural Networks in a Softcomputing Framework

K.-L. Du and M.N.S. Swamy

Neural Networks in a Softcomputing Framework

With 116 Figures

 Springer

K.-L. Du, PhD
M.N.S. Swamy, PhD, D.Sc (Eng)

Centre for Signal Processing and Communications
Department of Electrical and Computer Engineering
Concordia University
Montreal, Quebec
H3G 1M8
Canada

British Library Cataloguing in Publication Data
Du, K.-L.
    Neural networks in a softcomputing framework
    1.Neural networks (Computer science)
    I.Title II.Swamy, M. N. S.
    006.3'2

e-ISBN 1-84628-303-5
ISBN-13: 978-1-84996-574-3   e-ISBN-13: 978-1-84628-303-1

Printed in Germany

9 8 7 6 5 4 3 2 1

Springer Science+Business Media
springer.com

献给
我们的父母
和老师

मातृभ्यो नमः
पितृभ्यो नमः
गुरुभ्यो नमः

TO
OUR PARENTS
AND
TEACHERS

Preface

Softcomputing, a concept introduced by L.A. Zadeh in the early 1990s, is an evolving collection of methodologies for the representation of the ambiguity in human thinking. The core methodologies of softcomputing are fuzzy logic, neural networks, and evolutionary computation. Softcomputing targets at exploiting the tolerance for imprecision and uncertainty, approximate reasoning, and partial truth in order to achieve tractability, robustness, and low-cost solutions.

Research on neural networks dates back to the 1940s; the discipline of neural networks is well developed with wide applications in almost all areas of science and engineering. The powerful penetration of neural networks is due to their strong learning and generalization capability. After a neural network learns the unknown relation from given examples, it can then predict, by generalization, outputs for new samples that are not included in the learning sample set. The neural-network method is model free. A neural network is a black box that directly learns the internal relations of an unknown system. This takes us away from guessing functions for describing cause-and-effect relationships. In addition to function approximation, other capabilities of neural networks such as nonlinear mapping, parallel and distributed processing, associative memory, vector quantization, optimization, and fault tolerance also contribute to the widespread applications of neural networks.

The theory of fuzzy logic and fuzzy sets was introduced by L.A. Zadeh in 1965. Fuzzy logic provides a means for treating uncertainty and computing with words. This is especially useful to mimic human recognition, which skillfully copes with uncertainty. Fuzzy systems are conventionally created from explicit knowledge expressed in the form of fuzzy rules, which are designed based on experts' experience. A fuzzy system can explain its action by fuzzy rules. Fuzzy systems can also be used for function approximation. The synergy of fuzzy logic and neural networks generates neurofuzzy systems, which inherit the learning capability of neural networks and the knowledge-representation capability of fuzzy systems.

Evolutionary computation is a computational method for obtaining the best possible solutions in a huge solution space based on Darwin's survival-of-the-fittest principle. Evolutionary algorithms are a class of robust adaptation and global optimization techniques for many hard problems. Among evolutionary algorithms, the genetic algorithm is the best known and most studied, while evolutionary strategy is more efficient for numerical optimization. More and more biologically or nature-inspired algorithms are emerging. Evolutionary computation has been applied for the optimization of the structure or parameters of neural networks, fuzzy systems, and neurofuzzy systems. The hybridization between neural network, fuzzy logic, and evolutionary computation provides a powerful means for solving engineering problems.

At the invitation of Springer, we initially intended to write a monograph on neural-network applications in array signal processing. Since neural-network methods are general-purpose methods for data analysis, signal processing, and pattern recognition, we, however, decided to write an advanced textbook on neural networks for graduate students. More specifically, neural networks can be used in system identification, control, communications, data compression and reconstruction, audio and speech processing, image processing, clustering analysis, feature extraction, classification, and pattern recognition, etc. Conventional model-based data-processing methods require experts' knowledge for the modeling of a system. In addition, they are computationally expensive. Neural-network methods provide a model-free, adaptive, parallel-processing solution.

In this book, we will elaborate on the most popular neural-network models and their associated techniques. These include multilayer perceptrons, radial basis function networks, Hopfield networks, Boltzmann machines and stochastic neural-network models, many models and algorithms for clustering analysis and principal component analysis. The applications of these models constitute the majority of all neural-network applications. Self-contained fundamentals of fuzzy logic and evolutionary algorithms are introduced, and their synergies in the other two paradigms of softcomputing described.

We include in this book a thorough review of various models. Major research results published in the past decades have been introduced. Problems of array signal processing are given as examples to illustrate the applications of each neural-network model.

This book is divided into ten chapters and an appendix. Chapter 1 gives an introduction to neural networks. Chapter 2 describes some fundamentals of neural networks and softcomputing. A detailed description of the network architecture and the theory of operation for each softcomputing method is given in Chapters 3 through 9. Chapter 10 lists some other interesting or emerging neural-network and softcomputing methods and also mentions some topics that have received recent attention. Some mathematical preliminaries are given in the appendix. The contents of the various chapters are as follows.

- In Chapter 1, a general introduction to neural networks is given. This involves the history of neural-network research, the McCulloch–Pitts neuron, network topologies, learning methods, as well as properties and applications of neural networks.

- Chapter 2 introduces some topics of neural networks and softcomputing such as the statistical learning theory, learning and generalization, model selection, robust learning as well as feature selection and feature extraction.

- Chapter 3 is dedicated to multilayer perceptrons. Perceptron learning is first introduced. This is followed by the backpropagation learning algorithm and its numerous improvement measures. Many other learning algorithms including second-order algorithms are described.

- Hopfield networks and Boltzmann machines are described in Chapter 4. Some aspects of associative memory and combinatorial optimization are developed. Simulated annealing is introduced as a global optimization method. Some unsupervised learning algorithms for Hopfield networks and Boltzmann machines are also discussed.

- Chapter 5 treats competitive learning and clustering networks. Dozens of clustering algorithms, such as Kohonen's self-organizing map, learning vector quantization, adaptive resonance theory (ART), C-means, neural gas, and fuzzy C-means, are introduced.

- Chapter 6 systematically deals with radial basis function networks, which are fast alternatives to the multilayer perceptron. Some recent learning algorithms are also introduced. A comparison with the multilayer perceptron is made.

- Numerous neural networks and algorithms for principal component analysis, minor component analysis, independent component analysis, and singular value decomposition are described in Chapter 7.

- Fuzzy logic and neurofuzzy systems are described in Chapter 8. The relation between neural networks and fuzzy logic is addressed. Some popular neurofuzzy models including the ANFIS are detailed in this chapter.

- In Chapter 9, we elaborate on evolutionary algorithms with emphasis on genetic algorithms and evolutionary strategies. Applications of evolutionary algorithms to the optimization of the structure and parameters of a neural network or a fuzzy logic are also described.

- A brief summary of the book is given in Chapter 10. Some other useful or emerging neural-network models and softcomputing paradigms are briefly discussed. In Chapter 10, we also propose some foresights in this discipline.

This book is intended for scientists and practitioners who are working in engineering and computer science. The softcomputing paradigms are of general purpose in nature, thus this book is also useful to people who are interested in applications of neural networks, fuzzy logic, or evolutionary computation to their specific fields. This book can be used as a textbook for graduate students. Researchers interested in a particular topic will benefit from the appropriate chapter of the book, since each chapter provides a systematic introduction and survey on the respective topic. The book contains 1272 references. The state-of-the-art survey leads the readers to the most recent results, and this saves the readers enormous amounts of time in document retrieval.

In this book, all acronyms and symbols are explained at their first appearance. Readers may encounter some abbreviations or symbols not explained in a particular section, and in this case they can refer to the lists of abbreviations and symbols at the beginning of the book.

We would like to thank the editors of Springer for their support. We also would like to thank our respective families for their patience and understanding during the course of writing this book.

K.-L. Du
M.N.S. Swamy

Concordia University, Montreal, Canada
March, 2006

Contents

List of Abbreviations

ACO	ant-colony optimization	BAM	bidirectional associative memory
ACS	ant-colony system		
adaline	adaptive linear element	BCL	branching competitive learning
A/D	analog-to-digital		
AFC	adaptive fuzzy clustering	BER	bit error rate
AHN	adaptive Hamming net	BFGS	Broyden–Fletcher–Goldfarb–Shanno
AIC	Akaike information criterion		
ALA	adaptive learning algorithm	BIC	Bayesian information criterion
ANFIS	adaptive-network-based FIS		
AOSVR	accurate online SVR	BIRCH	balanced iterative reducing and clustering using hierarchies
APCA	asymmetric PCA		
APEX	adaptive principal components extraction		
		BP	backpropagation
AR	autoregressive	BPM	BP with momentum
ARBP	annealing robust BP	BSB	brain-states-in-a-box
ARC	adaptive resolution classifier	BSS	blind source separation
ARLA	annealing robust learning algorithm		
		CAM	content-addressable memory
ARRBFN	annealing robust RBFN	CDF	cumulative distribution function
ART	adaptive resonance theory		
ASIC	application-specific integrated circuit	CEM	classification EM
		CFA	clustering for function approximation
ASP	array signal processing		
ASSOM	adaptive-subspace SOM	CFHN	compensated fuzzy Hopfield network
		CG	conjugate gradient

CICA	constrained ICA	EKF	extended Kalman filtering
CMA	covariance matrix adaptation	ELSA	evolutionary local selection algorithm
CMA-ES	covariance matrix adaptation ES	EM	expectation-maximization
		EP	evolutionary programming
CMAC	cerebellar model articulation controller	ER	edge recombination
		ERM	empirical risk minimization
CNN	cellular neural network	ERR	error-reduction ratio
COP	combinatorial optimization problem	ES	evolutionary strategy
		ESFC	enhanced sequential fuzzy clustering
CPCA	constrainted PCA		
CSA	chaotic SA	E-step	expectation step
CURE	clustering using representation	ETF	elementary transcendental function
CX	cycle crossover	EVD	eigenvalue decomposition
DAC	digital-to-analog converter	FAGA	fuzzy adaptive GA
dART	distributed ART	FALVQ	fuzzy algorithms for LVQ
dARTMAP	distributed ARTMAP	FAM	fuzzy associative memory
DBSCAN	density-based spatial clustering of applications with noise	FBFN	fuzzy basis function network
		FBP	fuzzy BP
DCS	dynamic cell structures	FCL	fuzzy competitive learning
DCT	discrete cosine transform	FDA	functional data analysis
DEKF	decoupled EKF algorithm	FFA	fuzzy finite-state automaton
DFA	deterministic finite-state automaton	FFSCL	fuzzy FSCL
		FFT	fast Fourier transform
DFP	Davidon–Fletcher–Powell	FHN	fuzzy Hopfield network
DFT	discrete Fourier transform	FIR	finite impulse response
DFNN	dynamic fuzzy neural network	FIS	fuzzy inference system
		FKCN	fuzzy Kohonen clustering network
DHT	discrete Hartley transform		
DI	dependence identification	flop	floating-point operation
DoA	direction-of-arrival	FLVQ	fuzzy LVQ
DPD	dot-product-decorrelation	FMEKF	fading-memory EKF
DPE	dynamic parameter encoding	FNN	feedforward neural network
DWT	discrete wavelet transform	FOSART	fully self-organizing SART
		FPE	final prediction error
EA	evolutionary algorithm	FSCL	frequency-sensitive competitive learning
EART	efficient ART		
EBD	early brain damage	FuGeNeSys	fuzzy genetic neural system
ECAM	exponential correlation associative memory model		
		GA	genetic algorithm
ECFC	entropy-constrained fuzzy clustering	GAP-RBF	growing and pruning algorithm for RBF
		GART	Gaussian ART
ECLVQ	entropy-constrained LVQ	GAVaPS	GA with varying population size
EEBP	equalized error BP		
EHF	extended H_∞ filtering		

		LCMV	linear constrained minimum variance
GCS	growing cell structures		
GEFREX	genetic fuzzy rule extractor	LDA	linear discriminant analysis
GESA	guided evolutionary SA	LII	local identical index
GEVD	generalized EVD	LLCS	life-long learning cell structures
GFCM	generalized FCM		
GFP	generic fuzzy perceptron	LLLS	local linearized LS
GGAP-RBF	generalized GAP-RBF	LM	Levenberg–Marquardt
GHA	generalized Hebbian algorithm	LMAM	LM with adaptive momentum
GII	global identical index	LMS	least mean squares
GLVQ-F	generalized LVQ family algorithms	LMSE	least mean squared error
		LMSER	least mean square error reconstruction
GNG	growing neural gas		
GNG-U	GNG with utility criterion	LP	linear programming
GOTA	globally optimal training algorithm	LS	least-squares
		LSE	least-squares error
GP	genetic programming	LSSM	linear systems in a saturated mode
G-Prop	genetic backpropagation		
GSLN	generalized single-layer network	LTCL	lotto-type competitive learning
		LTG	linear threshold gate
GSO	Gram–Schmidt orthonormal	LTM	long-term memory
		LUT	look-up table
HFPNN	hybrid fuzzy polynomial neural network	LVQ	learning vector quantization
HWO	hidden weight optimization		
HSOM	hyperbolic SOM	MAD	median of the absolute deviation
HUFC	hierarchical unsupervised fuzzy clustering	MBCL	multiplicatively biased competitive learning
HUX	half-uniform crossover		
HyFIS	Hybrid neural FIS	MC	minor component
		MCA	minor component analysis
IART 1	improved ART 1	MCETL	multicore expand-and-truncate learning
ICA	independent component analysis		
		MCL	multicore learning
iff	if and only if	MCV	minimum cluster volume
iid	independently drawn and identically distributed	MDL	minimum description length
		MEKA	multiple extended Kalman algorithm
i-or	interactive-or		
IRprop	improved Rprop	MF	membership function
		MFCC	Mel frequency cepstral coefficient
KKT	Karush–Kuhn–Tucker		
KLT	Karhunen–Loeve transform	MFT	mean-field theory
k-NN	k-nearest neighbor	MIMO	multi-input multi-output
k-WTA	k-winners-take-all	ML	maximum-likelihood
		MLP	multilayer perceptron
LBG	Linde–Buzo–Gray	MMAS	min-max ant system
LBG-U	LBG with utility	MRA	multiresolution analysis

MSA	minor subspace analysis
MSE	mean squared error
MSOM	merge SOM
MST	minimum spanning tree
M-step	maximization step
MSV	minimum scatter volume
NARX	nonlinear autoregressive with exogenous input
NEFCLASS	neurofuzzy classification
NEFCM	non-Euclidean FCM
NEFCON	neurofuzzy controller
NEFLVQ	non-Euclidean FLVQ
NEFPROX	neuronfuzzy function approximation
NERFCM	non-Euclidean relational FCM
NFL	no free lunch
NG	neural gas
NG-ES	NG-type ES
NIC	novel information criterion
NLCPCA	nonlinear complex PCA
NLDA	nonlinear discriminant analysis
NOja	normalized Oja
NOOja	normalized orthogonal Oja
NOVEL	nonlinear optimization via external lead
NPGA	niched Pareto GA
NSGA	nondominated sorting GA
OBD	optimal brain damage
OBS	optimal brain surgeon
ODE	ordinary differential equation
OLS	orthogonal least squares
OmeGA	ordering messy GA
OOja	orthogonal Oja
OPTOC	one-prototype-take-one-cluster
OWO	output weight optimization
OX	order crossover
PAC	probably approximately correct
PAES	Pareto archived ES
PARC	pruning ARC
PART	projective ART

PAST	projection approximation subspace tracking
PASTd	PAST with deflation
PC	principal component
PCA	principal component analysis
PCB	printed circuit board
PCG	projected conjugate gradient
PCM	possibilistic C-means
PDF	probability density function
PESA	Pareto envelope-based selection algorithm
PGA	parallel GA
PMX	partial matched crossover
PNN	probabilistic neural network
PSA	principal subspace analysis
PSO	particle swarm optimization
PTG	polynomial threshold gate
PWM	pulse width modulation
QP	quadratic programming
QR-cp	QR with column pivoting
RAN	resource-allocating network
RBF	radial basis function
RBFN	radial basis function network
RCA	robust competitive agglomeration
RCAM	recurrent correlation associative memory
RCE	restricted Coulomb energy
RecSOM	recursive SOM
RLS	recursive least squares
RNN	recurrent neural network
ROLS	recursive OLS
rPCA	robust PCA
RPCL	rival penalized competitive learning
RProp	resilient propagation
RRLSA	robust RLS algorithm
RSOM	recurrent SOM
RTRL	real-time recurrent learning
SA	simulated annealing
SAM	standard additive model
SART	simplified ART
SCL	simple competitive learning
SCS	soft competition scheme
SER	average storage error rate

S-Fuzzy ART	symmetric fuzzy ART	TDRL	time-dependent recurrent learning
SISO	single-input single-output		
SLA	subspace learning algorithm	TLMS	total least mean squares
		TLS	total least squares
SLE	set of linear equations	TKM	temporal Kohonen map
SLP	single-layer perceptron	TNGS	theory of neuronal group selection
SMO	sequential minimal optimization		
		TREAT	trust-region-based error aggregated training
SNN	stochastic neural network		
SNR	signal-to-noise ratio	TRUST	terminal repeller unconstrained subenergy tunneling
SOFNN	self-organizing fuzzy neural network		
		TSK	Takagi–Sugeno–Kang
SOM	self-organization maps	TSP	traveling salesman problem
SOM-ES	SOM-type ES		
SOMSD	SOM for structured data	UD-FMEKF	UD factorization-based FMEKF
SPEA	strength Pareto EA		
SRM	structural risk minimization	UNBLOX	uniform block crossover
SSCL	self-splitting competitive learning		
		VC	Vapnik–Chervonenkis
SSE	sum-of-squares error	VEGA	vector-evaluated GA
STM	short-term memory	VHDL	very high level hardware description language
SURE	Stein's unbiased risk estimator		
		VLSI	very large scale integrated
SVC	support vector clustering	VQ	vector quantization
SVD	singular value decomposition		
		WAV	weighted averaging
SVM	support vector machine	WINC	weighted information criterion
SVR	support vector regression		
		WNN	wavelet neural network
TABP	terminal attractor-based BP	WP	wavelet packets
		WTA	winner-take-all
TDNN	time-delay neural network	WWW	world wide web

List of Symbols

$\lvert \cdot \rvert$	the cardinality of the set or region within; Also the absolute value of the scalar within
$\lVert \cdot \rVert$	the Euclidean norm
$\lVert \cdot \rVert_{\mathbf{A}}$	the weighted Euclidean norm
$\lVert \cdot \rVert_F$	the Frobenius norm
$\lVert \cdot \rVert_p$	the p-norm or L_p-norm
$\lVert \cdot \rVert_\varepsilon$	the ε-insensitive loss function
$\hat{[\cdot]}, \widehat{[\cdot]}$	the estimate of the parameter within
$\overline{[\cdot]}, \neg[\cdot]$	the complement of the set or fuzzy set within
$\overline{[\cdot]}$	the normalized form or unit direction of the vector within
$[\cdot]^\dagger$	the pseudoinverse of the matrix within
$[\cdot]^*$	the conjugate of the matrix within; Also the fixed point or optimum point of the variable within
$[\cdot]^{\mathrm{H}}$	the Hermitian transpose of the matrix within
$[\cdot]^{\mathrm{T}}$	the matrix transpose of the matrix within
$[\cdot]_{\varphi_0}$	the operator that finds in the interval $(-\pi, \pi]$ the quantization of the variable, which can be a discrete argument $m\varphi_0$
$[\cdot]_{\max}$	the maximal value of the quantity within
$[\cdot]_{\min}$	the minimal value of the quantity within
$[\cdot] \circ [\cdot]$	the max-min composition of the two fuzzy sets within
$[\cdot] \diamond [\cdot]$	the min-max composition of the two fuzzy sets within
$\nabla_{\mathbf{x}}$	$\frac{\partial}{\partial \mathbf{x}}$
$*$	the don't care symbol in the GA
$\oplus$	addition mod 2 or exclusive-or
$\otimes$	the i-or operator
$\wedge$	the logic AND operator; Also the intersection operator; Also the t-norm operator; Also the minimum operator
$\vee$	the union operator
$\subseteq$	the inclusion operator
$\emptyset$	the empty set
$\mathbf{1}$	a vector or matrix with all its entries being unity

α	the momentum factor in the BP algorithm; Also an annealing schedule parameter in the SA; Also a design parameter in the mountain and subtractive clustering; Also the parameter defining the size of neighborhood; Also a scaling factor; Also a pheromone decay parameter in the ant system; Also the inertia weight in the PSO
$\boldsymbol{\alpha}$	the diagonal damping coefficient matrix with the (i,i)th entry α_i; Also an eigenvector of the kernel matrix K
α_0	a positive constant
α_i	the damping coefficient of the ith neuron in the Hopfield model; Also the ith entry of the eigenvector of the kernel matrix K; Also the quantization of the phase of net_i in the complex Hopfield-like network; Also the Lagrange multiplier for the ith example in the SVM
$\boldsymbol{\alpha}_i$	the ith eigenvector of the kernel matrix K
$\alpha_{i,j}$	the jth entry of $\boldsymbol{\alpha}_i$
α_{ik}	a coefficient used in the GSO procedure
$\alpha_{ij}^{(m)}$	the momentum factor corresponding to $w_{ij}^{(m)}$
$\alpha_{\max}$	the upper bound for $\alpha_{ij}^{(m)}$ in the Quickprop
β	the gain of the sigmoidal function; Also a positive parameter that determines the relative importance of pheromone versus distance in the ant system; Also a deterministic annealing scale estimator; Also a scaling factor in the chaotic neural network; Also a design parameter in the mountain and subtractive clustering; Also the scale estimator, known as the cutoff parameter, in a loss function
$\boldsymbol{\beta}$	the variable vector containing the Lagrange multipliers for all the examples in the SVR
$\beta(t)$	a step size to decide $\mathbf{d}(t+1)$ in the CG method
β_0	positive constants
β_1, β_2	time-varying error cutoff points of Hampel's tanh estimator
β_i	the phase of x_i in the complex Hopfield-like network; Also a shape parameter associated with the ith dimension of the RBF; Also the weighting factor of the constraint term for the ith cluster; Also an annealing scaling factor for the learning rate of the ith neuron in the ALA algorithm; Also the ith entry of $\boldsymbol{\beta}$
$\beta_i^{(m)}$	the gain of the sigmoidal function of the ith neuron at the mth layer; Also a scaling factor for the weight vector to the ith neuron at the mth layer, $\mathbf{w}_i^{(m)}$
δ	a small positive constant; Also a threshold for detecting noise and outliers in the noise clustering method; Also a global step size in the CMA-ES
$\delta(H)$	the defining length of a schema H
δ_i	the approximation accuracy at the ith phase of the successive approximative BP
$\delta_i(t)$	an exponentially weighted estimate of the ith eigenvalue in the PASTd

δ_{ij}	the Kronecker delta
$\delta_{\min}$	the set distance
$\delta_{p,v}^{(m)}$	the delta function of the vth neuron in the mth layer for the pth pattern
δ_t	the radius of the trust region at the tth step
$\delta\sigma_i$	a parameter for mutating σ_i in the ES
$\delta\mathbf{y}_p(t)$	the training error vector for the pth example at the tth phase
$\Delta[\cdot]$	the change in the variable within
$\Delta(t, y)$	a function with domain $[0, y]$ whose probability of being close to 0 increases as t increases
$\Delta_{ij}^{(m)}(t)$	a parameter associated with $w_{ij}^{(m)}$ in the RProp
$\Delta_{\max}$	the upper bound on $\Delta_{ij}^{(m)}(t)$
$\Delta_{\min}$	the lower bound on $\Delta_{ij}^{(m)}(t)$
$\Delta_p[\cdot]$	the change in the variable within due to the pth example
$(\Delta E)_i$	the saliency of the ith weight
ϵ	the measurement noise; Also a perturbation related to $\mathbf{W}$; Also an error vector, whose ith entry corresponds to the L_2-norm of the approximation error of the ith example
ϵ_i	the ith entry of ϵ
$\boldsymbol{\epsilon}_i$	the error vector as a nonlinear extension to $\mathbf{e}_i$; Also the encoded complex memory state of $\mathbf{x}_i$; Also a perturbation vector for splitting the ith RBF prototype
$\boldsymbol{\epsilon}_i(t)$	an instantaneous representation error vector for nonlinear PCA
$\epsilon_{i,j}$	the jth entry of $\boldsymbol{\epsilon}_i$
ε	the decaying coefficient at each weight change in the weight-decaying technique; Also a positive constant in the delta-bar-delta; Also a threshold parameter in the mountain and subtractive clustering
$\bar{\varepsilon}, \underline{\varepsilon}$	the two thresholds used in the mountain and substractive clustering
$\varepsilon(t)$	a threshold in the RAN
$\varepsilon_0, \varepsilon_1$	predefined small positive numbers
$\varepsilon_{\max}$	the largest scale of the threshold $\varepsilon(t)$ in the RAN
$\varepsilon_{\min}$	the smallest scale of the threshold $\varepsilon(t)$ in the RAN
$\phi(\cdot)$	the activation function; Also the nonlinearity introduced in the nonlinear PCA and ICA
$\dot{\phi}(\cdot)$	the first-order derivative of $\phi(\cdot)$
$\phi^{-1}(\cdot)$	the inverse of $\phi(\cdot)$
$\boldsymbol{\phi}^{(m)}$	a vector comprising all the activation functions in the mth layer
$\phi_1(\cdot), \phi_2(\cdot), \phi_3(\cdot)$	nonlinear functions introduced in the ICA
ϕ_i	the azimuth angle of the ith point source in the space; Also the ith RBF in the RBFN;
$\boldsymbol{\phi}_i$	the ith column of $\boldsymbol{\Phi}$
$\phi_{i,j}$	the jth entry of $\boldsymbol{\phi}_i$
$\phi_i^{(m)}$	the activation function at the ith node of the mth layer, the ith entry of $\boldsymbol{\phi}^{(m)}$

$\overline{\overline{\phi}}_l$	the lth row of $\mathbf{\Phi}$
$\overline{\overline{\phi}}_i(\mathbf{x})$	the normalized form of the ith RBF node, over all the examples and all the nodes
$\overline{\overline{\phi}}_i$	the vector comprising all $\overline{\overline{\phi}}_i(\mathbf{x}_p)$, $p = 1, \cdots, N$
$\phi_\mu(net)$	an activation function defined according to $\phi(net)$
$\phi_I(\cdot)$	the imaginary part of a complex activation function
$\phi_R(\cdot)$	the real part of a complex activation function
$\langle \phi_i(\mathbf{x}), \phi_j(\mathbf{x}) \rangle$	the inner product of the two RBFs
Φ	a nonlinear mapping between the input and the output of the examples
$\mathbf{\Phi}$	the response matrix of the hidden layer of the RBFN
γ	a proportional factor; Also a constant in the LM method; Also a bias parameter in the chaotic neural network
γ_0	a non-negative constant
γ_i	a positive coefficient corresponding to the ith PC in the weighted SLA
γ_i^j	a value obtained by dividing the interval of the ith dimension of the input, x_i
$\gamma_G(f; \mathbf{c}, \sigma)$	the Gaussian spectrum of the known function
η	the learning rate or step size
η_0	the initial learning rate
η_0^-, η_0^+	two learning parameters in the RProp or the SuperSAB
η_{batch}	the learning rate for batch learning
η_f	the final learning rate
$\eta_{ij}^{(m)}$	the learning rate corresponding to $w_{ij}^{(m)}$
η_{inc}	the learning rate for incremental learning
η_k	the learning rate for the kth prototype
η_r	the learning rate for the rival prototype
η_w	the learning rate for the winning prototype
η_β	the learning rate for adapting the gain of the sigmoidal function
φ	a nonlinear mapping from R^{J_1} to R^{J_2}
$\varphi(\cdot)$	the influence function
$\varphi(x_i)$	the phase of the ith array element
φ_0	the Lth root of unity; Also the resolution of the phase quantizer
$\varphi_1(\cdot)$	a very robust estimate of the influence function in the τ-estimator
$\varphi_2(\cdot)$	a highly efficient estimate of the influence function in the τ-estimator
$\varphi_i(x_i)$	the ith factor of the RBF $\phi(\mathbf{x})$
$\widetilde{\varphi}_i(x_i)$	the compensating function for $\varphi_i(x_i)$
φ_S	the phase of $\mu_S(x)$
ϑ	a positive constant taking value between 0 and 1; Also a variable used in the OOja
$\kappa(\mathbf{x}_i, \mathbf{x}_j)$	the kernel function defined for kernel methods

$\kappa\left(y_i\right)$	the kurtosis of signal y_i
κ_0	a small positive number
λ	the number of offspring generated from the population in the ES
$\lambda(t)$	the exact step size to the local minimum of E along the direction of $\mathbf{d}(t)$
λ_c	the regularization parameter for E_c
λ_i	the wavelength of the radiation from the ith source; Also the ith eigenvalue of the Hessian $\mathbf{H}$; Also the ith eigenvalue of the auto-correlation matrix $\mathbf{C}$; Also the ith eigenvalue of the kernel matrix $\mathbf{K}$; Also the ith generalized eigenvalue in the GEVD problem
$\boldsymbol{\lambda}_i$	the prototype of a hyperspherical shell
$\tilde{\lambda}_{i,}$	the ith principal eigenvalue of $\mathbf{C_s}$
λ_i^{EVD}	the ith eigenvalue of $\mathbf{C}$, calculated by the EVD method
$\lambda_{\max}$	the largest eigenvalue of the Hessian matrix of the error function; Also the largest eigenvalues of $\mathbf{C}$
λ_o	the regularization parameter for E_o
$\boldsymbol{\Lambda}$	the diagonal matrix with all the eigenvalues of $\mathbf{C}$ as its diagonal entries, $\boldsymbol{\Lambda} = \mathrm{diag}\left(\lambda_1, \cdots, \lambda_{J_2}\right)$
μ	the mean of the data set $\{x_i\}$; Also the membership degree of a fuzzy set; Also a positive number; Also the forgetting factor in the RLS method; Also the population size in the ES
$\boldsymbol{\mu}$	the mean of the data set $\{\mathbf{x}_i\}$
μ_i	the degree of activation of the ith rule
$\mu_i^{(1)}$	an MF of the premise part of the NEFPROX
$\mu_i^{(2)}$	an MF of the consequence part of the NEFPROX
$\boldsymbol{\mu}_j$	the mean of all the data in class j
$\mu_{\mathcal{A}_i'^j}$	the association between the jth input of $\boldsymbol{\mathcal{A}}'$ and the ith rule
$\mu_{\mathcal{B}_i^k}$	the association between the kth input of $\boldsymbol{\mathcal{B}}$ and the ith rule
$\mu_{\mathcal{A}}(x)$	the membership degree of x to the fuzzy set $\mathcal{A}$
$\mu_{\mathcal{A}}[\alpha]$	the α-cut of the fuzzy set $\mathcal{A}$
$\mu_{\mathcal{A}'}(\mathbf{x})$	the membership degree of $\mathbf{x}$ to the fuzzy set $\mathcal{A}'$
$\mu_{\mathcal{A}_i}(\mathbf{x})$	the membership degree of $\mathbf{x}$ to the fuzzy set $\mathcal{A}_i$
$\boldsymbol{\mu}_{\mathcal{B}_i}(\mathbf{y})$	the membership degree of $\mathbf{y}$ to the fuzzy set $\mathcal{B}_i$
μ_{kp}	the connection weight assigned to prototype $\mathbf{c}_k$ with respect to $\mathbf{x}_p$, denoting the membership of pattern p into cluster k
$\mu_{\mathcal{R}}(x,y)$	the degree of membership for association between x and y
ν	an index for iteration
θ	the bias or threshold at a single neuron
$\boldsymbol{\theta}$	the bias vector; Also a vector of parameters to estimate
θ_i	the threshold for the ith neuron; Also the elevation angle of the ith point source in the space; Also the angle between $\mathbf{w}_i$ and $\mathbf{c}_i$ in PCA
$\boldsymbol{\theta}^{(m)}$	the bias vector at the mth layer

ρ	the condition number of a matrix; Also a vigilance parameter in ART models; Also a small positive constant, representing the power of the repeller in the global-descent method; Also a small positive tolerance
$\rho(E)$	a function of E
ρ^+, ρ^-	two positive constants in the bold driver technique
ρ_0	a positive constant; Also the initial neighborhood parameter for the NG
ρ_{dyn}	a dynamic factor
ρ_{f}	the final neighborhood parameter for the NG
ρ_i	a constant coefficient
$\rho_i^{(j)}$	the scaling factor for the weights connected to the ith neuron at the jth layer
ρ_{ij}^+	the (i,j)th entry of the correlation matrix of the state vector $\mathbf{x}$ in the clamped condition
ρ_{ij}^-	the (i,j)th entry of the correlation matrix of the state vector $\mathbf{x}$ in the free-running condition
ρ_t	the ratio of the actual reduction in error to the predicted reduction in error
σ	the variance parameter; Also the width of the Gaussian RBF; Also a shifting parameter in the global-descent method
$\boldsymbol{\sigma}$	the strategy parameters in the ES, the vector containing all standard deviations σ_i in the ES
$\sigma(\cdot)$	the loss function
$\sigma(t)$	a small positive value in the LM method, used to indirectly control the size of the trust region
$\sigma(\mathbf{c}_i)$	the standard deviation of cluster i
$\sigma(\mathcal{X})$	the standard deviation of dataset $\mathcal{X}$
σ^-, σ^+	the lower and upper thresholds of σ_i in the DDA algorithm
σ_-, σ_+	the left and right standard deviations used in the pseudo-Gaussian function
σ_0	a positive constant
σ_i	the standard deviation, width or radius of the ith Gaussian RBF; Also the ith singular value of $\mathbf{C}_{xy}$; Also the ith singular value of $\mathbf{X}$
$\boldsymbol{\sigma}_i$	the vector containing all the diagonal entries of $\boldsymbol{\Sigma}_i$
σ_i'	a quantity obtained by mutating σ_i in the ES
$\boldsymbol{\sigma}_k$	the variance vector of the kth cluster
$\sigma_{k,i}$	the ith entry of $\boldsymbol{\sigma}_k$
$\sigma_{\max}$	the maximum singular values of a matrix
$\sigma_{\min}$	the minimum singular values of a matrix
σ_{n}^2	the noise variance
$\boldsymbol{\Sigma}$	the covariance matrice for the Gaussian function; Also the singular value matrix arising from the SVD of $\mathbf{X}$
$\boldsymbol{\Sigma}_i$	the covariance matrice for the ith Gaussian RBF
$\boldsymbol{\Sigma}_{J_2}$	the singular value matrix with the J_2 principal singular values of $\mathbf{X}$ as diagonal entries

τ	the size of neighborhood used for estimating the step size of the line search; Also a decay constant in the RAN
$\boldsymbol{\tau}$	the circuit time constant matrix, which is a diagonal matrix with the (i,i)th entry being τ_i
τ_i	the circuit time constant of the ith neuron in the Hopfield network or an RNN
$\tau_{i,j}$	the intensity of the pheromone on edge $i \to j$
$\tau_{i,j}^k(t)$	the intensity of the pheromone on edge $i \to j$ contributed by ant k at generation t
$\tau_l\,(\phi_i, \theta_i)$	the time delay on the lth array element for the i source
$\boldsymbol{v}$	the observation noise
ς^j	the jth value obtained by dividing the interval of the output y
$\Omega(E)$	a non-negative continuous function of E
$\boldsymbol{\xi}$	the vector containing all the slack variables ξ_i
ξ_i	a zero-mean Gaussian white noise process for the regression of the ith RBFN weight; Also a slack variable
$\boldsymbol{\Xi}$	a matrix used in the EKF
$\psi(\cdot)$	a repulsive potential function
$\psi_{ij}(\cdot)$	a continuous function of one variable
ζ	a threshold
ζ_i	a slack variable used in the SVR
a	the radix of the activation function in the ECAM; Also a scaling factor; Also a shape parameter of the fuzzy MF
$\mathbf{a}$	the coefficient vector in the objective function of the LP problem; Also the output vector of the left part of the crosscorrelation APCA network
a_0	a positive constant
a_1, a_2	real parameters
a_i	a shape parameter associated with the ith dimension of the RBF; Also the ith entry of $\mathbf{a}$
$\mathbf{a}_i$	the ith column of the mixing matrix $\mathbf{A}$; Also the regression parameter vector for the ith RBF weight
a_{ij}	a constant integer in a class of COPs
$a_{i,j}$	the adjustable parameter corresponding to the ith rule and the jth input; Also the jth entry of $\mathbf{a}_i$
$a_{i,k}^j$	the adjustable parameter corresponding to the kth input, ith rule, and jth output
a_j^i	a premise parameter in the ANFIS model, corresponding to the ith input and jth rule
a_p'	a variable defined for the pth eigenvalue in the APCA network

A	the grid of neurons in the Kohonen network
$\mathbf{A}$	a general matrix; Also a matrix defined in the LMSER; Also the mixing matrix in the ICA data model
$\mathcal{A}$	a fuzzy set
$A(\cdot)$	a nonlinear function in the NOVEL method
$\mathcal{A}'$	an input fuzzy set of an FIS
$\mathcal{A}'^j$	an input fuzzy set of an FIS
$\mathcal{A}^i$	a fuzzy set obtained by fuzzifying x_i; Also the ith partition of the fuzzy set $\mathcal{A}$;
A_1, A_2	two algorithms; Also two weighting parameters in the cost function of the COP
$\mathbf{A}_i$	the transformed form of $\mathbf{A}$ for extracting the ith principal singular component of $\mathbf{A}$; Also a decorrelating matrix used in the LEAP
$\mathcal{A}_i$	the fuzzy set associated with the antecedent part of the ith rule; Also a fuzzy set corresponding to the antecedent part of the ith fuzzy rule
$\mathcal{A}_i^j$	the fuzzy subset associated with the ith fuzzy rule and the jth input
$\mathcal{A} \times \mathcal{B}$	the Cartesian product of fuzzy sets $\mathcal{A}$ and $\mathcal{B}$
b	a shape parameter for an activation function or a fuzzy MF; Also a positive number used in the nonuniform mutation
b_0	a positive constant
b_1	a shape parameter for a nonmonotonic activation function
$\mathbf{b}$	a vector; Also the output vector of the right part of the APCA network
b_i	the ith entry of $\mathbf{b}$; Also a numerical value associated with the consequent of the i rule, $\mathcal{B}_i$; Also the ith binary code bit in a binary code; Also a constant integer in a class of COPs
$\tilde{b}_{ij}$	the (i, j) entry of $\widetilde{\mathbf{B}}$
$\bar{b}_{ij}$	the (i, j) entry of $\overline{\mathbf{B}}$
b_j^i	a premise parameter in the ANFIS model, corresponding to the ith input and jth rule
b_p'	a variable defined for the pth eigenvalue in the APCA network
B	the size of a block in a pattern; Also a weighting parameter in the cost function of the COP
$\mathbf{B}$	the rotation matrix in the mutation operator in the CMA-ES
$\mathcal{B}$	a fuzzy subset
$\overline{\mathbf{B}}$	a matrix obtained during the ROLS procedure at the tth iteration; Also a matrix obtained during the batch OLS procedure
$\widetilde{\mathbf{B}}$	a matrix obtained during the batch OLS procedure
$B(\cdot)$	a nonlinear function in the NOVEL method
$B(t)$	a variable used in the OSS method
$\widetilde{\mathbf{B}}(t)$	a matrix obtained during the ROLS procedure at the tth iteration
$\mathcal{B}'$	an output fuzzy set by an FIS
$\mathcal{B}^{(k)}$	the fuzzy set corresponding to the consequent part of the rule $R^{(k)}$

$\mathbf{B}_i$	a matrix defined for the ith neuron in the LMSER; Also a decorrelating matrix used in the LEAP
$\mathcal{B}_i$	a fuzzy set corresponding to the consequent part of the ith fuzzy rule
$\mathcal{B}_i^k$	a fuzzy set associated with the ith fuzzy rule and the kth input of vector $\mathcal{B}$
$\mathcal{B}^l$	the lth fuzzy subset obtained by partitioning of the interval of the output, y
$\mathcal{B}^{l^p}$	the lth fuzzy subset obtained by partitioning of the interval of the output, y, corresponding to the pth pattern
c	speed of light; Also a center parameter for an activation function; Also a shape parameter of the fuzzy MF; Also the acceleration constant in the PSO, positive
$\mathbf{c}$	a prototype in VQ; Also an RBF prototype
$c(\cdot)$	the center parameter of an MF of a fuzzy variable
$c(t)$	a coefficient of self-coupling in the chaotic neural network
$\mathbf{c}^{\mathrm{in}}$	the center of the input space
c_0	a real constant
c_1	the cognitive parameter in the PSO, positive constant; Also a shape parameter in the π-shaped MF; Also a real constant
c_2	the social parameter in the PSO, positive constant; Also a shape parameter in the π-shaped MF; Also a real constant
$\mathbf{c}_i$	the eigenvectors of $\mathbf{C}$ corresponding to eigenvalue λ_i; Also the prototype of the ith cluster in VQ; Also the ith prototypes in the RBFN; Also the feedback weights from the F2 neuron i to all input nodes in the ART model
c_i^{in}	the ith entry of $\mathbf{c}^{\mathrm{in}}$
$\tilde{\mathbf{c}}_i$	the ith principal eigenvectors of the skewed autocorrelation matrix $\mathbf{C}_{\mathrm{s}}$
$c_{i,j}$	its jth entry of $\mathbf{c}_i$;
c_{ij}	the connectivity from nodes i to j; Also a constant integer coefficient in a class of COPs; Also the (i,j)th entry of $\mathbf{C}$
c_j^i	a premise parameter in the ANFIS model, corresponding to the ith input and jth rule
$\mathbf{c}_w$	the winning prototype
$\mathbf{c}_{x,j}$	the input part of the augmented cluster center $\mathbf{c}_j$ in supervised clustering
$\mathbf{c}_{y,j}$	the output part of the augmented cluster center $\mathbf{c}_j$ in supervised clustering
$\mathrm{co}(\cdot)$	the core of the fuzzy set within
$\mathrm{csign}(u)$	the multivalued complex-signum activation function
C	a weighting parameter in the cost function of the COP; Also the product of the gradients at time t and time $t+1$ in the RProp; Also the number of classes; Also a prespecified constant that trades off wide margin with a small number of margin failures in the SVM
$\mathbf{C}$	the autocorrelation of a set of vectors $\{\mathbf{x}\}$; Also a transform matrix in the kernel orthogonalization-based RBFN weight learning

$\mathcal{C}$	all concepts in a class, $\mathcal{C} = \{\mathcal{C}_n\}$; Also the complex plane
$C(t)$	a variable used in the OSS method
$\mathcal{C}(\mathcal{U})$	the set of all continuous real-valued functions on a compact domain $\mathcal{U}$
$C(\mathbf{W}, \mathbf{W}^*)$	the criterion function used in the global-descent method
$C^*(x, y)$	the t-conorm using the drastic union
$\mathcal{C}^J$	the J-dimensional complex space
$\mathbf{C}_1$	the autocorrelation matrix in the feature space
$C_b(x, y)$	the t-conorm using the bounded sum
C_i	the capacitance associated with neuron i in the Hopfield network
$\mathbf{C}_i$	a matrix defined for the ith neuron in the LMSER
$C_m(x, y)$	the t-conorm using the standard union
$\mathcal{C}_n$	a set of target concepts over the instance space $\{0, 1\}^n$, $n \geq 1$; Also the set of input vectors represented by $\mathbf{c}_n$ according to the nearest-neighbor paradigm, namely the nth cluster; Also a fuzzy set corresponding to the condition part of the nth fuzzy rule
$C_p(x, y)$	the t-conorm using the algebraic sum
$\mathbf{C}_s$	the skewed autocorrelation matrix
$\mathbf{C}_{xy}$	the crosscorrelation matrix of two sets of random vectors $\{\mathbf{x}_t\}$ and $\{\mathbf{y}_t\}$
$\mathbf{C}_{xy}^{(i)}$	a transformed form of $\mathbf{C}_{xy}$ for extracting the ith principal singular vector of $\mathbf{C}_{xy}$
d	the VC dimension of a machine; Also a shape parameter of the trapezoid MF
$\mathbf{d}(t)$	the update step of the weight vector $\overline{\mathbf{w}}$; Also the descent direction approximating Newton's direction
$d(\mathcal{C}_1, \mathcal{C}_2)$	the distance between clusters $\mathcal{C}_1$ and $\mathcal{C}_2$
$d(\mathbf{x}_1, \mathbf{x}_2)$	the distance between data points $\mathbf{x}_1$ and $\mathbf{x}_2$
d_0	the interelement separation
$\mathbf{d}_0$	the steering vector in the desired direction
$d_{BCS}(\mathbf{c}_k, \mathbf{c}_l)$	the between-cluster separation for cluster k and cluster l
$d_H(\cdot, \cdot)$	the Hamming distance between the two binary vectors within
$\mathbf{d}_i$	the steering vector associate with the ith source; Also the coefficient vector of the ith inequality constraint in LP problems
$d_{i,j}$	the jth entry of $\mathbf{d}_i$; Also the distance between nodes i and j in the ant system; Also the (i, j)th entry of $\mathbf{D}$; Also the distance between the ith pattern and the jth prototype
$\mathbf{d}_{i,j}$	the distance vector between pattern $\mathbf{x}_i$ and prototype $\boldsymbol{\lambda}_j$ in spherical shell clustering
$d_{WCS}(\mathbf{c}_k)$	the within-cluster scatter for cluster k
d_{max}	the maximum distance between the selected RBF centers
d_{min}	the shortest of the distances between the new cluster center and all the existing cluster centers in the mountain and subclustering clustering
$\text{defuzz}(\cdot)$	the defuzzification function of the fuzzy set within
$\det(\cdot)$	the determinant of the matrix within
$\dim_{BVC}(\mathcal{N})$	the Boolean VC dimension of the class of functions or the neural network within

$\mathrm{dim}_{\mathrm{VC}}(\cdot)$	the VC dimension of the class of functions or the neural network within
$\mathbf{D}$	the steering matrix, $\mathbf{D} = [\mathbf{d}_1, \cdots, \mathbf{d}_M]$; Also the constraint coefficient matrix
$\mathcal{D}_i, \overline{\mathcal{D}}_i$	the data subsets of the total pattern set obtained by the ith partitioning
$D_j^{(m)}$	the degree of saturation of the jth neuron at the mth layer
D_{in}	the maximum possible distance between two points of the input space
$D_i^{(j)}$	the dynamic range of the activation function of the ith neuron at the jth layer
$\mathbf{D}_f$	an approximation to $\mathbf{D}$ at the signal plane of the network in the nonideal case
$\mathbf{D}_g$	an approximation to $\mathbf{D}$ at the constraint plane of the network in the nonideal case
$D(\mathbf{s})$	the decoding transformation in the GA
$D(R_p)$	a degree of fulfillment of the rule associated with the pth example
$\mathbf{e}_i$	the error vector between the network output and the desired output for the ith example
$\mathbf{e}(t)$	the instantaneous representation error vector for the tth input for PCA
$\overline{e}_i$	the average of $e_{p,i}$ over all the patterns
$\mathbf{e}_i(t)$	the instantaneous representation error vector associated with the ith output node for the tth input in robust PCA
e_{max}	the maximum error at the output nodes for a given pattern
e_{min}	a threshold in the RAN
$e_{i,j}$	the jth entry of $\mathbf{e}_i$
err	the training error for a model
E	an objective function for optimization such as the MSE between the actual network output and the desired output
$\mathbf{E}$	a matrix whose columns are the eigenvectors of the covariance matrix $\mathbf{C}$
$\mathrm{E}[\cdot]$	the expectation operator
E^*	the optimal value of the cost function
E_0	an objective function used in the SVM
E_1, E_2	two objective functions used in the SVR
E_3	an objective function used in the SVC
E_c	the constraint term in the cost function
E_{coupled}	the information criterion for coupled PCA/MCA
E_o	the objective term in the cost function
E_p	the error contribution due to the pth pattern
E_{AFD}	the average fuzzy density criterion function
E_{AIC}	the AIC criterion function
E_{APCA}	the objective function for the APCA network
E_{BIC}	the BIC criterion function
$E_{\mathrm{CMP}}, E_{\mathrm{CMP1}}$	two cluster compactness measures
E_{CPCA}	the criterion function for the CPCA problem

E^*_{CPCA}	the minimum of E_{CPCA}
E_{CSA}	the extra energy term in the CSA
E_{CV}	the crossvalidation criterion function
E_{DoA}	the error function defined for the DoA problem
$E_{\text{DoA},l}$	the error function defined for the DoA problem, corresponding to the lth snapshot
E_{FHV}	the fuzzy hypervolume criterion function
E_{GEVD}	the criterion function for GEVD
E_{Hebb}	the instantaneous criterion function for Hebbian learning
$E_{\text{LDA},1}, E_{\text{LDA},2}, E_{\text{LDA},3}$	three criterion functions for LDA
E_{MDL}	the total description length
E_{NIC}	the NIC cost function
E^*_{NIC}	the global maximum of E_{NIC}
E_{OCQ}	the overall cluster quality measure
E_{PCA}	the criterion function for PCA
$E_{\text{SEP}}, E_{\text{SEP1}}$	two cluster separation measures
E_{SLA}	the criterion function for the SLA
E_{SVC}	the objective function for the SVC
E_{SVM}	the objective function for the SVM classification
E_{SVR}	the objective function for the SVR
E_{T}	the total optimization objective function comprising the objective and regularization terms
E_{THK}	the average shell thickness criterion function
E_{T}^j	the individual objective function corresponding to the jth cluster in the PCM
E_{WBR}	the ratio of the sum of the within-cluster scatters to the between-cluster separation
E_α, E_β	the energy levels of a physical system in states α and β
Err	the generalization error on the new data
E_{R}	the cost function of robust learning
ERR_k	the ERR due to the kth RBF neuron
$E_{\mathcal{S}}$	the expectation operation over all possible training sets
f	a function
$f(\cdot)$	the fitness function in EAs
$\mathbf{f}(\cdot)$	the operator to perform the function of the MLP, used in the EKF
$f(\mu_{ji})$	the fuzzy complement of μ_{ji}
$f(H)$	the average fitness of all strings in the population matched by the schema H
$\overline{f}(t)$	the average fitness of the whole population at time t
$\mathbf{f}(\mathbf{x})$	the vector containing multiple functions as entries
$\dot{f}(x)$	the first-order derivative of $f(x)$, that is, $\frac{df(x)}{dx}$
$f'(\mathbf{x})$	the objective function obtained by the penalty function method
$\mathbf{f}(\mathbf{z},t)$	a vector with functions as entries
$f : \mathcal{X} \to \mathcal{Y}$	a mapping from fuzzy sets $\mathcal{X}$ onto $\mathcal{Y}$
f_0	the carrier frequency
f_1, f_2	functions

f_c	the carrier frequency
$f_i(\cdot)$	the output function in the TSK model for the ith rule; Also the nonlinear relation characterized by the ith TSK system of the hierarchical fuzzy system
$f_i(\mathbf{x})$	the ith entry of the function vector $\mathbf{f}(\mathbf{x})$
$\mathbf{f}_i(\mathbf{x})$	the crisp vector function of $\mathbf{x}$, related to the ith rule and the output of the TSK model
$f_i^j(\mathbf{x})$	the jth entry of $\mathbf{f}_i(\mathbf{x})$, related to the jth output component of the TSK model
$f_p(\mathbf{x})$	the penalty term characterizing the constraints
$\mathrm{fuzz}(\cdot)$	a fuzzification operator
$\mathcal{F}$	the set of all functions; Also a set of real continuous functions
$F(\cdot)$	the fairness function in the FSCL; Also the CDF of the random variable within
$F(\mathbf{x})$	the weighted objective of all the entries of $\mathbf{f}(\mathbf{x})$
$\mathcal{F}_1$	a set of benchmark functions
$\mathbf{F}$	the Jacobian matrix used in the EKF method
$\mathbf{F}_i$	the fuzzy covariance matrix of the ith cluster
g	a function in $\mathcal{C}(\mathcal{U})$
$\mathbf{g}(t)$	the gradient vector of E with respect to $\vec{\mathbf{w}}(t)$
$\mathbf{g}^{(m)}$	the gradient vector of E with respect to $\vec{\mathbf{w}}^{(m)}$
g_i	the ith bit in a Gray code
$g_{ij}^{(m)}(t)$	the gradient of E with respect to $w_{ij}^{(m)}(t)$
$\bar{g}_{ij}^{(m)}(t)$	a gradient term decided by $g_{ij}^{(m)}(t)$
$\mathbf{g}_\tau(t)$	the gradient vector of $E\left(\vec{\mathbf{w}}(t) + \tau\mathbf{d}(t)\right)$ with respect to $\vec{\mathbf{w}}$
G_{ij}	the conductance of the jth resistor of neuron i in the Hopfield network
h	a hypothesis in the PAC theory; the tournament size in the GA
$h(\cdot)$	a function defined as the square root of the loss function $\sigma(\cdot)$
h_0	a positive constant
h_j	a constant term in the jth linear inequality in the COP and LP problems
$h_j(\cdot)$	a continuous function of one variable
$h_{kw}(t)$	the neighborhood function, defining the response of neuron k when $\mathbf{c}_w$ is the excitation center
$\mathrm{hgt}(\cdot)$	the height of the fuzzy set within
H	a schema of length l, defined over the three-letter alphabet $\{0, 1, *\}$
$\mathbf{H}$	the Hessian matrix of a network
$\mathcal{H}$	the hypothesis space, $\mathcal{H} = \{\mathcal{H}_i\}$
$H(\mathbf{y})$	the joint entropy of all the entries of $\mathbf{y}$
$H(y_i)$	the marginal entropy of component i
$\mathbf{H}_b$	the block diagonal Hessian for the MLP
$\mathbf{H}_b^{(m)}$	the (m, m)th diagonal partition matrix $\mathbf{H}_b$, corresponding to $\vec{\mathbf{w}}^{(m)}$
$\mathcal{H}_i$	a set of hypotheses over the instance space $\{0, 1\}^i$, $i \geq 1$

H_{ij}	the (i,j)th entry of $\mathbf{H}$
$\mathbf{H}_{\text{BFGS}}$	the Hessain obtained by the BFGS methods
$\mathbf{H}_{\text{DFP}}$	the Hessain obtained by the DFP method
$\mathbf{H}_{\text{GN}}$	the Hessian matrix obtained by the Gauss–Newton method
$\mathbf{H}_{\text{LM}}$	the Hessian matrix obtained by the LM method
i	an index for iteration
$i \rightarrow j$	an edge from nodes i to j
$\mathbf{I}$	the identity matrix
$\mathcal{I}$	the intersection of fuzzy sets $\mathcal{A}$ and $\mathcal{B}$
$I(i)$	a running length for summation for extracting the ith PC
$I(\mathbf{x};\mathbf{y})$	the mutual information between signal vectors $\mathbf{x}$ and $\mathbf{y}$
$I(\mathbf{y})$	the mutual information between the components of vector $\mathbf{y}$ in the ICA
I_i	the external bias current source for neuron i in the Hopfield network
$\mathbf{I}_k$	the identity matrix of size $k \times k$
$\text{Im}(\cdot)$	the operator taking the imaginary part of a complex number
j	an index for iteration
j	$\sqrt{-1}$
J	the dimensionality of the input data
$\mathbf{J}(\vec{\mathbf{w}})$	the Jacobian matrix
J_{H}	the number of hidden units
J_i	the number of nodes in the ith layer
J_{ij}	the (i,j)th entry of the Jacobian matrix $\mathbf{J}$
$\mathcal{J}_i^k$	the set of nodes that remain to be visited by ant k positioned at node i
J_{o}	the number of network outputs
k	an index for iteration; Also a scaling factor that controls the variance of the Gaussian machine
$\overline{\mathbf{k}}$	a vector in the set of all the index vectors of a fuzzy rule base, $\overline{\mathbf{k}} = (k, k_1, \cdots, k_n)^{\text{T}}$
k_i	the index of the partitioned fuzzy subsets of the interval of x_i
k_i^p	the index of the partitioned fuzzy subsets of the interval of x_i, corresponding to the pth pattern
k_{B}	Boltzmann's constant
K	the number of clusters; Also the number of prototypes in the RBFN; Also the number of rules in the ANFIS model
$\mathbf{K}$	the Kalman gain matrix; Also a kernel matrix
$\mathcal{K}$	the index set of a fuzzy rule base
K_{ij}	the (i,j)th entry of the kernel matrix $\mathbf{K}$
l	the string length of a chromosome
l^p	the index of the partitioned fuzzy subsets of the interval of the output, y, corresponding to the pth pattern
l_i	the bit-length of the gene x_i in the chromosome

L	an integer used as the quantization step for phase quantization; Also the number of array elements; Also a constant parameter in the ART 1
$\mathcal{L}$	the undesirable subspace in the CPCA problem
$\mathcal{L}^{J_2}$	a J_2-dimensional subspace that is constrained to be orthogonal to $\mathcal{L}$
$L(t)$	the Lipschitz constant at time t
$L^p\left(R^p, \mathrm{d}\mathbf{x}\right)$	the L^p space, where $(R^p, \mathrm{d}\mathbf{x})$ is a measure space and p a positive number
$L(\widehat{\mathbf{W}}(\overline{\mathcal{D}}_i)|\mathcal{D}_i)$	the likelihood evaluated on the data set $\mathcal{D}_i$
L_k	the length of the tour performed by ant k
$L_N(\widehat{\mathbf{W}}_N)$	the likelihood estimated for a training set of size N and the model parameters $\widehat{\mathbf{W}}_N$
$\mathrm{LT}[\cdot]$	the operator extracting the lower triangle of the matrix contained within
m	an index for iteration; Also an integer; Also the fuzzifier
$m(H,t)$	the number of examples of a particular schema H within a population at time t
m_i	the number of fuzzy subsets obtained by partitioning the interval of x_i
$m_i(t)$	the complex modulating function for the ith source
m_y	the number of fuzzy subsets obtained by partitioning the interval of y
$\max\left(\mathbf{x}_1, \mathbf{x}_2\right)$	the operation that gives a vector with each entry obtained by taking the maximum of the corresponding entries of $\mathbf{x}_1$ and $\mathbf{x}_2$
$\min\left(\mathbf{x}_1, \mathbf{x}_2\right)$	the operation that gives a vector with each entry obtained by taking the minimum of the corresponding entries of $\mathbf{x}_1$ and $\mathbf{x}_2$
M	the number of signal sources; Also the number of layers of FNNs; Also the effective size of a time window
$\mathbf{n}(t)$	an unbiased noisy term at a particular instant
$n_i^{(j)}$	the number of weights to the ith unit at the jth layer
n_y	the time window of a time series
net	the net input to a single neuron
$\mathbf{net}$	the net input vector to the SLP
net_i	the ith entry of $\mathbf{net}$
$\mathbf{net}_p^{(m)}$	the net input to the mth layer for the pth pattern
$net_{p,j}$	the net input of the jth neuron for the pth sample
$net_{p,v}^{(m)}$	the net input to the vth neuron of the mth layer for the pth pattern
N	the size of the training set
$N\left(0, \sigma\right)$	a random number drawn from a normal distribution with zero mean and standard deviation σ_i; Also a normal distribution with zero mean and standard deviation σ_i
N_{PAC}	the sample complexity of a learning algorithm
$\mathcal{N}, \mathcal{N}_k$	neural networks
$\mathcal{N}_1$	a neural network whose hidden neurons are LTGs
$\mathcal{N}_2$	a neural network whose hidden neurons are binary RBF neurons

$\mathcal{N}_3$	a neural network whose hidden neurons are generalized binary RBF neurons
N_i	the number of samples in the ith cluster or class
$N_{\max}$	the storage capability of an associative memory network; Also the maximum number of fundamental memories
N_n	the number of nodes in a network
NOT $[\cdot]$	the complement of the set or fuzzy set within
N_{phase}	the number of training phases in the successive approximative BP learning
N_P	the size of the population; the number of ants in the ant system
N_r	the number of rules in an FIS
N_w	the total number of weights (free parameters) of a network
$N_w^{(m)}$	the total number of weights (free parameters) at the mth layer of a network
N_y	the dimension of $\mathbf{y}$
$o(H)$	the order of a schema H, namely the number of fixed positions (the number of 0s or 1s) present in the template
$o_i^{(m)}$	the output of the ith node in the mth layer
$\mathbf{o}_p^{(m)}$	the output vector at the mth layer for the pth pattern
$o_{p,i}^{(m)}$	the ith entry of $\mathbf{o}_p^{(m)}$
$\mathcal{O}$	a J-dimensional hypercube, $\{-1,1\}^J$
$O(\cdot)$	in the order of the parameter within
OP	the degree of optimism inherent in a particular estimate
p	the index for iteration; Also the density of fundamental memories
$p(\mathbf{x})$	the marginal PDF of the vector variable $\mathbf{x} \in R^n$
$p(\mathbf{x},\mathbf{y})$	the joint PDF of $\mathbf{x}$ and $\mathbf{y}$
$p \rightarrow g$	if p then q
$p(\mathbf{y})$	the joint PDF of all the elements of $\mathbf{y}$
$\mathbf{p}_1, \mathbf{p}_2$	two points
$p_i(y_i)$	the marginal PDF of y_i
P	the number of hypothesis parameters in a model; Also the probability of a state change
$\mathbf{P}$	the conditional error covariance matrix; Also the mean output power matrix of the beamformer
$\mathcal{P}^*$	the Pareto optimal frontier
$\mathbf{P}(0)$	the initial value of the conditional error covariance matrix $\mathbf{P}$
$P(i)$	the potential measure for the ith data point, $\mathbf{x}_i$
$\overline{P}(k)$	the potential measure for the kth cluster center, $\mathbf{c}_k$
$\mathcal{P}(t)$	the population at generation t in an EA
$P(\mathbf{x})$	a data distribution
P_α, P_β	the probabilities of a physical system being in states α and β
P_c	the probability of recombination
P_i	the probability of state change of the ith neuron in the Boltzmann machine; Also the selection probability of the ith chromosome in a population
$P_i(t)$	the inverse of the covariance of the output of the ith neuron

$P_{i,j}^k(t)$	the probability for ant k at node i moving to node j at generation t
P_m	the probability of mutation
$P_{m,0}$	the initial value for P_m
$P_{m,f}$	the final values for P_m
P_{max}	an acceptable level of the error probability
$P_{MUSIC}(\cdot)$	the MUSIC spectrum function
$\mathbf{q}(t)$	the update step of $\boldsymbol{\epsilon}(t)$; Also the transformed vector obtained by applying the the orthonormal constraint matrix $\mathbf{Q}$ to the input vector $\mathbf{x}_t$
$\mathbf{q}_j$	the coefficient vector of the jth linear inequality; Also the jth column vector of the orthogonal matrix $\mathbf{Q}$
$q_{j,k}$	the kth entry of $\mathbf{q}_j$
$\mathbf{Q}$	an orthogonal matrix arising from the QR decomposition of $\boldsymbol{\Phi}^T$; Also an orthonormal constraint matrix
$\mathbf{Q}(t)$	the covariance matrix of the observation noise
$\mathbf{Q}_1(t)$	an orthogonal matrix arising from the QR decomposition in the ROLS at the tth iteration
r	a random number; Also the index of the rival prototypes in clustering
$\mathbf{r}$	the vector whose entries are the radii of all the clusters in the subtractive clustering; Also a position vector of a point
$\mathbf{r}^l$	the position vectors of the lth array element
r_0	a real positive constant
r_1, r_2	two uniform random numbers in $[0, 1]$
r_a, r_b	two design parameters in the mountain and subtractive clustering, normalized radii defining the neighborhood
r_i	a scaling resistance; Also the radius of a hyperspherical shell; Also the amplitude of the net_i of the complex Hopfield-like network; Also the reciprocal of $\|\mathbf{w}_i(t)\|$
$\mathbf{r}_i$	the position vector of the ith point source; Also the coefficient vector of the ith linear equality
$r_{i,j}$	the jth entry of $\mathbf{r}_i$
$r_{\mathcal{S}}(x)$	the amplitude of $\mu_{\mathcal{S}}(x)$, $r_{\mathcal{S}}(x) \in [0, 1]$
R	a quantity representing a tradeoff between a reasonable potential of a new cluster center and its distance to the existing cluster centers; Also the radius of a spherical cluster in the feature space
$\mathbf{R}$	the autocorrelation matrix of vector $\mathbf{x}$; Also an upper triangular matrix arising from the QR decomposition of $\boldsymbol{\Phi}^T$; Also a fuzzy matrix
$\mathcal{R}$	a relation between $\mathcal{X}$ and $\mathcal{Y}$, or a relation on $\mathcal{X} \times \mathcal{Y}$; Also an indiscernibility relation in the theory of rough sets
$R(\boldsymbol{\alpha})$	the expected risk
$\mathbf{R}(t)$	the covariance matrix of the measurement noise
$R^{(k)}$	the kth rule in a rule base
R^n	the n-dimensional vector space
$R^{m \times n}$	the $m \times n$-dimensional matrix space

$\mathbf{R}_1, \mathbf{R}_2$	two real, symmetric and positive-definite matrices
$\mathcal{R}_1 \circ \mathcal{R}_2$	the max-min composition between relations $\mathcal{R}_1$ and $\mathcal{R}_2$
$\mathcal{R}_1 \diamond \mathcal{R}_2$	the min-max composition between relations $\mathcal{R}_1$ and $\mathcal{R}_2$
$R_{\mathrm{emp}}(\boldsymbol{\alpha})$	the empirical risk
R_{ij}	the (i,j)th entry of $\mathbf{R}$; Also the jth resistance of the ith neuron in the Hopfield network
$\mathrm{Re}(\cdot)$	the operator taking the real part of a complex number
$\mathbf{s}$	the signal vector; Also the genetic coding of the network parameters
$s(\cdot)$	the shape parameter of an MF of a fuzzy variable
$s(t)$	a bipolar quantity specifying the classification correctness in the OLVQ1
$\mathbf{s}(t)$	the step size in the second-order optimization techniques
s_i	the ith component of the signal vector $\mathbf{s}$; Also a constant in the ith linear inequality constraint; Also a bit in the GA
$\mathbf{s}_{u,v,q}$	a hypothetical signal vector for DoA estimation using the Hopfield network
$\mathrm{sp}(\cdot)$	the support of the fuzzy set within
S	the sum, over all the output units and all the patterns, of the magnitude of the covariance between the output of the new candidate hidden node and the residual output error observed at an output unit
$\mathbf{S}$	the source correlation matrix; Also the error matrix for estimating the Hessain in the Gauss–Newton method
$\mathcal{S}$	a training set; Also a complex fuzzy set
S_1, S_2, S_3	three zones of the sigmoidal function
$\mathbf{S}_\mathrm{b}$	the between-class scatter matrix
$\widetilde{\mathbf{S}}_b$	the transformed form of $\mathbf{S}_\mathrm{b}$
S_i	the sensitivity of the outputs of the trained network with respect to the ith input; Also the sum of the membership degrees of only those members within a hyperellipsoid
$\mathbf{S}_\mathrm{m}$	the mixture scatter matrix
$\widetilde{\mathbf{S}}_\mathrm{m}$	the transformed form of $\mathbf{S}_\mathrm{m}$
$\mathcal{S}_\mathrm{prune}$	a subset of weights to be removed
$\mathbf{S}_\mathrm{w}$	the within-class scatter matrix
$\widetilde{\mathbf{S}}_\mathrm{w}$	the transformed form of $\mathbf{S}_\mathrm{w}$
S_w^E	the sensitivity of the the error function E with respect to weight w
t	an index for iteration
t_{kl}	the (k,l)th entry of $\mathbf{T}$
$\mathrm{tr}(\cdot)$	the trace operator
T	the annealing temperature; Also the number of generations in the GA
$\mathbf{T}$	the weight matrix in the MAXNET of the Hamming network; Also a nonsingular diagonal matrix acting as a scaling matrix
$T(x,y)$	the t-norm
$T^*(x,y)$	the t-norm using the drastic intersection

T_0	the initial annealing temperature
$T_b(x, y)$	the t-norm using the bounded sum
T_f	the final temperature; Also the maximum number of iterations
T_j	the fuzzy shell thickness of the jth cluster
$T_m(x, y)$	the t-norm using the standard intersection
$T_p(x, y)$	the t-norm using the algebraic product
T_s	the search time
T_w	the time-window size
$\mathbf{u}$	the vector comprising all the intermediate voltage u_i in the Hopfield network; Also a complex vector as the input to the neural network in the DoA problem; Also a vector used in the OOja
$u(\cdot)$	the Heaviside step function
u_i	the signal counter of the ith prototype, namely, the number of times the ith prototype has been the winner; Also an interconnection-point voltage of neuron i in the Hopfield network
$\mathbf{u}_i$	the ith column vector of $\mathbf{U}$
$\widetilde{\mathbf{u}}_i$	the transformed form of $\mathbf{u}_i$ in the FastICA
$\overline{\mathbf{u}}_i$	the ith column of $\overline{\mathbf{U}}$
$\underline{\mathbf{u}}_i$	the ith column of $\underline{\mathbf{U}}$
u_{ij}	the lateral weight from units i to j; Also the (i, j)th entry of $\mathbf{U}$
$\overline{u}_{ij}$	the (i, j)th entry of $\overline{\mathbf{U}}$
$\underline{u}_{ij}$	the (i, j)th entry of $\underline{\mathbf{U}}$
$u_k(\mathbf{x})$	the kth orthonormal RBF obtained from $\phi_j(\mathbf{x})$, $j = 1, \cdots, k$, by the GSO procedure
$\mathbf{U}$	an orthogonal separating matrix; Also the lateral weight matrix of the PCA network; Also the response matrix of the orthonormal RBFs; Also the left matrix arising from the SVD of $\mathbf{X}$, with its columns being all the left singular vectors; Also the membership matrix, whose element μ_{ji} denotes the membership of $\mathbf{x}_i$ into cluster j
$\mathcal{U}$	a compact domain; Also the union of fuzzy sets $\mathcal{A}$ and $\mathcal{B}$; Also the universe of discourse
$\overline{\mathbf{U}}$	the right lateral connection weight matrix of the APCA network
$\underline{\mathbf{U}}$	the left lateral connection weight matrix of the APCA network
$\mathbf{U}_{J_2}$	the left matrix arising from the SVD of $\mathbf{X}$, with its columns being the J_2 principal left singular vectors
$\mathbf{U}_n$	a matrix associated with the noise subspace of $\widehat{\mathbf{R}}$, with its $(L-M)$ columns being the eigenvectors corresponding to the $(L-M)$ smallest eigenvalues of $\widehat{\mathbf{R}}$
$\mathbf{v}$	an eigenvector of $\mathbf{C}_1$; Also a vector used in the OOja
$\overline{v}$	the average output of the candidate units over all the patterns in the cascaded-correlation technique
$\mathbf{v}(t)$	the whitened vector of the input $\mathbf{x}_t$
v_{cc}	the power voltage
v_i	the output voltage of neuron i in the Hopfield network; Also the output of the candidate unit for the ith pattern in the cascaded-correlation technique

$\mathbf{v}_i$	the ith eigenvector of $\mathbf{C}_1$; Also the velocity of the ith particle in the PSO
v_{ij}	the (i,j)th entry of $\mathbf{V}$
$\mathbf{v}_i^x$	the ith left principal singular vector of $\mathbf{C}_{xy}$
$\mathbf{v}_i^y$	the ith right principal singular vector of $\mathbf{C}_{xy}$
$\mathbf{V}$	a variable matrix with elements as 0 or 1; Also a whitening matrix; Also the inverse of $\mathbf{A}$; Also the right matrix arising from the SVD of $\mathbf{X}$, with its columns being all the right singular vectors
V_i	the volume of the ith cluster
$\mathbf{V}_{J_2}$	the right matrix arising from the SVD of $\mathbf{X}$, with its columns being the J_2 principal right singular vectors
w	the index of the winning prototype in clustering
$\mathbf{w}$	the vector collecting all the weights terminated at a single neuron
$\overrightarrow{\mathbf{w}}$	the vector created by concatenating all the components of $\mathbf{W}$
$\overrightarrow{\mathbf{w}}^{(m)}$	the vector obtained by concatenating all the components of $\mathbf{W}^{(m)}$
w_0	a free parameter for scaling weights
w_i	the ith entry of $\mathbf{w}$; Also the ith entry of $\overrightarrow{\mathbf{w}}$
$\mathbf{w}_i$	the vector collecting all the weights starting from the first layer and terminated at the jth neuron, $\mathbf{w}_i = (w_{1i}, w_{2i}, \cdots, w_{J_1 i})^{\mathrm{T}}$
$\overline{\mathbf{w}}_i$	the ith column of $\overline{\mathbf{W}}$
$\underline{\mathbf{w}}_i$	the ith column of $\underline{\mathbf{W}}$
$\widetilde{\mathbf{w}}_i$	the ith column vector of $\widetilde{\mathbf{W}}$
$\mathbf{w}_i^{\mathrm{EVD}}$	the ith eigenvector of $\mathbf{C}$, calculated by the EVD method
$\mathbf{w}_i^{\mathrm{ICA}}$	the ith independent direction
$\mathbf{w}_i^{\mathrm{PCA}}$	the ith principal direction
w_{ij}	connection weight from nodes i to j, the (i,j)th entry of $\mathbf{W}$
$\widetilde{w}_{ij}$	the (i,j)th entry of $\widetilde{\mathbf{W}}$
$\mathbf{w}_j^{(1)}$	the vector collecting all the weights starting from the first layer and terminating at the jth unit in the second layer
$w_{ij}^{(m)}$	the connection weights from the ith neuron at the mth layer to the jth neuron at the $m+1$ layer. When there is no confusion, the superscript (m) is usually omitted
$w_{ij,\mathrm{BP}}^{(m)}$	the value of $w_{ij}^{(m)}$, calculated by BP learning
$w_{ij}^{(m)*}$	the last local minima of $w_{ij}^{(m)}$
$w_{\max}^{(m)}$	the maximum magnitude of the weights from the mth to $(m+1)$th layers
$\mathbf{W}$	the connection weight matrix of a network; Also the feedforward weight matrix of the Hamming network or the PCA network; Also the weight matrix between the hidden and output layers in the RBFN; Also the beamforming weight matrix
$\overline{\mathbf{W}}$	the feedforward weight matrix in the right part of the APCA network
$\underline{\mathbf{W}}$	the feedforward weight matrix in the left part of the APCA network
$\widetilde{\mathbf{W}}$	an intermediate weight matrix in the kernel orthogonalization-based weight learning
$\mathbf{W}^{(m)}$	the connection weight matrix from the mth to $(m+1)$th layers

$\widehat{\mathbf{W}}(\overline{\mathcal{D}}_i)$	the ML estimates on $\overline{\mathcal{D}}_i$
$\mathbf{W}_{\text{LCMV}}$	the weight matrix of the LCMV beamformer
$\mathbf{x}$	an input vector for a network or a function; Also a chromosome in a population
$\mathbf{x}^-$	the vector having elements as the lower bounds for the corresponding parameters in $\mathbf{x}$
$\tilde{\mathbf{x}}$	an augmented form of $\mathbf{x}$
$\mathbf{x}^+$	the vector having elements as the upper bounds for the corresponding parameters in $\mathbf{x}$
$\mathbf{x}^{\text{g}}$	the global best value, $gbest$, obtained so far by any particle in the swarm
$\mathbf{x}^j$	the jth row of $\mathbf{X}$, $\mathbf{x}^j = (x_{1,j}, x_{2,j}, \cdots, x_{N,j})^{\text{T}}$
$\mathbf{x}^{(k)}(t)$	the trace derived from step k in the NOVEL method
x_1', x_2'	two crisp inputs
$\mathbf{x}', \mathbf{x}_1', \mathbf{x}_2', \mathbf{x}_1'',$ $\mathbf{x}_2'', \mathbf{x}_3'', \mathbf{x}_4''$	some offspring obtained by genetic operators
$\mathbf{x}_1 \succ \mathbf{x}_2$	$\mathbf{x}_1$ dominates $\mathbf{x}_2$
x_i	the ith entry of the input vector $\mathbf{x}$; Also the input on the ith array element; Also the ith gene of a chromosome; Also a member of the discrete fuzzy set $\mathcal{X}$; Also a value obtained by dividing the interval of x linearly or through α-cut
x_i^-, x_i^+	the lower and upper bounds of the variable x_i
$\mathbf{x}_i$	the ith input pattern; Also the ith individual in a population; Also the position of the ith particle in a swarm
$\mathbf{x}_i(t)$	the input for extracting the ith PC at time t in the PASTd
$\mathbf{x}_i^*$	the particle best, $pbest$, the best solution the ith particle in the swarm has achieved so far
x_i'	an offspring obtained by mutating x_i
$\widetilde{\mathbf{x}}_i$	the input in supervised clustering obtained by augmenting the input pattern with its output pattern
$\{\mathbf{x}_i\}$	the data set of vector $\mathbf{x}_i$
$\{x_i\}$	a one-dimensional data set of data x_i
$\{\widetilde{x}_i\}$	the data set with zero mean and unit standard deviation, obtained by linearly transforming $\{x_i\}$
$\mathbf{x}_i^{(j)}$	the ith sample of class j
$(\mathbf{x}_p, \mathbf{y}_p)$	the pth input-output pattern pair in the training set $\mathcal{S}$
$x_{i,j}$	the jth entry of the ith pattern $\mathbf{x}_i$
$\mathbf{X}$	the full input or feature matrix, whose columns are all the input patterns; Also the stored pattern matrix
$\mathcal{X}$	a set of input patterns; Also the universe of discourse
$\overline{\mathbf{X}}$	the associated pattern matrix corresponding to the stored pattern matrix $\mathbf{X}$
$\overline{\mathcal{X}}$	samples in $\mathcal{X}$ that are misclassified
$\mathcal{X} \times \mathcal{Y}$	the set of all ordered pairs (x, y) for $x \in \mathcal{X}$ and $y \in \mathcal{Y}$
$\mathbf{y}$	the output of a network or a function for an input vector $\mathbf{x}$; Also the measured signal vector on the antenna array
$\mathbf{y}(t)$	the beamformer output vector at time t

y'	a crisp output
$\mathbf{y}'$	the output vector of the TSK model when the input is $\mathcal{A}'$
y^-, y^+	the lower and upper bounds of the output y
$\overline{y}^{(k)}$	the center of $\mathcal{B}^{(k)}$
y_{AND}	the output of the AND neuron
y_i	a value obtained by linearly or α-cut dividing the interval of y; Also a member of the discrete fuzzy set $\mathcal{Y}$
$y_i(\mathbf{x})$	the ith output of the RBFN for the input $\mathbf{x}$
$\overline{y}_{i,j}$	a level of involvement of $\mathbf{x}_i$ in the constructed cluster j
$\overline{\mathbf{y}}_k$	the kth row of $\mathbf{Y}$
y_{OR}	the output of the OR neuron
$\mathbf{y}_t$	the desired output vector at the tth example; Also an input vector to the APCA network at time t
$y_{t,j}$	the desired output of the jth output neuron for the tth example
$\hat{y}_{t,j}$	the network output of the jth output neuron for the tth example
$\mathbf{Y}$	the output matrix, whose columns are the output patterns
$\mathcal{Y}$	a universal set
z	the complex variable
$\mathbf{z}$	a state vector; Also the normalized complex vector as the input to the neural network in the DoA problem; Also a random vector whose elements are drawn from a normal distribution $N(0,1)$
$\mathbf{z}(t)$	the step size of gradient, $\mathbf{z}(t) = \mathbf{g}(t+1) - \mathbf{g}(t)$
$\mathbf{z}_i$	the ith input complex vector
z_α	the value of the Z-statistic at $\alpha\%$ level of significance
Z	the partition function of the distribution of states of a physical system

1
Introduction

In this chapter, we first briefly go through the history of neural-network research. We then describe the conventional McCulloch–Pitts neuron and its analog VLSI implementation, structures of neural networks as well as various learning paradigms. Properties of neural networks are summarized and applications of neural networks enumerated. The neural-network model for array signal processing is described as simulation examples, since they are used in the subsequent chapters. Finally, the scope of the book and summary by chapters are given.

1.1 A Brief History of Neural Networks

The discipline of neural networks originates from an understanding of the human brain. The average human brain consists of 3×10^{10} neurons of various types, with each neuron connecting to up to 10^4 synapses [831]. Many neural-network models, also called connectionist models, have been proposed. Neural networks are attractive since they consist of many neurons, with each one processing information separately and simultaneously. All the neurons are connected by synapses with variable weights. Thus, neural networks are actually parallel distributed processing systems.

Research on neural networks dates back to the 1940s when McCulloch and Pitts found that the neuron can be modeled as a simple threshold device to perform logic function [783]. In the late 1940s, Hebb proposed the Hebbian rule in [473] to describe how learning affects the synaptics between two neurons. In the late 1950s and early 1960s, Rosenblatt [963] proposed the perceptron model, and Widrow and Hoff [1182] proposed the adaline (adaptive linear element) model, trained with a least mean squares (LMS) method.

Interest in neural networks waned for a decade due to Minsky and Papert's book [808], which proved mathematically that the perceptron cannot be used for complex logic function. During the same period, the adaline model was successfully applied to many problems, thanks to the rigorous mathematical foundation of the LMS method. However, the adaline model and its cascaded multilayer version called the madaline cannot solve linearly inseparable problems due to the linear activation function.

In the 1970s, Grossberg [434, 435], von der Malsburg [1133], and Fukushima [387] conducted pioneering work on competitive learning and self-organization, based on the connection patterns found in the visual cortex. Kohonen proposed his self-organization maps (SOM) [609, 610]. The SOM algorithm adaptively transforms incoming signal patterns of arbitrary dimensions into one- or two-dimensional discrete maps in a topologically ordered fashion. A Kohonen network is a structure of interconnected processing units that compete for the signal. Fukushima proposed his cognitron [387] and neocognitron models [388], under the competitive learning paradigm. The neocognition is a neural network specially designed for visual or character pattern recognition. It can recognize input patterns correctly with a high tolerance to character distortion, position, and size.

Grossberg and Carpenter contributed with the adaptive resonance theory (ART) model in the mid-1980s [435, 159]. The ART networks, also based on competitive learning, are capable of stable clustering of an arbitrary sequence of input data in real time. Each cluster is represented by the weight vector of a prototype unit, and clusters are allocated incrementally by the network. The dynamics of the model are characterized by first-order differential equations. The ART model is recurrent and self-organizing. It has two basic layers: the input layer called *comparing* and the output layer called *recognizing*.

The modern era of neural-network research is commonly deemed to have started with the publication of the Hopfield network in 1982 [502, 503, 505]. The model works at the system level rather than at a single neuron level. It is an RNN working with the Hebbian rule. This network can be used as an associative memory for information storage. As a dynamically stable network, it can also be used to solve optimization problems. Boltzmann machine was introduced in 1985 as an extension to the Hopfield network by incorporating stochastic neurons [9]. The Boltzmann learning is based on a method called simulated annealing (SA) [607]. These works revived the interest in neural networks. Kosko extended the ideas of Grossberg and Hopfield and proposed the adaptive bidirectional associative memory (BAM) [621].

The Hamming network was proposed by Lippman in the mid-1980s [717]. It is composed of a similarity feedforward subnet with an n-node input layer and an m-neuron memory layer and a winner-take-all (WTA) subnet with a fully connected m-neuron topology. The network is the most straightforward associative memory. The Hamming network calculates the Hamming distance between the input pattern and each memory pattern, and selects the memory with the smallest Hamming distance, which is declared as the *winner*.

The landmark of the field is the multilayer perceptron (MLP) model trained with the backpropagation (BP) learning algorithm published in 1986 by Rumelhart *et al.* [979]. Later it turned out that the BP algorithm had already been described in 1974 by Werbos [1166] when he studied social problems. What the fast Fourier transform (FFT) algorithm to the research on signal processing is, the MLP with the BP algorithm[1] is to the research on neural networks.

In 1988, Broomhead and Lowe proposed the radial basis function network (RBFN) model [133]. The RBFN has equivalent capabilities as the MLP model, but with a much faster training speed.

The cellular neural network (CNN) model, proposed in 1988 by Chua and Yang [239, 240], has a unique network architecture. CNNs are especially useful in image and video processing, and are most suitable for very large scale integrated (VLSI) implementation.

In the early 1980s, Vapnik invented a powerful supervised learning neural-network model called the *support vector machine (SVM)*. The SVM model is based on the statistical learning theory and is particularly useful for classification with small sample sizes [1118, 1120]. Research in SVM has become very active recently, and the SVM model has been used for classification, regression and clustering.

As the neurobiological research makes more progress, more and more neural network models are being invented. While most engineers are only interested in the problem-solving capabilities of neural networks, there are many researchers devoted to modeling biological neural systems by artificial neural networks.

We conclude this section by mentioning in passing fuzzy logic and evolutionary computation. Fuzzy logic [1248] can incorporate the human knowledge into a system by means of fuzzy rules. Evolutionary computation [496, 1011] originates from Darwin's theory of natural selection, and can optimize in a domain that is difficult to solve by other means. These techniques are now widely used to enhance the interpretability of the neural networks or select optimum architecture and parameters of neural networks. In 1978, the theory of neuronal group selection (TNGS), which incorporates Darwin's theory of natural selection into neural network [329, 944], was proposed. Based on the TNGS, some neural Darwinism models were also proposed in [944].

1.2 Neurons

A neuron is a processing unit in a neural network. It is a node that processes all fan-in from other nodes and generates an output according to a transfer function called the *activation function*. The activation function represents a linear or nonlinear mapping from the input to the output and is denoted by $\phi(\cdot)$. A neuron is linked to other neurons by variable synapses (weights).

Figure 1.1 illustrates the simple neuron model proposed by McCulloch and Pitts [783]. The McCulloch–Pitts neuron model is biologically plausible, and is used in most neural-network models. The output of the neuron is given by

---

[1] The MLP trained with the BP algorithm is also called the *BP network* in the literature.

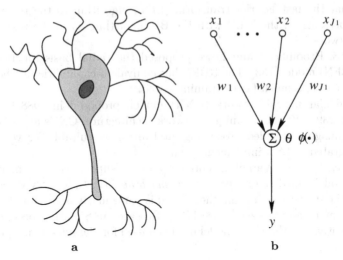

Fig. 1.1. The McCulloch–Pitts neuron model. (**a**) A biological neuron. (**b**) Mathematical model.

$$net = \sum_{i=1}^{J_1} w_i x_i - \theta = \mathbf{w}^{\mathrm{T}}\mathbf{x} - \theta \tag{1.1}$$

$$y = \phi(net) \tag{1.2}$$

where x_i is the ith input, w_i is the link weight from the ith input, $\mathbf{w} = (w_1, \cdots, w_{J_1})^{\mathrm{T}}$, $\mathbf{x} = (x_1, \cdots, x_{J_1})^{\mathrm{T}}$, θ is a threshold or bias, and J_1 is the number of inputs. The activation function $\phi(\cdot)$ is usually some continuous or discontinuous function mapping the real numbers into the interval $(-1, 1)$ or $(0, 1)$. We list below some functions that can be used as activation functions.

- Hard limiter (threshold)[2]:

$$\phi(x) = \begin{cases} 1, & x \geq 0 \\ -1, & x < 0 \end{cases} \tag{1.3}$$

- Semilinear function:

$$\phi(x) = \begin{cases} 1, & x > a \\ \frac{x}{2a} + \frac{1}{2}, & -a \leq x \leq a \\ 0, & x < -a \end{cases} \tag{1.4}$$

- Logistic function:

$$\phi(x) = \frac{1}{1 + e^{-\beta x}} \tag{1.5}$$

---

[2] It is sometimes given as $\phi(x) = \begin{cases} 1, & x \geq 0 \\ 0, & x < 0 \end{cases}$.

- Hyperbolic tangent function:

$$\phi(x) = \tanh(\beta x) \tag{1.6}$$

Other functions such as $\phi(x) = \frac{2}{\pi}\arctan(\beta x)$ and $\phi(x) = \frac{1}{2} + \frac{1}{\pi}\arctan(\beta x)$ are also used. In the above functions, β is a gain, typically selected as unity, and is used to control the steepness of the activation function.

All the above functions are monotonically increasing with the domain of output $(-1, 1)$ or $(0, 1)$. Sigmoidal functions are usually defined as those monotonically increasing functions satisfying $\lim_{x \to +\infty} \phi(x) = 1$, $\lim_{x \to -\infty} \phi(x) = 0$. Since all the above functions can satisfy this definition if stretched out, they are also treated as sigmoidal functions in the literature. Note that the hyperbolic tangent function is only a biased and scaled logistic function

$$\tanh(\beta x) = \frac{1 - e^{-\beta x}}{1 + e^{-\beta x}} = 2 \cdot \frac{1}{1 + e^{-\beta x}} - 1 \tag{1.7}$$

In practical implementations, all the neurons are typically assumed to have the same activation functions. Depending on the type of processor, the calculation of nonlinear activation functions such as (1.5) and (1.6) may consume considerable time [1167]. Piecewise-linear approximation to (1.5) along with look-up tables (LUTs) works very well [1167]. One can also use a Taylor-series approximation of (1.5) as the sigmoidal function [1167]

$$\phi(x) = \begin{cases} \frac{1}{b - x + 0.5x^2}, & x < 0 \\ 1 - \frac{1}{b + x + 0.5x^2}, & x \geq 0 \end{cases} \tag{1.8}$$

where $b \geq 2$ is a constant. When $b = 2$, (1.8) is a continuous function.

Piecewise-linear approximations of sigmoidal functions are also used in hardware implementation of a neural system [482]. Another sigmoid-like nonlinear activation function that has been found suitable for digital hardware implementation is [645]

$$\phi(x) = \begin{cases} \text{sign}(x), & |x| \geq c \\ -\frac{x \cdot |x|}{c^2} + \frac{2x}{c}, & \text{otherwise} \end{cases} \tag{1.9}$$

The above activation functions are illustrated in Fig. 1.2. Some new sigmoidal activation functions are introduced in [325].

The McCulloch–Pitts neuron model [783], which employs the sigmoidal activation function, was inspired biologically. This neuron model has been used in most neural-network models, including the MLP and the Hopfield network. Many other neural networks are also based on the McCulloch–Pitts neuron model, but use other activation functions. For example, in the adaline [1182] and Kohonen's SOM [610], linear activation functions are used, and the RBFN adopts a radial basis function (RBF). From an approximation viewpoint, the sigmoidal activation function is not always an optimal choice. Many different activation functions that are suitable for neural networks have been surveyed and defined in [326].

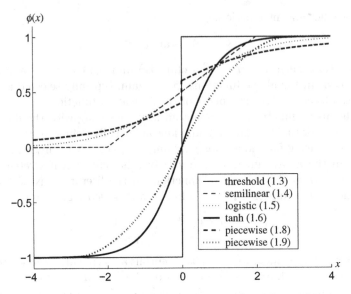

Fig. 1.2. Activation functions. $a = 2$, $b = 2.5$, $c = 3$.

1.3 Analog VLSI Implementation

Neural networks are suitable for VLSI circuit implementations. The analog approach is extremely attractive in terms of size, power, and speed. A neuron can be realized with a simple amplifier and the synapse is realized with a resistor. Corresponding to the neuron model given in Fig. 1.1, a VLSI model is given in Fig. 1.3. Since weights from the VLSI circuits can only be positive, an inverter can be applied to the input voltage so as to realize a negative synaptic weight.

According to Kirchhoff's current law, the output voltage of the neuron is derived as

$$y = \phi \left(\frac{\sum_{i=1}^{J_1} w_i x_i}{\sum_{i=0}^{J_1} w_i} - \theta \right) \tag{1.10}$$

where x_i is the ith input voltage, w_i is the conductance of the ith resistor, and θ the bias voltage. The bias voltage of a neuron in a VLSI circuit is caused by device mismatches, and is difficult to control. Due to practical limitations, design techniques based on CMOS technologies have been widely used in the hardware realizations of neural networks [659].

One limitation of VLSI implementation for general-purpose neural networks is the large number of interconnections and synapses. Since the number of synapses increases quadratically with that of neurons, silicon area is mainly occupied by the synaptic cells and the interconnection channels. For highly localized networks such as CNNs [239], the VLSI implementation is relatively simple.

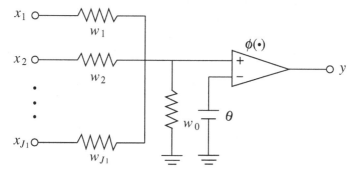

Fig. 1.3. VLSI model of a neuron. x_i, $i = 1, \cdots, J_1$, is the voltage at the ith input node, w_i, $i = 0, 1, \cdots, J_1$, is the conductance of the ith resistor, θ is the bias voltage, and $\phi(\cdot)$ is the transfer function of the amplifier.

Due to the functional regularities in neural networks, design automation is easy for the design of VLSI neural circuits. Analog VLSI circuits are sensitive to device mismatches, circuit layout, and parasitic elements; consequently, design automation in analog circuits is still quite primitive. In contrast, components in neural networks do not have to be of high precision or fast switching. The abundance of connection provides a powerful processing capability, though each neuron performs simple analog processing and the information transmission by the synapses is slow. The learning capability of neural networks can compensate initial device mismatches and long-term drift of the device characteristics. Out of the design methodologies for analog VLSI design, the analog standard-cell method is especially suitable for VLSI neural designs [659].

Pulse-stream-based architectures are recently gaining support in hardware design of neural networks [482]. This is a full digital implementation using analog circuitry. In a pulse-stream-based architecture, signals are encoded by using pulse amplitude, width, density, frequency, or phase. Pulse-mode architecture has a number of advantages over analog and conventional digital implementations. For instance, signal multiplication can be realized by using a very simple digital circuit like an AND gate, and nonlinear activation function can also be easily implemented.

1.4 Architecture of Neural Networks

The connection weight matrix $\mathbf{W} = [w_{ij}]$, where w_{ij} denotes the connection weight from node i to node j, is used to describe the network architecture. When $w_{ij} = 0$, there is no connection from node i to node j. By setting the connection weights between nodes as zero, one can realize different network topologies. According to the architecture, neural networks can be grossly classified into feedforward neural networks (FNNs), recurrent neural networks (RNNs), and their combinations. Some popular network topologies include fully connected layered FNNs, RNNs, lattice networks, layered FNNs with lateral connections, and CNNs, as shown in Fig. 1.4. The nonzero elements of $\mathbf{W}$ can be adapted by a learning algorithm.

In an FNN, the connections between neurons are in a feedforward manner. The network is usually arranged in the form of layers. In layered FNNs, there is no connection between the neurons within each layer, and no feedback between layers. A fully connected layered FNN is a network such that every node in any layer is connected to every node in its adjacent forward layer, see Fig. 1.4a. When some of the connections are missing, it becomes a partially connected layered FNN. FNNs exhibit no dynamic properties and the networks are simply a nonlinear mapping. The popular MLP and RBFN are fully connected layered FNNs.

In an RNN, there is at least one feedback connection that corresponds to an integration operation or unit delay. Thus, an RNN actually represents a nonlinear dynamic system, see Fig. 1.4b. The Hopfield model and the Boltzmann machine are RNNs.

A lattice network consists of one-, two- or higher-dimensional array of neurons, as shown in Fig. 1.4c. Each array has a corresponding set of input nodes. The Kohonen network [610] uses a one- or two-dimensional lattice architecture.

A layered FNN with lateral connections is a neural network that has lateral connections between the units at the same layer of its layered FNN architecture, as shown in Fig. 1.4d. A competition learning network has such an architecture. It is a two-layered network, and the output layer is also called the *competition layer*. The feedforward connections are excitatory, while the lateral connections in the same layer are inhibitive. Some networks for principal component analysis (PCA) using the Hebbian/anti-Hebbian learning rules [974] also employ this kind of network topology.

A CNN consists of regularly spaced neurons, called *cells*, which communicate only with the neurons in its immediate neighborhood. Adjacent cells are connected by mutual interconnections. Each cell is excited by its own signals and by signals flowing from its adjacent cells [239, 240]. The architecture of a CNN is shown in Fig. 1.4e.

Convention

When representing a neural network with a layered architecture, we use the notation J_1-J_2-$\cdots$-J_M, where M denotes the number of layers and J_i the number of nodes in the ith layer. Note that a node is sometimes not a neuron. For example, nodes at the input layer have no numerical processing function and they merely assign external inputs to other neurons, thus they are not neurons.

As a convention in this book, we count the input layer as the first layer. The notation 3-4-5-1 represents a four-layer neural network with 3 nodes at

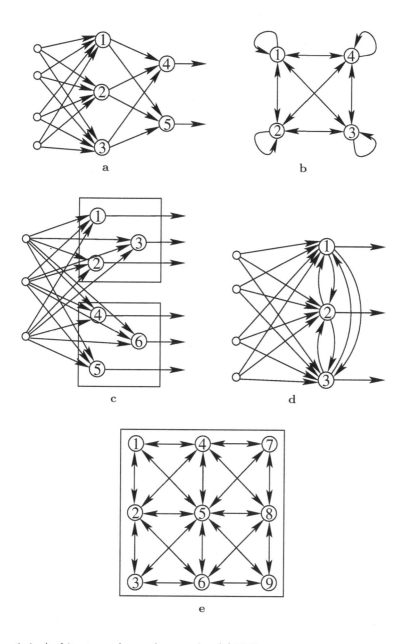

Fig. 1.4. Architecture of neural networks. (**a**) Fully connected layered FNN. (**b**) RNN. (**c**) Two-dimensional lattice network. (**d**) Layered FNN with lateral connections. (**e**) CNN. The numbered circles denotes neurons, and the small circles denote input nodes.

the input layer, 4 nodes at the second layer, 5 nodes at the third layer, and 1 node at the output layer. The second and the third layers are hidden layers. A neural network is sometimes simply called a network in this book.

1.5 Learning Methods

Learning is an important capability of neural networks. Learning rules are algorithms for finding suitable weights $\mathbf{W}$ and/or other network parameters. Learning of a neural network can be viewed as a nonlinear optimization problem in which the goal is to find a set of network parameters minimizing the cost function for given examples. Putting it in another way, learning is an optimization process that produces an output that is as close as possible to the desired output by adjusting network parameters such as $\mathbf{W}$. This kind of parameter estimation is also called a *learning* or *training algorithm*.

Neural networks are usually trained by epoch. An epoch is a complete run when all the training examples are presented to the network and are processed using the learning algorithm only once. After learning, a neural network represents a complex relationship, and possesses the ability for generalization. When a new input is presented to the trained neural network, a reasonable output is produced.

The computational complexity of an algorithm is usually characterized by the number of floating-point operations (flops)[3]. Each operation of multiplication or addition is counted as a flop. The complexity of an algorithm is usually denoted as $O(m)$, indicating that the order of floating-point operations is m.

Learning methods are conventionally divided into supervised, unsupervised, reinforcement, and evolutionary learning, and these schemes are illustrated in Fig. 1.5. Supervised learning is widely used in pattern recognition, approximation, control, modeling and identification, signal processing, and optimization. Reinforcement learning is usually used in control. Unsupervised learning schemes are mainly used for pattern recognition, clustering, vector quantization, signal coding, and data analysis. Evolutionary computation is a class of optimization techniques, which can be used to search for the global optimum of an objective function. Evolutionary learning is used for adjusting

---

[3] A floating-point number is a digital representation for a rational number and often used to approximate an arbitrary real number on a computer. A floating-point number a can be represented by two numbers m and e, such that $a - m \times b^e$, where b is the base of numeration called the *radix*, m is a p-digit number of the form $\pm d_1.d_2d_3 \cdots d_{p-1}d_p$, each digit d_i being an integer between 0 and $b - 1$ inclusive, and e is called the *exponent*. The cases of $b = 2$ and $b = 10$, respectively, correspond to the binary and decimal systems. The binary floating-point arithmetic is defined by the standard IEEE 174. For example, the decimal -0.001875 can be represented by a decimal floating-point number -0.1875×10^{-2}, a binary number -0.0011_2, and a binary floating-point number -0.11×2^{-2}.

neural network architecture and parameters using an evolutionary algorithm (EA), and can also be used to optimize the control parameters in a supervised or unsupervised learning algorithm.

1.5.1 Supervised Learning

Supervised learning is based on a direct comparison between the actual network output and the desired output. Network parameters are adjusted by a combination of the training pattern set and the corresponding errors between the desired output and the actual network response. Supervised learning is a closed-loop feedback system, where the error is the feedback signal. The trained network is used to emulate the system.

To control a learning process, a criterion is needed to decide the time for terminating the process. For supervised learning, an error measure, which shows the difference between the network output and the output from the training samples, is used to guide the learning process. The error measure is usually defined by the mean squared error (MSE)

$$E = \frac{1}{N} \sum_{p=1}^{N} \|\mathbf{y}_p - \hat{\mathbf{y}}_p\|^2 \tag{1.11}$$

where N is the number of pattern pairs in the sample set S, $\mathbf{y}_p$ is the output part of the pth pattern pair, and $\hat{\mathbf{y}}_p$ is the network output corresponding to the pattern pair p. This is also the objective function to optimize. The error E is calculated anew after each epoch. The learning process is terminated when E is sufficiently small or a failure criterion is met.

To make E decrease toward zero, a gradient-descent procedure is usually applied. The gradient-descent method always converges to a local minimum in a neighborhood of the initial solution of network parameters. The LMS [1182] and BP algorithms [979, 1166] are two early, but most popular, supervised learning algorithms. Both of them are derived using a gradient-descent procedure. Second-order methods, which are based on the computation of the Hessian matrix, are also used. The Levenberg–Marquardt (LM) method [764], which optimizes all the parameters of a model regardless of the parameter features, is very successful for nonlinear least-squares optimization problems. Some recent advances in supervised learning have been reviewed in [744].

1.5.2 Unsupervised Learning

Unsupervised learning involves no target values. It tries to autoassociate information from the inputs with an intrinsic reduction of data dimensionality or total amount of input data. Unsupervised learning is solely based on the correlations among the input data, and is used to find the significant patterns or features in the input data without the help of a teacher. Unsupervised

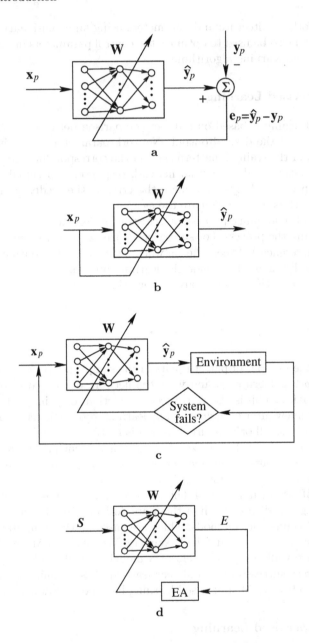

Fig. 1.5. Learning methods. (**a**) Supervised learning. (**b**) Unsupervised learning.
(**c**) Reinforcement learning. (**d**) Evolutionary learning. $\mathbf{x}_p$ and $\mathbf{y}_p$ are the input and
output of the pth pattern in the training set $\mathcal{S}$, $\hat{\mathbf{y}}_p$ is the neural network output for
the pth input, and E is an error function.

learning is particularly suitable for biological learning in that it does not rely on a teacher and it uses intuitive primitives like neural competition and co-operation.

A criterion is needed to terminate the learning process. Without a stop criterion, a continuous learning process continues even when a pattern, which does not belong to the training patterns set, is presented to the network. The network is adapted according to a constantly changing environment. Hebbian learning [473], competitive learning [435], and Kohonen's SOM [610] are the three mostly used unsupervised learning approaches. Generally speaking, unsupervised learning is slow to settle into stable conditions.

In Hebbian learning [473], learning is a purely local phenomenon, involving only two neurons and a synapse. The synaptic weight change is proportional to the correlation between the pre- and postsynaptic signals. Many neural networks for PCA and associative memory are based on Hebbian learning.

In competitive learning, the output neurons of a neural network compete for the right to respond. In a simple competitive learning, the neural network has a single layer of output neurons, called the competition layer, operating by the WTA rule. Kohonen's SOM is also based on competition learning. Competitive learning is directly related to clustering, and the latter maps similar input patterns into output patterns that are also similar. Clustering is valuable for reducing the amount of input data. The C-means algorithm is a popular competitive learning-based clustering method [746]. By using the correlation of the input vectors, the learning rule changes the network weights to group the input vectors into clusters.

The Boltzmann machine [9] uses a kind of stochastic training technique known as SA [607], which can been treated as a special type of unsupervised learning based on the inherent property of a physical system.

1.5.3 Reinforcement Learning

Reinforcement learning [75] is a special case of supervised learning, where the exact desired output is unknown. It is based only on the information as to whether or not the actual output is close to the estimate. Explicit computation of derivatives is not required. This, however, presents a slower learning process.

Reinforcement learning is a learning procedure that *rewards* the neural network for its *good* output result and *punishes* it for the *bad* output result. It is used in the case when the correct output for an input pattern is not available and there is need for developing a certain output.

The evaluation of an output as *good* or *bad* depends on the specific problem and the environment. For a control system, if the controller still works properly after an input, the output is judged as *good*; Otherwise, it is considered as *bad*. The evaluation of the output is binary, and is called *external reinforcement*. Thus, reinforcement learning is a kind of supervised learning with the external reinforcement as the error signal. Reinforcement learning can learn the system structure by trial-and-error, and is suitable for online learning [75, 74, 567].

The binary external reinforcement provides very limited information for the learning algorithm. An additional adaptive critic network [1168] is usually used to predict the future reinforcement signal, called *internal reinforcement*. This assures avoiding *bad* states from happening.

1.5.4 Evolutionary Learning

The evolutionary learning approach is attractive since it can handle the global search problem better on a vast, complex, multimodal, and nondifferentiable surface. It is not dependent on the gradient information of the error (or fitness) function, and thus is particularly appealing when this information is unavailable or very costly to obtain or estimate.

EAs can be used to search for the optimal control parameters in supervised as well as unsupervised learning by optimizing their respective objective functions. It can also be used as an independent training method for network parameters by optimizing the error function.

EAs are widely used for training neural networks and tuning fuzzy systems, and are generally much less sensitive to the initial conditions. They always search for a globally optimal solution, while supervised and unsupervised learning algorithms can only find a local optimum in a neighborhood of the initial solution [1225].

1.6 Operation of Neural Networks

Neural networks are characterized by the network architecture, node characteristics, and learning rules. They are roughly divided into feedforward and recurrent architectures according to their connectivity. RNNs cannot guarantee stability, and Lyapunov functions are usually constructed to prove their stability.

The operation of neural networks is divided into two stages: learning (training) and generalization (recalling). Network training is typically accomplished by using examples, and network parameters are adapted using a learning method. This can be done in an online or offline manner. Once the network is trained to accomplish the desired performance, the learning process is terminated and it can then be used directly to replace the complex system dynamics. The trained network can be used to operate in a static manner: to simulate an unknown dynamics or nonlinear relationship.

For real-time applications, a neural network is required to have a constant processing delay regardless of the number of input nodes, and a minimum number of layers. As the number of input nodes increases, the size of the network layers should grow at the same rate without additional layers.

1.6.1 Adaptive Neural Networks

Adaptive neural networks are a class of neural networks that do not need to be trained by a training pattern set. They can learn when they are performing. For adaptive neural networks, unsupervised learning methods are usually used. Unsupervised learning naturally adapts to whatever input is supplied to it.

For example, the Hopfield model [502] uses a generalized Hebbian learning rule for implementation as associative memory. Any time a pattern is presented to it, the Hopfield network always updates the connection weights. After the network is trained with standard patterns and is prepared for generalization, the learning capability should be disabled; otherwise, when an incomplete or noisy pattern is presented to the network, it will search the closest matching, meanwhile the memorized pattern is replaced by this new pattern.

Reinforcement learning is also naturally adaptive, where the environment is treated as a teacher. Supervised learning is not adaptive in nature. However, when environment is treated as a teacher, supervised learning can also be used online.

1.7 Properties of Neural Networks

Neural networks are usually biologically motivated. Each neuron is a computational node, which represents a nonlinear function. Neural networks possess the following advantages:

1. **Adaptive learning**
   Neural networks have strong learning capability. They can adapt themselves by changing the network parameters in a surrounding environment. Powerful learning algorithms make this capability possible. Learning and generalization are the most salient features of neural networks.
2. **Generalization**
   A well-trained neural network has superior generalization capability. This can be attributed to the bounded and smooth nature of the hidden-unit responses. The bounded-unit response localizes the nonlinear effects of the individual hidden units in a neural network and allows for the approximations in different regions of the input space to be independently tuned [461]. In contrast, in conventional curve-fitting methods, the polynomials and other functions have a potential divergence nature.
3. **General-purpose nonlinear nature**
   A neuron has a typical nonlinear property. Neural networks aim at solving problems using their universal approximation capability. They perform like a black box. This negates the necessity of analyzing the structure or looking for a knowledge of the problems. Only the inputs and outputs have physical meaning.

4. **Self-organizing**

   Some neural networks such as the SOM [610] and competitive learning-based neural networks have a self-organization property. The training of these networks is based on the unsupervised learning algorithms.

5. **Massive parallelism**

   Neural networks possess parallel-processing structure. Each basic processing unit usually has a uniform property, which uses simple addition, multiplication, division, and threshold operations. This parallel structure allows for highly parallel software and hardware implementations.

6. **Robustness and fault tolerance**

   Neural networks have robustness and fault-tolerant capability. A neural network can easily handle imprecise, fuzzy, noisy, and probabilistic information. It is a distributed information system, where information is stored in the whole network in a distributed manner by the network structure such as **W**. Thus, the overall performance does not degrade significantly when the information at some node is lost or some connections in the network are damaged. The network will immediately improve the performance by updating the connection weights using the learning rule and the current result. Therefore, the network can repair itself, and possesses a strong fault-tolerant capability.

   Some neural networks such as the Hopfield network can be used as associative storage of information. When a noisy or incomplete pattern is presented to a trained network, it will help to find the correct pattern; that is, the trained network is fault tolerant.

7. **Simple VLSI implementations**

   The massive parallelism using simple uniform units can be readily implemented in analog VLSI or optical hardware [345, 130], or be implemented on special-purpose massively parallel hardware [392].

1.8 Applications of Neural Networks

Neural networks find applications in almost all disciplines of science and engineering. The applications are generally in modeling and system identification [1042, 838], classification [1252], pattern recognition, optimization, control [838], industrial application [793], communications [528], and signal processing [321].

1.8.1 Function Approximation

Most applications of neural networks utilize the function approximation capability of neural networks. These include modeling and system identification, regression and prediction, control, signal processing, pattern recognition and classification, and associative memory. Image restoration is also a function approximation problem. The neural networks are trained using supervised learning. The input and output layers are defined according to the dimensions of the input and output patterns. The input-output pattern pairs are fed to the network, and network parameters are adjusted accordingly. A trained network represents the learned nonlinear relation between the input-output pairs, and can generalize when an unknown input pattern is presented to the network.

FNNs such as the MLP [273, 389] and the RBFN [887] are universal approximators for nonlinear functions. Some RNNs are universal approximators of dynamical systems [390].

1.8.2 Classification

A conventional statistical classifier is the optimal, parametric Bayes classifier. Bayesian design assumes functional forms for the densities in the mixture model and estimates the parameters. The maximum likelihood (ML) estimation is a popular statistical approach for parametric estimation and it treats the best estimate as the one that maximizes the joint probability density function (PDF), from which all the observations are generated. The k-nearest neighbor (k-NN) rule is a supervised, nonparametric approximation to the Bayes classifier. The k-NN rule presumes to have a large number of correctly labeled examples from each class, and is based on a simple majority voting. The C4.5 algorithm [930] is a popular classifier that creates a decision tree based on a labeled data set.

Classification is the most fundamental application of neural networks. Classification can be based on the function approximation capability of neural networks. Neural networks used for classification are usually trained by supervised methods. Both the MLP and the RBFN are widely used for classification. Although the network can have a single output node, with discrete values representing different classes, usually the network architecture is slightly different.

The architecture of a neural-network classifier is usually defined according to the convention from a competitive learning-based classifier. To classify a pattern set with J_1-dimensional inputs into K classes, the neural network is selected usually as having J_1 input nodes, K output nodes with each corresponding to one class, and zero or multiple hidden layers. During training, the target value of the kth output node corresponding to the kth target class is set to 1, while the target values of all the other nodes are set to 0. A pattern is considered to be correctly classified if the target output node has the highest output among all the output nodes.

1.8.3 Clustering and Vector Quantization

Clustering groups together similar objects. Unlike in classification problems, the classmembership of a pattern is not known *a priori*. Clustering is based on some distance measure. Vector quantization (VQ) is similar to clustering. It performs the division of the input space into several connected regions, called Voronoi regions[4]. Each point in the input space is mapped to its nearest codebook vector.

Clustering is particularly useful in data analysis and machine learning, including pattern classification, system modeling, data mining, document retrieval, and image segmentation. Many clustering neural networks and competitive learning-based algorithms such as SOM [610] and ART [159] models are also widely used for clustering and VQ.

1.8.4 Associative Memory

An association is an input-output pair. Associative memory, also known as *content-addressable memory*, is a memory organization that accesses memory by its content instead of its address. It picks up a desirable match from all stored prototypes. An associative memory can be autoassociative or heteroassociative.

The architecture of an associative memory network may be feedforward, bidirectional, or recurrent. The Hopfield model and the Boltzmann machine are popular autoassociative memories. An MLP trained with the equivalent input and output patterns also functions as autoassociative memory.

Associative memories are useful for pattern recognition and pattern association. When an incomplete or corrupted sample is presented to the neural network, the network is required to recall the stored correct pattern.

1.8.5 Optimization

Many significant engineering and scientific problems involve optimization of some criteria over a combinatorial or constrained configuration space. Some neural-network models, such as the Hopfield model and the Boltzmann machine, can be used to solve combinatorial optimization problems (COPs). The traveling salesman problem (TSP) is a classical COP. A clustering problem can also been transformed into a combinatorial optimization problem.

SA [607] is an effective global optimization algorithm for neural-network learning and general optimization problems. Many other global optimization methods including the popular EAs are also introduced in this book.

1.8.6 Feature Extraction and Information Compression

Coding and information compression is an essential processing task in the transmission and storage of speech, image, audio, video, and other information. The objective of data compression is to reduce the amount of data and

---

[4] Also known as *Voronoi polygons*.

to achieve low bit rate digital representation without perceived loss of information quality.

PCA networks can be used for feature extraction, which is necessary for pattern recognition. PCA is widely used in signal preprocessing or feature extraction. By compressing the high-dimensional data into low-dimensional data, the signals are compressed, and PCA networks thus accomplish data compression. VQ also achieves the objective of feature extraction and information compression.

1.9 Array Signal Processing as Examples

In many signal-processing applications such as biomedical, industrial, speech, and sonar/radar, and wireless communications, signals come from multichannels and are received by several sensors. Array signal processing (ASP) is an important research area [417], which focuses on two problems: direction finding of the sources and beamforming of the desired sources. A sensor array serves as a nonlinear mapping from signal sources to array measurement. Both problems, namely, finding the direction of the sources and beamforming the desired sources, are inverse problems in that the information regarding the signal sources is being determined from the array measurement. The objective of direction finding is to estimate the direction-of-arrival (DoA) of the source signals from the measurement of the array output, while the task of beamforming is to recover the desired signals from the received mix of signals.

1.9.1 Array Signal Model

Most ASP algorithms are developed for applications in the wireless scenario, where signal sources are usually assumed to be in the far field and as a result, the received signals are assumed to have planar wavefronts.

Consider a sensor array of L omnidirectional elements, surrounded by M point sources with position $\mathbf{r}_i = (r_i, \phi_i, \theta_i)$, $i = 1, \cdots, M$, in the spherical coordinate system, as shown in Fig. 1.6. For narrowband signals, the induced signal on the reference element due to the ith source can be expressed as $s_i = m_i(t)e^{j2\pi f_0 t}$, where $m_i(t)$ is the complex modulating function and $j = \sqrt{-1}$. The total measured signal on the lth element is

$$x_l = \sum_{i=1}^{M} s_i e^{j2\pi f_0 \tau_l(\phi_i, \theta_i)} + n_l(t)$$

$$= \sum_{i=1}^{M} d_{i,l} s_i + n_l(t) = \mathbf{d}_i^{\mathrm{T}} \mathbf{s} + n_l \qquad (1.12)$$

where $d_{i,l} = e^{j2\pi f_0 \tau_l(\phi_i, \theta_i)}$ is the steering direction of the ith source on the lth element, $\mathbf{d}_i = (d_{i,1}, \cdots, d_{i,L})^{\mathrm{T}}$ is the steering vector associated with the ith

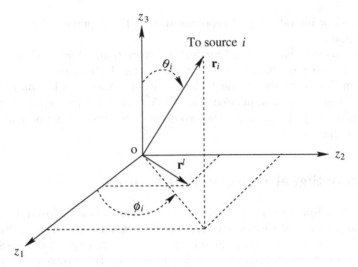

Fig. 1.6. Coordinate system for the signal model in the wireless communication scenario

source, $\mathbf{s} = (s_1, \ldots, s_M)^{\mathrm{T}}$, $\tau_l(\phi_i, \theta_i) = \frac{\mathbf{r}^l \cdot \hat{\mathbf{r}}_i}{c}$, $\mathbf{r}^l$ is the position vector of the lth element, $\hat{\mathbf{r}}_i$ is the unit position vector of the ith source, c is the speed of wave propagation, and $n_l(t)$ is the random noise component on the lth element, which is assumed to be zero-mean white Gaussian with variance σ_{n}^2.

For all the sources and array elements, the signal model can be written by

$$\mathbf{x} = \mathbf{D}^{\mathrm{T}}\mathbf{s} + \mathbf{n} \qquad (1.13)$$

where $\mathbf{x} = (x_1, \ldots, x_L)^{\mathrm{T}}$, $\mathbf{D} = [\mathbf{d}_1 \ldots \mathbf{d}_M]$, and $\mathbf{n} = (n_1, \ldots, n_L)^{\mathrm{T}}$. For a uniform linear array, $\mathbf{d}_i$ can be expressed as

$$\mathbf{d}_i = \left(1, e^{-\mathrm{j}\bar{\tau}_i}, e^{-\mathrm{j}2\bar{\tau}_i}, \cdots, e^{-\mathrm{j}(n-1)\bar{\tau}_i}\right)^{\mathrm{T}} \qquad (1.14)$$

where $\bar{\tau}_i = \frac{2\pi d_0}{\lambda_i}\sin\theta_i$ is the differential propagation delay between two adjacent elements for the ith source, d_0 is the interelement separation, and λ_i and θ_i are, respectively, the wavelength and the DoA of the radiation from the ith source.

The architecture of the narrowband beamformer is shown in Fig. 1.7. The output from the beamformer is defined by

$$\mathbf{y}(t) = \mathbf{W}^{\mathrm{H}}\mathbf{x}(t) \qquad (1.15)$$

where $\mathbf{y}(t) = (y_1(t), \cdots, y_M(t))^{\mathrm{T}}$, y_i denotes the restored signal for the i source, $\mathbf{W} = [\mathbf{w}_1 \ldots \mathbf{w}_M]$ is an $L \times M$ weight matrix, and superscript H denotes the Hermitian transpose.

If the components of $\mathbf{x}(t)$ can be modeled as a zero-mean stationary process, the mean output power of the beamformer is given by

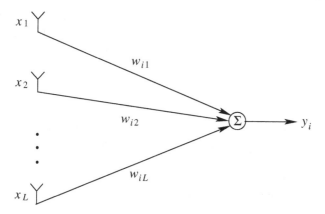

Fig. 1.7. Architecture of the narrowband beamformer. The inputs from sensor measurement are combined to restore the signal from the ith source.

$$P(\mathbf{W}) = E\left[\mathbf{y}(t)\mathbf{y}^H(t)\right] = \mathbf{W}^H \mathbf{R} \mathbf{W} \tag{1.16}$$

where $E[\cdot]$ denotes the expectation operator and $\mathbf{R}$ is the array correlation matrix

$$\mathbf{R} = E\left[\mathbf{x}(t)\mathbf{x}^H(t)\right] = \mathbf{D}\mathbf{S}\mathbf{D}^H + \sigma_n^2 \mathbf{I}_L \tag{1.17}$$

and $\mathbf{S} = E\left[\mathbf{s}\mathbf{s}^H\right]$, an $M \times M$ matrix, denotes the source correlation. For uncorrelated sources, $\mathbf{S}$ is a diagonal matrix.

$\mathbf{R}$ can be eigen-decomposed as

$$\mathbf{R} = \mathbf{V}\boldsymbol{\Lambda}\mathbf{V}^H \tag{1.18}$$

where $\boldsymbol{\Lambda} = \mathrm{diag}\left(\lambda_1, \ldots, \lambda_M, \lambda_{M+1}, \ldots, \lambda_L\right)$, with M signal eigenvalues and $L - M$ noise eigenvalues $\lambda_1 \geq \ldots \geq \lambda_M \geq \lambda_{M+1} \geq \ldots \geq \lambda_L$, and $\mathbf{V}$ is composed of L eigenvectors.

Based on different optimization objectives such as maximizing the output signal-to-noise ratio (SNR) or minimizing the output interference, many ASP algorithms have been formulated [417].

1.9.2 Direction Finding and Beamforming

ASP mainly involves two classes of problems, namely, direction finding and beamforming. Some systems, such as the radar or sonar systems, need only direction finding to detect the signals. In other systems, such as the wireless communications, biomedical, and speech-processing systems, signals have to be acquired by using beamforming techniques.

Direction Finding

The ML method is the optimum method for DoA estimation. It estimates the DOAs from the array measurement by maximizing the log-likelihood function, which is the joint PDF of the sampled data given the DOAs and viewed as a function of the DOAs. Many iterative methods are available in the literature for this nonlinear optimization problem [417]. The ML method gives a performance superior to other methods, particularly when the SNR is small, the number of samples is small, or the sources are correlated [417]. The ML method is typically computationally expensive.

MUSIC [1006] is a simple and efficient direction-finding method. In its standard form, the MUSIC estimates the noise subspace from the available samples. This can be done by either the eigenvalue decomposition (EVD) of the estimated array correlation matrix $\widehat{\mathbf{R}}$ or the singular value decomposition (SVD) of the data matrix $\mathbf{X}$, with its columns being samples $\mathbf{x}_i$. The SVD method provides a better numerical stability. After the noise subspace is estimated, the M source directions can be determined by finding those steering vectors that are as orthogonal to the noise subspace as possible. This is achieved by locating peaks in the MUSIC spectrum, $P_{\text{MUSIC}}(\phi, \theta)$, defined by [1006, 417]

$$P_{\text{MUSIC}}(\phi, \theta) = \frac{1}{|\mathbf{d}^{\text{H}}(\phi, \theta)\mathbf{U}_{\text{n}}|^2} \tag{1.19}$$

where $\mathbf{U}_{\text{n}}$ denotes an $L \times (L - M)$ matrix with its $(L - M)$ columns being the eigenvectors corresponding to the $(L - M)$ smallest eigenvalues of $\widehat{\mathbf{R}}$, and $\mathbf{d}(\phi, \theta)$ denotes the steering vector in the direction (ϕ, θ). MUSIC achieves a high resolution for signals with small angular separation and a good performance under low SNR, but it suffers from a high sensitivity to the structure of the covariance matrix.

Beamforming

Mobile communication systems need beamforming to acquire the signals. Beamforming methods are generally based on the DoA estimated by a calibrated array, or based on a known training signal transmitted by the user. There are also some blind beamforming methods, which make use of the inherent properties of signals such as cyclostationarity [323] and do not require a knowledge of the DoA or the training sequence.

Conventional beamforming algorithms such as the minimum-variance distortionless response (MVDR) and recursive least squares (RLS) algorithms [417] require a knowledge of the DoAs of the sources. Once the DoAs of the sources are available, the beamforming algorithms can be used to track, in real time, those sources that are of interest, and null out the other sources as interference by controlling the beampattern of the antenna array in an adaptive way. The MVDR method is derived by minimizing the output signal power while maintaining unity response in the desired direction. The MVDR method gives the beamforming weight vector as

$$\mathbf{w} = \frac{\mathbf{R}^{-1}\mathbf{d}_0}{\mathbf{d}_0^{\text{H}}\mathbf{R}^{-1}\mathbf{d}_0} \tag{1.20}$$

where $\mathbf{d}_0$ is the steering vector in the desired direction. The MVDR is an optimal method in the ML sense and it maximizes the output SNR as well. A review of direction-finding and beamforming algorithms can be found in [417].

Neural-network Methods

Conventional methods are typically linear algebra based, requiring computationally intensive matrix inversion, and cannot meet the requirements for real-time and/or multisource tracking. They also require calibrated antennas with uniform features, and are sensitive to the manufacturing fault and other physical uncertainties. These techniques make use of the first- and second-moment information of the data, and the higher-order statistics (cumulant) arising from the correlation between the signal sources and imperfect array geometry are missed, thus reducing the quality of performance. Cumulants are useful to describe non-Gaussian signals, and are also insensitive to additive Gaussian noise.

These limitations can be alleviated by the use of the properties of neural networks. Neural-network methods are adaptive, general-purpose methods. The neural-network processors for the two ASP problems are shown in Fig. 1.8. Some preprocessing and postprocessing are necessary. For example, for the purpose of direction finding, the initial phase contains no information about the DoA and can be eliminated at the preprocessing phase. The input of the network can also be normalized, since the signal gain does not affect the detection of the DoA. For ASP, a neural network is first trained, and it then performs direction finding or beamforming. The schematic of a neural-network beamformer is shown in Fig. 1.9.

The ASP problems are in essence complex-valued optimization problems, and thus one needs to separate the real and the imaginary parts of the complex computation. Currently, existing applications of neural networks to ASP are mainly focused on direction finding, albeit some on beamforming. A review of neural network methods for ASP is given in [321].

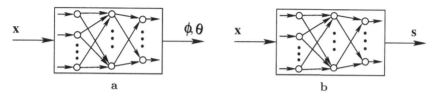

Fig. 1.8. ASP neural-network models. (**a**) Direction finding. (**b**) Beamforming. $\mathbf{x}$ is the array measurement, ϕ and θ are the azimuthal and elevation angle vectors, respectively, and $\mathbf{s}$ is the extracted signal vector.

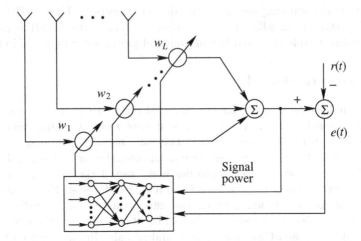

Fig. 1.9. Schematic of a neural-network beamformer. The neural network first undergoes supervised training with data generated from existing algorithms. After the training is completed, the neural network acts as a beamformer. $r(t)$ and $e(t)$ are, respectively, the reference and error signals.

1.10 Scope of the Book

This book is devoted to a detailed exposition of some of the most popular neural-network models. These include the MLP, the Hopfield model, the Boltzmann machine, clustering networks, the RBFN, PCA networks, and independent component analysis (ICA) networks. We include the major research results concerning these neural-network models that have been published in the past decade. These models suffice for most of the applications that employ neural networks, and in fact, the majority of the existing applications of neural networks actually employ one of these models.

Fuzzy logic and some neurofuzzy models are also introduced. Like neural networks, they are also universal approximators, and can be used as black-box models for nonlinear systems represented by examples. The difference is that fuzzy systems have the capability of incorporating knowledge and explaining the action of a nonlinear system by fuzzy rules.

EAs and some other heuristics are also described as general-purpose global optimization methods, with an emphasis on the genetic algorithm (GA) [496], the evolutionary strategy (ES) [1011], and SA [607]. These methods can be used for training neural networks and adapting fuzzy and neurofuzzy systems.

We select the problems of ASP, namely, direction finding and beamforming, as simulation examples in some chapters. They are typical signal-processing problems and are gaining significant interest as well.

1.10.1 Summary by Chapters

The contents of the subsequent chapters are described below.

- Chapter 2 describes some algorithm-independent topics on neural networks and softcomputing. These include the statistical learning theory, no free lunch theorem, universal function approximation of FNNs, learning and generalization, model selection, robust learning as well as some feature-selection and feature-extraction techniques.
- The rediscovery of the BP algorithm by Rumelhart *et al.* in 1986 was a landmark in the history of neural-network research. Due to the BP algorithm, the MLP has become one of the most widely used neural-network models. Chapter 3 is devoted to the MLP. The BP algorithm and second-order learning algorithms are described. Applications of the MLP to dynamic systems are also discussed.
- The emergence of the Hopfield model in 1982 aroused great interest in and gave impetus to neural networks. It is the most important dynamic neural-network model. In Chapter 4, the Hopfield model is first introduced as associative memory, and then the fixed-point property is used for solving COPs. The SA algorithm is described as a global optimization method, and the Boltzmann machine is introduced as an extension of the Hopfield model. Some deterministic SA algorithms are also introduced.
- Clustering is an unsupervised classification technique used in identifying some inherent structure present in a set of objects. In Chapter 5, we describe some popular clustering and competitive learning networks. Clustering algorithms are most useful for the prototype selection of kernel-based neural networks such as the RBFN and rule extraction for neurofuzzy systems.
- In Chapter 6, we elaborate on the RBFN model. The RBFN is considered as a faster alternative to the MLP, and is now widely used for function approximation. Supervised and unsupervised learning algorithms are described. RBFNs for dynamical systems are mentioned. A comparison between the RBFN and the MLP is also made.
- PCA is a statistical method that is widely used in data analysis. It is an orthogonalization technique that can be used for feature extraction and information compression. In Chapter 7, many PCA networks and algorithms are described. The minor component analysis (MCA) is a variant of the PCA, and some MCA algorithms are also discussed. ICA was originally presented for blind signal separation, and has been generalized for feature extraction. ICA, which is directly associated with the PCA, is also described in this chapter. Finally, crosscorrelation PCA networks, which are related to SVD of the crosscorrelation matrix of two stochastic signals, are introduced.
- Fuzzy logic provides a means to represent knowledge. It is also a universal approximator. A neurofuzzy system is a synergy of fuzzy logic and neural networks. In addition to its function as a black-box model to an unknown

system, a neurofuzzy system can also represent the system by means of fuzzy rules. In Chapter 8, we first give some fundamentals of fuzzy logic. Transformations between fuzzy logic and neural networks, as well as some popular neurofuzzy models, are also discussed.

- EAs are excellent general-purpose global optimization methods. In Chapter 9, we first describe EAs with emphasis on GAs and ESs. Then, applications of EAs to optimizing the parameters and architectures of neural networks or fuzzy systems are described.
- Chapter 10 contains a short summary of the material discussed in the earlier chapters. This is followed by a brief description of some interesting or emerging methods of neural networks, numerical optimization, and softcomputing. Finally, we point out some perspectives on neural networks and softcomputing, and propose some possible future research directions.
- Some mathematical preliminaries are included in the appendix.

2

Fundamentals of Machine Learning and Softcomputing

This chapter introduces some fundamentals of machine learning and softcomputing. These includes topics on the Vapnik–Chervonenkis (VC) theory, probably approximately correct (PAC) learning, no free lunch (NFL) theorem for search and optimization, universal function approximation of FNNs, learning and generalization, model selection, robust learning, neural-network processor as well as some feature-selection and feature-extraction techniques.

2.1 Computational Learning Theory

Machine learning makes predictions about the unknown underlying model based on a training set drawn from hypotheses. Due to the finite training set, learning theory cannot provide absolute guarantees of performance of the algorithms. The performance of learning algorithms is commonly bounded by probabilistic terms.

Computational learning theory is a statistical tool for the analysis of machine-learning algorithms, that is, for characterizing learning and generalization. There are a number of approaches to computation learning theory. Among them, two popular formalisms are the *VC theory* [1121] and the probably approximately correct (PAC) learning [1115]. Both approaches are nonparametric or distribution-free learning models.

The VC theory [1121], known as the statistical learning theory, is a dependency-estimation method with finite data. Necessary and sufficient conditions for consistency and fast convergence are obtained based on the empirical risk minimization (ERM) principle. Uniform convergence for a given class of approximating functions is associated with the capacity of the function class considered [1121]. The capacity and complexity of the function class is measured in terms of the VC dimension. The ERM principle has been practically applied in the SVM [1119].

The PAC learning [1115] aims to find a hypothesis that is a good approximation to an unknown target concept with a high probability. The PAC learning paradigm is intimately associated with the ERM principle [1121]. A hypothesis that minimizes the empirical error, based on a sufficiently large sample, will approximate the target concept with a high probability. The generalization ability of network training can be established estimating the VC dimension of neural architectures. Boosting [1003] is a PAC learning-inspired method for supervised learning.

2.1.1 Vapnik–Chervonenkis Dimension

The VC dimension [1120] is a combinatorial characterization of the diversity of functions that can be computed by a given neural architecture. The VC dimension can be viewed as a generalization of the concept of capacity first introduced by Cover [267]. For linear models, the VC dimension is equivalent to the number of model parameters.

A subset S of the domain $\mathcal{X}$ is shattered by a class of functions or neural network $\mathcal{N}$ if every function $f : S \rightarrow \{0,1\}$ can be computed on $\mathcal{N}$. The VC dimension of $\mathcal{N}$ is defined as the maximal size of a set $S \subseteq \mathcal{X}$ that is shattered by $\mathcal{N}$

$$\dim_{\mathrm{VC}}(\mathcal{N}) = \max \left\{ |S| \big| S \subseteq \mathcal{X} \text{ is shattered by } \mathcal{N} \right\} \qquad (2.1)$$

where the cardinality of S is denoted $|S|$.

For example, for a neural network with the relation $f(\mathbf{x}, \mathbf{w}, \theta) = \mathrm{sgn}\left(\mathbf{w}^{\mathrm{T}}\mathbf{x} + \theta\right)$, it can shatter at most any three points in $\mathcal{X}$, thus its VC dimension is 3. This is shown in Fig. 2.1.

A hard-limiter function with threshold θ_0 is typically used as the activation function for binary neurons. The basic function of the McCulloch–Pitts neuron has a linear relation applied by a threshold operation, hence called a *linear threshold gate (LTG)*.

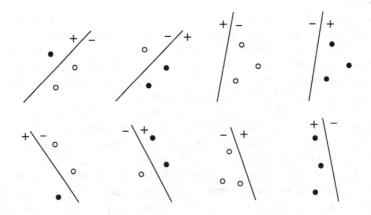

Fig. 2.1. Shatter any three points in $\mathcal{X}$ into two classes. The points are in general positions, that is, they are linearly independent.

The neural network with LTG has a VC dimension of $O\left(N_w \log N_w\right)$ [73], where N_w is the number of weights in a network. The VC dimension has been generalized for neural networks with real-valued output. The VC dimension of various neural networks have been studied in [73].

It can be used to estimate the number of training examples for a good generalization capability. The Boolean VC dimension of a neural network $\mathcal{N}$, written $\dim_{BVC}(\mathcal{N})$, is defined as the VC dimension of the class of Boolean functions that is computed by $\mathcal{N}$. The VC dimension can be regarded as a measure of the capacity or expressive power of a network.

2.1.2 Empirical Risk-minimization Principle

Given a set of N samples, $\{(\mathbf{x}_i, y_i)\}$. Assume that these data are independently drawn and identically distributed (iid) samples from some unknown probability distribution $P(\mathbf{x}, y)$. Assume a machine defined by a set of possible mappings $\mathbf{x} \to f(\mathbf{x}, \boldsymbol{\alpha})$, where $\boldsymbol{\alpha}$ contains adjustable parameters. When $\boldsymbol{\alpha}$ is selected, the machine is called a *trained machine.*

The expected risk is the expectation of the generalization error for a trained machine, and is given by

$$R(\boldsymbol{\alpha}) = \int L\left(y, f(\mathbf{x}, \boldsymbol{\alpha})\right) \mathrm{d}P(\mathbf{x}, y) \tag{2.2}$$

where $L\left(y, f(\mathbf{x}, \boldsymbol{\alpha})\right)$ is the loss function, measuring the discrepancy between the output pattern y and the output of the learning machine $f(\mathbf{x}, \boldsymbol{\alpha})$. The loss function can be defined in different forms for different purposes:

For classification

$$L(y, f(\mathbf{x}, \boldsymbol{\alpha})) = \begin{cases} 0, & y = f(\mathbf{x}, \boldsymbol{\alpha}) \\ 1, & y \neq f(\mathbf{x}, \boldsymbol{\alpha}) \end{cases} \tag{2.3}$$

For regression

$$L(y, f(\mathbf{x}, \boldsymbol{\alpha})) = (y - f(\mathbf{x}, \boldsymbol{\alpha}))^2 \tag{2.4}$$

For density estimation

$$L(p(\mathbf{x}, \boldsymbol{\alpha})) = -\ln p(\mathbf{x}, \boldsymbol{\alpha}) \tag{2.5}$$

The empirical risk $R_{\text{emp}}(\boldsymbol{\alpha})$ is defined to be the measured mean error on a given training set

$$R_{\text{emp}}(\boldsymbol{\alpha}) = \frac{1}{N} \sum_{i=1}^{N} L\left(y_i, f\left(\mathbf{x}_i, \boldsymbol{\alpha}\right)\right) \tag{2.6}$$

The ERM principle aims to approximate the loss function by minimizing the empirical risk (2.6) instead of the risk (2.2), with respect to model parameters.

When the loss function takes the value 0 or 1, with probability $1 - \delta$, there is the upper bound [1119]

$$R(\alpha) \leq R_{\text{emp}}(\alpha) + \sqrt{\frac{d \left(\ln \frac{2N}{d} + 1 \right) - \ln \frac{\delta}{4}}{N}} \qquad (2.7)$$

where d is the VC dimension of the machine. The second term on the right-hand side is called the *VC confidence*.

The VC confidence term in (2.7) monotonically increases with increasing d. Reducing d leads to a better upper bound on the actual error. The VC confidence depends on the class of functions, whereas the empirical risk and actual risk depend on the particular function obtained by the training procedure.

The principle of structural risk minimization (SRM) purports to minimize the risk functional with respect to both the empirical risk and the VC dimension of the set of functions, thus it aims to find the subset of functions that minimizes the bound on the actual risk.

2.1.3 Probably Approximately Correct (PAC) Learning

The PAC learning paradigm is concerned with learning from examples of a target function called *concept*, by choosing from a set of functions known as the *hypothesis space*, a function meant to be a good approximation to the target.

Let $\mathcal{C}_n$ and $\mathcal{H}_n$, $n \geq 1$, respectively, be a set of target concepts and a set of hypotheses over the instance space $\{0,1\}^n$, where $\mathcal{C}_n \subseteq \mathcal{H}_n$ for $n \geq 1$. When there exists a polynomial-time learning algorithm that achieves low error with high confidence in approximating all concepts in a class $\mathcal{C} = \{\mathcal{C}_n\}$ by the hypothesis space $\mathcal{H} = \{\mathcal{H}_n\}$ if enough training data is available, the class of concepts $\mathcal{C}$ is said to be *PAC learnable* by $\mathcal{H}$ or simply PAC learnable. Uniform convergence of the empirical error of a function towards the real error on all possible inputs guarantees that all training algorithms that yield a small training error are PAC. A function class is PAC learnable if and only if the capacity in terms of the VC dimension is finite.

In this framework, we are given a set of inputs and a hypothesis space of functions that maps the inputs onto $\{0,1\}$. Assume that there is an unknown but usually fixed probability distribution on the inputs, and the aim is to find a good approximation to a particular target concept from the hypothesis space, given only a random sample of the training examples and the value of the target concept on these examples.

The sample complexity of a learning algorithm, N_{PAC}, is defined as the smallest number of samples required for learning $\mathcal{C}$ by $\mathcal{H}$, that achieve a given approximation accuracy ϵ with a probability $1 - \delta$. Any consistent algorithm that learns $\mathcal{C}$ by $\mathcal{H}$ has a sample complexity with the upper bound [41, 467]

$$N_{\text{PAC}} = \frac{1}{\epsilon \left(1 - \sqrt{\epsilon} \right)} \left(2 \dim_{\text{VC}} (\mathcal{H}_n) \ln \frac{6}{\epsilon} + \ln \frac{2}{\delta} \right) \qquad (2.8)$$

for any $0 < \delta < 1$. In other words, with probability of at least $1 - \delta$, the algorithm returns a hypothesis $h \in \mathcal{H}_n$ with the error of h less than ϵ.

In terms of the cardinality of $\mathcal{H}_n$, denoted $|\mathcal{H}_n|$, it can be shown [1118, 839, 467] that the sample complexity is at most

$$N_{\text{PAC}} = \frac{1}{\epsilon} \left(\ln |\mathcal{H}_n| + \ln \frac{1}{\delta} \right) \tag{2.9}$$

For most hypothesis spaces on Boolean domains, the second bound gives a better bound. On the other hand, most hypothesis spaces on real-valued attributes are infinite, so only the first bound is applicable. The PAC learning is particularly useful for obtaining upper bounds on sufficient training sample size. Linear threshold concepts (perceptrons) are PAC learnable on both Boolean and real-valued instance spaces [467].

2.2 No Free Lunch Theorem

Before the NFL theorem [1192] was proposed, people intuitively believed that there exists some universally beneficial algorithms for search, and many people actually made efforts to design some algorithms. The NFL theorem asserts that there is no universally beneficial algorithm.

The *NFL theorem* states that no search algorithm is better than another in locating an extremum of a cost function when averaged over the set of all possible discrete functions. That is, all search algorithms achieve the same performance as random enumeration, when evaluated over the set of all functions. Given the set of all functions $\mathcal{F}$ and a set of benchmark functions $\mathcal{F}_1$, if algorithm A_1 is better on average than algorithm A_2 on $\mathcal{F}_1$, then algorithm A_2 must be better than algorithm A_1 on $\mathcal{F} - \mathcal{F}_1$.

The performance of any algorithm is determined by the knowledge concerning the cost function. Thus, it is meaningless to evaluate the performance of an algorithm without specifying the prior knowledge. Practical problems always contain priors such as smoothness, symmetry, and iid samples.

For example, although neural networks are usually deemed a powerful approach for classification, they cannot solve all classification problems. For some arbitrary classification problems, other methods may be efficient.

The NFL theorem was later extended to coding methods, crossvalidation, early stopping, and avoidance of overfitting by some authors. Again, it has been asserted that no one method is better than the others for all problems.

2.3 Neural Networks as Universal Machines

The power of neural networks depends on their representation capability. On the one hand, FNNs are proved to offer the capability of universal function approximation. This lays the theoretical foundation for the wide application of FNNs. On the other hand, RNNs using the sigmoidal activation function are Turing equivalent [1033]. There exists an RNN that simulates a universal Turing machine. Thus, RNNs can compute whatever function any digital computer can compute. Some aspects of the representation capability of FNNs are described in this section.

2.3.1 Boolean Function Approximation

FNNs with binary neurons can be used to represent logic or Boolean functions. In binary neural networks, the input and output values for each neuron are Boolean variables, denoted by binary (0 or 1) or bipolar (-1 or $+1$) representation. For J_1 independent Boolean variables, there are 2^{J_1} combinations of these variables. This leads to a total of $2^{2^{J_1}}$ different Boolean functions of J_1 variables. An LTG can discriminate between two classes. Any Boolean function is realizable using a network of LTGs.

The function counting theorem [267] gives the number of linearly separable dichotomies of m points in general position in R^n. It essentially estimates the separating capability of an LTG. It is given as follows [461].

Theorem 2.1 (Function counting theorem). *The number of linearly separable dichotomies of m points in general position in R^n is*

$$C(m,n) = \begin{cases} 2\sum_{i=0}^{n} \binom{m-1}{i}, & m > n+1 \\ 2^m, & m \leq n+1 \end{cases} \tag{2.10}$$

A set of m points in R^n is said to be in a *general position* if every subset of n or fewer points is linearly independent.

The total number of possible dichotomies of m points is 2^m. Under the assumption of 2^m equiprobable dichotomies, the probability of a single LTG with n inputs to separate m points in general position is given by

$$P(m,n) = \frac{C(m,n)}{2^m} = \begin{cases} \frac{2}{2^m}\sum_{i=0}^{n} \binom{m-1}{i}, & m > n+1 \\ 1, & m \leq n+1 \end{cases} \tag{2.11}$$

The fraction $P(m,n)$ is said to be the probability of linear dichotomy. This relation is plotted in Fig. 2.2.

Thus, if $\frac{m}{n+1} \leq 1$, $P = 1$; if $1 < \frac{m}{(n+1)} < 2$ and $n \to \infty$, $P \to 1$. At $\frac{m}{(n+1)} = 2$, $P = \frac{1}{2}$. Usually, $m = 2(n+1)$ is used to characterize the statistical capability of a single LTG.

A three-layer (J_1-2^{J_1}-1) feedforward LTG network can represent any Boolean function with J_1 arguments [300, 831]. To realize an arbitrary function $f : R^{J_1} \to \{0,1\}$ defined on N arbitrary points in R^{J_1}, the lower bound

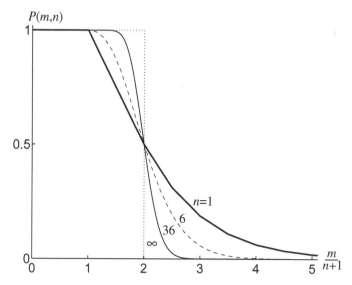

Fig. 2.2. The probability of linear dichotomy of m points in n dimensions, as given by (2.11)

for the number of hidden nodes is derived as $O\left(\dfrac{N}{J_1 \log_2 \frac{N}{J_1}}\right)$ for $N \geq 3J_1$ and $J_1 \to \infty$ [461, 86]; for N points in general position, the lower bound is $\dfrac{N}{2J_1}$ when $J_1 \to \infty$ [461]. Networks with two or more hidden layers are found to be potentially more size efficient than networks with a single hidden layer [461].

2.3.2 Linear Separability and Nonlinear Separability

Given a set $\mathcal{X}$ of N patterns $\mathbf{x}_i$ of J_1 dimensions, each belonging to one of two classes $\mathcal{C}_1$ and $\mathcal{C}_2$; if there is a hyperplane that separates all the samples of $\mathcal{C}_1$ from $\mathcal{C}_2$, then such a classification problem is said to be *linearly separable*. A single LTG can realize linearly separable dichotomy function, characterized by a linear separating surface (hyperplane)

$$\mathbf{w}^{\mathrm{T}}\mathbf{x} + w_0 = 0 \qquad (2.12)$$

where $\mathbf{w}$ is a J_1-dimensional vector, and w_0 is a bias toward the origin. For a pattern, if $\mathbf{w}^{\mathrm{T}}\mathbf{x} + w_0 > 0$, it belongs to $\mathcal{C}_1$; if $\mathbf{w}^{\mathrm{T}}\mathbf{x} + w_0 < 0$, it belongs to $\mathcal{C}_2$. Some examples of linearly separable classes and linearly inseparable classes in two-dimensional space are illustrated in Fig. 2.3.

A linearly inseparable dichotomy can become nonlinearly separable. A dichotomy $\{\mathcal{C}_1, \mathcal{C}_2\}$ of set $\mathcal{X}$ is said to be *φ-separable* if there exists a mapping $\varphi : R^{J_1} \to R^{J_2}$ that satisfies a separating surface [267]

$$\mathbf{w}^{\mathrm{T}}\varphi(\mathbf{x}) = 0 \qquad (2.13)$$

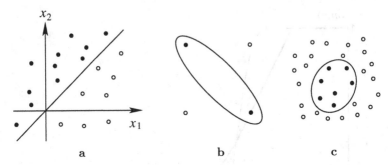

Fig. 2.3. Linearly separable, linearly inseparable, and nonlinearly separable classification in two-dimensional space. Dots and circles denote patterns of different classes. **(a)** Two linearly separable classes with $x_1 - x_2 = 0$ as the delimiter. **(b)** and **(c)** are linearly inseparable classes, where **(b)** is the exclusive-or problem. Note that the linearly inseparable classification in cases **(b)** and **(c)** become nonlinearly separable, where the separating surfaces are ellipses.

such that $\mathbf{w}^{\mathrm{T}}\varphi(\mathbf{x}) > 0$ if $\mathbf{x} \in \mathcal{C}_1$ and $\mathbf{w}^{\mathrm{T}}\varphi(\mathbf{x}) < 0$ if $\mathbf{x} \in \mathcal{C}_2$. Here $\mathbf{w}$ is a J_2-dimensional vector. As shown in Fig. 2.3, the two linearly inseparable dichotomies become φ-separable.

The nonlinearly separable problem can be realized by using a polynomial threshold gate (PTG), which changes the linear term in the LTG into high-order polynomials. The function counting theorem is also applicable to PTGs.

The function counting theorem still holds true if the set of m points is in general position in φ-space, that is, the set of m points is in φ-general position.

2.3.3 Binary Radial Basis Function

For three-layer FNNs, if the activation function of the hidden neurons is selected as the binary RBF or generalized binary RBF and the output neurons are selected as LTGs, one obtains binary or generalized binary RBFNs. Binary or generalized binary RBFN can be used for the mapping of Boolean functions.

The parameters of the generalized binary RBF neuron are the center $\mathbf{c} \in R^n$ and the radius $r \geq 0$. The activation function $\phi : R^n \to \{0,1\}$ is defined by

$$\phi(\mathbf{x}) = \begin{cases} 1, & \|\mathbf{x} - \mathbf{c}\|_{\mathbf{A}} \leq r \\ 0, & \text{otherwise} \end{cases} \tag{2.14}$$

where $\mathbf{A}$ is any real, symmetric, and positive-definite matrix and $\| \cdot \|_{\mathbf{A}}$ is the weighted Euclidean norm. When $\mathbf{A}$ is the identity matrix $\mathbf{I}$, the neuron becomes a binary RBF neuron.

Every Boolean function computed by the LTG can also be computed by any generalized binary RBF neuron, and generalized binary RBF neurons are more powerful than LTGs [375]. As an immediate consequence, in any neural network, any LTG that receives only binary inputs can be replaced by a generalized binary RBF neuron having any norm, without any loss of the computational power of the neural network.

Given a J_1-J_2-1 FNN, whose output neuron is an LTG; we denote the network as $\mathcal{N}_1$, $\mathcal{N}_2$, and $\mathcal{N}_3$, when the J_2 hidden neurons are respectively selected as LTGs, binary RBF neurons, and generalized binary RBF neurons. The VC dimensions of the three networks have the relation [375]

$$\text{dim}_{\text{BVC}}\left(\mathcal{N}_1\right) = \text{dim}_{\text{BVC}}\left(\mathcal{N}_2\right) \leq \text{dim}_{\text{BVC}}\left(\mathcal{N}_3\right) \tag{2.15}$$

When $J_1 \geq 3$ and $J_2 \leq \frac{2^{J_1+1}}{J_1^2+J_1+2}$, the lower bound for the three neural networks is given as [72, 375]

$$\text{dim}_{\text{BVC}}\left(\mathcal{N}_1\right) = J_1 J_2 + 1 \tag{2.16}$$

2.3.4 Continuous Function Approximation

A three-layer FNN with a sufficient number of hidden units can approximate any continuous function to any degree of accuracy. This is guaranteed by Kolmogorov's theorem [620, 474, 389].

Theorem 2.2 (Kolmogorov). *Any continuous real-valued function $f(x_1, \cdots, x_n)$ defined on $[0,1]^n$, $n \geq 2$, can be represented in the form*

$$f\left(x_1, \cdots, x_n\right) = \sum_{j=1}^{2n+1} h_j \left(\sum_{i=1}^{n} \psi_{ij}\left(x_i\right)\right) \tag{2.17}$$

where h_j and ψ_{ij} are continuous functions of one variable, and ψ_{ij} are monotonically increasing functions independent of f.

According to Kolmogorov's theorem, a continuous multivariate function on a compact set can be expressed using superpositions and compositions of a finite number of single-variable functions.

Based on Kolmogorov's theorem, Hecht-Nielsen provided a theorem that is directly related to neural networks [474]

Theorem 2.3 (Hecht-Nielsen). *Any continuous real-valued mapping $f : [0,1]^n \to R^m$ can be approximated to any degree of accuracy by an FNN with n input nodes, $2n+1$ hidden units, and m output units.*

To date, numerous attempts have been made in searching for suitable forms of activation functions and proving the corresponding network's universal approximation capabilities.

The Weierstrass theorem asserts that any continuous real-valued multivariate function can be approximated to any accuracy using a polynomial. The Stone–Weierstrass theorem [971] is a generalization of the Weierstrass theorem, and is usually used for verifying a model's approximation capability to dynamic systems.

Theorem 2.4 (Stone–Weierstrass). *Let $\mathcal{F}$ be a set of real continuous functions on a compact domain $\mathcal{U}$ of n dimensions. Let $\mathcal{F}$ satisfy the following criteria*

1. Algebraic closure: $\mathcal{F}$ *is closed under addition, multiplication, and scalar multiplication. That is, for any two $f_1, f_2 \in \mathcal{F}$, we have $f_1 f_2 \in \mathcal{F}$ and $a_1 f_1 + a_2 f_2 \in \mathcal{F}$, where a_1 and a_2 are any real numbers.*
2. Separability on $\mathcal{U}$: *for any two different points $\mathbf{x}_1, \mathbf{x}_2 \in \mathcal{U}$, $\mathbf{x}_1 \neq \mathbf{x}_2$, there exists $f \in \mathcal{F}$ such that $f(\mathbf{x}_1) \neq f(\mathbf{x}_2)$;*
3. Not constantly zero on $\mathcal{U}$: *for each $\mathbf{x} \in \mathcal{U}$, there exists $f \in \mathcal{F}$ such that $f(\mathbf{x}) \neq 0$.*

Then $\mathcal{F}$ is a dense subset of $C(\mathcal{U})$, the set of all continuous real-valued functions on $\mathcal{U}$. In other words, for any $\varepsilon > 0$ and any function $g \in C(\mathcal{U})$, there exists $f \in \mathcal{F}$ such that $|g(\mathbf{x}) - f(\mathbf{x})| < \varepsilon$ for any $\mathbf{x} \in \mathcal{U}$

Universal approximation to a given nonlinear functional under certain conditions can be realized by using the classical Volterra series or the Wiener series.

2.4 Learning and Generalization

From an approximation viewpoint, learning is a hypersurface reconstruction based on existing examples, while generalization means estimating the value on the hypersurface where there is no example, requiring approximation. Mathematically, the learning process is a nonlinear curve-fitting process, while generalization is the interpolation and extrapolation of the input data. Neural networks have the capabilities of learning and generalization.

The problem of reconstructing the mapping is said to be *well-posed* if an input always generates a unique output, and the mapping is continuous [470]. Learning is an ill-posed inverse problem. Given examples of an input-output mapping, an approximate solution is required to be found for the mapping. The input data may be noisy or imprecise, and also may be insufficient to uniquely construct the mapping. The regularization technique aims to transform an ill-posed problem into a well-posed one so as to stabilize the solution by adding some auxiliary non-negative functional that embeds prior information such as the smoothness constraints for approximation [1089, 923].

When a network is overtrained with too many examples, parameters or epochs, it may produce good results for the training data, but has a poor generalization capability. This is the overfitting phenomenon, and is illustrated in Fig. 2.4. In statistics, overfitting applies to the situation wherein a model possesses too many parameters, and fits the noise in the data rather than the underlying function. A simple network with smooth input-output mapping usually has a better generalization capability. Generally, the generalization capability of a network is jointly determined by the size of the training pattern set, the complexity of the problem, and the architecture of the network.

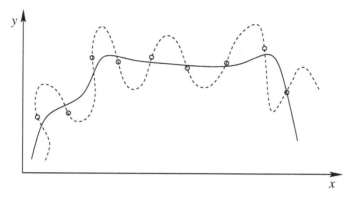

Fig. 2.4. Proper fitting and overfitting. Solid line corresponds to proper fitting, and dashed line corresponds to overfitting.

2.4.1 Size of Training Set

For a given network topology, we can estimate the minimal size of the training set for successfully training the network. For conventional curve-fitting techniques, the required number of examples usually grows with the dimensionality of the input space, namely, the *curse of dimensionality*. Feature extraction can reduce input dimensionality and thus improve the generalization capability of the network.

The training set should be sufficiently large and diverse so that it could represent the problem well. For good generalization, the size of the training set should be at least several times larger than the network's capacity, *i.e.* $N \gg \frac{N_w}{N_y}$, where N is the size of training set, N_w the total number of weights or free parameters, and N_y the number of output components [1183].

Computational learning theory [1121, 1115] addresses the problem of optimal generalization capability for supervised learning. The VC dimension is a measure for the capacity of a neural network to learn from a training set. FNNs with threshold and logistical activation functions have VC dimensions of $O\left(N_w \ln N_w\right)$ [85] and $O\left(N_w^2\right)$ [617], respectively. Sufficient sample sizes are, respectively, estimated by using the PAC paradigm and the VC dimension for FNNs with sigmoidal neurons [1025] and FNNs with LTGs [85]. These bounds on sample sizes are dependent on the error rate of hypothesis ϵ and the probability of failure δ. A practical size of the training set for a good generalization is $N = O\left(\frac{N_w}{\varepsilon}\right)$ [470], where ε specifies the accuracy. For example, for an accuracy level of 90%, $\varepsilon = 0.1$.

2.4.2 Generalization Error

The generalization error of a trained network can be decomposed into two parts, namely, an *approximation error* that is due to the finite number of parameters of the approximation scheme used, and an *estimation error* that is due to the finite number of data available [853, 852].

For an FNN with J_1 input nodes and a single output node, a bound for the generalization error is given by [853, 852]

$$O\left(\frac{1}{P}\right) + O\left(\left[\frac{PJ_1\ln(PN) - \ln\delta}{N}\right]^{1/2}\right) \tag{2.18}$$

with a probability greater than $1 - \delta$, where N is the number of examples, $\delta \in (0,1)$ is the confidence parameter, and P is proportional to the number of parameters, such as P centers in an RBFN, or P sigmoidal hidden units or P threshold units in an MLP.

The first term in (2.18) corresponds to the bound on the approximation error, and the second on the estimation error. Thus, a bound on the total generalization error is a function of the order of the number of hypothesis parameters P and the number of examples N.

As P, the order of the model, increases, the approximation error decreases since we are using a larger model; however, the estimation error increases due to overfitting (or alternatively, more data). Thus, one cannot reduce the upper bounds on both the error components simultaneously. When P is selected as [853]

$$P \propto N^{\frac{1}{3}} \tag{2.19}$$

then the tradeoff between the approximation and estimation errors is best maintained, and this is the optimal size of the model for the amount of data available. After suitably selecting P and N, the generalization error for FNNs should be $O\left(\frac{1}{P}\right)$. This result is similar to that for an MLP with sigmoidal functions [71].

2.4.3 Generalization by Stopping Criterion

Generalization can be controlled during training. Overtraining can be avoided by stopping the training before the absolute minimum is reached. Neural networks trained with iterative gradient-based methods tend to learn a mapping in the hierarchical order of its increasing components of frequency. When the training is stopped at an appropriate point, the network will not learn the high-frequency noise. While the training error will always decrease, the generalization error will decrease to a minimum and then begins to rise again as the network is being overtrained, as is illustrated in Fig. 2.5. The training should stop at the optimum stopping point. The generalization error is defined in the same form as the learning error, but on a separate validation set of data.

Early stopping is implemented with crossvalidation to decide when to stop. Three early-stopping criteria are defined and empirically compared in [927]. As pointed out in [927], slower stopping criteria, which stop later than others, on average lead to small improvements in generalization, but result in a much longer training time.

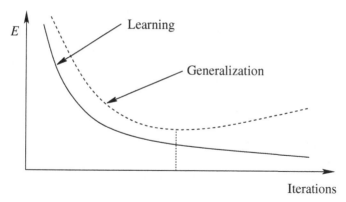

Fig. 2.5. Learning and generalization. When the network is overtrained, its generalization performance degrades.

Statistical analysis for the three-layer MLP has been performed in [30]. As far as the generalization performance is concerned, exhaustive learning is satisfactory when $N > 30N_w$. When $N < N_w$, early stopping can really prevent overtraining. When $N < 30N_w$, overtraining may also occur. In the latter two cases, crossvalidation can be used to stop training.

2.4.4 Generalization by Regularization

Regularization is a reliable method for improving generalization. The target function is assumed to be smooth, and small changes in the input do not cause large changes in the output. A constraint term E_c, which penalizes poor generalization, is added to the standard training cost function E

$$E_T = E + \lambda_c E_c \qquad (2.20)$$

where λ_c is a positive value that balances the tradeoff between error minimization and smoothing. In contrast to the early-stopping criterion method, the regularization method is applicable to both the iterative gradient-based training techniques and the one-step linear optimization such as the SVD technique.

Network-pruning techniques such as the weight-decay technique also help to improve generalization [943, 118]. At the end of training, there are some weights significantly different from zero, while some other weights are close to zero. Those connections with small weights can be removed from the network. During the implementation of the weight-decay technique, the biases should be excluded from the penalty term so that the network yields an unbiased estimate of the true target mean. Many network-pruning techniques can be derived under this regularization framework.

Training with a small amount of jitter in the input while keeping the same output can improve generalization. With jitter, the learning problem is equivalent to a smoothing regularization with the noise variance playing the role of the regularization parameter [943, 117]. Training with jitter thus allows regularization within the conventional layered FNN architecture. Although large networks are generally trained rapidly, they tend to generalize poorly due to insufficient constraints. Training with jitter helps to prevent overfitting by providing additional constraints.

In [488], each weight is encoded with a short bit-length to decrease the complexity of the model. The amount of information in a weight can be controlled by adding Gaussian noise and the noise level can be adapted during learning to optimize the tradeoff between the expected squared error of the network and the amount of information in the weights.

Early stopping has a behavior similar to that of a simple weight-decay technique in the case of the MSE function [117]. The quantity $\frac{1}{\eta t}$, where η is the learning rate and t is the iteration index, plays the role of λ_c. The effective number of weights, that is, the number of weights whose values differ significantly from zero, grows as training proceeds.

Weight sharing is to control several weights by a single parameter [979]. This reduces the number of free parameters in a network, and thus improves generalization. A soft weight-sharing method is implemented in [856] by adding a regularization term to the error function, where the learning algorithm decides which of the weights should be tied together.

Regularization decreases the representation capability of the network, but increases the bias (bias–variance dilemma [405]). The principle of regularization is to choose a well-defined regularizer to decrease the variance by affecting the bias as little as possible [117].

2.5 Model Selection

The objective of model selection is to find a model that is as simple as possible that fits a given data set with sufficient accuracy, and has a good generalization capability to unseen data. The generalization performance of a network gives a measure of the quality of the chosen model. Existing model-selection approaches can be generally grouped into four categories [684]: crossvalidation, complexity criteria, regularization, and network pruning/growing. Some existing model-selection methods have been reviewed in [684].

In crossvalidation methods, many networks of different complexity are trained and then tested on an independent validation set. The procedure is computationally demanding and/or requires additional data withheld from the total pattern set. In complexity criterion-based methods, training of many networks is required and hence, computationally demanding, though a validation set is not required. Regularization methods are described in the preceding section; they are more efficient than crossvalidation techniques, but the results may be suboptimal since the penalty terms damage the representation capability of the network. Pruning/growing methods can be under the framework of regularization, which often makes restrictive assumptions, resulting in networks that are suboptimal. In addition, the evolutionary approach is also used to select an optimum architecture of neural networks.

2.5.1 Crossvalidation

Crossvalidation, a standard tool in statistics, is a classical model-selection method [548]. The total pattern set is randomly partitioned into a training set and a validation (test) set. The major part of the total pattern set is included in the training set, which is used to train the network. The remaining, typically, 10 to 20 per cent, is included in the validation set and is used for validation. When only one sample is used for validation, the method is called *leave-one-out* crossvalidation. Methods on conducting crossvalidation are given in [1142, 927].

Let $\mathcal{D}_i$ and $\overline{\mathcal{D}}_i$, $i = 1, \cdots, m$, be the data subsets of the total pattern set arising from the ith partitioning, which are, respectively, used for training and testing. The crossvalidation process is actually to find a suitable model by minimizing the log-likelihood function

$$E_{\mathrm{CV}} = -\frac{1}{m} \sum_{i=1}^{m} \ln \left(L \left(\widehat{\mathbf{W}} \left(\overline{\mathcal{D}}_i \right) \middle| \mathcal{D}_i \right) \right) \tag{2.21}$$

where $\widehat{\mathbf{W}} \left(\overline{\mathcal{D}}_i \right)$ denotes the ML parameter estimates on $\overline{\mathcal{D}}_i$, and $L \left(\widehat{\mathbf{W}} \left(\overline{\mathcal{D}}_i \right) \middle| \mathcal{D}_i \right)$ is the likelihood evaluated on the data set $\mathcal{D}_i$.

Sometimes it is not optimal if we train the network to perfection on a given pattern set due to the ill-posedness of the finite training pattern set. Crossvalidation helps to generate good generalization of the network. When N, the size of the training set, is too large, it will cause overfitting (overtraining). In this case, crossvalidation can be used to select a good network architecture. Crossvalidation is effective for finding a large network with a good generalization performance.

Validation uses data different from the training set, thus the validation set is independent from the estimated model. This helps to select the best one among the different model parameters. To avoid overfitting or underfitting, the optimal model parameters should be selected so as to have the best performance measure associated with the validation set. Since this data set is independent from the estimated model, the generalization error obtained is a fair estimate.

2.5.2 Complexity Criteria

An efficient approach for improving the generalization performance is to construct a small network using a parsimonious principle. Statistical model selection with information criteria such as Akaike's final prediction error (FPE) criterion [17], Akaike information criterion (AIC) [18], Schwartz's Bayesian information criterion (BIC) [1010], and Rissanen's minimum description length (MDL) principle [951] are popular and have been widely used for model selection of neural networks.

Although the motivations and approaches for these criteria may be very different from one another, most of them can be expressed as a function with two components, one for measuring the training error and the other for penalizing the complexity [951, 396]. For example, the AIC and BIC criteria can be, respectively, represented by [396]

$$E_{\text{AIC}} = -\frac{1}{N} \ln \left(L_N \left(\widehat{\mathbf{W}}_N \right) \right) + \frac{P}{N} \qquad (2.22)$$

$$E_{\text{BIC}} = -\frac{1}{N} \ln \left(L_N \left(\widehat{\mathbf{W}}_N \right) \right) + \frac{P}{2N} \ln N \qquad (2.23)$$

where $L_N \left(\widehat{\mathbf{W}}_N \right)$ is the likelihood estimated for a training set of size N and model parameters $\widehat{\mathbf{W}}_N$, and P is the number of parameters in the model. These criteria penalize large-size models.

The description length of the model characterizes the information needed for simultaneously encoding a description of the model and a description of the prediction errors of the model. The best model according to the MDL principle is the one with the minimum description length [951, 952]. The total description length E_{MDL} has three terms: code cost for coding the input vectors, model cost for defining the reconstruction method, and reconstruction error due to reconstruction of the input vector from its code. The description length is described by the number of bits. Most existing unsupervised learning algorithms such as the competitive learning and PCA can be explained using the MDL principle [488]. Good generalization can be achieved by encoding the weights with short bit-lengths by penalizing the amount of information they contain using the MDL principle [488].

Generalization error Err is characterized by the sum of the training (approximation) error err and the degree of optimism OP inherent in a particular estimate [410], that is, $Err = err + OP$. Complexity criteria such as BIC can be used for estimating OP. A theoretically well-motivated criterion for describing the generalization error is developed by using Stein's unbiased risk estimator (SURE) [600, 1056, 410], which, for the additive Gaussian model, is given as [410]

$$Err = err - \sigma^2 + \frac{2\sigma^2}{N} \sum_{i=1}^{N} \frac{\mathrm{d}\hat{f}(x_i)}{\mathrm{d}y_i} \qquad (2.24)$$

where $\hat{f}(\cdot)$ is the prediction model, and σ^2 is the noise variance of the model. $\hat{f}(\cdot)$, err and σ can be estimated from the training data.

2.6 Bias and Variance

The generalization error can also be represented by the sum of the *bias* squared plus the *variance* [405]. Most existing supervised learning algorithms suffer from the bias–variance dilemma [405]. That is, the requirements for small bias and small variance are conflicting and a tradeoff must be made.

Let $f(\mathbf{x}; \hat{\mathbf{w}})$ be the best model in model space. Thus, $\hat{\mathbf{w}}$ does not depend on the training data. The bias and variance can be defined by [117]

$$\text{Bias} = E_{\mathcal{S}}(f(\mathbf{x})) - f(\mathbf{x}; \hat{\mathbf{w}}) \tag{2.25}$$

$$\text{Var} = E_{\mathcal{S}}\left((f(\mathbf{x}) - E_{\mathcal{S}}(f(\mathbf{x})))^2\right) \tag{2.26}$$

where $f(\mathbf{x})$ is the function to be estimated, and $E_{\mathcal{S}}$ denotes the expectation operation over all possible training sets. Bias is caused by an inappropriate choice of the size of a class of models when the number of training samples is assumed infinite, while the variance is the error caused by the finite number of training samples. An illustration of the concepts of bias and variance in the two-dimensional space is shown in Fig. 2.6.

The generalized error can be decomposed into a sum of the bias and variance

$$
\begin{aligned}
E_{\mathcal{S}}&\left([f(\mathbf{x}) - f(\mathbf{x}, \hat{\mathbf{w}})]^2\right) \\
&= E_{\mathcal{S}}\left(\{[f(\mathbf{x}) - E_{\mathcal{S}}(f(\mathbf{x}))] + [E_{\mathcal{S}}(f(\mathbf{x})) - f(\mathbf{x}, \hat{\mathbf{w}})]\}^2\right) \\
&= E_{\mathcal{S}}\left([f(\mathbf{x}) - E_{\mathcal{S}}(f(\mathbf{x}))]^2\right) + E_{\mathcal{S}}\left([E_{\mathcal{S}}(f(\mathbf{x})) - f(\mathbf{x}, \hat{\mathbf{w}})]^2\right) \\
&\quad +2E_{\mathcal{S}}\left([f(\mathbf{x}) - E_{\mathcal{S}}(f(\mathbf{x}))]\,[E_{\mathcal{S}}(f(\mathbf{x})) - f(\mathbf{x}, \hat{\mathbf{w}})]\right) \\
&= (\text{Bias})^2 + \text{Var}
\end{aligned}
\tag{2.27}
$$

A network with a small number of adjustable parameters gives poor generalization on new data, since the model has very little flexibility and thus yields underfitting with a high bias and low variance. In contrast, a network with too many adjustable parameters also gives a poor generalization performance, since it is too flexible and fits too much of the noise on the training data, thus yielding overfitting with a low bias but high variance. The best generalization performance is achieved by balancing bias and variance, which optimizes the complexity of the model through either finding a model with an optimal size or by adding a regularization term in an objective function.

For nonparametric methods, most complexity criteria-based techniques operate on the variance term in order to get a good compromise between the contributions made by the bias and variance to the error. When the number of hidden cells is increased, the bias term is likely to be reduced, whereas the variance would increase.

For three-layer FNNs with P hidden sigmoidal units, the bias and variance are bounded explicitly [71]: $O\left(\frac{1}{P}\right)$ and $O\left(\frac{PJ_1 \ln N}{N}\right)$ are the upper bounds for

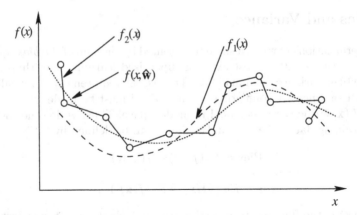

Fig. 2.6. Bias and variance. Circles denote examples from a training set. $f(x; \hat{\mathbf{w}})$ is the underlying function; $f_1(x)$ and $f_2(x)$ are used to approximate $f(x; \hat{\mathbf{w}})$: $f_1(x)$ is a fixed function independent of the data points, while $f_2(x)$ is an exact interpolation of the data points. For $f_1(x)$, the bias is high while the variance is zero; for $f_2(x)$, the bias is zero at the data points and is small in the neighborhood of the data points, while the variance is the variance of the noise on the data, which could be significant.

the bias and variance, respectively, where N is the size of the training set and J_1 is the dimensionality of the feature vectors. Thus when P is large, the bias is small. However, when N is finite, a network with an excessively large space complexity will overfit the training set. The average performance can decrease as P gets larger. As a result, a tradeoff needs to be made between the bias and variance.

2.7 Robust Learning

When the training data is corrupted by large noise, such as outliers[1], conventional least mean squares (LMS)-based learning algorithms may not yield acceptable performance since a small number of outliers have a large impact on the MSE. For nonlinear regression, the techniques of robust statistics [514] can be applied to deal with the outliers. M-estimators are derived from the ML estimators to deal with situations, where the exact probability model is unknown. The M-estimator replaces the conventional squared error term by the so-called *loss functions*. The loss function is used to degrade the effects of

---

[1] An outlier is an observation that deviates significantly from the other observations; this may be due to erroneous measurements or noisy data from the tail of the noise distribution functions. When noise becomes large or outliers exist, the networks may try to fit those improper data and thus, the learned systems are corrupted.

those outliers in learning. One difficulty is the selection of the scale estimator of the loss function in the M-estimator.

The cost function of a robust learning algorithm is defined by

$$E_R = \sum_{i=1}^{N} \sigma\left(\epsilon_i; \beta\right) \tag{2.28}$$

where $\sigma(\cdot)$ is the loss function, which is a symmetric function with a unique minimum at zero, $\beta > 0$ is the scale estimator, known as the *cutoff* parameter, ϵ_i is the estimated error for the ith training pattern, and N is the size of the training set. The loss function can be typically selected as one of the following functions:

- The logistic function [514]

$$\sigma(\epsilon_i; \beta) = \frac{\beta}{2} \ln\left(1 + \frac{\epsilon_i^2}{\beta}\right) \tag{2.29}$$

- Huber's function [514]

$$\sigma(\epsilon_i; \beta) = \begin{cases} \frac{1}{2}\epsilon_i^2, & |\epsilon_i| \leq \beta \\ \beta\,|\epsilon_i| - \frac{1}{2}\beta^2, & |\epsilon_i| > \beta \end{cases} \tag{2.30}$$

- Talwar's function [247]

$$\sigma\left(\epsilon_i; \beta\right) = \begin{cases} \frac{1}{2}\epsilon_i^2, & |\epsilon_i| \leq \beta \\ \frac{1}{2}\beta^2, & |\epsilon_i| > \beta \end{cases} \tag{2.31}$$

- Hampel's tanh estimator [196, 992]

$$\sigma\left(\epsilon_i; \beta_1, \beta_2\right) = \begin{cases} \frac{1}{2}\epsilon_i^2, & |\epsilon_i| \leq \beta_1 \\ \frac{1}{2}\beta_1^2 - \frac{2c_1}{c_2}\ln\frac{1+e^{c_2(\beta_2-|\epsilon_i|)}}{1+e^{c_2(\beta_2-\beta_1)}} - c_1\left(|\epsilon_i| - \beta_1\right), & \beta_1 < |\epsilon_i| \leq \beta_2 \\ \frac{1}{2}\beta_1^2 - \frac{2c_1}{c_2}\ln\frac{2}{1+e^{c_2(\beta_2-\beta_1)}} - c_1\left(\beta_2 - \beta_1\right), & |\epsilon_i| > \beta_2 \end{cases} \tag{2.32}$$

In the tanh estimator, β_1 and β_2 are two cutoff points, and c_1 and c_2 are constants used to adjust the shape of the influence function[2]. When $c_1 = \frac{\beta_1}{\tan(c_2(\beta_2-\beta_1))}$, the influence function is continuous. In the interval of the two cutoff points, the influence function can be represented by a hyperbolic tangent relation.

Using the gradient-descent method, the weights are updated by

$$\Delta w_{jk} = -\eta\frac{\partial E_R}{\partial w_{jk}} = -\eta\sum_{i=1}^{N}\varphi\left(\epsilon_i; \beta\right)\frac{\partial \epsilon_i}{\partial w_{jk}} \tag{2.33}$$

---

[2] This will be defined in (2.34).

where η is a learning rate or step size, and $\varphi(\cdot)$, called the *influence function*, is given by

$$\varphi(\epsilon_i; \beta) = \frac{\partial \sigma(\epsilon_i; \beta)}{\partial \epsilon_i} \tag{2.34}$$

The conventional MSE function corresponds to $\sigma(\epsilon_i) = \frac{1}{2}\epsilon_i^2$ and $\varphi(\epsilon_i; \beta) = \epsilon_i$. To suppresses the effect of large errors, loss functions used for robust learning are defined such that $\varphi(\epsilon_i; \beta)$ is sublinear. The loss functions given above and their respective influence functions are illustrated in Fig. 2.7. Some other loss functions can be designed according to this principle [703].

When the initial weights are not properly selected, the loss functions may not be able to correctly discriminate against the outliers. The selection of β is also a problem, and one approach is to select β as the median of the absolute deviation (MAD)

$$\beta = c \times \text{median}\left(|\epsilon_i - \text{median}(\epsilon_i)|\right) \tag{2.35}$$

with constant c chosen as 1.4826 [468, 514]. Some other methods for selecting β are based on using the median of all errors [468, 514], or counting out a fixed percentage of points as outliers [196].

τ-estimator [1072] can be viewed as an M-estimator with an adaptive bounded influence function $\varphi(\cdot)$ given by the weighted average of two functions $\varphi_1(\cdot)$ and $\varphi_2(\cdot)$, with $\varphi_1(\cdot)$ corresponding to a very robust estimate and $\varphi_2(\cdot)$ to a highly efficient estimate. τ-estimator simultaneously has a high breakdown point and a high efficiency under Gaussian errors.

2.8 Neural-network Processors

A typical architecture of a neural-network processor is illustrated in Fig. 2.8. It is composed of three components: input preprocessing, a neural network for performing inversion, and output postprocessing. Input preprocessing is used to remove redundant and/or irrelevant information in order to achieve a small network and to reduce the dimensionality of the signal parameter space, thus improving the generalization capability of the network. Postprocessing the output of the network generates the desired information.

It is well known that some neural networks, such as the MLP and RBFN, can approximate arbitrary functions to any degree of accuracy. Theoretically, we can use such a network to directly map the raw data onto the required output variables. In practice, it is always beneficial if we apply preprocessing to the input data and postprocessing to the output data, since this significantly improves learning efficiency and generalization.

2.8.1 Preprocessing and Postprocessing

Preprocessing is to transform the raw data into a new representation before being presented to a neural network. If the input data is preprocessed at the

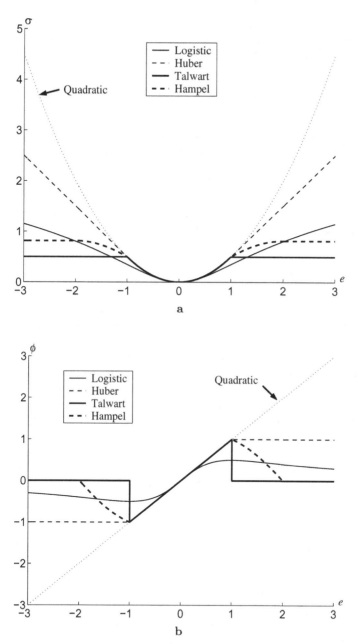

Fig. 2.7. Loss functions and their respective influence functions. For logistic, Huber's, and Talwart's functions, $\beta = 1$. For Hampel's tanh estimator, $\beta_1 = 1$, $\beta_2 = 2$, $c_2 = 1$, and $c_1 = 1.313$. (a) Loss functions σ. (b) Influence functions φ.

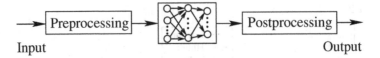

Fig. 2.8. Architecture of a neural-network processor

training stage, accordingly at the generalizing stage, the input data also needs to be preprocessed before being passed to the neural network. Similarly, if the output data is preprocessed at the training stage, the network output at the generalization stage is also required to be postprocessed. The postprocessing of the network output, as the inverse of the preprocessing of raw output patterns, generates the target output corresponding to the raw output patterns.

Preprocessing has a significant influence on the generalization performance of a neural network. This process removes the redundancy in the input space and reduces the space of the input data, thus usually resulting in a reduction in the amount or the dimensionality of the input data. This helps to alleviate the problem of the curse of dimensionality. A network with preprocessed inputs may be constrained by a smaller data set, and thus one needs only to train a small network, which also achieves a better generalization capability.

The preprocessing usually take the form of linear or nonlinear transformation of the raw input data to generate input data for the network. The preprocessing can also be based on the prior knowledge of the network architecture, or the problem itself. When the preprocessing removes redundant information in the input data, it also results in a loss of information. Thus, preprocessing should retain as much relevant information as possible.

The raw data to neural networks are sometimes called *features*, and the preprocessing for input data can be feature selection or feature extraction. Feature selection concentrates on selecting from the original set of features a smaller subset of salient features, while feature extraction is to combine the original features in such a way as to produce a new reduced set of salient features. There are many feature-selection and feature-extraction methods.

When some examples of the raw data suffer from missing components, one simple treatment is to discard those examples from the data set. This, however, is applicable only when the data set is sufficiently large, the percentage of examples with missing components is small, and the mechanism for loss of data is independent of the data itself. In other cases, one has to make use of these deficient examples and various heuristics are available in the literature for this purpose. For function approximation problems, one can represent any variable with a missing value as a regression over the other variables using the available data, and then find the missing value by interpolating the regression function. For density estimation problems, the ML solution to problems with the missing data can be found by applying an expectation–maximization (EM) algorithm [299, 409]. Some strategies for dealing with incomplete data in the context of clustering are given in [466].

2.8.2 Linear Scaling and Data Whitening

The raw data may be orders of magnitude in range, and linear scaling of the raw data is usually employed as a preprocessing step. By linear normalization, all the raw data can be brought in the vicinity of an average value. For a one-dimensional data set of size N, $\{x_i\}$, the mean $\hat{\mu}$ and variance $\hat{\sigma}^2$ are estimated as[3]

$$\hat{\mu} = \frac{1}{N} \sum_{i=1}^{N} x_i \tag{2.36}$$

$$\hat{\sigma}^2 = \frac{1}{N-1} \sum_{i=1}^{N} (x_i - \hat{\mu})^2 \tag{2.37}$$

The transformed data are now defined by

$$\widetilde{x}_i = \frac{x_i - \hat{\mu}}{\hat{\sigma}} \tag{2.38}$$

It is seen that the transformed data set $\{\widetilde{x}_i\}$ has zero mean and unit standard deviation.

When the raw data set $\{\mathbf{x}_i\}$ are composed of vectors, accordingly, the mean vector $\boldsymbol{\mu}$ and covariance matrix $\boldsymbol{\Sigma}$ are calculated as

$$\hat{\boldsymbol{\mu}} = \frac{1}{N} \sum_{i=1}^{N} \mathbf{x}_i \tag{2.39}$$

$$\hat{\boldsymbol{\Sigma}} = \frac{1}{N-1} \sum_{i=1}^{N} (\mathbf{x}_i - \hat{\boldsymbol{\mu}})(\mathbf{x}_i - \hat{\boldsymbol{\mu}})^{\mathrm{T}} \tag{2.40}$$

New input vectors can be defined by the linear transformation

$$\widetilde{\mathbf{x}}_i = \boldsymbol{\Lambda}^{-\frac{1}{2}} \mathbf{U}^{\mathrm{T}} (\mathbf{x}_i - \hat{\boldsymbol{\mu}}) \tag{2.41}$$

where $\mathbf{U} = [\mathbf{u}_1 \cdots \mathbf{u}_M]$, $\boldsymbol{\Lambda} = \mathrm{diag}(\lambda_1, \ldots, \lambda_M)$, M is the dimension of data vectors, and λ_i and $\mathbf{u}_i$ are the eigenvalue and the corresponding eigenvectors of $\boldsymbol{\Sigma}$, which satisfies

$$\boldsymbol{\Sigma}\mathbf{u}_i = \lambda_i \mathbf{u}_i \tag{2.42}$$

The new data set $\{\widetilde{\mathbf{x}}_i\}$ has zero mean and its covariance matrix is the identity matrix [117]. The above process is also called *data whitening*.

---

[3] Equations (2.37) and (2.40) are, respectively, the unbiased estimates of the variance and the covariance matrix. When the factor $\frac{1}{N-1}$ is replaced by $\frac{1}{N}$, the estimates for μ and Σ are the ML estimates. The ML estimate for variance and covariance are biased.

2.8.3 Feature Selection and Feature Extraction

Selection of training data plays a vital role in the performance of learning. The quality and representational nature of the data set are more important than the volume of the data set.

Feature Selection

Feature-selection techniques select the best subset or the best subspace of the features out of the original set, since irrelevant features degrade the performance. Feature selection is a task where the optimum salient characteristics are retained, and hence the dimension of the measurement space is reduced.

A criterion is required to evaluate each subset of the features so that an optimum subset can be selected. The selection criterion should be the same as that for assessing the complete system, such as the MSE criterion for function approximation and the misclassification rate for classification. Theoretically, the global optimum subset of the features can only be selected by an exhaustive search of all the possible subsets of the features. Simplified selection criteria as well as nonexhaustive search procedures are usually used in order to reduce the computational complexity of the search process. Some nonexhaustive search methods such as the branch and bound procedure, sequential forward selection, and sequential backward elimination are discussed in [117].

Usually, feature selection is used to select an optimum subset of the features, which results in a reduction in the size of the feature set. When a specific dimension of the feature set has little contribution to the defined criterion, the dimension for the whole feature set can be discarded. This type of feature selection leads to a reduction in the dimension of the feature space. Feature selection can be a combination of the two situations.

Mutual information-based feature selection [80] is a common method for feature selection. The mutual information measures the arbitrary dependence between random variables, whereas linear relations, such as the correlation-based methods, are prone to mistakes. By calculating the mutual information, the importance levels of the features are ranked based on their ability to maximize the evaluation criterion. Relevant inputs are found by estimating the mutual information between the inputs and the desired outputs.

Mutual information between two signals $\mathbf{x}$ and $\mathbf{y}$ is characterized by calculating the crossentropy, known as *Kullback–Leibler divergence*, between the joint PDF of $\mathbf{x}$ and $\mathbf{y}$ given by $p(\mathbf{x}, \mathbf{y})$ and the product of the marginal PDFs $p(\mathbf{x})$ and $p(\mathbf{y})$

$$I(\mathbf{x}; \mathbf{y}) = \int p(\mathbf{x}, \mathbf{y}) \ln \frac{p(\mathbf{x}, \mathbf{y})}{p(\mathbf{x})p(\mathbf{y})} d\mathbf{x}d\mathbf{y} \qquad (2.43)$$

This may be implemented by estimating the PDFs in terms of the cumulants of the signals. This approach requires the numerical estimation of the joint and marginal densities.

In [80], a mutual information-based feature selection algorithm has been proposed; it is based on a greedy selection of the features and takes into account both the mutual information with respect to the output class and the mutual information with respect to the already-selected features. Neural networks are also used for feature selection, for example, the MLP-based algorithm for feature ranking [975], where the sensitivity of the output of the network to its input is used to rank the input features. Other feature-selection methods include neurofuzzy approaches for supervised feature selection [882] as well as unsupervised feature selection [76], those based on Kohonen's SOM [760] and the RBFN [77].

Feature Extraction

Feature extraction is a dimensionality-reduction technique, mapping high-dimensional patterns onto lower-dimensional patterns, by extracting the most prominent features. The extracted features do not have any physical meaning. Feature extraction reduces the dimension of the features by orthogonal transforms, while feature selection decreases the size of the feature set or reduces the dimension of the features by discarding the raw information according to a criterion, though the distinction between them is somewhat blurred in the literature.

Feature extraction is usually conducted by using orthogonal transforms, though the Gram–Schmidt orthonormalization (GSO) is more suitable for feature selection. This is due to the fact that the physically meaningless features in the Gram–Schmidt space can be linked back to the same number of variables of the measurement space, thus resulting in no dimensionality reduction. In situations where the features are used for pattern understanding and analysis, the GSO transform provides a good option.

The advantage of employing an orthogonal transform is that the correlations among the candidate features are decomposed so that the significance of the individual features can be evaluated independently. PCA is a well-known orthogonal transform that is used for dimensionality reduction in pattern recognition.

Two of the most popular techniques for feature extraction are the PCA and linear discriminant analysis (LDA). LDA is due to Fisher, and is also known as *Fisher's discriminant analysis* [327, 832]. Taking all the data into account, the PCA computes vectors that have the largest variance associated with them. The generated PCA features may not have clear physical meanings. In contrast, the LDA searches for those vectors in the underlying space that best discriminate among the classes (rather than those that best describe the data).

Recently, ICA has emerged as a new statistical signal-processing technique for feature extraction. ICA can extract the statistically independent components from the input data set. It is to estimate the mutual information between the signals by adjusting the estimated matrix to give outputs that are maximally independent [52]. Many ICA algorithms are based on the minimization of the Kullback–Leibler divergence. By applying ICA to estimate the independent input data from raw data, a statistical test can be derived to reduce the input dimension. The dimensions to remove are those that are independent of the output. In contrast, in PCA the input dimensionality reduction is achieved by removing those with a low variance.

For time/frequency-continuous signal systems such as speech-recognition systems, the fixed time-frequency resolution FFT power spectrum, and the multiresolution discrete wavelet transform (DWT) and wavelet packets (WP) [348] are usually used for feature extraction. The features used are chosen from the Fourier or wavelet coefficients having high energy. The cepstrum and its time derivative remain a most commonly used feature set [917]. These features are calculated by taking the discrete cosine transform (DCT) of the logarithm of the energy at the output of a Mel filter and are commonly called *Mel frequency cepstral coefficients (MFCC)*. In order to have the temporal information, the first and second time derivatives of the MFCC are taken.

2.9 Gram–Schmidt Orthonormalization Transform

Ill-conditioning is usually measured for a data matrix $\mathbf{A}$ by its condition number ρ, defined as $\rho(\mathbf{A}) = \frac{\sigma_{\max}}{\sigma_{\min}}$, where $\sigma_{\max}$ and $\sigma_{\min}$ are respectively the maximum and minimum singular values of $\mathbf{A}$. In the batch least-squares (LS) algorithm, the information matrix $\mathbf{A}^{\mathrm{T}}\mathbf{A}$ needs to be manipulated. Since $\rho\left(\mathbf{A}^{\mathrm{T}}\mathbf{A}\right) = \rho(\mathbf{A})^2$, the effect of ill-conditioning on the parameter estimation will be more severe. Orthogonal decomposition is a well-known technique to eliminate ill-conditioning.

The GSO procedure starts with the QR decomposition of the full feature matrix. Denote

$$\mathbf{X} = [\mathbf{x}_1, \mathbf{x}_2, \cdots, \mathbf{x}_N] \tag{2.44}$$

where the ith pattern $\mathbf{x}_i = (x_{i,1}, x_{i,2}, \cdots, x_{i,J})^{\mathrm{T}}$, $x_{i,j}$ denotes the jth component of $\mathbf{x}_i$, and J is the dimensions of the raw data. We then represent $\mathbf{X}^{\mathrm{T}}$ by

$$\mathbf{X}^{\mathrm{T}} = [\mathbf{x}^1, \mathbf{x}^2, \cdots, \mathbf{x}^J] \tag{2.45}$$

where $\mathbf{x}^j = (x_{1,j}, x_{2,j}, \cdots, x_{N,j})^{\mathrm{T}}$.

The QR decomposition is performed on $\mathbf{X}^{\mathrm{T}}$

$$\mathbf{X}^{\mathrm{T}} = \mathbf{QR} \tag{2.46}$$

where $\mathbf{Q}$ is an orthonormal matrix, that is, $\mathbf{Q}^{\mathrm{T}}\mathbf{Q} = \mathbf{I}_J$, $\mathbf{Q} = [\mathbf{q}_1, \mathbf{q}_2, \cdots, \mathbf{q}_J]$, $\mathbf{q}_i = (q_{i,1}, q_{i,2}, \cdots, q_{i,N})^{\mathrm{T}}$, $q_{i,j}$ denotes the jth component of $\mathbf{q}_i$, and $\mathbf{R}$ is an upper triangular matrix. The QR decomposition can be performed by the

Householder transform or Givens rotation [424], which is suitable for hardware implementation.

The GSO procedure is given as

$$\mathbf{q}_1 = \mathbf{x}^1 \tag{2.47}$$

$$\mathbf{q}_k = \mathbf{x}^k - \sum_{i=1}^{k-1} \alpha_{ik} \mathbf{q}_i \tag{2.48}$$

$$\alpha_{ik} = \begin{cases} \frac{(\mathbf{x}^k)^{\mathrm{T}} \mathbf{q}_i}{\mathbf{q}_i^{\mathrm{T}} \mathbf{q}_i}, & \text{for} \quad i = 1, 2, \cdots, k-1 \\ 1, & \text{for} \quad i = k \\ 0, & \text{for} \quad i > k \end{cases} \tag{2.49}$$

Thus $\mathbf{q}_k$ is a linear combination of $\mathbf{x}^1, \cdots, \mathbf{x}^k$, and the Gram–Schmidt features $\mathbf{q}_1, \cdots, \mathbf{q}_k$ and the vectors $\mathbf{x}^1, \cdots, \mathbf{x}^k$ are one-to-one mappings, for $1 \leq k \leq J$. The GSO transform can be used for feature subset selection; it inherits the compactness of the orthogonal representation and at the same time provides features retaining their original meaning.

2.10 Principal Component Analysis

For a J_1-dimensional data set $\{\mathbf{x}_i\}$ of size N, PCA [733] generates a J_2-dimensional feature set $\{\mathbf{y}_i\}$ of the same size, $J_1 > J_2$, by using the linear transformation

$$\mathbf{y}_i = \mathbf{W}^{\mathrm{T}} \mathbf{x}_i \tag{2.50}$$

The weight matrix $\mathbf{W}$ can be solved under different criteria such as the output variance maximization or MSE minimization for the objective of dimensionality reduction.

In comparison with the GSO transform, PCA generates each of its features based on the covariance matrix of all vectors $\mathbf{x}_i$, $i = 1, \cdots, N$. Dimensionality reduction is achieved by dropping the variables with insignificant variances.

PCA is often used to select inputs, but it is not always useful, since the variance of a signal is not always related to the importance of the variable, for example, for non-Gaussian signals. An improvement on the PCA is provided by nonlinear generalizations of the PCA, which extend the ability of the PCA to incorporate nonlinear relationships in the data. Two-dimensional PCA [1220] is designed for image feature extraction. ICA [258] has been shown to be particularly suitable for demixing noise from measurements. PCA and ICA will be dealt with in Chapter 7.

2.11 Linear Discriminant Analysis

LDA creates a linear combination of the given independent features that yield the largest mean differences between the desired classes [327]. Given the data set $\{\mathbf{x}_i\}$ of size N, which is composed of J_1-dimensional vectors, for all the samples of all the C classes, the within-class scatter matrix $\mathbf{S}_w$, the between-class scatter matrix $\mathbf{S}_b$, and the mixture scatter matrix $\mathbf{S}_m$ are, respectively, defined by [760]

$$\mathbf{S}_w = \frac{1}{N} \sum_{j=1}^{C} \sum_{i=1}^{N_j} \left(\mathbf{x}_i^{(j)} - \boldsymbol{\mu}_j\right) \left(\mathbf{x}_i^{(j)} - \boldsymbol{\mu}_j\right)^{\mathrm{T}} \tag{2.51}$$

$$\mathbf{S}_b = \frac{1}{N} \sum_{j=1}^{C} N_j \left(\boldsymbol{\mu}_j - \boldsymbol{\mu}\right) \left(\boldsymbol{\mu}_j - \boldsymbol{\mu}\right)^{\mathrm{T}} \tag{2.52}$$

$$\mathbf{S}_m = \frac{1}{N} \sum_{j=1}^{N} \left(\mathbf{x}_j - \boldsymbol{\mu}\right) \left(\mathbf{x}_j - \boldsymbol{\mu}\right)^{\mathrm{T}} \tag{2.53}$$

where $\mathbf{x}_i^{(j)}$ is the ith sample of class j, $\boldsymbol{\mu}_j$ is the mean of class j, N_j is the number of samples in class j, and $\boldsymbol{\mu}$ represents the mean of all classes. Note that $\sum_{j=1}^{C} N_j = N$. All the scatter matrices are of size $J_1 \times J_1$, and are related by

$$\mathbf{S}_m = \mathbf{S}_w + \mathbf{S}_b \tag{2.54}$$

It is easily seen that the criterion based on the trace of $\mathbf{S}_w$ is the same as the MSE criterion. The minimization of the MSE criterion is equivalent to the minimization of the trace of $\mathbf{S}_w$ or maximizing the trace of $\mathbf{S}_b$ [1207].

The objective for LDA is to maximize the between-class measure while minimizing the within-class measure after applying a $J_1 \times J_2$ transform matrix $\mathbf{W}$, $J_1 > J_2$, which transforms the $J_1 \times J_1$ scatter matrices into $J_2 \times J_2$ matrices $\tilde{\mathbf{S}}_w$, $\tilde{\mathbf{S}}_b$, and $\tilde{\mathbf{S}}_m$

$$\tilde{\mathbf{S}}_w = \mathbf{W}^{\mathrm{T}} \mathbf{S}_w \mathbf{W}$$
$$\tilde{\mathbf{S}}_b = \mathbf{W}^{\mathrm{T}} \mathbf{S}_b \mathbf{W}$$
$$\tilde{\mathbf{S}}_m = \mathbf{W}^{\mathrm{T}} \mathbf{S}_m \mathbf{W} \tag{2.55}$$

The $\mathrm{tr}(\mathbf{S}_w)$ measures the closeness of the samples within the clusters and $\mathrm{tr}(\mathbf{S}_b)$ measures the separation between the clusters, where $\mathrm{tr}(\cdot)$ denotes the trace operator. An optimal $\mathbf{W}$ should preserve the given cluster structure, and simultaneously maximize $\mathrm{tr}\left(\tilde{\mathbf{S}}_b\right)$ and minimize $\mathrm{tr}\left(\tilde{\mathbf{S}}_w\right)$. This is equivalent to maximizing [760]

$$E_{\mathrm{LDA},1}(\mathbf{W}) = \mathrm{tr}(\tilde{\mathbf{S}}_w^{-1} \tilde{\mathbf{S}}_b) \tag{2.56}$$

when $\tilde{\mathbf{S}}_w$ is a nonsingular matrix.

Assuming that $\mathbf{S}_w$ is a nonsingular matrix, one can maximize the Rayleigh coefficient [832]

$$E_{\mathrm{LDA},2}(\mathbf{w}) = \frac{\mathbf{w}^{\mathrm{T}} \mathbf{S}_b \mathbf{w}}{\mathbf{w}^{\mathrm{T}} \mathbf{S}_w \mathbf{w}} \tag{2.57}$$

to find the principal projection direction $\mathbf{w}_1$. Conventionally, the following Fisher's determinant ratio criterion is maximized for finding the projection directions [269, 509]

$$E_{\text{LDA},3}(\mathbf{W}) = \frac{\det\left(\widetilde{\mathbf{S}}_b\right)}{\det\left(\widetilde{\mathbf{S}}_w\right)} = \frac{\det\left(\mathbf{W}^{\mathrm{T}}\mathbf{S}_b\mathbf{W}\right)}{\det\left(\mathbf{W}^{\mathrm{T}}\mathbf{S}_w\mathbf{W}\right)} \tag{2.58}$$

where the column vectors $\mathbf{w}_i$, $i = 1, \cdots, J_2$, of the projection matrix $\mathbf{W}$, are the first J_2 principal eigenvectors of $\mathbf{S}_w^{-1}\mathbf{S}_b$.

Under the assumption that the class distributions are identically distributed Gaussians, LDA is Bayes optimal [832]. An illustration of PCA and LDA for a two-dimensional data set is shown in Fig. 2.9. It is clearly seen that PCA is purely descriptive, while LDA is discriminative.

It is noted that there are at most $C-1$ nonzero generalized eigenvalues and thus an upper bound on J_2 is $C-1$, and that at least J_1+C samples are needed to guarantee $\mathbf{S}_w$ to be nonsingular. This requirement on the number of samples may be severe for some problems like image processing. For example, for an image of 100-by-100 pixels, a 10 000-dimensional vector is needed to represent it; thus, the number of samples required is so large that it is impractical.

When the dimension of the feature space is equal to or higher than the number of training samples, $\mathbf{S}_w$ will be singular and regularization may be necessary. By introducing kernel into the linear $\mathbf{w}$, nonlinear discriminant analysis (NLDA) is obtained [804, 832, 1200]. A multiple of the identity or the kernel matrix can be added to $\mathbf{S}_w$ or its reformulated matrix $\widetilde{\mathbf{S}}_w$ after introducing the kernels to penalize $\|\mathbf{w}\|^2$ or $\|\widetilde{\mathbf{w}}\|^2$, respectively [804]. One can also first compress by PCA the high-dimensional features into intermediate-dimensional

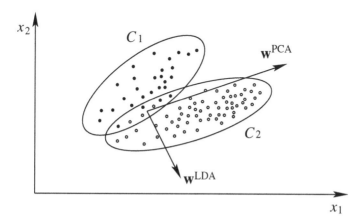

Fig. 2.9. The projections of PCA and LDA for a data set. Two Gaussian classes $\mathcal{C}_1$ and $\mathcal{C}_2$ are represented by two ellipses. The principal direction obtained from PCA, namely, $\mathbf{w}^{\text{PCA}}$, cannot discriminate the two classes, while $\mathbf{w}^{\text{LDA}}$, the principal direction obtained from LDA, can discriminate the two classes.

space, which is further projected by LDA onto the low-dimensional space. The overall performance of the two-stage approach is sensitive to the reduced dimension in the first stage. A generalization of LDA by using generalized SVD [509] can be used to solve the problem of singularity of S_w. The generalized SVD method has numerical advantages over the two-stage approach. Two-dimensional LDA algorithms [696, 564, 1201] provide an efficient approach to image feature extraction and can overcome the singularity problem.

In [269], an NLDA network with the MLP as the architecture and Fisher's determinant ratio as the criterion function has been proposed. The motivation is to combine the universal approximation properties of the MLP with the target-free nature of the LDA. A layered lateral network-based LDA network and an MLP-based NLDA network have also been proposed in [760]. Based on a single-layer linear feedforward network, LDA algorithms are also given in [190, 297].

3

Multilayer Perceptrons

The perceptron is the earliest and the simplest neural network model [962].
Rosenblatt used a single-layer perceptron for the classification of linearly separable patterns. The rediscovery of the backpropagation (BP) algorithm for
training the multilayer perceptron (MLP) in 1986 heralded a new era of neural-
network research. The MLP can be used for the classification of linearly inseparable patterns, and can also work as universal approximators. In this chapter,
the MLP and its learning algorithms are described. Applications of the MLP
to temporal learning and complex-valued signal processing are also addressed.

3.1 Single-layer Perceptron

The perceptron model is based on the McCulloch–Pitts neuron model [783]
introduced in Sect. 1.2. For a one-neuron perceptron, the network topology is
the same as that shown in Fig. 1.1, and the net input to the neuron is given
by

$$net = \sum_{i=1}^{J_1} w_i x_i - \theta = \mathbf{w}^{\mathrm{T}} \mathbf{x} - \theta \tag{3.1}$$

where all the symbols are as explained in Sect. 1.2. The one-neuron perceptron
using the hard-limiter activation function is useful for classification of vector
$\mathbf{x}$ into two classes. The two decision regions are separated by a hyperplane

$$\mathbf{w}^{\mathrm{T}} \mathbf{x} - \theta = 0 \tag{3.2}$$

where the threshold θ is a parameter used to shift the decision boundary away
from the origin.

When more neurons with the hard-limiter activation function are used,
we have a single-layer perceptron (SLP), as illustrated in Fig. 3.1. The SLP
can be used to classify input vector data $\mathbf{x}$ into more classes. For a J_1-J_2
perceptron, the system state is updated by

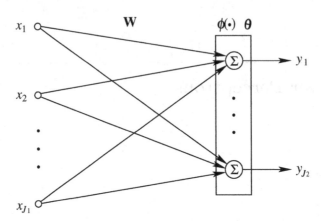

Fig. 3.1. Architecture of the single-layer perceptron. $\phi = (\phi_1, \cdots, \phi_{J_2})^{\mathrm{T}}$ and $\boldsymbol{\theta} = (\theta_1, \cdots, \theta_{J_m})^{\mathrm{T}}$ correspond to all the activation functions and all the biases in the second layer, respectively.

$$\mathbf{net} = \mathbf{W}^{\mathrm{T}}\mathbf{x} - \boldsymbol{\theta} \tag{3.3}$$

$$\hat{\mathbf{y}} = \phi(\mathbf{net}) \tag{3.4}$$

where the net input vector $\mathbf{net} = (net_1, \cdots, net_{J_2})^{\mathrm{T}}$, the output vector $\hat{\mathbf{y}} = (\hat{y}_1, \cdots, \hat{y}_{J_2})^{\mathrm{T}}$, and $\phi(\mathbf{net}) = \left(\phi_1 (net_1)^{\mathrm{T}}, \cdots, \phi_{J_2} (net_{J_2}) \right)$, ϕ_i being the activation function of the ith neuron.

The adaptation of $\mathbf{W}$ is error driven, which can be according to Rosenblatt's perceptron learning algorithm [962, 963] or according to the LMS algorithm based on the adaline model [1182].

3.1.1 Perceptron Learning Algorithm

Rosenblatt proved the perceptron convergence theorem for classification problems [963, 808]

Theorem 3.1 (Perceptron Convergence). *Given a one-neuron perceptron and input patterns* $\mathbf{x} \in \mathcal{X}$ *from two linearly separable classes. Let the patterns be presented in an arbitrary sequence in each epoch. Then, starting from an arbitrary initial state, the perceptron learning procedure always converges and yields a decision hyperplane between the two classes in finite time.*

The perceptron convergence theorem can be proved by minimizing the perceptron criterion function

$$E(\mathbf{w}) = \sum_{\mathbf{x} \in \overline{\mathcal{X}}} (-\mathbf{w}^{\mathrm{T}}\mathbf{x}) \tag{3.5}$$

using the gradient-descent method, where $\overline{\mathcal{X}}$ is the set of samples misclassified by $\mathbf{w}$. Thus, the weights are modified in such a manner as to reduce the number of misclassifications.

The perceptron convergence theorem can be easily extended to the SLP by extending the perceptron learning algorithm from one neuron to multiple neurons. The perceptron learning algorithm is given as

$$net_{t,j} = \sum_{i=1}^{J_1} x_{t,i} w_{ij}(t) - \theta_j = \mathbf{w}_j^{\mathrm{T}} \mathbf{x}_t - \theta_j \tag{3.6}$$

$$\hat{y}_{t,j} = \begin{cases} 1, & net_{t,j} > 0 \\ 0, & \text{otherwise} \end{cases} \tag{3.7}$$

$$e_{t,j} = y_{t,j} - \hat{y}_{t,j} \tag{3.8}$$

$$w_{ij}(t+1) = w_{ij}(t) + \eta x_{t,i} e_{t,j} \tag{3.9}$$

for $i = 1, \cdots, J_1$, $j = 1, \cdots, J_2$, where $net_{t,j}$ is the net input of the jth neuron for the tth example, $\mathbf{w}_j = (w_{1j}, w_{2j}, \cdots, w_{J_1 j})^{\mathrm{T}}$ is the vector collecting all weights terminated at the jth neuron, θ_j is the threshold for the jth neuron, $x_{t,i}$ is the ith input of the tth example, $\hat{y}_{t,j}$ and $y_{t,j}$ are, respectively, the network output and the desired output of the jth neuron for the tth example, with values zero or unity representing classmembership, $e_{t,j}$ is the difference between $y_{t,j}$ and $\hat{y}_{t,j}$, and η is the learning rate. All the network weights w_{ij} are randomly initialized. The choice of η does not affect the stability of the perceptron learning, and affects the convergence speed only if the initial weight vector is nonzero. The learning rate η is typically selected as 0.5. The learning process stops when the errors are sufficiently small.

When used for classification, the perceptron learning algorithm can operate only for linearly separable patterns, and does not terminate for linearly inseparable patterns. The limitations of perceptron learning has been mathematically analyzed in [808]. The failure of Rosenblatt's and similar methods to converge for linearly inseparable problems is caused by the inability of the methods to detect the minimum of the error function [330].

Two improvements on the perceptron learning are the pocket algorithm [395] and the thermal perceptron learning [374]. Both these algorithms can be applied for the classification of linearly inseparable patterns. The pocket algorithm is a variant of perceptron learning derived by adding a checking amendment to stop the algorithm, and was introduced to optimally dichotomize the given patterns in the sense of minimizing the erroneous classification rate [395]. The weight vector with the longest unchanged run is identified as the best solution so far and is stored in the *pocket*. The content of the pocket is replaced by any new weight vector with a longer successful run. The pocket convergence theorem guarantees the optimal convergence of the pocket algorithm, if the inputs in the training set are integers or rational [395, 834]. In [834], the pocket algorithm with ratchet is asserted to find an optimal weight vector with probability one within a finite number of iterations, independently of the given training set. Thermal perceptron learning [374] is a simple extension to perceptron learning and is obtained by multiplying the second term of (3.9) by a temperature annealing factor $e^{-\frac{|net_j|}{T}}$, where T is an annealing temperature. It finds stable weights for inseparable problems as well as for separable ones.

The perceptron convergence theorem has been extended for the MLP, stating that the pattern mode BP algorithm converges to an optimal solution for linearly separable patterns with no upper bound on the learning rate [428].

3.1.2 Least Mean Squares Algorithm

The LMS algorithm [1182] achieves a robust separation between the patterns of different classes by minimizing the MSE rather than the number of misclassified patterns, and the gradient-descent method is applied. Like the perceptron learning, it can only be used for the classification of linearly separable patterns. In the LMS algorithm, the activation function is linear, and the error is defined by

$$e_{t,j} = y_{t,j} - net_{t,j} \qquad (3.10)$$

where $net_{t,j}$ is defined by (3.6). The weight update rule is the same as (3.9), and is reproduced here for easy presentation

$$w_{ij}(t+1) = w_{ij}(t) + \eta x_{t,i} e_{t,j} \qquad (3.11)$$

For classification problems, a threshold activation function is further applied to the linear output so as to render the final output to $\{0, 1\}$ or $\{+1, -1\}$

$$\hat{y}_{t,j} = \begin{cases} 1, & net_{t,j} > 0 \\ 0, & \text{otherwise} \end{cases} \qquad (3.12)$$

The whole unit including a linear combiner and the following threshold operation is called an *adaptive linear element (adaline)*.

The above LMS rule is also called the *μ-LMS rule*. For most practical purposes, η can be selected as $0 < \eta < \frac{2}{\max_t \|\mathbf{x}_t\|^2}$ to ensure its convergence. The Widrow–Hoff delta rule, known as the *α-LMS*, is a modification to the LMS rule obtained by normalizing the input vector so that the weights change independently of the magnitude of the input vector [1183]

$$w_{ij}(t+1) = w_{ij}(t) + \eta \frac{x_{t,i} e_{t,j}}{\|\mathbf{x}_t\|^2} \qquad (3.13)$$

The selection of η is important for the stability of the algorithm. For the convergence of the α-LMS rule, η should be selected as $0 < \eta < 2$, and a practical range for η is $0.1 < \eta < 1.0$ [1183]. Unlike the perceptron learning, the LMS method can also be used for function approximation. In this case, the threshold operation in the Adaline is dropped, and the behavior of the Adaline is identical to that of linear regression. The Adaline model is still a widely used model today, especially in adaptive filtering. There are also various madaline models using layered multiple adalines [1183]. The Widrow–Hoff delta rule has become the foundation of modern adaptive signal processing [398].

3.1.3 Other Learning Algorithms

There are many other valid learning rules such as the Mays' rule [782, 1183, 461], the Ho–Kashyap rule [493, 327], and adaptive Ho–Kashyap rules [462, 461]. Like the perceptron learning algorithm, these algorithms converge only in the case of linearly separable data sets. The one-shot Hebbian learning [1116] and nonlinear Hebbian learning [123] have also been applied for perceptron learning.

A complex perceptron learning algorithm was proposed in [843]. As in the perceptron, input and output patterns are composed of the values 1 and -1; only the weights take complex values. The complex perceptron learning has a better separating power than the perceptron learning. The dynamics of the network is defined by

$$y_i = \text{sgn}\left(|\mathbf{w}_i\mathbf{x}| - \theta_i\right) \tag{3.14}$$

where $\text{sgn}(x)$ is the signum function, which takes the value 1 for $x \geq 0$ and -1 otherwise. The use of a decision circle in the complex plane for the output function increases the separating power by a factor greater than the increase of the degree of freedom [843].

Separating Both Linearly Separable and Linearly Inseparable Data Sets

There are a number of SLP learning algorithms that are suitable for both linearly separable and linearly inseparable classification problems. Examples are the convex analysis and nonsmooth optimization-based method [330], the linear programming (LP) method [327, 759], the constrained steepest descent algorithm [907], fuzzy perceptron [597, 200], and the conjugate-gradient (CG) method [836, 835].

The problem of training an SLP is to find a solution to a set of linear inequalities, thus it is known as an LP problem. LP techniques have been applied to SLP learning [327, 759]. They can solve linearly inseparable problems. When the training vectors are from $\{-1, +1\}^{J_1}$, the method requires $O\left(J_1^3 \log_2 J_1\right)$ learning cycles in the worst case, while the perceptron convergence procedure may require $O\left(2^{J_1}\right)$ learning cycles [759].

Unlike the perceptron learning [962], the constrained steepest-descent algorithm [907] has no free learning parameters and therefore no heuristics is involved. Learning proceeds by iteratively lowering the perceptron cost function following the direction of steepest descent, under the constraint that patterns already correctly classified are not to be affected. A decrease in the error is achieved at each iteration by employing the projection search direction when needed. The training task is decomposed into a succession of small-scale quadratic programming (QP) problems, whose solutions determine the appropriately constrained direction of steepest descent. For linearly separable problems, it always finds a hyperplane that completely separates the patterns belonging to different categories in a finite number of steps. In the case of linearly inseparable problems, the algorithm detects the inseparability in a finite number of steps and terminates, having usually found a good separation hyperplane. Thus, it provides a natural criterion for linear separability or inseparability.

Fuzzy set techniques were introduced into the SLP learning algorithm for two-class classification problems [597]. The algorithm assigns fuzzy membership functions to the input data to reflect their geometrical proximity to the means of class 1 and class 2 before training the perceptron. This fuzzy perceptron learning scheme can improve the convergence significantly, especially when the crisp data are overlapping. The fuzzy perceptron network described in [200] is a Type-I neurofuzzy system. The input to the network can be either fuzzy IF-THEN rules or numerical data. The learning scheme is derived based on the α-cut concept, which extends the perceptron learning [962] to fuzzy input vectors. Moreover, the fuzzy pocket algorithm was derived and then further incorporated into the fuzzy perceptron learning scheme to tackle inseparable cases.

3.2 Introduction to Multilayer Perceptrons

MLPs are FNNs with one or more layers of units between the input and output layers. The output units represent a hyperplane in the space of the input patterns. The architecture of the MLP is illustrated in Fig. 3.2. Assume that there are M layers, each layer having J_m, $m = 1, \cdots, M$, nodes. The weights from the $(m-1)$th layer to the mth layer are described by $\mathbf{W}^{(m-1)}$; also, the bias, output and activation function of the ith neuron in the mth layer are, respectively, denoted as $\theta_i^{(m)}$, $o_i^{(m)}$ and $\phi_i^{(m)}(\cdot)$.

From Fig. 3.2, we have the following relations[1]

$$\hat{\mathbf{y}}_p = \mathbf{o}_p^{(M)}, \quad \mathbf{o}_p^{(1)} = \mathbf{x}_p \tag{3.15}$$

$$\mathbf{net}_p^{(m)} = \left[\mathbf{W}^{(m-1)}\right]^{\mathrm{T}} \mathbf{o}_p^{(m-1)} + \boldsymbol{\theta}^{(m)} \tag{3.16}$$

$$\mathbf{o}_p^{(m)} = \phi^{(m)}\left(\mathbf{net}_p^{(m)}\right) \tag{3.17}$$

for $m = 2, \cdots, M$, where the subscript p corresponds to the pth example, $\mathbf{net}_p^{(m)} = \left(net_{p,1}^{(m)}, \cdots, net_{p,J_m}^{(m)}\right)^{\mathrm{T}}$, $\mathbf{W}^{(m-1)}$ is a J_{m-1}-by-J_m matrix,

---

[1] Unlike in previous sections, a plus sign precedes the bias vector for easy presentation

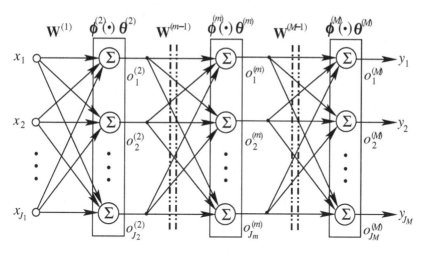

Fig. 3.2. Architecture of the MLP. The MLP is an M-layer FNN with the mth layer having J_m nodes. $\boldsymbol{\phi}^{(m)} = \left(\phi_1^{(m)}, \cdots, \phi_{J_m}^{(m)}\right)^{\mathrm{T}}$, $\boldsymbol{\theta}^{(m)} = \left(\theta_1^{(m)}, \cdots, \theta_{J_m}^{(m)}\right)^{\mathrm{T}}$, $m = 2, \cdots, M$, are the activation functions and the biases in the mth layer, respectively. $\mathbf{W}^{(m-1)}$, $m = 2, \cdots, M$, are the weights from the $(m-1)$th layer to the mth layer. $o_i^{(m)}$, $m = 2, \cdots, M$, is the output of the ith node in the mth layer.

$\mathbf{o}_p^{(m-1)} = \left(o_{p,1}^{(m-1)}, \cdots, o_{p,J_{m-1}}^{(m-1)}\right)^{\mathrm{T}}$, $\boldsymbol{\theta}^{(m)} = \left(\theta_1^{(m)}, \cdots, \theta_{J_m}^{(m)}\right)^{\mathrm{T}}$ is the bias vector, and $\boldsymbol{\phi}^{(m)}(\cdot)$ applies $\phi_i^{(m)}(\cdot)$ to the ith component of the vector within.

For simple implementation, all $\phi_i^{(m)}(\cdot)$ are typically selected to be the same sigmoidal function; one can also select all $\phi_i^{(m)}(\cdot)$ in the first $M - 1$ layers as the same sigmoidal function, and all $\phi_i^{(m)}(\cdot)$ in the Mth layer as another continuous yet differentiable function.

3.2.1 Universal Approximation

The MLP has a strong classification capability, and is also a universal approximator. The universal approximation capability of the MLP stems from the nonlinearities used in the nodes. A four-layer MLP can approximate any nonlinear relation to any accuracy as long as the number of nodes in the hidden layers are sufficiently large [272]. According to Kolmogorov's theorem [620], a four-layer MLP with $J_1 (2J_1 + 1)$ nodes using continuously increasing nonlinearities can compute any continuous function of J_1 variables. In [1078], a four-layer MLP with $\frac{N}{2} + 3$ hidden neurons has been designed and it can represent N distinct examples, with negligibly small error. In [510] a four-layer MLP with $2\sqrt{(m+2)N}$ hidden neurons can learn any distinct examples with any arbitrarily small error, where m is the required number of output neurons.

It has been mathematically proved that a three-layer MLP using sigmoidal activation function can approximate well any continuous multivariate function to any accuracy [273, 508, 389, 1197]. Cybenko's proof [273] is based on the Hahn–Banach theorem [971], and is concise. The proof of Hornik *et al.* [508] is based on the Stone–Weierstrass theorem, and Funahashi [389] proved the same problem using an integral formula. Xiang *et al.*'s proof [1197] is most elegant and simple, and is derived from a piecewise-linear approximation of the sigmoidal activation function [1197]. Readers interested in the proofs are referred to the original papers.

The MLP is very efficient for function approximation in high-dimensional spaces. The error convergence rate of the MLP is independent of the input dimensionality, while conventional linear regression methods suffer from the curse of dimensionality, which results in a decrease of the convergence rate with an increase of the input dimensionality [70, 71]. The necessary number of MLP neurons for approximating a target function depends only upon the basic geometrical shape of the target function, and not on the dimensionality of the input space. Based on a geometrical interpretation of the MLP, the minimal number of line segments or hyperplanes that can construct the basic geometrical shape of the target function is suggested as the first trial for the number of hidden neurons of a three-layer MLP [1197]. A similar result is given in [1264], where the optimal network size can be selected according to the number of extrema and the number of hidden nodes should be selected as the number of extrema.

Since both the three-layer and four-layer MLPs can be used as universal approximators, one may wonder as to which one is more effective. Usually, a four-layer network can approximate the target with fewer connection weights, but this may, however, introduce extra local minima [222, 1078, 1197]. Xiang *et al.* provided a geometrical interpretation of the MLP on the basis of the special geometrical shape of the activation function. For the target function with a flat surface located in the domain, a small four-layer MLP can generate better results [1197].

3.2.2 Sigma-Pi Networks

The Sigma-Pi network [979] is a generalization of the MLP. Unlike the MLP, the Sigma-Pi network uses product units as well as summation units to build higher-order terms. The BP learning rule can be applied to the learning of the network. The Sigma-Pi network is known to provide inherently more powerful mapping capabilities than first-order models such as the MLP [808]. It is a universal approximator [508]. However, the Sigma-Pi network has a combinatorial increase in the number of product terms and weights. This limits its applications.

3.3 Backpropagation Learning Algorithm

The backpropagation (BP) learning algorithm is currently the most popular learning rule for performing supervised learning tasks [979, 1166, 1167]. It is not only used to train FNNs such as the MLP, it has also been adapted to RNNs [918].

The BP algorithm is a generalization of the delta rule, known as the *LMS algorithm* [1182]. Thus, it is also called the *generalized delta rule*. The BP overcomes the limitations of the perceptron learning enumerated by Minsky and Papert [808]. It uses a gradient-search technique to minimize a cost function equivalent to the MSE between the desired and actual network outputs. Due to the BP algorithm, the MLP can be extended to many layers. In contrast, the adaline network [1182] is a linear neural network. For a linear neural network, it makes no sense to introduce multilayer topology, since consecutive layers can be simplified to a single layer by multiplying the respective weight matrices. The Adaline network cannot solve linearly inseparable problems.

The BP algorithm propagates backward the error between the desired signal and the network output through the network. After providing an input pattern, the output of the network is then compared with a given target pattern and the error of each output unit calculated. This error signal is propagated backward, and a closed-loop control system is thus established. The weights can be adjusted by a gradient-descent-based algorithm.

In order to implement the BP algorithm, a continuous, nonlinear, monotonically increasing, differentiable activation function is required. The two most-used activation functions are the logistic function (1.5) and the hyperbolic tangent function (1.6), and both are sigmoidal functions. In the following, we derive the BP algorithm for the MLP. BP algorithms for other neural-network models can be derived in a similar manner.

The objective function for optimization is defined as the MSE between the actual network output $\hat{\mathbf{y}}_p$ and the desired output $\mathbf{y}_p$ for all the training pattern pairs $(\mathbf{x}_p, \mathbf{y}_p) \in \mathcal{S}$

$$E = \frac{1}{N} \sum_{p \in \mathcal{S}} E_p = \frac{1}{2N} \sum_{p \in \mathcal{S}} \|\hat{\mathbf{y}}_p - \mathbf{y}_p\|^2 \tag{3.18}$$

where N is the size of the sample set, and[2]

$$E_p = \frac{1}{2} \|\hat{\mathbf{y}}_p - \mathbf{y}_p\|^2 = \frac{1}{2} \mathbf{e}_p^{\mathrm{T}} \mathbf{e}_p \tag{3.19}$$

$$\mathbf{e}_p = \hat{\mathbf{y}}_p - \mathbf{y}_p \tag{3.20}$$

where the ith element of $\mathbf{e}_p$ is $e_{p,i} = \hat{y}_{p,i} - y_{p,i}$.

All the network parameters $\mathbf{W}^{(m-1)}$ and $\boldsymbol{\theta}^{(m)}$, $m = 2, \cdots, M$, can be combined and represented by the matrix $\mathbf{W} = [w_{ij}]$. The error function E

---

[2] The factor $\frac{1}{2}$ is used for simple derivation.

or E_p can be minimized by applying the gradient-descent procedure. When minimizing E_p, we have

$$\Delta_p \mathbf{W} = -\eta \frac{\partial E_p}{\partial \mathbf{W}} \tag{3.21}$$

where η is the learning rate or step size, provided that it is a sufficiently small positive number. Note that the gradient term $\frac{\partial E_p}{\partial \mathbf{W}}$ is a matrix whose (i,j)th entry is $\frac{\partial E_p}{\partial w_{ij}}$.

Applying the chain rule, the derivative in (3.21) can be expressed as

$$\frac{\partial E_p}{\partial w_{uv}^{(m)}} = \frac{\partial E_p}{\partial net_{p,v}^{(m+1)}} \frac{\partial net_{p,v}^{(m+1)}}{\partial w_{uv}^{(m)}} \tag{3.22}$$

The second factor of (3.22) is derived from (3.16)

$$\frac{\partial net_{p,v}^{(m+1)}}{\partial w_{uv}^{(m)}} = \frac{\partial}{\partial w_{uv}^{(m)}} \left(\sum_{\omega=1}^{J_m} w_{\omega v}^{(m)} o_{p,\omega}^{(m)} + \theta_v^{(m+1)} \right) = o_{p,u}^{(m)} \tag{3.23}$$

The first factor of (3.22) can again be derived using the chain rule

$$\frac{\partial E_p}{\partial net_{p,v}^{(m+1)}} = \frac{\partial E_p}{\partial o_{p,v}^{(m+1)}} \frac{\partial o_{p,v}^{(m+1)}}{\partial net_{p,v}^{(m+1)}} = \frac{\partial E_p}{\partial o_{p,v}^{(m+1)}} \dot{\phi}_v^{(m+1)} \left(net_{p,v}^{(m+1)} \right) \tag{3.24}$$

where the dot denotes differentiation. In the derivation of (3.24), we make use of (3.17). To solve the first factor of (3.24), we need to consider two situations

- For the output units, $m = M - 1$,

$$\frac{\partial E_p}{\partial o_{p,v}^{(m+1)}} = e_{p,v} \tag{3.25}$$

- For the hidden units, $m = 1, \cdots, M - 2$,

$$\begin{aligned}
\frac{\partial E_p}{\partial o_{p,v}^{(m+1)}} &= \sum_{\omega=1}^{J_{m+2}} \left(\frac{\partial E_p}{\partial net_{p,\omega}^{(m+2)}} \frac{\partial net_{p,\omega}^{(m+2)}}{\partial o_{p,v}^{(m+1)}} \right) \\
&= \sum_{\omega=1}^{J_{m+2}} \left[\frac{\partial E_p}{\partial net_{p,\omega}^{(m+2)}} \frac{\partial}{\partial o_{p,v}^{(m+1)}} \left(\sum_{u=1}^{J_{m+1}} w_{u\omega}^{(m+1)} o_{p,u}^{(m+1)} + \theta_\omega^{(m+2)} \right) \right] \\
&= \sum_{\omega=1}^{J_{m+2}} \frac{\partial E_p}{\partial net_{p,\omega}^{(m+2)}} w_{v\omega}^{(m+1)} \tag{3.26}
\end{aligned}$$

Define the delta function by

$$\delta_{p,v}^{(m)} = -\frac{\partial E_p}{\partial net_{p,v}^{(m)}} \tag{3.27}$$

for $m = 2, \cdots, M$. By substituting (3.22), (3.26), and (3.27) into (3.24), we finally obtain the following.

- For the output units, $m = M - 1$,

$$\delta_{p,v}^{(M)} = -e_{p,v}\dot{\phi}_v^{(M)}\left(net_{p,v}^{(M)}\right) \tag{3.28}$$

- For hidden units, $m = 1, \cdots, M - 2$,

$$\delta_{p,v}^{(m+1)} = \dot{\phi}_v^{(m+1)}\left(net_{p,v}^{(m+1)}\right) \sum_{\omega=1}^{J_{m+2}} \delta_{p,\omega}^{(m+2)} w_{v\omega}^{(m+1)} \tag{3.29}$$

Equations (3.28) and (3.29) provide a recursive method to solve $\delta_{p,v}^{(m+1)}$ for the whole network. Thus, $\mathbf{W}$ can be adjusted by

$$\frac{\partial E_p}{\partial w_{uv}^{(m)}} = -\delta_{p,v}^{(m+1)} o_{p,u}^{(m)} \tag{3.30}$$

For the activation functions, we have the following relations

- For the logistic function

$$\dot{\phi}(net) = \beta\phi(net)\left[1 - \phi(net)\right] \tag{3.31}$$

- For the tanh function

$$\dot{\phi}(net) = \beta\left[1 - \phi^2(net)\right] \tag{3.32}$$

The update for the biases can be in two ways. The biases in the $(m+1)$th layer $\boldsymbol{\theta}^{(m+1)}$ can be expressed as the expansion of the weight $\mathbf{W}^{(m)}$, that is, $\boldsymbol{\theta}^{(m+1)} = \left(w_{0,1}^{(m)}, \cdots, w_{0,J_{m+1}}^{(m)}\right)^{\mathrm{T}}$. Accordingly, the output $\mathbf{o}^{(m)}$ is expanded into $\mathbf{o}^{(m)} = \left(1, o_1^{(m)}, \cdots, o_{J_m}^{(m)}\right)^{\mathrm{T}}$. Another way is to use a gradient-descent method with regard to $\boldsymbol{\theta}^{(m)}$, by following the above procedure. Since biases can be treated as special weights, these are usually omitted in practical applications.

The BP algorithm is defined by (3.21), and is rewritten below

$$\Delta_p\mathbf{W}(t) = -\eta\frac{\partial E_p}{\partial \mathbf{W}} \tag{3.33}$$

The algorithm is convergent in the mean if $0 < \eta < \frac{2}{\lambda_{\max}}$, where $\lambda_{\max}$ is the largest eigenvalue of the autocorrelation of the vector $\mathbf{x}$, denoted as $\mathbf{C}$ [1184]. When η is too small, the possibility of getting stuck at a local minimum of the error function is increased. In contrast, the possibility of falling into oscillatory traps is high when η is too large. By statistically preprocessing the input patterns, namely, decorrelating the input patterns, the excessively large eigenvalues of $\mathbf{C}$ can be avoided and thus, increasing η can effectively speed up the convergence. PCA preconditioning speeds up the BP in most cases, except when the pattern set consists of sparse vectors. In practice, η is usually

---

Algorithm 3.3 (BP) BP algorithm for a three-layer MLP. All units have the same activation function $\phi(\cdot)$, and all biases are assumed to be zero.

1. Initialize $\mathbf{W}^{(1)}$ and $\mathbf{W}^{(2)}$.
2. Calculate E using (3.18).
3. If E is satisfactory, return.
4. For each input pattern $\mathbf{x}_p$, $1 \leq p \leq N$
   a) Forward pass
      i. Compute $\mathbf{net}_p^{(2)}$ by (3.16) and $\mathbf{o}_p^{(2)}$ by (3.17).
      ii. Compute $\mathbf{net}_p^{(3)}$ by (3.16) and $\hat{\mathbf{y}}_p = \mathbf{o}_p^{(3)}$ by (3.17).
      iii. Compute $\mathbf{e}_p$ by (3.20).
   b) Backward pass, for all neurons
      i. Compute
$$\delta_{p,v}^{(3)} = -e_{p,v}\dot{\phi}\left(net_{p,v}^{(3)}\right)$$
      ii. Update $\mathbf{W}^{(2)}$ by
$$\Delta w_{uv}^{(2)} = \eta\delta_{p,v}^{(3)}o_{p,u}^{(2)}$$
      iii. Compute
$$\delta_{p,v}^{(2)} = \left(\sum_{\omega=1}^{J_3}\delta_{p,\omega}^{(3)}w_{v\omega}^{(2)}\right)\dot{\phi}\left(net_{p,v}^{(2)}\right)$$
      iv. Update the weights $\mathbf{W}^{(1)}$ by
$$\Delta w_{uv}^{(1)} = \eta\delta_{p,v}^{(2)}o_{p,u}^{(1)}$$
5. Go to Step 2.

---

chosen to be $0 < \eta < 1$ so that successive weight changes do not overshoot the minimum of the error surface. The flowchart of the BP for a three-layer MLP is shown in Algorithm 3.3.

The BP algorithm can be improved by adding a momentum term [979]

$$\Delta_p\mathbf{W}(t) = -\eta\frac{\partial E_p}{\partial \mathbf{W}} + \alpha\Delta\mathbf{W}(t-1) \tag{3.34}$$

where α is the momentum factor, usually $0 < \alpha \leq 1$. The typical value for α is 0.9. This method is usually called the *BP with momentum (BPM)* algorithm. The momentum term can effectively magnify the descent in almost-flat steady downhill regions of the error surface by $\frac{1}{1-\alpha}$. In regions with high fluctuations, the momentum has a stabilizing effect. The momentum term actually inserts second-order information in the training process that performs like the conjugate-gradient (CG) method. A momentum term in some cases helps to overshoot a local minimizer. As a result, the momentum term effectively smoothes oscillations and accelerates convergence.

The essential storage requirement for the BP algorithm consists of all the N_w weights of the network. The computational complexity per iteration of the BP is around N_w multiplications for the forward pass, around $2N_w$ multiplications for the backward pass, and N_w multiplications for multiplying the gradient with η. Thus, four multiplications are required per iteration per weight [570].

In the above derivation, the optimization objective is E_p and the weights are updated after the presentation of each pattern. Thus, the learning is termed as incremental learning, online learning, pattern learning, or adaptive learning. When optimizing the average error E, we get the batch learning algorithm, where weights are updated only after all the training patterns are presented.

The BP algorithm is a supervised gradient-descent technique, wherein the MSE between the actual output of the network and the desired output is minimized. It is prone to local minima in the cost function. The performance can be improved and the occurrence of local minima reduced by allowing extra hidden units, lowering the gain term, and by training with different initial random weights.

The process of presenting all the examples in the pattern set, with each example being presented once, is called an *epoch*. Neural networks are trained by presenting all the examples cyclically by epoch, until the convergence criteria is reached. The training examples should be presented to the network in a random order during each epoch. When the algorithm converges to a local or global minimum, the gradient of the error function should be zero, or the change in the error function is sufficiently small. Thus, the algorithm can be terminated when the gradient of the error function is sufficiently small, or the change in the error function is sufficiently small per epoch, or the error function itself is sufficiently small.

3.4 Criterion Functions

The MSE is by far the most popular measure of error. This error measure ensures that a large error receives much greater attention than a small error. The MSE criterion is optimal and results in an ML estimation of the weights if the distributions of the feature vectors are Gaussian [978]. This is desired for most applications. Apart from the MSE function, other criterion functions can be used for the BP learning. In some situations, other error measures such as the mean absolute error, maximum absolute error, and median squared error, may be preferred.

The logarithmic error function, which takes the form of the instantaneous relative entropy or Kullback–Leibler divergence criterion, has some merits over the MSE function [86]

$$E_p(\mathbf{W}) = \frac{1}{2} \sum_{i=1}^{J_M} \left[(1 + y_{p,i}) \ln \left(\frac{1 + y_{p,i}}{1 + \hat{y}_{p,i}} \right) + (1 - y_{p,i}) \ln \left(\frac{1 - y_{p,i}}{1 - \hat{y}_{p,i}} \right) \right] \quad (3.35)$$

for the tanh activation function, where $y_{p,i} \in (-1,1)$. For the logistic activation function, the criterion can be written as [779]

$$E_p(\mathbf{W}) = \frac{1}{2} \sum_{i=1}^{J_M} \left[y_{p,i} \ln \left(\frac{y_{p,i}}{\hat{y}_{p,i}} \right) + (1 - y_{p,i}) \ln \left(\frac{1 - y_{p,i}}{1 - \hat{y}_{p,i}} \right) \right] \qquad (3.36)$$

where $y_{p,i} \in (0,1)$. In the latter case, $y_{p,i}$, $\hat{y}_{p,i}$, $1-y_{p,i}$, and $1-\hat{y}_{p,i}$ are regarded as probabilities. These criteria take the value zero only when $y_{p,i} = \hat{y}_{p,i}$, $i = 1, \cdots, J_M$, and are strictly positive otherwise. Another criterion function obtained by simplifying (3.36) via omitting the constant terms related to the patterns is [1048]

$$E_p(\mathbf{W}) = -\frac{1}{2} \sum_{i=1}^{J_M} [y_{p,i} \ln \hat{y}_{p,i} + (1 - y_{p,i}) \ln (1 - \hat{y}_{p,i})] \qquad (3.37)$$

The BP algorithm derived from the entropy criteria can partially solve the flat-spot problem. These criteria do not add computation load to calculate the error function. They, however, remarkably reduce the training time, and alleviate the problem of getting stuck at local minima by reducing the density of local minima [779]. Besides, the entropy-based BP is well suited to probabilistic training data, since it can be viewed as learning the correct probabilities of a set of hypotheses represented by the outputs of the neurons. The rationale for these cost functions is derived from a probabilistic viewpoint, with the goal of finding a network that is the most likely explanation of the observed data sequence [978].

The MSE criterion can be generalized into the Minkowski-r metric [455]

$$E_p = \frac{1}{r} \sum_{i=1}^{J_M} |\hat{y}_{p,i} - y_{p,i}|^r \qquad (3.38)$$

When $r = 1$, the metric is called the *city block* metric. The Minkowski-r metric corresponds to the MSE criterion for $r = 2$. A small value of r ($r < 2$) reduces the influence of large deviations, thus it can be used in the case of outliers. In contrast, a large r weights large deviations, and generates a better generation surface when the noise is absent in the data or when the data clusters in the training set are compact.

A modified BP algorithm is derived in [7] based on a criterion with an additional linear quadratic error term

$$E_p = \frac{1}{2} \| \hat{\mathbf{y}}_p - \mathbf{y}_p \|^2 + \frac{1}{2} \lambda_c \left\| \mathbf{net}^{(M)} - \phi^{-1}(\mathbf{y}_p) \right\|^2 \qquad (3.39)$$

where $\phi^{-1}(\cdot)$, the inverse of $\phi(\cdot)$, applies to each component of the vector within, and λ_c is a small positive number, usually $0 \leq \lambda_c \leq 1$. For each pattern, the modified BP is slightly more complex than the BP, while it always has a significantly faster convergence than the BP in the number of training iterations and in the computation time for a suitably selected λ_c. The parameter λ_c can be selected as a decreasing parameter from one to zero during the learning process.

A new term embodying the saturation degree is added to the conventional criterion function to prevent premature saturation [1157]

$$E_p = \frac{1}{2} \|\hat{\mathbf{y}}_p - \mathbf{y}_p\|^2 + \frac{1}{2} \|\hat{\mathbf{y}}_p - \mathbf{y}_p\|^2 \cdot \sum_{j=1}^{J_H} (o_{p,j} - 0.5)^2 \qquad (3.40)$$

where J_H is the number of hidden units, and 0.5 is the average value of a sigmoidal activation function. $\sum_{j=1}^{J_H} (o_{p,j} - 0.5)^2$ is defined as the saturation degree for all the hidden neurons for pattern p. The modified BP algorithm for the three-layer MLP is derived in [1157]. This modification significantly improves the BP in both the accuracy and the convergence speed. Besides, η can be selected as a large value without the worry of saturation.

Many other criterion functions can be used for deriving BP algorithms, such as those based on robust statistics [514] or regularization [923].

3.5 Incremental Learning versus Batch Learning

Incremental learning and batch learning are two methods for the BP learning. For incremental learning, the training patterns are presented to the network sequentially. It is a stochastic optimization method. For each training example, the weights are updated by the gradient-descent method

$$\Delta_p w_{ij}^{(m)} = -\eta_{inc} \frac{\partial E_p}{\partial w_{ij}^{(m)}} \qquad (3.41)$$

The learning algorithm has been proved to minimize the global error E when η_{inc} is sufficiently small [979].

In batch learning, the optimization objective is E, and the weight update is performed at the end of an epoch [979]. It is a deterministic optimization method. The weight incrementals for each example are accumulated over all the training examples before the weights are actually adapted

$$\Delta w_{ij}^{(m)} = -\eta_{batch} \frac{\partial E}{\partial w_{ij}^{(m)}} = \sum_p \Delta_p w_{ij}^{(m)} \qquad (3.42)$$

For sufficiently small learning rates, incremental learning approaches batch learning and the two methods produces the same results [354].

Incremental learning can be used when the complete training set is not available, and it is especially effective when the training set is very large, which necessitates large additional storage in the case of batch learning. For small constant learning rates, the randomness introduced provides incremental learning with a quasiannealing property, and allows for a wider exploration of the search space, which often helps in escaping from local minima [247]. However, incremental learning is hard to parallelize.

Gradient-descent algorithms are only truly gradient descent when their learning rates approach zero; thus, both the batch and incremental training are using approximations of the true gradient as they move through the weight space. When η_{batch} is sufficiently small, batch learning follows incremental learning quite closely.

Incremental learning tends to be orders of magnitude faster than batch learning, and is at least as accurate as batch training, especially for large training sets [1191]. For large training sets, batch training is often completely impractical due to the minuscule η_{batch} required. Incremental training can safely use a larger η_{inc}, and can thus train more quickly. As explained in [1191], for a training set with 20 000 examples, if η is selected as 0.1 and the average gradient is of the order of ± 0.1 for each weight per example, then the total accumulated weight change for batch learning will be of the order of $\pm 0.1 \times 0.1 \times 20\,000 = \pm 200$. These changes are unreasonably big and will result in wild oscillations across the weight space. When using incremental learning, each weight change will be of the order of $\pm 0.1 \times 0.1 = \pm 0.01$. Thus, for a converging batch learning with η_{batch}, the corresponding incremental learning algorithm can take $\eta_{inc} = N\eta_{batch}$, where N is the size of the training set. It is recommended in [1191] that $\eta_{inc} = \sqrt{N}\eta_{batch}$. As soon as η is small enough to avoid drastic overshooting of curves and local minima, there is a linear relationship between η and the number of epochs required for learning.

3.6 Activation Functions for the Output Layer

Usually, all neurons in the MLP use the same sigmoidal activation function. This limits the outputs of the network to be in the range of $(0, 1)$ or $(-1, 1)$. For classification problems, the representation is suitable. However, for function approximation problems the output may be far from the desired output, and the training algorithm is actually invalid. One solution is to scale the output patterns in the training set to within the range of the output of the sigmoidal function, and then to postprocess the output of the network for new input patterns.

3.6.1 Linear Activation Function

The preprocessing and postprocessing procedures are not necessary if the activation function for the neurons in the output layer is selected as a linear function $\phi(x) = x$ to increase the dynamical range of the network output. The approximation capability of such networks is discussed in [751]. A three-layer MLP with J_2 hidden units has a lower bound for the degree of approximation

[751]. By suitably selecting an analytic, strictly monotonic, sigmoidal activation function, this lower bound is essentially attainable. Using this same
activation function, a four-layer MLP using a fixed finite number of units in
each layer can approximate arbitrarily well any continuous function on any
compact domain [751].

When the activation function of the output layer is selected as $\phi(x) = x$,
the relation between the last hidden layer and the output layer is linear, and
the network can thus be trained in two steps. With the linearity property of
the output units, there is the relation

$$\left[\mathbf{W}^{(M-1)}\right]^{\mathrm{T}} \mathbf{o}_p^{(M-1)} = \mathbf{y}_p \tag{3.43}$$

where $\mathbf{W}^{(M-1)}$, a J_{M-1}-by-J_M matrix, can be optimized by the LS method
such as the SVD, RLS, or CG method [758]. The CG method converges to the
exact solution in J_{M-1} or J_M steps, whichever is larger. The BP algorithm is
then used to update the remaining weights.

3.6.2 Generalized Sigmoidal Function

The generalized sigmoidal function is introduced in [978, 837] for neurons in
the output layer of an MLP used for 1-of-n classification. The output of the
ith output neuron is defined by[3]

$$o_i^{(M)} = \phi\left(net_i^{(M)}\right) = \frac{e^{net_i^{(M)}}}{\sum_{j=1}^{J_M} e^{net_j^{(M)}}} \tag{3.44}$$

where the summation in the denominator is over all the neurons in the output
layer. With reference to (3.44), the derivative of the generalized sigmoidal
function is $o_i^{(M)}\left(1 - o_i^{(M)}\right)$, which is identical to the derivative for the logistic
sigmoidal function.

The generalized sigmoidal function introduces a behavior that resembles
in some respects the behavior of WTA networks. The sum of the outputs
of the neurons in the output layer is always equal to unity. The use of the
generalized sigmoidal function introduces additional flexibility into the MLP
model. Since the response of each output neuron is tempered by the responses
of all the output neurons, the competition actually fosters cooperation among
the output neurons. An SLP using the generalized sigmoidal function can
solve linearly inseparable classification problems [837].

3.7 Optimizing Network Structure

In analogy to curve fitting, smaller networks that use fewer parameters usually
have better generalization capability. When training an MLP, the optimal

---

[3] This unit is sometimes called the *soft-max* or *Potts unit* [978].

number of neurons in the hidden layers is unknown and is estimated usually by trial-and-error. Two strategies, namely, network pruning and network growing, are used to determine the size of the hidden layers.

3.7.1 Network Pruning

Network-pruning strategy first selects a network with a large number of hidden units, then removes the redundant units during the learning process. Pruning approaches usually fall into two broad groups [942, 187]. In sensitivity-based methods, one estimates the sensitivity of the error function E to the removal of a weight or unit, and removes the least important element. In penalty-based methods, additional terms are added to the error function E so that the new objective function rewards the network for choosing efficient solutions. The BP algorithm derived from this objective function drives unnecessary weights to zero and removes them during training. These two groups overlap if the objective function includes sensitivity terms. Some network pruning algorithms are surveyed in [942].

Network Pruning Using Sensitivity Analysis

Network pruning can be performed based on the relevance or sensitivity analysis of the error function E with respect to a weight w. The relevance or sensitivity measure is usually used to quantify the contribution that individual weights or nodes make in solving the network task. The less relevant weights or units can be selected as those to be removed. Mathematically, the normalized sensitivity is defined by

$$S_w^E = \lim_{\Delta w \to 0} \frac{\frac{\Delta E}{E}}{\frac{\Delta w}{w}} = \frac{\partial \ln E}{\partial \ln w} = \frac{w}{E} \frac{\partial E}{\partial w} \tag{3.45}$$

A sensitivity-based method utilizing retraining is described in [1034]. The output of each hidden unit is monitored and analyzed for all the training set after the network converges. If the output of a hidden unit is approximately constant for all the training set, this unit actually functions as a bias to all the neurons it feeds, and hence can be removed. Similarly, if two hidden units produce the same or proportional outputs for all the training set, one of the units can be removed. Small weights are assumed to be irrelevant and are pruned. After some units are removed, the network is required to be retrained. This technique unfortunately leads to a prohibitively long training process for large networks.

A pruning procedure, which iteratively removes hidden units and then adjusts the remaining weights in such a way as to preserve the overall network behavior, is described in [170]. The pruning problem is formulated as solving a set of linear equations (SLE), which is solved by a CG algorithm in the LS sense. A simple criterion for choosing the units to be removed according to sensitivity analysis is also provided.

In the skeletonization technique [826], the sensitivity of E with respect to w is defined as $S_w^E = -w\frac{\partial E}{\partial w}$. This definition of sensitivity has been applied in Karnin's pruning method [592]. In Karnin's method, during the normal course of the training process, the sensitivity for each connection is calculated by making use of the available terms. Upon completion of the training process, those connections that have low sensitivities are pruned, and no retraining procedure is necessary. This method has been further improved in [418] by devising some pruning rules to prevent an input being removed from the network or a particular hidden layer being totally removed. A fast training algorithm is also included to retrain the network after a weight is removed. In [925], Karnin's method has been extended by introducing the local relative sensitivity index within each subgroup or layer of the network. This enables parallel pruning of weights that are relatively redundant in different layers of an FNN.

A sensitivity-based method that uses linear models for hidden units is developed in [551]. If a hidden unit can be well approximated as a linear model of its net input, then it can be eliminated and replaced by adding biases in subsequent layers and by changing weights that bypass the unit. Thus, such units are useless and can be pruned. No retraining of the network is necessary. In [187], an effective hidden unit-pruning algorithm called linear-dependence pruning utilizing SLEs is presented; this algorithm improves upon the linear models given in [551] and includes the network retraining. Redundant hidden units are well modeled as linear combinations of the outputs of the other units. Hidden units are modeled as linear combinations of nonlinear units in the same layer and in the earlier layers. The least useful hidden unit is identified as that which is predicted to increase the training error the least when replaced by its model. After this hidden unit is found, the pruning algorithm replaces it with its model and retrains the weights connecting to the output layer by one iteration of training.

In [575], orthogonal transforms such as the SVD and QR with column pivoting (QR-cp) are used for pruning neural networks. The SVD serves as the null space detector, and can cure the numerical ill-conditioning problems, without reducing the parameter set. The QR-cp coupled with SVD is used for subset selection and elimination of the redundant set. Based on the transforms on the training set, one can select the optimal sizes of the input and hidden nodes, leading to a reduction in the network size. The reduced-size network is then reinitialized and retrained to the desired convergence.

The principal components pruning technique [691] is based on PCA of the node activations of successive layers of trained FNNs for a validation set. The node activation correlation matrix at each layer is required, while the calculation of the full Hessian of the error function is avoided. This method prunes the least salient eigen-nodes, and network retraining is not necessary.

Optimal Brain Damage and Optimal Brain Surgeon

The optimal brain damage (OBD) [656] and optimal brain surgeon (OBS) [460] procedures are two network-pruning methods based on the perturbation analysis of the second-order Taylor expansion of the error function.

In the following, we use $\vec{w}$ to represent the vector generated by concatenating all entries of $\mathbf{W}$. When the training process converges, the gradient is close to zero, and thus the increase in E due to a change in $\vec{w}$ is given by

$$\Delta E \simeq \frac{1}{2} \Delta \vec{w}^{\mathrm{T}} \mathbf{H} \Delta \vec{w} \qquad (3.46)$$

where $\mathbf{H}$ is the Hessian matrix, $\mathbf{H} = \frac{\partial^2 E}{\partial \vec{w}^2}$.

Removing a weight amounts to equating this weight to zero. Thus, removing a subset of weights, $\mathcal{S}_{\mathrm{prune}}$, results in a change in E by setting $\Delta w_i = w_i$, if $i \in \mathcal{S}_{\mathrm{prune}}$, otherwise $\Delta w_i = 0$. The OBS is based on the saliency (3.46). The OBD is a special case of the OBS, where the Hessian $\mathbf{H}$ is assumed to be a diagonal matrix; in this case, each weight has a saliency

$$(\Delta E)_i \simeq \frac{1}{2} w_i^2 H_{ii} \qquad (3.47)$$

In the procedure, a weight with the smallest saliency is selected for deletion. The calculation of the Hessian $\mathbf{H}$ is fundamental to the OBS procedure.

Optimal cell damage (OCD) [246] extends the OBD to remove irrelevant input and hidden units. The unit-OBS [1053] improves the OBS by removing one whole unit in each step. The unit-OBS can also conduct feature extraction on the input data by removing unimportant input units. As an intermediate between the OBD and the OBS, the principal components pruning [691] is based on a block-diagonal approximation of the Hessian.

In the case of early stopping, the OBD and the OBS are not suitable since the network is not in a local minimum and the first-order term in the Taylor-series expansion is not zero. Early brain damage (EBD) is an extension to OBD and the OBS in connection with early stopping. In addition, the EBD allows the revival of the already pruned weights [1100].

A pruning procedure similar to the OBD approach [656] is constructed using the error covariance matrix $\mathbf{P}$ obtained during the RLS training [689]. As $\mathbf{P}$ is obtained along with the RLS algorithm, pruning becomes much easier. The RLS-based pruning has a computational complexity of $O\left(N_{\mathrm{w}}^3\right)$, which is much smaller than that of the OBD method, namely, $O\left(N_{\mathrm{w}}^2 N\right)$, while the performance of the RLS-based pruning is very close to that of the OBD method in terms of the number of pruning weights and generalization ability. In addition, the RLS-based pruning is also suitable for the online situation.

Another network pruning technique, based on the training results from the extended Kalman filtering (EKF) technique, is given in [1066]. The method prunes a neural network based solely on the obtained error covariance matrix

P and the state (weight) vector, which are used to evaluate the saliency of the weights.

The variance nullity pruning [334] is based on the sensitivity analysis of the output, rather than on that of the error function. If the gradient search and the MSE function are used, then the OBD [656] and the output sensitivity analysis are conceptually the same under the assumptions that the Hessain **H** is diagonal. Parameter relevance is measured as the variance in sensitivity over the training set, and those hidden or input nodes that are irrelevant are removed. The pruned network is then retrained. The method achieves much smaller architectures than those obtained from the OBD, the OBS [460], and the method given in [1034], but with a similar or better generalization performance.

Network Pruning Using Regularization

The regularization technique is also used for network pruning. The optimization objective is defined as

$$E_{\mathrm{T}} = E + \lambda_{\mathrm{c}} E_{\mathrm{c}} \qquad (3.48)$$

where E is the error function, E_{c} is a penalty for the complexity of the structure, and $\lambda_{\mathrm{c}} > 0$ is the regularization parameter. Extra local minima are introduced to the optimization process by the penalty term, and λ_{c} needs to be appropriately determined for a particular problem.

In the weight-decay technique [484, 1165, 533, 442], E_{c} is defined as a function of the weights. In [484], E_{c} is defined as the sum of the squares of all the weights

$$E_{\mathrm{c}} = \sum_{i,j} w_{ij}^2 \qquad (3.49)$$

As a result, the change of each weight is proportional to its value. In [1165], E_{c} is defined as

$$E_{\mathrm{c}} = \sum_{i,j} \frac{\frac{w_{ij}^2}{w_0^2}}{1 + \frac{w_{ij}^2}{w_0^2}} \qquad (3.50)$$

where w_0 is a free parameter. For small w_0, the network prefers large weights. For example, in [442], w_0 is taken as unity, and this penalty term decays the small weights more rapidly than the large weights. For large w_0, the network prefers small weights. In [533], E_{c} is defined as the sum of the absolute values of the weights

$$E_{\mathrm{c}} = \sum_{i,j} |w_{ij}| \qquad (3.51)$$

Thus, all the weights are decaying at a constant step to zero.

The BP algorithm derived from E_{T} using the weight-decay term is a structural learning algorithm

$$\Delta w_{ij}^{(m)} = -\eta \frac{\partial E_{\mathrm{T}}}{\partial w_{ij}^{(m)}} = \Delta w_{ij,\mathrm{BP}}^{(m)} - \varepsilon \frac{\partial E_{\mathrm{c}}}{\partial w_{ij}^{(m)}} \qquad (3.52)$$

where $\Delta w_{ij,\mathrm{BP}}^{(m)} = -\eta \frac{\partial E}{\partial w_{ij}^{(m)}}$ is the weight change corresponding to the conventional BP learning, and $\varepsilon = \eta \lambda_{\mathrm{c}}$ is the decaying coefficient at each weight change. The amplitudes of the weights decrease continuously towards zero, unless they are reinforced by the BP rule. At the end of the training, only the essential weights deviate significantly from zero. By pruning the weights that are close to zero, a skeleton network is obtained. This effectively increases generalization and reduces the danger of overtraining as well. For example, in the modified BP with forgetting [533, 626], the weight-decay term (3.51) is used, and $\frac{\partial E_{\mathrm{c}}}{\partial w_{ij}^{(m)}} = \mathrm{sgn}\left(w_{ij}^{(m)}\right)$, where $\mathrm{sgn}(\cdot)$ is the signum function. Neural networks trained by weight-decay algorithms are not sensitive to the initial choice of the network.

The weight-decay technique given in [442] is an implementation of a robust network that is insensitive to noise. The weight-decay technique decays the weights in a network towards zero by weakening the small weights more rapidly. Because small weights can be used by the network to code noisy patterns, this weight-decay mechanism is considered to be especially important in the case of noisy data. The weight-decay technique is shown to converge as fast as BP, if not faster, and shows some significant improvement over BP in noisy situations [442].

The conventional RLS algorithm is essentially a weight-decay algorithm [689]. This conclusion is made by comparing the objective function for the RLS method with that for the weight-decay technique using (3.49). The error covariance matrix $\mathbf{P}$ obtained during the RLS training possesses properties similar to the Hessian matrix $\mathbf{H}$ of the error function. The initial value of $\mathbf{P}$, namely, $\mathbf{P}(0)$ can be used to control the generalization ability.

The weight-smoothing regularization introduces the constraint of Jacobian profile smoothness during the learning step [15]. For functional approximation with inputs resulting from the discretization of a continuous function, the weight-smoothing technique smoothes the neural Jacobian profiles with respect to the input index. Such approximations are confronted with the compensation phenomenon, namely, a lower contribution of one input can be compensated by a larger one of its neighboring inputs. Solving the compensation phenomenon, this weight-smoothing algorithm makes it possible to estimate a physically acceptable and more robust solution. Other regularization methods include neural Jacobians like the double BP [319], the input perturbation [118], or generalized regular network [923] that minimize the neural Jacobian amplitude to smooth the neural-network behavior, and the approximate smoother [819].

3.7.2 Network Growing

Another method for training the MLP is the *constructive approach*, which starts with a small network and then gradually adds hidden units until a given performance is achieved. This helps us in finding a minimal network. Constructive algorithms search for small network solutions first. They are computationally more economical than pruning algorithms, in which the majority of the training time is spent on networks larger than necessary. Constructive algorithms are likely to find smaller network solutions that are more efficient in forward computation.

The constructive approach also helps in escaping a local minimum by adding a new hidden unit [490]. When the error E does not decrease or decreases too slowly, the network may be trapped in a local minimum, and a new hidden unit is added to change the shape of the error function and thus to escape from the local minimum. The weights of the newly added neurons can be set randomly as in the standard BP algorithm.

Cascade-correlation Learning

Cascade-correlation learning is a well-known constructive learning approach. It is an efficient technique both computationally and in terms of modeling performance [347]. In the cascaded architecture, as shown in Fig. 3.3, each newly recruited unit is connected both to the input nodes and to every pre-existing unit. Network construction is based on the one-by-one training and addition of hidden units. The training starts with no hidden unit. If the minimal network cannot solve the problem after a certain number of training cycles, a set of candidate hidden units with random initial weights are generated, from which an additional hidden unit is selected and added to the network. The constructed network has direct connections between the input and output units. Previously trained units are frozen for the sake of computational efficiency.

To select the hidden unit from the candidate set, we need to evaluate the sum, over all the output units and all the patterns, of the magnitude of the covariance between the output of the new candidate and the residual output error observed at an output unit [347]

$$S = \sum_{i=1}^{J_o} \left| \sum_{p=1}^{N} (v_p - \bar{v})(e_{p,i} - \bar{e}_i) \right| \qquad (3.53)$$

where J_o is the number of network outputs, v_p is the output of the candidate unit for the pth pattern, $\bar{v}$ and $\bar{e}_i$ are the values of v_p and $e_{p,i}$ averaged over all patterns. This sum can be maximized by using a gradient-ascent procedure until S stops improving. The gradient-ascent procedure can be replaced by the Quickprop update rule for a faster convergence. The candidate with the maximum covariance is selected as a hidden unit. The network is then trained

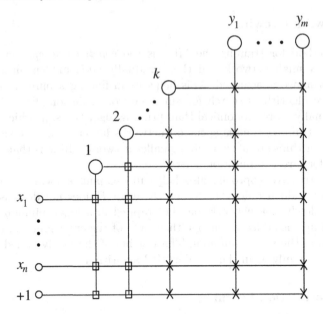

Fig. 3.3. Illustration of the cascaded architecture. For each newly added hidden unit k, all the weights connected to the previously trained units are frozen, and are denoted by boxes. All the weights connected to the newly added unit are trained, and all the weights connected to the output units are continuously updated. The crossed connections denote all the weights that are to be trained when the kth hidden unit is added.

by using the Quickprop algorithm. There is no need to backpropagate the error signals through the network.

Many ideas from the cascade-correlation learning are employed in the constructive algorithms developed in [647], where a number of objective functions for training new hidden units are also proposed. The constructive BP learning technique proposed in [675] is related to the cascade-correlation learning. Constructive BP is computationally as efficient as cascade-correlation learning, even though the error is needed to backpropagate through no more than one hidden layer, and they share the same constructive features. Constructive BP benefits from simpler implementation and the ability to utilize stochastic optimization routines. Moreover, constructive BP can be extended to allow the addition of multiple new units simultaneously instead of only one unit at a time. Multiple units trained together can reduce the modeling error more than multiple units trained independently. It can be used to perform continuous automatic structure adaptation, including both addition and deletion of units.

The cascade-correlation learning, however, is not suitable for VLSI implementation. This is due to the irregularity in connections and the unbounded fan-in of the hidden and output units. Moreover, the depth or the propagation delay through the network is directly proportional to the number of hidden units and can be excessive. In [915], the cascaded-correlation learning has been modified so as to generate networks with restricted fan-in and small depth by controlling the connectivity. The strictly layered architecture is one way of limiting the depth and fan-in.

Other Constructive Methods

A quasi-Newton method for constructing a three-layer MLP is reported in [1017]. Training starts with a network topology having a small number of hidden units and automatically grows by adding one hidden unit at a time until a network that can solve the application problem is found. The quasi-Newton method is used to minimize the sequence of error functions associated with the growing network. The construction process is fast due to the fast convergence property of the quasi-Newton method. In addition, the small number of hidden units results in an acceptable computational cost for the memory-intensive quasi-Newton method.

The dependence identification (DI) algorithm constructs and trains an MLP by transforming the training problem into a set of quadratic optimization problems, which are then solved by a succession of SLEs [818]. It is a batch learning process, and can quickly determine the MLP architectures for a given problem. The DI method uses the concept of linear dependence to group patterns. The overall convergence speed is orders of magnitude faster than the BP, although the resulting network is usually large [818]. The DI is a faster and more systematic method for developing initial network architectures than the trial-and-error or gradient-based pruning techniques. The resulting network architecture and weights can be further refined with the BP. It works well for creating neural-network approximations of continuous functions as well as providing a starting point for further BP training.

A constructive learning algorithm for the MLP using an incremental training procedure has been proposed in [722], which may be useful for real-time learning. Training patterns are learned one by one. The algorithm starts with a single training pattern and a single hidden neuron. During training, when the algorithm gets stuck at a local minimum, the weight-scaling technique [386, 950] is applied for the algorithm to escape from the local minimum. If the algorithm fails in escaping from a local minimum after several consecutive attempts, the network is allowed to grow by adding a hidden neuron. An optimization procedure based on the QP and LP techniques is employed to select initial weights for the newly added neuron, which tends to make the network reach the error tolerance with no or little training after adding a hidden neuron.

The MLP iterative construction algorithm (MICA) is a constructive training algorithm for three-layer MLPs for classification problems [938]. The Ho–Kashyap algorithm [327] is central to training both the hidden layer nodes and the output layer nodes. A pruning procedure that removes the least important hidden node, one at a time, can be included to increase the generalization ability of the MICA. MICA can also be used to determine the number of hidden nodes needed, seed the lower-level weights or seed all the weights for improving the use of BP.

Examples of early constructive methods for training FNNs with LTG neurons are the tower algorithm [395], the tiling algorithm [798] and the upstart algorithm [373]. These algorithms are based on the pocket algorithm [395], and are used for classification. For a review on constructive approaches to structural learning of FNNs including the MLP, readers are referred to [648].

3.8 Speeding Up Learning Process

The BP algorithm is a gradient-descent method and has a slow convergence speed. Numerous measures have been reported in the literature in order to speed up the convergence of the BP algorithm. These methods are introduced in this section.

3.8.1 Preprocessing of Data Set

As discussed in Chapter 2, preprocessing of a training pattern set relieves the curse of dimensionality, and also improves the generalization ability of the network. This method is efficient when the training set is very large.

A popular objective of network pruning is to detect irrelevant weights and neurons. This can be achieved through an evaluation of the sensitivities of the error function to the weights. Rather than to weights or neurons, clear and practical measures of sensitivities to inputs are developed, and utilized towards deletion of redundant inputs in [1270]. Some measures of sensitivity are given in [1270]. When one or more dimensions of the input vectors have relatively small sensitivity in comparison to others, that dimension of the input vectors can be removed, and a smaller-size neural network can be successfully retrained in most cases.

A feature-selection method particularly suited for FNNs has been proposed in [975]. A saliency metric describing the sensitivity of the outputs of the trained network with respect to the jth input is defined by

$$S_j = \sum_{\mathbf{x} \in \mathcal{X}} \sum_i \left| \frac{\partial o_i^{(M)}}{\partial x_j} \right| \tag{3.54}$$

where the first sum denotes inclusion of the entire training set. The FNN is preliminarily trained using all the available features, and a saliency metric S_j of each feature is subsequently computed. The partial derivatives can be derived according to the chain rule. Pretraining may be repeated a number of times, *e.g.*, with different initial weights or different partitions of the training set. Input features for training the FNN are arranged in a descending order of saliency. Once the saliency metrics with saliencies exceeding a certain threshold have been evaluated, the FNN is trained again. Only the most salient features are then used in the final training process, thus reducing the dimensionality of the feature space. In [95], the noise-injection method has been proposed to determine the most appropriate threshold.

In [906], a feature-extraction method that exhibits some similarity to the PCA has been proposed. Unlike the L_1-norm saliency metric defined in (3.54), the L_2-norm is used in the saliency metric. Following a pretraining stage of an FNN with the original features, linear combinations of these features, which locally maximize the response of the network's outputs to small perturbations of the inputs, are extracted. The method also takes into account the supervised character of the learning process. The method generally provides a significant increase in generalization ability with considerable reduction in the number of required input features.

3.8.2 Eliminating Premature Saturation

One major reason for slow convergence is the occurrence of premature saturation of the output of the sigmoidal functions. This can be seen from Fig. 3.4, where the sigmoidal functions and their derivatives are plotted. When the absolute value of *net* is large, $\dot{\phi}(net)$ is so small that the weight change approaches zero and the learning takes an excessively long time. This is the flat-spot problem.

Once trapped at saturation, the outputs of saturated units preclude any significant improvement in the training weights directly connected to the units. The premature saturation leads to an increase in the number of training cycles required to release the trapped weights. In order to combat premature

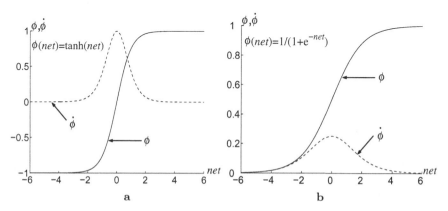

Fig. 3.4. Sigmoidal functions and their derivatives, $\beta = 1$. (**a**) Hyperbolic tangent function and its derivative. (**b**) Logistic function and its derivative.

saturation, one can modify the slope of the sigmoidal function, or modify the error function E so that when the output unit is saturated, *i.e.* the slope of the sigmoidal function approaches zero, the backpropagated error is finite.

In [673], premature saturation of network output units has been analyzed as a static phenomenon that occurs at the beginning of the training stage as a consequence of random initialization of weights. The probability of premature saturation at the first training cycle is derived as a function of the initial weights, the number of nodes in each layer, and the slope of the sigmoidal function.

A dynamic mechanism for premature saturation is analyzed in [1131]. The momentum term is identified as a leading role in the occurrence of premature saturation. The entire premature saturation process is partitioned into three distinct stages, namely, the beginning of the premature saturation, saturation plateau, and the complete recovery from saturation. For the onset of premature saturation to occur, a set of four necessary conditions must be simultaneously satisfied and usually remain satisfied for a number of consecutive iterations. In view of the four necessary conditions, a method for preventing premature saturation is to temporarily modify the momentum factor α. Once the four conditions are satisfied at iteration t, a value for α is calculated so that the second condition is not satisfied at iteration $t + 1$. The parameter α is then used to update $\Delta \mathbf{W}(t + 1)$. If more than one output unit satisfies the four conditions, α is calculated for each of these units and the smallest α used to update $\Delta \mathbf{W}(t + 1)$. The original α is used again after the $(t + 1)$th iteration. The algorithm works like the standard BP algorithm unless the four conditions are satisfied simultaneously.

In [666], the BP update equation is revised by adding a term embodying the degree of saturation to prevent premature saturation. This turns out to be adding an additional term embodying the degree of saturation in the energy function. Similarly, in [845, 844], the partial derivatives of the logistic activation function are generalized to $\left[o_{p,i}^m(1 - o_{p,i}^m)\right]^{\frac{1}{\rho_0}}$ with $\rho_0 \geq 1$ so that error signals are significantly enlarged when $o_{p,i}^m$ approaches saturation. An EA for weight update is also proposed whenever the modified energy function is trapped into local minima [844]. Other authors also avoid premature saturation by using modified energy functions [862, 1157]. These energy functions make the learning term reasonable regardless of the distribution of the actual output value.

3.8.3 Adapting Learning Parameters

The performances of the BP and BPM algorithms are highly dependent upon a suitable selection for η and α, which, unfortunately, are usually selected by trial-and-error. For different tasks and at different stages of training, some heuristics are needed for optimally adjusting η and α to speed up the convergence of the algorithms. According to [979], the process of starting with a large η and gradually decreasing it is similar to that in SA [607]. The algorithm escapes from a shallow local minimum in early training and converges into a deeper, possibly global minimum. Learning parameters are typically adapted once for each epoch.

Globally Adapted Learning Parameters

All the weights in the network are typically updated using the global learning parameters η and α. The optimal η is the inverse of the largest eigenvalue, *i.e.* $\lambda_{\max}$, of the Hessian matrix $\mathbf{H}$ of the error function, [658]. An online algorithm for estimating $\lambda_{\max}$ has been proposed in [658] that does not even require the calculation of the Hessian.

A simple and popular method for accelerating the learning is to use the search-then-converge schedule [276], which starts with a large η and gradually decreases it as the learning proceeds. Typically, η is selected as [276]

$$\eta(t) = \frac{\eta_0}{1 + \frac{t}{T_s}} \tag{3.55}$$

where T_s is the search time, during which η is relatively large and the weights may evolve to a state in the vicinity of a good minimum. When $t \gg T_s$, $\eta \to \frac{T_s}{t}\eta_0$, and the algorithm converges.

The bold-driver technique is a heuristic for optimal network performance [1132, 78], where at the $(t+1)$th epoch η is updated by

$$\eta(t+1) = \begin{cases} \rho^+\eta(t), & \Delta E(t) < 0 \\ \rho^-\eta(t), & \Delta E(t) > 0 \end{cases} \tag{3.56}$$

In the above, ρ^+ is chosen to be slightly larger than unity, typically 1.1, ρ^- is chosen significantly less than unity, typically 0.5, and $\Delta E(t) = E(t) - E(t-1)$. If the error decreases ($\Delta E < 0$), the training is approaching the minimum, and we can increase η to speed up the search process. If, however, the error increases ($\Delta E > 0$), the algorithm must have overshot the minimum and thus η is too large. In this case, the weights are not updated. This process is repeated until a decrease in the error is found. The parameter α can be selected as a fixed value. However, at each occurrence of an error increase, the next weight update is needed to be along the negative gradient direction to speed up the convergence and α is selected as zero temporarily. There are many variants of this heuristic method, including those with fixed increment of η.

The gradient-descent rule can be reformulated as

$$\vec{\mathbf{w}}(t+1) = \vec{\mathbf{w}}(t) - \eta(t)\nabla E\left(\vec{\mathbf{w}}(t)\right) \tag{3.57}$$

where $\vec{\mathbf{w}}$ is a vector formed by concatenating all the columns of $\mathbf{W}$. In [888], the learning rate of the batch BP algorithm is adapted according to the instantaneous value of $E(t)$

$$\eta(t) = \rho_0 \frac{\rho\left(E(t)\right)}{\|\nabla E\left(\overrightarrow{\mathbf{w}}(t)\right)\|} \tag{3.58}$$

where ρ_0 is a positive constant, $\rho(E)$ is a function of E, typically $\rho(E) = E$, and

$$\nabla E\left(\overrightarrow{\mathbf{w}}(t)\right) = \left.\frac{\partial E(\overrightarrow{\mathbf{w}})}{\partial \overrightarrow{\mathbf{w}}}\right|_t \tag{3.59}$$

This adaptation leads to fast convergence. However, $\eta(t)$ is a very large number in the neighborhood of a local or global minimum, leading to jumpy behavior of the weights. The method converges faster than the Quickprop algorithm [346].

In [1213], both η and α are changed adaptively according to the topography of the error function. A heuristic based on the correlation coefficient between the negative gradient and the last weight update is used to adjust η, and α is also suitably revised to retain its acceleration effect. The method can significantly reduce the overall convergence time, though it results in a small increase in the computational effort in each epoch. This algorithm achieves an exponential increase or decrease of η. The method outperforms the locally adapted learning rate algorithm proposed in [1035].

In [750], η is updated according to the local approximation of the Lipschitz constant $L(t)$ based on Armijo's condition for line search. The utilization of the Lipschitz constant guarantees the algorithm to be globally convergent[4]. $L(t)$ is approximated by

$$L(t) = \frac{\|\nabla E\left(\overrightarrow{\mathbf{w}}(t)\right) - \nabla E\left(\overrightarrow{\mathbf{w}}(t-1)\right)\|}{\|\overrightarrow{\mathbf{w}}(t) - \overrightarrow{\mathbf{w}}(t-1)\|} \tag{3.60}$$

Select $\eta(t) = 0.5\frac{1}{L(t)}$. If $\eta(t) < \eta_{LB}$, a user-specified stepsize lower bound, then $\eta(t) = 2\eta(t)$. If $\eta(t)$ is too large and successive weight updates for different patterns do not satisfy the criterion

$$E\left(\overrightarrow{\mathbf{w}}(t+1)\right) - E\left(\overrightarrow{\mathbf{w}}(t)\right) \leq -\frac{1}{2}\eta(t)\|\nabla E\left(\overrightarrow{\mathbf{w}}(t)\right)\|^2 \tag{3.61}$$

then $\eta(t)$ is set as η_0, an arbitrary positive value, and the criterion tested. If it is not satisfied, $\eta(t)$ is decreased by half in order to prevent the updated weights overshooting the minimum of the error surface. This procedure is continued until the criterion is satisfied. The algorithm terminates when the error is sufficiently small. The algorithm is robust against oscillations due to large η and avoids the phenomenon of the nearly constant E value, by ensuring that E is decreased with every weight update. The algorithm results in an improvement in the performance when compared with the BP, the delta-bar-delta [536], and the bold-driver technique [1132].

---

[4] A globally convergent algorithm is an algorithm that converges to a local minimum of the objective function from almost any starting point. Global convergence is different from global optimization.

The fuzzy inference system (FIS) is also used to adapt the learning parameters for an MLP with the BP learning [234]. The fuzzy system incorporates Jacobs' heuristics [536] about the unknown learning parameters using fuzzy IF-THEN rules. The heuristics are driven by the behavior of $E(t)$. Change in $E(t)$, denoted by $\Delta E(t)$, is an approximation to the gradient of E, and change in $\Delta E(t)$ is an approximation to the second-order derivatives of E. FISs are constructed for adjusting η and α, respectively. This fuzzy BP learning is much faster than the BP, with a significantly smaller MSE [234].

Locally Adapted Learning Parameters

Each weight $w_{ij}^{(m)}$ can have its own learning rate $\eta_{ij}^{(m)}(t)$ so that

$$\Delta w_{ij}^{(m)}(t) = -\eta_{ij}^{(m)}(t) g_{ij}^{(m)}(t) \qquad (3.62)$$

where the gradient

$$g_{ij}^{(m)}(t) = \nabla E\left(w_{ij}^{(m)}\right)\bigg|_t = \frac{\partial E}{\partial w_{ij}^{(m)}}\bigg|_t \qquad (3.63)$$

There are many locally adaptive learning algorithms using weight-specific learning rates such as [1084, 1035], the SuperSAB [1094], the delta-bar-delta algorithm [536], the Quickprop [346], and the equalized error BP (EEBP) [765], and the globally convergent strategy [749].

In [1084], the learning rates for all input weights to a neuron is selected to be inversely proportional to the fan-in of the neuron, namely

$$\eta_{ij}^{(m-1)}(t) = \frac{\kappa_0}{net_j^{(m)}(t)} \qquad (3.64)$$

where κ_0 is a small positive number. This can maintain a balance among the learning speed of units with different fan-in. Otherwise, units with high fan-in have their net input changed by a larger amount than units with low fan-in. Due to the nature of the sigmoidal function, a large input may result in saturation that will slow down the adaptation process. The increase in convergence speed is theoretically justified by studying the eigenvalue distribution of $\mathbf{H}$ [657].

The heuristic proposed in [1035] and in the SuperSAB [1094] is to adapt $\eta_{ij}^{(m)}$ by

$$\eta_{ij}^{(m)}(t+1) = \begin{cases} \eta_0^+ \eta_{ij}^{(m)}(t), & g_{ij}^{(m)}(t) \cdot g_{ij}^{(m)}(t-1) > 0 \\ \eta_0^- \eta_{ij}^{(m)}(t), & g_{ij}^{(m)}(t) \cdot g_{ij}^{(m)}(t-1) < 0 \end{cases} \qquad (3.65)$$

where $\eta_0^+ > 1, 0 < \eta_0^- < 1$. In the SuperSAB, $\eta_0^+ \simeq \frac{1}{\eta_0^-}$. Since $\eta_{ij}^{(m)}$ grows and decreases exponentially, too many successive acceleration steps may generate too large or too small $\eta_{ij}^{(m)}$ and thus slow down the learning process. To avoid this, a momentum term is included in the SuperSAB.

In [1245, 1244], $\eta_{ij}^{(m)}$ and $\alpha_{ij}^{(m)}$ are optimally tuned using three approaches, namely, the second-order-based, first-order-based, and CG-based approaches. These approaches make use of the derivatives of E with respect to $\eta_{ij}^{(m)}$ and $\alpha_{ij}^{(m)}$. The algorithms make use of the information gathered from the forward and backward procedures, and do not need explicit computation of the first- and second-order derivatives in the weight space. The computational and storage burdens are at most triple that of the standard BP, with an order of magnitude faster speed.

The EEBP algorithm [765] is an incremental learning algorithm. The relative magnitudes of $\eta_{ij}^{(m)}$ can be fixed prior to the training by just taking simple properties of the network and the input data into account. In addition, the EEBP incorporates a simple statistically motivated input normalization, and hidden-unit substitution, a new technique that further improves the convergence rate. EEBP usually yields a dramatic speed-up of the convergence.

A general theoretical result has been derived for developing first-order batch training algorithms with local learning rates based on Wolfe's conditions for linear search and the Lipschitz condition [749]. This result provides conditions under which global convergence is guaranteed. This globally convergent strategy can be equipped with algorithms of this class to adapt the overall search direction to a descent one at each training iteration. When the Quickprop [346] and the Silva and Almeida's algorithms [1035] are equipped with this strategy, they exhibit a significantly better percentage of success in reaching local minima than their original versions, although they may require additional evaluations on the error function and gradients depending on the algorithm.

Delta-bar-delta

The delta-bar-delta algorithm [536] is similar to the method given by (3.65) [1035], but eliminates its problems by making linear acceleration and exponential deceleration of the learning rates.

Individual $\eta_{ij}^{(m)}(t)$ are updated based on a local optimization method

$$\eta_{ij}^{(m)}(t+1) = \eta_{ij}^{(m)}(t) + \Delta\eta_{ij}^{(m)}(t) \tag{3.66}$$

and $\Delta\eta_{ij}^{(m)}(t)$ is given as

$$\Delta\eta_{ij}^{(m)}(t) = \begin{cases} \kappa_0, & \bar{g}_{ij}^{(m)}(t-1)g_{ij}^{(m)}(t) > 0 \\ -\beta\eta_{ij}^{(m)}(t), & \bar{g}_{ij}^{(m)}(t-1)g_{ij}^{(m)}(t) < 0 \\ 0, & \text{otherwise} \end{cases} \tag{3.67}$$

where

$$\overline{g}_{ij}^{(m)}(t) = (1 - \varepsilon)g_{ij}^{(m)}(t) + \varepsilon\overline{g}_{ij}^{(m)}(t - 1) \tag{3.68}$$

and ε, κ_0, β are positive constants specified by the user. All $\eta_{ij}^{(m)}$s are initial-ized with small values. Basically, $g_{ij}^{(m)}(t)$ is an exponentially decaying trace of gradient values. The inclusion of the momentum term sometimes causes the delta-bar-delta to diverge, and as such an adaptively changing momen-tum has been introduced to improve the delta-bar-delta [807]. However, the delta-bar-delta algorithm requires a careful selection of the parameters.

Quickprop

In the Quickprop method [346, 347, 1123], $\alpha_{ij}^{(m)}(t)$ are heuristically adapted. The Quickprop algorithm is given by

$$\Delta w_{ij}^{(m)}(t) = \begin{cases} \alpha_{ij}^{(m)}(t)\Delta w_{ij}^{(m)}(t - 1), & \Delta w_{ij}^{(m)}(t - 1) \neq 0 \\ \eta_0 g_{ij}^{(m)}(t), & \Delta w_{ij}^{(m)}(t - 1) = 0 \end{cases} \tag{3.69}$$

where

$$\alpha_{ij}^{(m)}(t) = \min\left\{ \frac{g_{ij}^{(m)}(t)}{g_{ij}^{(m)}(t - 1) - g_{ij}^{(m)}(t)}, \alpha_{max} \right\} \tag{3.70}$$

α_{max} is typically 1.75 and $0.01 \leq \eta_0 \leq 0.6$; η_0 is only used at the start or restart of the training. Quickprop can suffer from the flat-spot problems, and Fahlman [346] improves it by adding 0.1 to the derivative of the sigmoidal function. The use of error gradient at two consecutive time steps is a discrete approximation to second-order derivatives, and the method is actually a quasi-Newton method that uses the so-called secant steps. α_{max} is used to avoid very large Quickprop updates. Benchmark experiments show that the Quickprop typically performs very reliably and converges very fast [913]. However, the simplification of the Hessian to a diagonal matrix used in Quickprop has not been theoretically justified and convergence problems may occur for certain tasks.

3.8.4 Initializing Weights

The initial weights of a network play a significant role in the convergence of a training method. Without *a priori* knowledge of the final weights, it is com-mon practice to initialize all the weights with random small absolute values, or with small zero-mean random numbers [979]. Starting from large weights may prematurely saturate the units and slow down the learning process. Theoreti-cally, the probability of prematurely saturated neurons in the MLP increases with the maximal value of the weights [673]. Thus, a smaller initial weight range increases the convergence speed of the MLP [673]. By statistical ana-lysis, the maximum amplitude for the initial weights is derived in [318]. For the three-layer MLP, a weight range of $[-0.77, 0.77]$ empirically gives the best mean performance over many existing random weight initialization techniques [1086].

Randomness also helps to break the symmetry of the system and thus prevents redundancy in the network. Symmetric weight distributions severely restrict the performance of gradient-based learning algorithms, which are unable to break weight symmetry. Initial weights may drastically affect the learning behavior regardless of the learning algorithms. Poor weight values may result in slow convergence or the network gets stuck at a local minimum. The objective of weight initialization is to find weights that are as close as possible to a global minimum before training.

Heuristics for Weight Initialization

In [1169], the initial weights of the ith unit at the jth layer are selected based on the order of $\frac{1}{\sqrt{n_i^{(j)}}}$, where $n_i^{(j)}$ is the number of weights to the ith unit at the jth layer. When the weights to a unit are uniformly distributed in $\left[-\frac{3}{\sqrt{n_i^{(j)}}}, \frac{3}{\sqrt{n_i^{(j)}}} \right]$, the total input to that unit, namely, $net_i^{(j)}$, is a random variable with zero mean and a standard deviation of unity. This is an empirical optimal initialization of the weights, which has been verified by the extensive experiments performed in [1086].

In [787], the weights are first randomly initialized to the range $[-a_0, a_0]$, a_0 being a positive constant, and are then individually scaled to ensure that each neuron is active over its full dynamic range. The scaling factor for the weights connected to the ith neuron at the jth layer is given by

$$\rho_i^{(j)} = \frac{D_i^{(j)}}{a_0 n_i^{(j)}} \tag{3.71}$$

where $D_i^{(j)}$ is the dynamic range of the activation function.

The optimal magnitudes of the initial weights and biases can be determined based on multidimensional geometry [1212]. This method ensures that the outputs of the hidden and output layers are well within the active region, while the dynamic range of the activation function is fully utilized. For a three-layer MLP, the maximum magnitude of the weights between the input and hidden layers is given as

$$w_{\max}^{(1)} = \frac{8.72}{D_{\text{in}}} \sqrt{\frac{3}{J_1}} \tag{3.72}$$

where D_{in} is the maximum possible distance between two points of the input space

$$D_{\text{in}} = \left(\sum_{i=1}^{J_1} [\max(x_i) - \min(x_i)]^2 \right)^{\frac{1}{2}} \tag{3.73}$$

The maximum magnitude of the weights between the hidden and output layers is given as

$$w_{\max}^{(2)} = \frac{15.10}{J_2} \tag{3.74}$$

The biases of the hidden and output nodes are, respectively, given by

$$\theta_j^{(1)} = -\sum_{i=1}^{J_1} c_i^{\text{in}} w_{ij}^{(1)} \tag{3.75}$$

$$\theta_j^{(2)} = -0.5 \sum_{i=1}^{J_2} w_{ij}^{(2)} \tag{3.76}$$

where $\mathbf{c}^{\text{in}} = \left(c_1^{\text{in}}, c_2^{\text{in}}, \cdots, c_{J_1}^{\text{in}} \right)^{\text{T}}$ is the center of the input space, and

$$c_i^{\text{in}} = \frac{1}{2} \left[\max\left(x_i \right) + \min\left(x_i \right) \right] \tag{3.77}$$

The maximum magnitude of the weights between the hidden and output layers is shown to be inversely proportional to the hidden layer fan-in.

The hidden-layer weights can be initialized in such a way that each hidden node is assigned to approximate a portion of the range of the desired function based on the piecewise-linear approximation of a sigmoidal function at the start of the network training [847, 874]. For a three-layer MLP using tanh activation function, Nguyen and Widrow [847] randomly selected $w_{ij}^{(1)}$, $w_{jk}^{(2)}$ between -0.5 and 0.5, and further scaled $w_{ij}^{(1)}$ by $w_{ij}^{(1)} = \frac{b_0 w_{ij}^{(1)}}{\left\| \mathbf{w}_j^{(1)} \right\|}$, where $\mathbf{w}_j^{(1)} = \left(w_{1j}^{(1)}, \cdots, w_{J_1 j}^{(1)} \right)^{\text{T}}$, and $b_0 = 0.7 \left(J_2 \right)^{\frac{1}{J_1}}$. The biases $\theta_j^{(1)}$ are set as random numbers in $[-b_0, b_0]$.

The mini-max initialization technique can efficiently approximate functions [1264]. The number of extrema can be used to estimate the complexity of the function, and a compact network constructed to guarantee the approximation at the extremum points. Some heuristics based on interpolation and extrapolation are adopted to locate the extrema of the mapping function from the given sample data points. Based on this, the mini-max initialization can construct the network of optimal size with the initial point located very likely within the promising area. Thus, local minima are avoided, and fast convergence is achieved.

Weight Initialization Using Parametric Estimation

Random initialization is ineffective because of the lack of prior information on the mapping function between the input and output of the examples. The sensitivity of the BP to the initial weights is discovered to be a complex fractal-like structure for convergence as a function of the initial weights [619]. In addition to the heuristics, such as random initialization, there are also various weight-estimation techniques, where a nonlinear mapping between pattern and target is introduced [301, 318, 678, 680].

Clustering is useful for weight initialization of three-layer MLPs. A three-layer MLP with prototypes is initialized in [301]. The input patterns are first augmented with the output patterns and are then transformed into vectors of unit length and increased size. Cluster analysis is used for finding prototypes. Substantial improvements are achieved in convergence speed, avoidance of local minima, and generalization [301]. In [1044], the clustering and nearest-neighbor methods are utilized to initialize hidden-layer weights, and the output-layer weights are then initialized by solving an SLE using the SVD. These ideas are also employed in the initialization method given in [1172], which incorporates the clustering and nearest-neighbor classification technique for a number of cluster sets, each representing the training examples with a different degree of accuracy. A network construction algorithm uses these cluster sets for constructing an initial network, and the network optimization stage searches for the best networks by evaluating all the networks that can be constructed this way.

The orthogonal least squares (OLS) method [208, 210] is used as a practical weight initialization algorithm for the MLP in [678, 680]. The maximum covariance initialization method [676] uses a procedure similar to that of the cascade-correlation algorithm [347].

An optimal weight initialization for the three-layer MLP is derived based on ICA [1215]. The algorithm is able to initialize the hidden-layer weights that extract the salient feature components from the input data. The initial output-layer weights are evaluated in such a way that the output neurons are kept inside the active region.

To increase the convergence speed, the outputs of the hidden neurons can be assigned in the nonsaturation region, and the optimal initial weights evaluated using the LS and linear algebraic method [1214, 1211]. In [1214], the optimal initial weights between layers are evaluated using the LS method by assigning the outputs of hidden neurons with random numbers in the range between 0.1 and 0.9. The actual outputs of the hidden neurons are obtained by propagating the input patterns through the network. The optimal weights between the hidden and output layers can then be evaluated by using the LS method. In [1211], the weights connected to the hidden layers are determined by the Cauchy's inequality and the weights connected to the output layer are determined by the LS method.

Weight Initialization Based on Taylor-series Expansion

MLP learning is to estimate a nonlinear mapping between the input and the output of the examples, Φ, by superposition of the sigmoidal functions. By using a Taylor-series development of Φ and the nonlinearity of the sigmoidal function, two weight initialization strategies for the three-layer MLP are obtained based on the first- and second-order identification of Φ [263]. These techniques effectively avoid local minima, significantly speed up the convergence, obtain a better generalization, and estimate the size of the network. The first-order identification strategy is described below. These ideas are also embodied in the two weight initialization techniques that combine a random weight initialization scheme with the pseudoinverse method [195].

For a J_1-J_2-J_3 MLP with linear output neurons, we have

$$\mathbf{Y} = \left[\mathbf{W}^{(2)}\right]^{\mathrm{T}} \phi \left(\left[\mathbf{W}^{(1)}\right]^{\mathrm{T}} \mathbf{X} + \boldsymbol{\theta}^{(2)} \mathbf{1}^{\mathrm{T}}\right) + \boldsymbol{\theta}^{(3)} \mathbf{1}^{\mathrm{T}} \tag{3.78}$$

where $\mathbf{X}$ is the $J_1 \times N$ input matrix, $\mathbf{Y}$ is the $J_3 \times N$ output matrix, and $\mathbf{1}$ is an N-dimensional vector whose entries are all unity.

The training error is defined as

$$E = \left\|\mathbf{Y} - \widehat{\mathbf{Y}}\right\|_{\mathrm{F}}^2 \tag{3.79}$$

where $\|\cdot\|_{\mathrm{F}}$ is the Frobenius norm, and $\widehat{\mathbf{Y}}$ is the network output corresponding to $\mathbf{Y}$.

If the linear model is used, the training pattern pairs $(\mathbf{X}, \mathbf{Y})$ can be related by

$$\mathbf{Y} = \mathbf{A}\mathbf{X} + \boldsymbol{\theta}\,\mathbf{1}^{\mathrm{T}} \tag{3.80}$$

Assuming that the network operates over the linear part of the sigmoidal function with a slope of unity, the bias $\boldsymbol{\theta}^{(2)}$ is naturally taken as 0. For the training set, the approximation according to the network is given by

$$\mathbf{Y} = \left[\mathbf{W}^{(2)}\right]^{\mathrm{T}} \left[\mathbf{W}^{(1)}\right]^{\mathrm{T}} \mathbf{X} + \boldsymbol{\theta}^{(3)} \mathbf{1}^{\mathrm{T}} \tag{3.81}$$

Applying the LS method to (3.81), we obtain

$$\left[\mathbf{W}^{(2)}\right]^{\mathrm{T}} \left[\mathbf{W}^{(1)}\right]^{\mathrm{T}} = \left(\mathbf{Y} - \boldsymbol{\theta}^{(3)} \mathbf{1}^{\mathrm{T}}\right) \mathbf{X}^{\dagger} \tag{3.82}$$

where $\mathbf{X}^{\dagger}$ denotes the pseudoinverse of $\mathbf{X}$. SVD is used to solve the pseudoinverse, $\mathbf{X} = \mathbf{U}\boldsymbol{\Sigma}\mathbf{V}^{\mathrm{T}}$; hence, $\mathbf{X}^{\dagger} = \mathbf{V}\boldsymbol{\Sigma}^{-1}\mathbf{U}^{\mathrm{T}}$. For a hidden layer of J_2 units, only the first J_2 principal singular values and their corresponding left and right singular vectors are retained, and singular values arranged in the descending order. A dynamic factor $\rho_{\mathbf{dyn}}$ is introduced so as to determine the size of the hidden layer, $\sigma_1 \geq \sigma_2 \geq \cdots \sigma_{J_2} > \frac{\sigma_1}{\rho_{\mathrm{dyn}}}$. Thus

$$\left[\mathbf{W}^{(2)}\right]^{\mathrm{T}} \left[\mathbf{W}^{(1)}\right]^{\mathrm{T}} = \left(\mathbf{Y} - \boldsymbol{\theta}^{(3)} \mathbf{1}^{\mathrm{T}}\right) \mathbf{V}_{J_2} \boldsymbol{\Sigma}_{J_2}^{-1} \mathbf{U}_{J_2}^{\mathrm{T}} \tag{3.83}$$

One can select $\mathbf{W}^{(1)}$ and $\mathbf{W}^{(2)}$ by

$$\left[\mathbf{W}^{(1)}\right]^{\mathrm{T}} = \mathbf{T}\boldsymbol{\Sigma}_{J_2}^{-1}\mathbf{U}_{J_2}^{\mathrm{T}} \tag{3.84}$$

$$\left[\mathbf{W}^{(2)}\right]^{\mathrm{T}} = \left(\mathbf{Y} - \boldsymbol{\theta}^{(3)} \mathbf{1}^{\mathrm{T}}\right) \mathbf{V}_{J_2} \mathbf{T}^{-1} \tag{3.85}$$

where $\mathbf{T}$, a $J_2 \times J_2$ matrix, is an arbitrary nonsingular diagonal matrix acting as a scaling matrix. By calculating the network output $\hat{\mathbf{Y}}$ according to (3.81), the error E can be calculated. The above procedure can be easily extended to three-layer MLPs with sigmoidal output neurons.

3.8.5 Adapting Activation Function

During the training, if a unit has a large net input, net, the output of this unit is close to a saturation region of its sigmoidal function. Thus, if the target value is substantially different from that of the saturated one, the unit has entered a flat spot. Since the first-order derivative of the sigmoidal function $\dot{\phi}(net)$ is very small when net is large in magnitude, the weight update is very slow. Fahlman [346] developed a simple solution by adding a bias, typically 0.1, to $\dot{\phi}(net)$. Hinton [483] suggested the design of an error function that goes to infinity at points where $\dot{\phi}(net) \to 0$. This leads to a finite nonzero error update.

One way to solve the flat-spot problem is to define an activation function such that

$$\phi_\mu(net) = \mu net + (1 - \mu)\phi(net) \tag{3.86}$$

with $\mu \in [0, 1]$ [1221]. At the beginning, $\mu = 1$ and all the nodes have linear activation, and the BP is used to obtain a local minimum in E. Then μ is decreased gradually and the BP is applied until $\mu = 0$. The flat-spot problem does not occur since $\dot{\phi}_\mu(net) > \mu$ and $\mu > 0$ for most of the training time. When $\mu = 1$, $E(\mathbf{W}, \mu)$ is a polynomial of $\mathbf{W}$ and thus has few local minima. This process can be viewed as an annealing process, which helps us in finding a global or good minimum.

For sigmoidal functions, such as the logistic and hyperbolic tangent functions, the gain β represents the steepness (slope) of the activation function. In the BP algorithm, β is fixed and typically $\beta = 1$. The modified BP algorithm with an adaptive β significantly increases the learning speed and improves the generalization [636, 1051]. Each neuron has its own variable gain $\beta_i^{(m)}$, which is adapted by using the gradient-descent method

$$\Delta\beta_i^{(m)} = -\eta_\beta \frac{\partial E}{\partial \beta_i^{(m)}} \tag{3.87}$$

where η_β is a small positive learning rate.

A large gain yields results similar to those with a high learning rate. Changing the gain β is equivalent to changing the learning rate η, the weights and the biases. This is asserted by Theorem 3.2 [335].

Theorem 3.2. *An MLP with the logistic activation function $\phi(\cdot)$, gain β, learning rates $\boldsymbol{\eta}$, weights $\mathbf{W}$, and biases $\boldsymbol{\theta}$ is equivalent to a network of identical topology with the activation function $\phi(\cdot)$, gain 1, learning rates $\beta^2\boldsymbol{\eta}$, weights $\beta\mathbf{W}$, and biases $\beta\boldsymbol{\theta}$, in the sense of the BP learning.*

A fuzzy system for automatically tuning the gain β has been proposed in [335] to improve the performance of the BP algorithm. The inputs of the fuzzy system are the sensitivities of the error with respect to the output and hidden layers, and the output is the appropriate gain of the activation function.

An adaptation rule for the gain β is derived using the gradient-descent method based on a sigmoidal function such as [186]

$$\phi(x) = \left(\frac{1}{1 + e^{-x}} \right)^{\beta} \tag{3.88}$$

where $\beta \in (0, \infty)$. For $\beta \neq 1$, the derivative $\dot{\phi}(x)$ is skewed and its maxima shift from the point corresponding to $x = 0$ for $\beta = 1$ and the envelope of the derivatives is also sigmoidal. The method is an order of magnitude faster than the standard BP algorithm.

A sigmoidal activation function with a wide linear part is derived in [325] by integrating an input distribution of the soft trapezoidal shape[5] to generate a probability of fullfilment

$$\phi(x) = \frac{1}{2\beta b} \ln \left(\frac{1 + e^{\beta[(x-c)+b]}}{1 + e^{\beta[(x-c)-b]}} \right) \tag{3.89}$$

where β is the gain parameter, c decides the center of the shape, and $b > 0$ decides the slope at $x = 0$. A larger b leads to a smaller slope, while a larger β generates a longer linear part. When $b \to 0$, the log-exp function approaches the logistic function $\phi(x) = \frac{1}{1+e^{-\beta x}}$. For the same β, the log-exp function always has a longer linear part and a smaller slope at $x = 0$.

An illustration of the log-exp function as well as the logistic function is shown in Fig. 3.5. The extended linear central part of the log-exp function makes it more suitable as the activation function for MLPs and MLPs with such activation functions can learn quickly. This is due to the fact that the wide nonsaturation zone prevents premature saturation. The extended linear center part is also desirable for implementation.

3.8.6 Other Acceleration Techniques

A simple gradient reuse strategy is applied in [521] to improve the convergence speed. Gradients computed using the BP algorithm are reused several times until the resulting weight updates no longer lead to a reduction in error. To enhance the reuse rate, batch mode must be used. Batch mode generates a more accurate estimate of the true gradient. The reuse rate actually controls η in the BP algorithm. This strategy can be viewed as a line search in the gradient direction. However, according to the benchmark experiments [913], this algorithm is only faster than the BP for simple problems, and it is either slower than the BP or does not converge at all for more complex problems.

---

[5] This is given by (6.13).

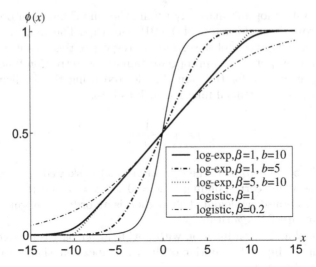

Fig. 3.5. Illustration of the log-exp and logistic functions. For the log-exp function, $c = 0$. For the same β, the log-exp function always has a larger nonsaturation zone and a wider linear part than the logistic function. For the log-exp function, the slope at $x = 0$ is decided by b and the width of the linear part is determined by β. The logistic function can extend its nonsaturation zone by decreasing β, but the width of its linear part is still limited.

The BP algorithm can be accelerated by extrapolation of each individual weight [570]. This extrapolation procedure is easy to implement and is activated only a few times in between iterations of the BP algorithm. It leads to significant savings in computation time of the BP algorithm and the solution is always located in close proximity to the one obtained by the BP procedure. BP by weight extrapolation reduces the required number of iterations at the expense of only a few extrapolation steps embedded in the BP algorithm and some additional storage for computed weights and gradient vectors.

In the BP with selective training [1113], a selective training procedure is appended to the BP. When the MLP is sufficiently trained by the BP algorithm, if we continue training on only some inputs in a new epoch, the network usually forgets other learnt input-output associations. Thus, when the network has learnt most of the vector mappings and the training procedure has slowed down, *i.e.* the average error becomes smaller than a threshold value, the BP is stopped and selective training is continued. A stability margin for correct classification in the training step is considered, and training on marginally learnt inputs are carried out. This can effectively prevent overtraining, de-emphasize the overtrained patterns, and enable the network to learn the last per cent of unlearned associations in a short period of time. However, selective training cannot be used on conflicting data, or on a data set with large overlapping areas of classes.

Three-term Backpropagation

In addition to the usual gradient and momentum terms, a third term, namely, a proportional term, can be added to the BP update equation [1271, 1272]. The algorithm can be applied for both the batch and incremental learning. For incremental learning, the learning rule can be written as

$$\Delta_p \mathbf{W}(t) = -\eta \frac{\partial E_p(t)}{\partial \mathbf{W}(t)} + \alpha \Delta_p \mathbf{W}(t-1) + \gamma E_p(\mathbf{W}(t)) \mathbf{1} \qquad (3.90)$$

where the matrix $\mathbf{1}$ has the same size as $\mathbf{W}$ but with all the entries being unity, and γ is a proportional factor. This three-term BP algorithm is analogous to the common PID control algorithm used in feedback control. Three-term BP, having a complexity similar to the BP, significantly outperforms the BP in terms of the convergence speed and the ability to escape from local minima. The three-term BP algorithm is more robust to the choice of the initial weights, especially when relatively high values for the learning parameters are selected.

Successive Approximative Backpropagation

The successive approximative BP algorithm [702] can effectively avoid local minima. Given a set of N pattern pairs $\{(\mathbf{x}_p, \mathbf{y}_p)\}$, all the training patterns are normalized so that $|x_{p,j}| \leq 1$, $|y_{p,k}| \leq 1$, where $p = 1, \cdots, N$, $j = 1, \cdots, J_1$, $k = 1, \cdots, J_M$. The training is composed of N_{phase} successive BP training phases, each phase being terminated when a predefined accuracy δ_i, $i = 1, \cdots, N_{\text{phase}}$, is achieved.

At the first phase, the network is trained using BP on the training set. After accuracy δ_1 is achieved, the output of the network for the N input $\{\mathbf{x}_p\}$ are $\{\hat{\mathbf{y}}_p(1)\}$ and the weights are $\mathbf{W}(1)$. Calculate output errors $\delta \mathbf{y}_p(1) = \mathbf{y}_p - \hat{\mathbf{y}}_p(1)$ and normalize each $\delta \mathbf{y}_p(1)$ so that $|\delta y_{p,k}(1)| \leq 1$. In the second phase, the N training patterns are $\{(\mathbf{x}_p, \delta \mathbf{y}_p(1))\}$. The training terminates at accuracy δ_2, with weights $\mathbf{W}(2)$, and output $\{\hat{\mathbf{y}}_p(2)\}$. Calculate $\delta \mathbf{y}_p(2) = \delta \mathbf{y}_p(1) - \hat{\mathbf{y}}_p(2)$ and normalized $\delta \mathbf{y}_p(2)$. This process continues up to phase N_{phase} with accuracy $\delta_{N_{\text{phase}}}$ and weights $\mathbf{W}(N_{\text{phase}})$.

The final training error is given by

$$E < 2^{N_{\text{phase}}} \prod_{i=1}^{N_{\text{phase}}} \delta_i \qquad (3.91)$$

If all $\delta_i < \frac{1}{2}$, as $N_{\text{phase}} \to \infty$, $E \to 0$. Successive approximative BP is empirically shown to significantly outperform the BP in terms of the convergence speed and generalization performance.

3.9 Backpropagation with Global Descent

The gradient-descent method is a stochastic dynamical system whose stable points only locally minimize the energy (error) function. The global-descent method, which is based on a global optimization technique called *terminal repeller unconstrained subenergy tunneling (TRUST)* [179, 68], is a deterministic dynamic system consisting of a single vector differential equation. The global-descent rule replaces the gradient-descent rule for MLP learning [179, 970]. TRUST was introduced for general optimization problems, and it formulates optimization in terms of the flow of a special deterministic dynamical system.

3.9.1 Global Descent

Global descent is a gradient-descent method using the criterion function

$$C\left(\mathbf{W}, \mathbf{W}^*\right) = \ln\left(\frac{1}{1 + e^{-[E(\mathbf{W}) - E(\mathbf{W}^*) + \sigma]}}\right)$$

$$-\frac{3}{4}\rho\sum_{ij}\left(w_{ij} - w_{ij}^*\right)^{\frac{4}{3}} u\left(E(\mathbf{W}) - E\left(\mathbf{W}^*\right)\right) \qquad (3.92)$$

where $\mathbf{W}^* = \left[w_{ij}^*\right]$ is a fixed weight matrix corresponding to a local minimum of $E(\mathbf{W})$ or an initial weight matrix, $u(\cdot)$ is the Heaviside step function, σ is a shifting parameter whose typical value is taken as 2, and $\rho > 0$ is a small constant, such as 0.001, representing the power of the repeller. The first term on the right-hand side of (3.92) is a nonlinear, monotonic transformation of $E(\mathbf{W})$ that preserves all the critical points of $E(\mathbf{W})$ and the same relative ordering of the local and global minima. It flattens the portions of $E(\mathbf{W})$, which lie above $E\left(\mathbf{W}^*\right)$, and leaves it nearly unmodified elsewhere. The term $\sum_{ij}\left(w_{ij} - w_{ij}^*\right)^{\frac{4}{3}}$ is the *repeller energy* term, which generates a convex surface with a unique minimum at $\mathbf{W} = \mathbf{W}^*$. This is schematically represented in Fig. 3.6 for a one-dimensional criterion function $E(\mathbf{W})$.

Applying gradient descent to $C\left(\mathbf{W}, \mathbf{W}^*\right)$, we get the global-descent update

$$\Delta w_{ij} = -\eta\frac{\partial E(\mathbf{W})}{\partial w_{ij}}\frac{1}{1 + e^{E(\mathbf{W}) - E(\mathbf{W}^*) + \sigma}}$$

$$+\eta\rho\left(w_{ij} - w_{ij}^*\right)^{\frac{1}{3}} u\left(E(\mathbf{W}) - E\left(\mathbf{W}^*\right)\right) \qquad (3.93)$$

where the first term on the right-hand side is a *subenergy gradient* and the second term is a *non-Lipschitzian terminal repeller*. The BP with global-descent update can escape from local minima of $E(\mathbf{W})$.

The update (3.93) automatically switches between two phases: the tunneling phase and the local-search phase.

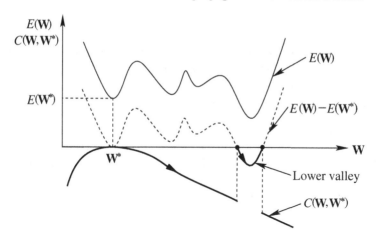

Fig. 3.6. Schematic of $C(\mathbf{W}, \mathbf{W}^*)$ for one-dimensional criterion function $E(\mathbf{W})$. $C(\mathbf{W}, \mathbf{W}^*)$ transforms the current local minimum of $E(\mathbf{W})$ into a unique maximum so that the gradient-descent method can escape from it to a lower valley.

- The tunneling phase is characterized by $E(\mathbf{W}) \geq E(\mathbf{W}^*)$.
  For this condition, the subenergy gradient term is nearly zero in the vicinity of the local minimum $\mathbf{W}^*$, and thus the terminal repeller term dominates

$$\Delta w_{ij} \simeq \eta \rho \left(w_{ij} - w_{ij}^* \right)^{\frac{1}{3}} \tag{3.94}$$

  The system has a repelling unstable equilibrium point at $w_{ij} = w_{ij}^*$ [1251]. When initialized with a small perturbation from $\mathbf{W}^*$, the system will be repelled from the local minimum until it reaches a lower basin of attraction, where $E(\mathbf{W}) < E(\mathbf{W}^*)$.
- The local-search phase is characterized by $E(\mathbf{W}) < E(\mathbf{W}^*)$.
  The repeller term is identically zero, and the multiplier term for the gradient in the first term is close to unity. Thus

$$\Delta w_{ij} \simeq \eta \frac{\partial E(\mathbf{W})}{\partial w_{ij}} \tag{3.95}$$

This phase implements gradient descent.

The two phases are repeated until a stopping criterion is achieved.

3.9.2 Backpropagation with Tunneling

The BP with tunneling for training the MLP is similar to the global descent [179] and can find the the global minimum from arbitrary initial choice in the weight space in polynomial time [970]. The algorithm consists of two phases. The first phase is a local search that implements the BP learning. The second phase implements dynamic tunneling in the weight space avoiding the local trap and thereby generates the point of next descent. Alternately, repeating the two phases forms a training procedure that leads to the global minimum. A good initial point results in the algorithm converging to the global minimum very rapidly, while an improper starting point results in it taking more time. This algorithm is computationally efficient.

A random point $\mathbf{W}$ in the weight space is selected. For a new point $\mathbf{W}+\epsilon$, where ϵ is a perturbation related to $\mathbf{W}$, if $E(\mathbf{W} + \epsilon) \leq E(\mathbf{W})$, the BP is applied until a local minimum is found; otherwise, the tunneling technique takes place until a point at a lower basin is found. The algorithm automatically enters the two phases alternately and weights are modified according to their respective update rules.

The tunneling is implemented by solving the differential equation

$$\frac{\mathrm{d}w_{kj}^{(l)}}{\mathrm{d}t} = \eta_1 \left(w_{kj}^{(l)} - w_{kj}^{(l)*} \right)^{1/3} \tag{3.96}$$

where η_1 is the learning rate, $w_{kj}^{(l)*}$ is the last local minima of $w_{kj}^{(l)}$, and $w_{kj}^{(l)} = w_{kj}^{(l)*} + \epsilon_{kj}^{(l)}$. Equation (3.96) is integrated for a fixed amount of time, with a small time-step Δt. After every Δt, $E(\mathbf{W})$ is computed with the new value of $w_{kj}^{(l)}$ keeping the remaining components of $\mathbf{W}$ unchanged. Tunneling comes to a halt when $E(\mathbf{W}) \leq E(\mathbf{W}^*)$, and initiates the next gradient descent. If this condition of descent is not satisfied, then this process is repeated with all the components of $w_{kj}^{(l)*}$ until the above condition of descent is reached. If the above condition is not satisfied for all $w_{kj}^{(l)}$, then the last local minimum is the global minimum.

3.10 Robust Backpropagation Algorithms

Since the BP algorithm is a special case of stochastic approximation, the techniques of robust statistics [514] can be applied to the BP [1173]. In the presence of outliers, M-estimator-based robust learning, which has been introduced in Sect. 2.7, can be applied. The rate of convergence is improved since the influence of the outliers is suppressed. Robust BP algorithms using the M-estimator-based criterion functions are a typical class of robust algorithms, such as the robust BP using Hampel's tanh estimator with time-varying error cutoff points β_1 and β_2 [196], and the annealing robust BP (ARBP) algorithm [241].

The ARBP algorithm [241] adopts the annealing concept into robust learning. A deterministic annealing process is applied to the scale estimator. The cost function of the ARBP learning has the same form as (2.28), with $\beta = \beta(t)$ as a deterministic annealing scale estimator. The algorithm is based on gradient-descent method. As $\beta(t) \to \infty$, the ARBP becomes a BP algorithm. The basic idea of using an annealing schedule is to use a larger scale estimator in the early training stage and then to use a smaller scale estimator in the later training stage.

When $\beta(t) \to 0+$ for $t \to \infty$, the Huber M-estimator is equivalent to the linear L_1-norm estimator [697]. Since the L_1-norm estimator is robust against outliers [247], the Huber M-estimator equipped with such an annealing schedule is equivalent to the robust mixed-norm learning algorithm [247], where the L_2-norm is used at the beginning, then gradually tending to the L_1-norm according to the total error. The annealing schedule $\beta(t) = \frac{\gamma_1}{t}$ is found to achieve a better performance than the schedule of the form $\beta(t) = \gamma_2 - \frac{t}{\gamma_3}$ and the commonly used exponential form $\beta(t) = 1.5^{\frac{\gamma_4}{t}} - 1$, where $\beta(t)$ is always positive and γ_i, $i = 1, 2, 3, 4$, are positive constants [241].

M-estimator-based robust methods have difficulties in the selection of the scale estimator β. Tao-robust BP algorithm [910] overcomes this problem by using a τ-estimator. Tao-robust BP algorithm also achieves two important properties: robustness with a high breakdown point and a high efficiency for normal distributed data. The annealing schedule used in [241] is also integrated into the Tao-robust BP algorithm.

3.11 Resilient Propagation

The motivation behind the resilient propagation (RProp) [949] is to eliminate the influence of the magnitude of the partial derivative on the step size of the weight update. The update of each weight is according to the sequence of signs of the partial derivatives in each dimension of the weight space.

The update for each weight or bias $w_{ij}^{(m)}$ is given according to the following procedure [949]

$$C = g_{ij}^{(m)}(t-1) \cdot g_{ij}^{(m)}(t) \tag{3.97}$$

$$\Delta_{ij}^{(m)}(t) = \begin{cases} \min\left\{\eta_0^+ \Delta_{ij}^{(m)}(t-1), \Delta_{\max}\right\}, & C > 0 \\ \max\left\{\eta_0^- \Delta_{ij}^{(m)}(t-1), \Delta_{\min}\right\}, & C < 0 \\ \Delta_{ij}^{(m)}(t-1), & C = 0 \end{cases} \tag{3.98}$$

$$\Delta w_{ij}^{(m)}(t) = \begin{cases} -\text{sign}\left(g_{ij}^{(m)}(t)\right) \cdot \Delta_{ij}^{(m)}(t), & C \geq 0 \\ -\Delta w_{ij}^{(m)}(t-1), & C < 0 \end{cases} \tag{3.99}$$

$$g_{ij}^{(m)}(t) = 0, \quad C < 0 \tag{3.100}$$

$$w_{ij}^{(m)}(t+1) = w_{ij}^{(m)}(t) + \Delta w_{ij}^{(m)}(t) \tag{3.101}$$

where $0 < \eta_0^- < 1 < \eta_0^+$, and typically $\eta_0^+ = 1.2$ and $\eta_0^- = 0.5$. The value of $\Delta_{ij}^{(m)}(0)$ is not critical to the algorithm, and is selected as a positive constant

Δ_0. The upper and lower bounds, denoted by Δ_{max} and Δ_{min}, respectively, are used to restrict overflow/underflow problems of floating-point variables. For example, one can select $\Delta_{max} = 50.00$ and $\Delta_{min} = 10^{-6}$ [949]. A smaller value of Δ_{max} such as 1.0 may result in a smoothened behavior of the decrease in error.

The RProp is robust against the choice of its initial parameters. In comparison with the BP, the Quickprop [346], and the SuperSAB [1094], the number of learning steps is significantly reduced and computational complexity of RProp at each step is considerably smaller [949, 452]. The RProp has a performance comparable to that of the CG method [452].

The RProp [949] is one of the best performing first-order learning methods for neural networks. It is suitable for hardware implementation and is not susceptible to numerical problems. The RProp algorithm has also been used for training RBFNs [88], and recurrent fuzzy neural networks [773].

Variants of Resilient Propagation

Due to its effectiveness in practical applications, some variants aiming at improving the RProp have been proposed. The QRprop and the diagonal estimation Rprop (DERprop) [914] are two similar hybrids of the Rprop and second-order search steps. They adaptively switch between the two methods by using the strategy of Rprop and switching to second-order approximation only when the search is in the vicinity of a local minimum. The QRprop makes use of local one-dimensional secant steps, which are used in the Quickprop. The DERprop directly computes the diagonal elements of the Hessian matrix.

The SASS [451] uses the same update rule as the RProp, but the update of $\Delta w_{ij}^{(m)}(t)$ is based on the bisection method for minimization in one dimension and uses two previous signs

$$\Delta_{ij}^{(m)}(t) = \begin{cases} 2.0\Delta_{ij}^{(m)}(t-1), & C \geq 0 \text{ and } g_{ij}^{(m)}(t-2)\,g_{ij}^{(m)}(t) \geq 0 \\ 0.5\Delta_{ij}^{(m)}(t-1), & \text{otherwise} \end{cases} \tag{3.102}$$

The SASS is shown to provide a performance comparable to that of the RProp [452].

The addition of SA [607] in the form of noise and weight decay to the RProp algorithm yields the SARProp algorithm [1098]. The SARProp may be used with a restart training phase, which allows a more thorough search of the error surface and provides an automatic annealing schedule.

The RProp applies weight-backtracking, which retracts a previous update for weights if the weight updates cause changes to the signs of the corresponding partial derivatives, that is, $C < 0$. In the improved Rprop (IRprop) algorithm [529], weight updates that cause $C < 0$ are reverted only in the case of an increase in the error, that is, if $E(t) > E(t-1)$. The IRprop algorithm shows a performance better than that of the Rprop as well as that of the CG, and has a performance comparable to that of the BFGS method.

The GRprop [33] is a globally convergent modification of the Rprop algo-rithm. It is built on a mathematical framework for the convergence analysis [749], which ensures that the adaptive local learning rates of the Rprop's schedule generate a descent-search direction at each iteration. The GRprop exhibits a better convergence speed and stability than the Rprop or the IR-prop.

3.12 Second-order Learning Methods

The training of FNNs can be viewed as an unconstrained optimization prob-lem. The BP algorithm is slow to converge when the error surface is flat along a weight dimension. Second-order optimization techniques have a strong theo-retical basis and provide significantly faster convergence. Second-order meth-ods make use of the Hessian matrix $\mathbf{H}$, that is, the second-order derivative of the error E with respect to the N_{w}-dimensional weight vector $\vec{\mathbf{w}}$ [6]

$$\mathbf{H}(t) = \left. \frac{\partial^2 E}{\partial \vec{\mathbf{w}}^2} \right|_t \tag{3.103}$$

It is an $N_{\mathrm{w}} \times N_{\mathrm{w}}$ matrix. This matrix contains information as to how the gra-dient changes in different directions of the weight space. The calculation of $\mathbf{H}$ can implemented into the BP algorithm [116]. For FNNs, $\mathbf{H}$ is ill-conditioned [986].

Second-order algorithms can either be of matrix or vector type. Matrix-type algorithms require the storage for the Hessian and its inverse. The Broyden–Fletcher–Goldfarb–Shanno (BFGS) method [362] and a class of New-ton's methods belong to matrix-type algorithms. Matrix-type algorithms are typically two orders of magnitude faster than the BP. The computational com-plexity is at least $O\left(N_{\mathrm{w}}^2\right)$ floating-point operations, when used for supervised learning of the MLP. This limits their applications in large-scale problems.

Vector-type algorithms, on the other hand, require the storage of a few vectors. Examples of such algorithms include the limited-memory BFGS [81], one-step secant (OSS) [79, 82], scaled CG [812], and CG methods [556, 1117]. They are typically one order of magnitude faster than the BP algorithm. Vector-type algorithms require iterative computation of the Hessian or im-plicitly exploit the structure of the Hessian. They are based on line-search or trust-region-search methods.

In the BP algorithm, the selection of the learning parameters by trial-and-error is a daunting task for a large training set. In second-order methods, learning parameters can be automatically adapted. However, second-order methods are required to be used in batch mode due to the numerical sensitivity of the computation of second-order gradients. Besides, second-order methods

---

[6] In this section, $\vec{\mathbf{w}}$ is a vector obtained by concatenating all the weights and biases of a network.

become trapped in a local minimum more frequently than the BP algorithm [470].

3.12.1 Newton's Methods

Newton's methods [79, 69] require explicit computation and storage of the Hessian. They are variants of the classical Newton's method, and are matrix-type algorithms. These include the classical Newton's, the Gauss–Newton, and the Levenberg–Marquardt (LM) methods. Newton's methods achieves the quadratic convergence. They are less sensitive to the learning constant, and the choice of a proper learning constant is not difficult.

Classical Newton's Method

At step $t + 1$, we expand $E(\overrightarrow{\mathbf{w}})$ into a Taylor series

$$E\left(\overrightarrow{\mathbf{w}}\right)|_{t+1} = E\left(\overrightarrow{\mathbf{w}}\right)|_t + [\overrightarrow{\mathbf{w}}(t+1) - \overrightarrow{\mathbf{w}}(t)]^{\mathrm{T}} \mathbf{g}(t)$$
$$+ \frac{1}{2}[\overrightarrow{\mathbf{w}}(t+1) - \overrightarrow{\mathbf{w}}(t)]^{\mathrm{T}} \mathbf{H}(t)[\overrightarrow{\mathbf{w}}(t+1) - \overrightarrow{\mathbf{w}}(t)] + \cdots \quad (3.104)$$

where the gradient vector is given by

$$\mathbf{g}(t) = \nabla E\left(\overrightarrow{\mathbf{w}}(t)\right) = \nabla E(\overrightarrow{\mathbf{w}})|_t \quad (3.105)$$

Equating $\mathbf{g}(t + 1)$ to zero

$$\mathbf{g}(t+1) = \mathbf{g}(t) + \mathbf{H}(t)\left(\overrightarrow{\mathbf{w}}(t+1) - \overrightarrow{\mathbf{w}}(t)\right) + \cdots = 0 \quad (3.106)$$

By ignoring the third- and higher-order terms, the classical Newton's method is obtained:

$$\overrightarrow{\mathbf{w}}(t+1) = \overrightarrow{\mathbf{w}}(t) + \mathbf{d}(t) \quad (3.107)$$

$$\mathbf{d}(t) = -\mathbf{H}^{-1}(t)\mathbf{g}(t) \quad (3.108)$$

For the MLP, the Hessian is a singular matrix [1159], and thus (3.108) cannot be used. Nevertheless, we can make use of (3.106) and solve the following SLE for the step $\mathbf{d}(t)$

$$\mathbf{g}(t) = -\mathbf{H}(t)\mathbf{d}(t) \quad (3.109)$$

This SLE can be solved by using SVD or QR decomposition.

From second-order conditions, $\mathbf{H}(t)$ must be positive for searching a minimum. At each iteration, E is approximated locally by a second-order Taylor polynomial, which is minimized subsequently. This minimization is computationally prohibitive, since the computation of $\mathbf{H}(t)$ needs global information and the solution of an SLE is also required [117]. In the classical Newton's method, $O\left(N_w^3\right)$ floating-point operations are needed for computing the search direction. However, the classical Newton's method is not suitable when $\overrightarrow{\mathbf{w}}(t)$ is remote from the solution, since $\mathbf{H}(t)$ may not be positive-definite.

Gauss–Newton Method

Denote $E(\overrightarrow{\mathbf{w}})$ as

$$E(\overrightarrow{\mathbf{w}}) = \frac{1}{2} \sum_{i=1}^{N} \epsilon_i^2(\overrightarrow{\mathbf{w}}) = \frac{1}{2} \epsilon^T \epsilon \qquad (3.110)$$

where $\epsilon(\overrightarrow{\mathbf{w}}) = (\epsilon_1(\overrightarrow{\mathbf{w}}), \epsilon_2(\overrightarrow{\mathbf{w}}), \cdots, \epsilon_N(\overrightarrow{\mathbf{w}}))^T$, $\epsilon_i = \|\mathbf{e}_i\|$, and $\mathbf{e}_i = \hat{\mathbf{y}}_i - \mathbf{y}_i$. Thus, $\epsilon_i^2 = \mathbf{e}_i^T \mathbf{e}_i$.

Taking the derivative of $E(\overrightarrow{\mathbf{w}})$ with respect to $\overrightarrow{\mathbf{w}}$, we get the gradient vector as

$$\mathbf{g} = \mathbf{J}^T(\overrightarrow{\mathbf{w}}) \epsilon(\overrightarrow{\mathbf{w}}) \qquad (3.111)$$

where the Jacobian matrix $\mathbf{J}(\overrightarrow{\mathbf{w}})$, an $N \times N_w$ matrix, is defined by

$$\mathbf{J}(\overrightarrow{\mathbf{w}}) = \frac{\partial \epsilon(\overrightarrow{\mathbf{w}})}{\partial \overrightarrow{\mathbf{w}}} = [J_{ij}] \qquad (3.112)$$

$$J_{ij} = \frac{\partial \epsilon_i}{\partial w_j} \qquad (3.113)$$

Further, taking the derivative of the gradient with respect to $\overrightarrow{\mathbf{w}}$ results in the Hessian

$$\mathbf{H} = \mathbf{J}^T(\overrightarrow{\mathbf{w}})\mathbf{J}(\overrightarrow{\mathbf{w}}) + \mathbf{S}(\overrightarrow{\mathbf{w}}) \qquad (3.114)$$

where

$$\mathbf{S}(\overrightarrow{\mathbf{w}}) = \sum_{i=1}^{N} \epsilon_i(\overrightarrow{\mathbf{w}}) \nabla^2 \epsilon_i(\overrightarrow{\mathbf{w}}) \qquad (3.115)$$

Assuming that $\mathbf{S}(\overrightarrow{\mathbf{w}})$ is small, we approximate the Hessian using

$$\mathbf{H}_{\mathrm{GN}}(t) = \mathbf{J}^T(t)\mathbf{J}(t) \qquad (3.116)$$

where $\mathbf{J}(t)$ denotes $\mathbf{J}(\overrightarrow{\mathbf{w}}(t))$.

In view of (3.107), (3.108) and (3.111), we obtain

$$\overrightarrow{\mathbf{w}}(t+1) = \overrightarrow{\mathbf{w}}(t) + \mathbf{d}(t) \qquad (3.117)$$

$$\mathbf{d}(t) = -\mathbf{H}_{\mathrm{GN}}^{-1}(t)\mathbf{J}^T(t)\epsilon(t) \qquad (3.118)$$

where $\epsilon(t)$ denotes $\epsilon(\overrightarrow{\mathbf{w}}(t))$. The above procedure is the Gauss–Newton method.

The Gauss–Newton method approximates the Hessian using information from first-order derivatives only. However, far away from the solution, the term $\mathbf{S}$ is not negligible and thus the approximation to the Hessian $\mathbf{H}$ is poor, resulting in slow convergence. The Gauss–Newton method may have an ill-conditioned Jacobian matrix and $\mathbf{H}$ may be noninvertible. In this case, like in the classical Newton's method, one can instead solve $\mathbf{H}_{\mathrm{GN}}(t)\mathbf{d}(t) = -\mathbf{J}^T(t)\epsilon(t)$ for $\mathbf{d}(t)$.

An iterative Gauss–Newton method based on the generalized secant method using Broyden's approach is given as [362]

$$\mathbf{q}(t) = \boldsymbol{\epsilon}(t+1) - \boldsymbol{\epsilon}(t) \tag{3.119}$$

$$\mathbf{J}(t+1) = \mathbf{J}(t) + \frac{[\mathbf{q}(t) - \mathbf{J}(t)\mathbf{d}(t)]\mathbf{d}^{\mathrm{T}}(t)}{\mathbf{d}^{\mathrm{T}}(t)\mathbf{d}(t)} \tag{3.120}$$

The method uses the same update given by (3.117) and (3.118).

Levenberg–Marquardt Method

The LM method [821] eliminates the possible singularity of $\mathbf{H}$ by adding a small identity matrix to it. This method is derived by minimizing the quadratic approximation to $E\left(\overrightarrow{\mathbf{w}}\right)$ subject to the constraint that the step length $\|\mathbf{d}(t)\|$ is within a trust region at step t. At given $\overrightarrow{\mathbf{w}}(t)$, the second-order Taylor approximation of $E\left(\overrightarrow{\mathbf{w}}\right)$ is given by

$$\hat{E}\left(\overrightarrow{\mathbf{w}}(t) + \mathbf{d}(t)\right) = E\left(\overrightarrow{\mathbf{w}}(t)\right) + \mathbf{g}(t)^{\mathrm{T}}\mathbf{d}(t) + \frac{1}{2}\mathbf{d}^{\mathrm{T}}(t)\mathbf{H}(t)\mathbf{d}(t) \tag{3.121}$$

The search step $\mathbf{d}(t)$ is computed by solving the trust-region subproblem

$$\min_{\mathbf{d}(t)} \hat{E}\left(\overrightarrow{\mathbf{w}}(t) + \mathbf{d}(t)\right) \quad \text{subject to} \quad \|\mathbf{d}(t)\| \leq \delta_t \tag{3.122}$$

where δ_t is a positive scalar and $\left\{\mathbf{d}(t)\big|\|\mathbf{d}(t)\| \leq \delta_t\right\}$ is the trust region around $\overrightarrow{\mathbf{w}}(t)$.

This inequality constrained optimization problem can be solved by using the Karush–Kuhn–Tucker (KKT) theorem[7] [362], which leads to

$$\mathbf{H}_{\mathrm{LM}}(t) = \mathbf{H}(t) + \sigma(t)\mathbf{I} \tag{3.123}$$

where $\sigma(t)$ is a small positive value, which indirectly controls the size of the trust region.

The LM modification to the Gauss–Newton method is given as [444]

$$\mathbf{H}_{\mathrm{LM}}(t) = \mathbf{H}_{\mathrm{GN}}(t) + \sigma(t)\mathbf{I} \tag{3.124}$$

Thus, $\mathbf{H}_{\mathrm{LM}}$ is always invertible. The LM method given in (3.124) can be treated as a trust-region modification to the Gauss–Newton method [79].

The LM method is based on the assumption that such an approximation to the Hessian is valid only inside a trust region of small radius, controlled by σ. If the eigenvalues of $\mathbf{H}$ are $\lambda_1 \geq \lambda_2 \geq \cdots \geq \lambda_{N_{\mathrm{w}}}$, then the eigenvalue of $\mathbf{H}_{\mathrm{LM}}$ are $\lambda_i + \sigma$, $i = 1, \cdots, N_{\mathrm{w}}$, with the same corresponding eigenvectors. σ is selected so that $\mathbf{H}_{\mathrm{LM}}$ is positive-definite, that is, $\lambda_{N_{\mathrm{w}}} + \sigma > 0$. As a result, the LM method eliminates the singularity of $\mathbf{H}$ for the MLP.

---

[7] It is also called the *Kuhn–Tucker theorem*.

The LM method is therefore given by

$$\vec{\mathbf{w}}(t+1) = \vec{\mathbf{w}}(t) + \mathbf{d}(t) \tag{3.125}$$

$$\mathbf{d}(t) = -\mathbf{H}_{\mathrm{LM}}^{-1}(t)\mathbf{J}^{\mathrm{T}}(t)\boldsymbol{\epsilon}(t) \tag{3.126}$$

When σ is large, the algorithm becomes the BP with $\eta = \frac{1}{\sigma}$. However, when σ is small, the algorithm reduces to the Gauss–Newton. Thus, there is a tradeoff between the fast learning speed of the classical Newton's method and the guaranteed convergence of the gradient descent. One can adapt σ by [444]

$$\sigma(t) = \begin{cases} \sigma(t-1)\gamma, & \text{if } E(t) \geq E(t-1) \\ \frac{\sigma(t-1)}{\gamma}, & \text{if } E(t) < E(t-1) \end{cases} \tag{3.127}$$

where $\gamma > 1$ is a constant. Typical selections of $\sigma(0)$ and γ are, respectively, 0.01 and 10 [444]. The computation of the Jacobian is based on a simple modification to the BP algorithm [444]. There are some other methods for finding a value for $\sigma(t)$, such as the hook step [821], Powell's dogleg method [362], and other rules of thumb [362].

Newton's methods for the MLP lack iterative implementation of $\mathbf{H}$, and the computation of $\mathbf{H}^{-1}$ is also expensive. Besides, they also suffer from the ill-representability of the diagonal terms of $\mathbf{H}$ and the requirement of a good initial estimate of the weights. The LM method is a trust-region method with a hyperspherical trust region. It is an efficient algorithm for medium-sized neural networks [444]. The LM method demands large memory space to store the Jacobian matrix, the approximated Hessian matrix, and the inversion of a matrix of size $N_{\mathrm{w}} \times N_{\mathrm{w}}$ at each iteration.

Variants of the LM Method

The trust-region-based error aggregated training (TREAT) algorithm [218] is similar to the LM method, but uses a different Hessian matrix approximation based on the Jacobian matrix derived from aggregated error vectors. The new Jacobian matrix is significantly smaller. The size of the matrix to be inverted at each iteration is also reduced by using the matrix inversion lemma.

A modified LM method [1186] is obtained by modifying the error function and using the slope between the desired and actual outputs in the activation function to replace the standard derivative at the point of the actual output. This method gives a better convergence rate with less computational complexity and reduces the memory from the N_{w}^2 to J_M^2.

The disadvantages of the LM as well as of Newton's methods can be alleviated by the block Hessian-based Newton's method [1159], where a block Hessian matrix $\mathbf{H}_{\mathrm{b}}$ is defined to approximate and simplify $\mathbf{H}$. Each $\mathbf{W}^{(m)}$, or its vector form $\vec{\mathbf{w}}^{(m)}$, corresponds to a diagonal partition matrix $\mathbf{H}_{\mathrm{b}}^{(m)}$, and

$$\mathbf{H}_{\mathrm{b}} = \mathrm{blockdiag}\left(\mathbf{H}_{\mathrm{b}}^{(1)}, \mathbf{H}_{\mathrm{b}}^{(2)}, \cdots, \mathbf{H}_{\mathrm{b}}^{(M-1)}\right) \tag{3.128}$$

$\mathbf{H_b}$ is proved to be a singular matrix [1159]. In the LM implementation, the inverse of $\mathbf{H_b} + \sigma\mathbf{I}$ can be decomposed into the inverse of each diagonal block $\mathbf{H}_b^{(m)} + \sigma\mathbf{I}$, and the problem is decomposed into $M - 1$ subproblems

$$\Delta\overrightarrow{\mathbf{w}}^{(m)} = -\left(\mathbf{H}_b^{(m)} + \sigma\mathbf{I}\right)^{-1}\mathbf{g}^{(m)} \tag{3.129}$$

for $m = 1, \cdots, M - 1$, where the gradient partition $\mathbf{g}^{(m)} = \frac{\partial E}{\partial\overrightarrow{\mathbf{w}}^{(m)}}$. The inverse in each subproblem can be computed recursively according to the matrix inversion lemma.

The LM with adaptive momentum (LMAM) and the optimized LMAM [32] combine the merits of both the LM and CG techniques, and help the LM escape from the local minima. The LMAM is derived by optimizing the mutually conjugate property of the two steps subject to a constraint on the error change as well as a different trust-region condition

$$\mathbf{d}^{\mathrm{T}}(t)\mathbf{H}(t)\mathbf{d}(t) \leq \delta_t \tag{3.130}$$

This results in two free parameters to be tuned. The optimized LMAM is adaptive, requiring minimal input from the end user. The LMAM is globally convergent. Their implementations require minimal additional computations when compared to the LM iteration, and this is, however, compensated by their excellent convergence properties. Both the methods generate better results than the LM, BFGS, and Polak–Ribiere CG with restarts [556][8].

A recursive LM algorithm for online training of neural networks is given in [846] for nonlinear system identification. The recursive LM algorithm is shown to have advantages similar to its corresponding offline algorithm.

Attractor-based Trust-region Method

The natural way to implement Newton's methods is to confine a quadratic approximation of the objective function $E\left(\overrightarrow{\mathbf{w}}\right)$ to a trust region. This idea is credited to Levenberg and Marquart in deriving the LM method. The trust-region subproblem is then solved to obtain the next iteration.

The attractor-based trust-region method is an alternating two-phase algorithm for MLP learning [667]. The first phase is a trust-region-based local search for fast training of the network and global convergence [362], while the second phase is an attractor-based global search for escaping local minima utilizing a quotient gradient system [668]. The repeated iteration of the two phases results in a fast convergence to global optimum.

---

[8] MATLAB source code of the LMAM, OLMAM, LM, BFGS, and Polak–Ribiere CG with restarts is available at *http://www.iit.demokritos.gr/ $\sim$ abazis/toolbox*.

Phase 1: Trust-region-based local search

The trust-region subproblem is to minimize $\hat{E}\left(\overrightarrow{\mathbf{w}}(t) + \mathbf{d}(t)\right)$ given in (3.121) and (3.122). Powell's dogleg trust-region method [362] is applied to solve the problem.

Define the ratio of the actual reduction in error to the predicted reduction in error

$$\rho_t = \frac{E\left(\overrightarrow{\mathbf{w}}(t)\right) - E\left(\overrightarrow{\mathbf{w}}(t) + \mathbf{d}(t)\right)}{\hat{E}\left(\overrightarrow{\mathbf{w}}(t)\right) - \hat{E}\left(\overrightarrow{\mathbf{w}}(t) + \mathbf{d}(t)\right)} \tag{3.131}$$

When the ratio is close to unity, the actual reduction in the error is similar to the predicted reduction in error, and δ_t can be increased. When the similarity ratio is small or negative, δ_t should be decreased. One popular strategy is given by [362, 667]

$$\delta_{t+1} = \begin{cases} \frac{\delta_t}{4}, & \rho_t < 0.25 \\ 2\delta_t, & \rho_t > 0.75 \\ \delta_t, & \text{otherwise} \end{cases} \tag{3.132}$$

where $\delta_t = \|\mathbf{d}(t)\|$. The weights are adapted by

$$\overrightarrow{\mathbf{w}}(t+1) = \begin{cases} \overrightarrow{\mathbf{w}}(t) + \mathbf{d}(t), & \rho_t \geq 0 \\ \overrightarrow{\mathbf{w}}(t), & \text{otherwise} \end{cases} \tag{3.133}$$

The trial step computation is repeated until a local minimum is found.

Phase 2: Attractor-based global search

This step is to escape from the local minima by using a quotient gradient system [668]. The adaptation is given by

$$\frac{d\overrightarrow{\mathbf{w}}}{dt} = \begin{cases} -\left(E\left(\overrightarrow{\mathbf{w}}\right) - E\left(\overrightarrow{\mathbf{w}}^*\right)\right)\nabla E\left(\overrightarrow{\mathbf{w}}\right), & E\left(\overrightarrow{\mathbf{w}}\right) > E\left(\overrightarrow{\mathbf{w}}^*\right) \\ \mathbf{0}, & \text{otherwise} \end{cases} \tag{3.134}$$

where $\overrightarrow{\mathbf{w}}^*$ is a locally optimal weight vector amongst the previously obtained weight vectors.

The algorithm outperforms the BPM, BP with tunneling [970], and LM algorithms with respect to the average number of epochs, average time of training, and the mean squared training error, according to some benchmark tests [667].

3.12.2 Quasi-Newton Methods

Quasi-Newton methods approximate Newton's direction without evaluating second-order derivatives of the cost function. The approximation of the Hessian or its inverse is computed in an iterative process. They are a class of gradient-based methods whose descent direction vector $\mathbf{d}(t)$ approximates the Newton's direction[9]

$$\mathbf{d}(t) = -\mathbf{H}^{-1}(t)\mathbf{g}(t) \qquad (3.135)$$

Thus, one can obtain $\mathbf{d}(t)$ by solving the following SLE

$$\mathbf{H}(t)\mathbf{d}(t) = -\mathbf{g}(t) \qquad (3.136)$$

The Hessian is always symmetric and is often positive-definite. Quasi-Newton methods with positive-definite Hessian are called *variable-metric methods*. Secant methods are a class of variable-metric methods that use differences to obtain an approximation to the Hessian. The storage requirement for quasi-Newton methods is $\frac{1}{2}N_w^2 + O(N_w)$, which is the same as that for Newton's methods. These methods approximate the classical Newton's method, thus the convergence is very fast.

There are two globally convergent strategies available, namely, the line-search and trust-region methods. The line-search method tries to limit the step size along the Newton's direction until it is unacceptably large, whereas in the trust-region method the quadratic approximation of the cost function can be trusted only within a small region in the vicinity of the current point. Both the methods retain the rapid-convergence property of Newton's methods and are generally applicable [362].

In quasi-Newton methods, a line search is applied such that

$$\lambda(t) = \arg\min_{\lambda \geq 0} E\left(\overrightarrow{\mathbf{w}}(t) + \lambda\mathbf{d}(t)\right) \qquad (3.137)$$

and

$$\overrightarrow{\mathbf{w}}(t + 1) = \overrightarrow{\mathbf{w}}(t) + \lambda(t)\mathbf{d}(t) \qquad (3.138)$$

Line search is used to guarantee that at each iteration the objective function decays, which is dictated by the convergence requirement. The optimal $\lambda(t)$ can be theoretically derived from

$$\frac{\partial}{\partial \lambda} E\left(\overrightarrow{\mathbf{w}} + \lambda\mathbf{d}(t)\right) = 0 \qquad (3.139)$$

and this yields a representation using the Hessian. The second-order derivatives are approximated by the difference of the first-order derivatives at two neighboring points, and thus λ is calculated by

$$\lambda(t) = \frac{-\tau\mathbf{g}(t)^{\mathrm{T}}\mathbf{d}(t)}{\mathbf{d}(t)^{\mathrm{T}}\left[\mathbf{g}_\tau(t) - \mathbf{g}(t)\right]\mathbf{d}(t)} \qquad (3.140)$$

where $\mathbf{g}_\tau(t) = \nabla_{\overrightarrow{\mathbf{w}}} E\left(\overrightarrow{\mathbf{w}}(t) + \tau\mathbf{d}(t)\right)$, and the size of neighborhood τ carefully selected. Some inexact line-search and line-search-free optimization methods

---

[9] In this subsection, $\mathbf{d}(t)$ denotes the descent direction, and $\mathbf{s}(t)$ the step size. In the preceding subsection, both vectors are equal and are represented by $\mathbf{d}(t)$.

are applied to quasi-Newton methods, which are further used for training FNNs [91].

There are many secant methods of rank one or rank two. The Broyden family is a family of rank-two and rank-one methods generated by taking [362]

$$\mathbf{H}(t) = (1 - \vartheta)\mathbf{H}_{\mathrm{DFP}}(t) + \vartheta\mathbf{H}_{\mathrm{BFGS}}(t) \tag{3.141}$$

where $\mathbf{H}_{\mathrm{DFP}}$ and $\mathbf{H}_{\mathrm{BFGS}}$ are, respectively, the Hessain obtained by the Davidon–Fletcher–Powell (DFP) and BFGS methods, and ϑ a positive constant between 0 and 1. By giving different values for ϑ, one can get the DFP ($\vartheta = 0$), the BFGS ($\vartheta = 1$), or other rank-one or rank-two formulae. The DFP and BFGS methods are two dual rank-two secant methods, and the BFGS emerges as a leading variable-metric contender in theory and practice [842]. Many of the properties of the DFP and BFGS methods are common to the whole family.

The BFGS Method

The BFGS method [81, 362, 842] is implemented as follows. Inexact line search can be applied to the BFGS, and this significantly reduces the number of evaluations of the error function. The Hessian $\mathbf{H}$ or its inverse is updated by

$$\mathbf{H}(t + 1) = \mathbf{H}(t) - \frac{\mathbf{H}(t)\mathbf{s}(t)\mathbf{s}^{\mathrm{T}}(t)\mathbf{H}(t)}{\mathbf{s}^{\mathrm{T}}(t)\mathbf{H}(t)\mathbf{s}(t)} + \frac{\mathbf{z}(t)\mathbf{z}^{\mathrm{T}}(t)}{\mathbf{s}^{\mathrm{T}}(t)\mathbf{z}(t)} \tag{3.142}$$

$$\mathbf{H}^{-1}(t + 1) = \mathbf{H}^{-1}(t) + \left(1 + \frac{\mathbf{z}^{\mathrm{T}}(t)\mathbf{H}^{-1}(t)\mathbf{z}(t)}{\mathbf{s}^{\mathrm{T}}(t)\mathbf{z}(t)}\right)\frac{\mathbf{s}(t)\mathbf{s}^{\mathrm{T}}(t)}{\mathbf{s}^{\mathrm{T}}(t)\mathbf{z}(t)}$$
$$- \left(\frac{\mathbf{s}(t)\mathbf{z}^{\mathrm{T}}(t)\mathbf{H}^{-1}(t) + \mathbf{H}^{-1}(t)\mathbf{z}(t)\mathbf{s}^{\mathrm{T}}(t)}{\mathbf{s}^{\mathrm{T}}(t)\mathbf{z}(t)}\right) \tag{3.143}$$

where

$$\mathbf{z}(t) = \mathbf{g}(t + 1) - \mathbf{g}(t) \tag{3.144}$$

$$\mathbf{s}(t) = \overline{\mathbf{w}}(t + 1) - \overline{\mathbf{w}}(t) \tag{3.145}$$

For its implementation, $\overline{\mathbf{w}}(0)$, $\mathbf{g}(0)$, and $\mathbf{H}^{-1}(0)$ are needed to be specified. $\mathbf{H}^{-1}(0)$ is typically selected as the identity matrix. The computational complexity is $O\left(N_{\mathrm{w}}^2\right)$ floating-point operations. The method requires storage of the matrix $\mathbf{H}^{-1}$. By interchanging $\mathbf{H} \leftrightarrow \mathbf{H}^{-1}$, $\mathbf{s} \leftrightarrow \mathbf{z}$ in (3.142) and (3.143), one can obtain the DFP method [362, 842].

All the secant methods including the BFGS method are derived to satisfy the so-called *quasi-Newton condition* or *secant relation* [362, 842]

$$\mathbf{H}^{-1}(t + 1)\mathbf{z}(t) = \mathbf{s}(t) \tag{3.146}$$

The One-step Secant Method

The OSS method [79, 82] is a memoryless BFGS method, and is obtained by resetting $\mathbf{H}^{-1}(t)$ as the identity matrix in the BFGS update equation (3.143) at the $(t+1)$th iteration, and multiplying both sides of the update by $-\mathbf{g}(t+1)$ to obtain the search direction

$$\mathbf{d}(t + 1) = -\mathbf{g}(t + 1) + B(t)\mathbf{z}(t) + C(t)\mathbf{s}(t) \qquad (3.147)$$

where

$$B(t) = \frac{\mathbf{s}^{\mathrm{T}}(t)\mathbf{g}(t + 1)}{\mathbf{s}^{\mathrm{T}}(t)\mathbf{z}(t)} \qquad (3.148)$$

$$C(t) = -\left(1 + \frac{\mathbf{z}^{\mathrm{T}}(t)\mathbf{z}(t)}{\mathbf{s}^{\mathrm{T}}(t)\mathbf{z}(t)}\right) B(t) + \frac{\mathbf{z}^{\mathrm{T}}(t)\mathbf{g}(t + 1)}{\mathbf{s}^{\mathrm{T}}(t)\mathbf{z}(t)} \qquad (3.149)$$

This algorithm does not store the Hessian, and the new search direction can be calculated without computing a matrix inverse. The OSS method reduces the computational complexity to $O(N_{\mathrm{w}})$. However, it results in a considerable reduction of second-order information, and thus yields a slow convergence compared to the BFGS. When exact line search is applied, the OSS generates conjugate directions.

The BFGS and OSS methods are two efficient methods for MLP training. Parallel implementations of the two algorithms are discussed in [788]. A parallel secant method of Broyden's family with parallel inexact searches is developed and applied for the training of FNNs [916].

Limited-memory BFGS methods implement parts of the Hessian approximation by using second-order information from the most recent iterations [81]. A number of limited-memory BFGS algorithms, which have a memory complexity of $O(N_{\mathrm{w}})$ and do not require accurate line searches, are listed in [1022]. In the trust-region implementation of the BFGS method [908], Powell's dogleg trust-region method is used to solve the constrained optimization subproblems. Other variants of quasi-Newton methods are the variable-memory BFGS [789] and memory-optimal BFGS methods [791].

A class of limited-memory quasi-Newton methods is given in [126]. These methods utilize an iterative scheme of a generalized BFGS-type method, and suitably approximate the whole Hessian matrix with a rank-two formula determined by a fast unitary transform such as the Fourier, Hartley, Jacobi type, or trigonometric transform. It has a computational complexity of $O(N_{\mathrm{w}} \log(N_{\mathrm{w}}))$ and requires $O(N_{\mathrm{w}})$ memory allocations, and is thereby especially suitable for the training of MLPs with large size.

The close relationship between the BFGS and CG methods is important for formulating algorithms with variable storage or limited memory [842]. Memoryless or limited-memory quasi-Newton algorithms can be viewed as a tradeoff between the CG and quasi-Newton algorithms, and are closely related to the CG.

3.12.3 Conjugate-gradient Methods

The CG method [481, 79, 812, 189] is a popular alternative to the BP algorithm. It has many tried and tested, linear and nonlinear variants, each using a different search direction and line-search method. Mathematically, the CG method is closely related to the quasi-Newton method. The CG method conducts a series of line searches along noninterfering directions that are constructed to exploit the Hessian structure without explicitly storing it. The storage requirement is $O(N_w)$, and is about four times that for the BP algorithm [570]. The computation time per weight update cycle is significantly increased due to the line search for an appropriate step size, involving several evaluations of either the error E or its derivative, which requires the presentation of the complete training set.

The CG method conducts a special kind of gradient descent. The CG method constructs a set of N_w linearly independent, nonzero search directions $\mathbf{d}(t)$, $\mathbf{d}(t+1)$, $\cdots$, $\mathbf{d}(t + N_w - 1)$, for $t = 0, 1, \cdots$. These search directions are derived from (3.139), and, at the minimum of the line search, satisfy

$$\mathbf{g}^T(t+1)\mathbf{d}(t) = 0 \qquad (3.150)$$

Based on this, one can construct a sequence of N_w successive search directions that satisfy the so-called $\mathbf{H}$-conjugate property

$$\mathbf{d}(t+i)\mathbf{H}\mathbf{d}(t+j) = 0, \qquad i \neq j \qquad (3.151)$$

for all $0 < |i - j| < N_w - 1$. The CG method is updated by

$$\overrightarrow{\mathbf{w}}(t+1) = \overrightarrow{\mathbf{w}}(t) + \lambda(t)\mathbf{d}(t) \qquad (3.152)$$

$$\mathbf{d}(t+1) = -\mathbf{g}(t+1) + \beta(t)\mathbf{d}(t) \qquad (3.153)$$

where $\lambda(t)$ is the exact step to the minimum of $E(\overrightarrow{\mathbf{w}}(t+1))$ along the direction of $\mathbf{d}(t)$ and is found by a linear search as given in (3.137), and $\beta(t)$ is a step size to decide $\mathbf{d}(t+1)$. A comparison of the search directions of the gradient descent, Newton's methods, and the CG are illustrated in Fig. 3.7.

A practical implementation of the line-search method is to increase λ until $E(\overrightarrow{\mathbf{w}}(t) + \lambda\mathbf{d}(t))$ stops being strictly monotonically decreasing and begins to increase. Thus, a minimum is bracketed. The search in the interval between the last two values of λ is then repeated several times until $E(\overrightarrow{\mathbf{w}}(t) + \lambda\mathbf{d}(t))$ is sufficiently close to a minimum. Exact line search is computationally expensive, and the speed of the CG method depends critically on the line-search efficiency. Faster CG algorithms with inexact line searches [310, 1022] are used to train the MLP [430]. The scaled CG algorithm [812] avoids the line search by introducing a scalar to regulate the positive-definiteness of the Hessian $\mathbf{H}$ as used in the LM method. This is achieved by automatically scaling the weight update vector magnitude using Powell's dogleg trust-region method [362].

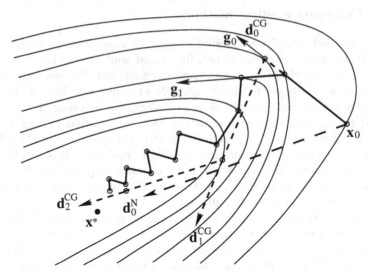

Fig. 3.7. An illustration of the search directions of the gradient-descent, Newton's, and CG methods. $\mathbf{x}_0$ and $\mathbf{x}^*$ are, respectively, the starting point and the local minimum. $\mathbf{g}_0, \mathbf{g}_1, \cdots$ are negative gradient directions, and are the search directions for the gradient-descent method. $\mathbf{d}_0^{\mathrm{N}}$ is the Newton direction, and it generally points towards the local minimum. $\mathbf{d}_0^{\mathrm{CG}}, \mathbf{d}_1^{\mathrm{CG}}, \cdots$ are the conjugate directions. Note that $\mathbf{d}_0^{\mathrm{CG}}$ and $\mathbf{g}_0$ are the same. The contours denote constant E.

There are many choices of $\beta(t)$. Usually $\beta(t)$ is selected as one of the following five.

$$\beta(t) = \frac{\mathbf{g}^{\mathrm{T}}(t+1)\mathbf{z}(t)}{\mathbf{d}^{\mathrm{T}}(t)\mathbf{z}(t)} \qquad \text{(Hestenes–Stiefel [481])} \qquad (3.154)$$

$$\beta(t) = \frac{\mathbf{g}^{\mathrm{T}}(t+1)\mathbf{g}(t+1)}{\mathbf{g}^{\mathrm{T}}(t)\mathbf{g}(t)} \qquad \text{(Fletcher–Reeves [363])} \qquad (3.155)$$

$$\beta(t) = \frac{\mathbf{g}^{\mathrm{T}}(t+1)\mathbf{z}(t)}{\mathbf{g}^{\mathrm{T}}(t)\mathbf{g}(t)} \qquad \text{(Polak–Ribiere [924])} \qquad (3.156)$$

$$\beta(t) = \frac{\mathbf{g}^{\mathrm{T}}(t+1)\mathbf{g}(t+1)}{\mathbf{d}^{\mathrm{T}}(t)\mathbf{z}(t)} \qquad \text{(Dai–Yuan [275])} \qquad (3.157)$$

$$\beta(t) = -\frac{\mathbf{g}^{\mathrm{T}}(t+1)\mathbf{g}(t+1)}{\mathbf{g}^{\mathrm{T}}(t)\mathbf{s}(t)} \qquad \text{(Conjugate descent [362])} \qquad (3.158)$$

where $\mathbf{z}(t)$ is defined as in (3.144). In the implementation, $\overline{\mathbf{w}}(0)$ is set as a random vector, and we set $\mathbf{d}_0 = -\mathbf{g}(0)$. When $\|\mathbf{g}(t)\|$ is small enough, we terminate the process. The computational complexity of the CG method is $O(N_{\mathrm{w}})$.

When the objective function is strict convex quadratic and an exact line search is applied, $\beta(t)$ is identical for all the five choices, and termination occurs at most in N_w steps [362, 842]. With periodic restarting, all the above nonlinear CG algorithms are well known to be globally convergent[10] [842]. The Polak–Ribiere CG with a suitable restart strategy, proposed by Powell, is considered to be one of the most efficient methods [1117, 842]. The Polak–Ribiere CG with restarts forces $\beta = 0$ whenever $\beta < 0$. This is equivalent to forgetting the last search direction and restarting it from the direction of steepest descent.

Empirically, in general, the local minimum achieved with the BP algorithm will in fact be a global minimum, or at least a solution that is good enough for most purposes. In contrast, the CG method is easy to be trapped at a bad local minimum, since the CG method moves towards the bottom of whatever valley it reaches. Escaping a local minimum requires an increase in E, and this is excluded by the line-search procedure. Consequently, the convergence condition can never be reached [570, 1097]. The CG method is usually applied several times with different random $\overline{\mathbf{w}}(0)$, and the final $\overline{\mathbf{w}}$ that gives the minimum error level, is taken as the result [570].

The CG method can be regarded as an extension of the BPM by automatically selecting an appropriate learning rate $\eta(t)$ and momentum factor $\alpha(t)$ in each epoch [189, 1117, 1245, 109]. The CG algorithm can be considered as the BPM, which has adjustable $\eta(t)$ and $\alpha(t)$, $\eta(t) = \lambda(t)$, $\alpha(t) = \lambda(t)\beta(t)$ [189]. By an adaptive selection of both $\eta(t)$ and $\alpha(t)$ for a quadratic error function, referred to as *optimally tuned*, the BPM is proved to be exactly equivalent to the CG method [109].

In [721], the MLP is first decomposed into a set of Adalines, each having its own local MSE function. The desired local output at each Adaline is estimated based on error backpropagation. Each local MSE function has a unique optimum, which can be found within finite steps by using the CG method. By using a modified CG that avoids the line search, the local training method achieves a significant reduction in the number of iterations and the computation time. Given the approximation accuracy, the local method requires a computation time that is typically one order of magnitude less than that of the CG-based global method. The local method is particularly suited for parallel implementation.

3.12.4 Extended Kalman Filtering Methods

The EKF method belongs to second-order methods. The EKF is an optimum filter for a linear system resulting from the linearization of a nonlinear system. It attempts to estimate the state of a system that can be modeled as a linear system driven by an additive white Gaussian noise. It always estimates the optimum step size, namely, the Kalman gain, for weight updating and thus the rate of convergence is significantly increased when compared to that of

---

[10] A globally convergent algorithm is an iterative algorithm that converges to a local minimum from almost any starting point.

the gradient-descent method. The EKF approach is an incremental training method in that the weights are updated immediately after the presentation of a training pattern. The method is of general purpose and can be used for training any FNN. The RLS method is a reduced form of the EKF method, which is more widely used in adaptation.

Extended Kalman Filtering

When using the EKF, the training of an MLP is viewed as a parametric identification problem of a nonlinear system, where the weights are unknown and have to be estimated for the given set of training patterns. The weight vector $\overrightarrow{\mathbf{w}}$ is now a state vector, and an operator $\mathbf{f}(\cdot)$ is defined to perform the function of an MLP that maps the state vector and the input onto the output. The training of an MLP can be posed as a state estimation problem with the following dynamic and observation equations

$$\overrightarrow{\mathbf{w}}(t+1) = \overrightarrow{\mathbf{w}}(t) + \boldsymbol{v}(t) \tag{3.159}$$

$$\mathbf{y}_t = \mathbf{f}\left(\overrightarrow{\mathbf{w}}(t), \mathbf{x}_t\right) + \boldsymbol{\epsilon}(t) \tag{3.160}$$

where $\mathbf{x}_t$ is the input to the network at time t, $\mathbf{y}_t$ is the observed (or desired) output of the network, and $\boldsymbol{v}$ and $\boldsymbol{\epsilon}$ are the observation and measurement noise, assumed white and Gaussian with zero mean and covariance matrices $\mathbf{Q}(t)$ and $\mathbf{R}(t)$, respectively. The state estimation problem is then to determine $\widehat{\overrightarrow{\mathbf{w}}}$, the estimated weight vector, that minimizes the sum of squared prediction errors of all prior observations.

The EKF is a minimum variance estimator based on the Taylor-series expansion of $\mathbf{f}\left(\overrightarrow{\mathbf{w}}\right)$ in the vicinity of the previous estimate. The EKF method for estimating $\overrightarrow{\mathbf{w}}(t)$ is given by [976, 530, 1066]

$$\widehat{\overrightarrow{\mathbf{w}}}(t+1) = \widehat{\overrightarrow{\mathbf{w}}}(t) + \mathbf{K}(t+1)\left[\mathbf{y}_{t+1} - \widehat{\mathbf{y}}_{t+1}\right] \tag{3.161}$$

$$\boldsymbol{\Xi}(t+1) = \mathbf{F}(t+1)\mathbf{P}(t)\mathbf{F}^{\mathrm{T}}(t+1) + \mathbf{R}(t+1) \tag{3.162}$$

$$\mathbf{K}(t+1) = \mathbf{P}(t)\mathbf{F}^{\mathrm{T}}(t+1)\boldsymbol{\Xi}^{-1}(t+1) \tag{3.163}$$

$$\mathbf{P}(t+1) = \mathbf{P}(t) - \mathbf{K}(t+1)\mathbf{F}(t+1)\mathbf{P}(t) + \mathbf{Q}(t) \tag{3.164}$$

where $\mathbf{K}$ is the $N_w \times N_y$ Kalman gain, N_y is the dimension of $\mathbf{y}$, $\mathbf{P}$ is the $N_w \times N_w$ conditional error covariance matrix, $\widehat{\mathbf{y}}$ is estimated output, and

$$\mathbf{F}(t+1) = \left.\frac{\partial \mathbf{f}}{\partial \overrightarrow{\mathbf{w}}}\right|_{\overrightarrow{\mathbf{w}} = \widehat{\overrightarrow{\mathbf{w}}}(t)} = \left.\frac{\partial \widehat{\mathbf{y}}}{\partial \overrightarrow{\mathbf{w}}}\right|_{\overrightarrow{\mathbf{w}} = \widehat{\overrightarrow{\mathbf{w}}}(t)} \tag{3.165}$$

is an $N_y \times N_w$ matrix.

In the absence of *a priori* information, the initial state vector can be set randomly and $\mathbf{P}$ can be set as a diagonal matrix: $\mathbf{P}(0) = \frac{1}{\varepsilon}\mathbf{I}$, $\overrightarrow{\mathbf{w}}(0) = \mathcal{N}(0, \mathbf{P}(0))$, where ε is a small positive number and $\mathcal{N}(0, \mathbf{P}(0))$ denotes a zero-mean Gaussian distribution with covariance matrix $\mathbf{P}(0)$. A relation between the Hessian $\mathbf{H}$ and $\mathbf{P}$ is estabilished in [1066].

The method given above is the global EKF method for MLP training [1040, 530]. As compared to the BP algorithm, it needs far fewer training cycles to reach convergence and the quality of the solution is better, at the expense of a much higher computational complexity at each cycle. The BP is proved to be a degenerate form of the EKF [976]. The fading-memory EKF (FMEKF) and a UD factorization-based FMEKF (UD-FMEKF), which use an adaptive forgetting factor, are two fast algorithms for FNN learning [1262]. The EKF variant reported in [955] performs as efficiently as the nonrecursive LM method.

In order to reduce the complexity of the EKF method, one can partition the global problem into many small-scaled, separate, localized identification subproblems for each neuron in the network so as to solve the individual subproblems. Examples of the localized EKF methods are the multiple extended Kalman algorithm (MEKA) [1019], and the decoupled EKF algorithm (DEKF) [929]. The major problem with the localized algorithms is that the offdiagonal terms of the covariance matrix in the global EKF method are set to zero, thereby ignoring the natural coupling of the weights. The UD-FMEKF algorithm [1262] can be applied to the decoupled EKF formulation, with each of the decoupled filters using the UD-FMEKF so as to reduce the computational requirement in the training of large-scale systems.

The Kalman filtering method is based on the assumption of the noise being Gaussian, and is thus sensitive to noises of other distributions. According to the H_∞ theory, the maximum energy gain that the Kalman filtering algorithm contributes to the estimation error due to disturbances has no upper bound [849]. The extended H_∞ filtering (EHF) method, in the form of global and local algorithms, can be treated as an extension to the EKF method for enhancing the robustness to disturbances. The computational complexity of the EHF method is typically twice that of the EKF method.

Recursive Least Squares

When $\mathbf{R}(t+1)$ in (3.163) and $\mathbf{Q}(t)$ in (3.164), respectively, reduce to the identity matrix $\mathbf{I}$ and zero matrix $\mathbf{O}$ of the same size, the EKF method is reduced to the RLS method. RLS methods are applied for the learning of layered FNNs in [1040, 51, 114, 689]. The RLS algorithm is typically an order of magnitude faster than the LMS algorithm, which is equivalent to the BP in one-layer networks. For a given accuracy, the RLS is shown to require tenfold fewer epochs than the BP [114].

The RLS method is derived from the optimization of the energy function [689]

$$E\left(\overrightarrow{\mathbf{w}}(t)\right) = \sum_{p=1}^{t} \left[\mathbf{y}_p - \mathbf{f}\left(\overrightarrow{\mathbf{w}}(t), \mathbf{x}_p\right)\right]^{\mathrm{T}} \left[\mathbf{y}_p - \mathbf{f}\left(\overrightarrow{\mathbf{w}}(t), \mathbf{x}_p\right)\right]$$

$$+ \left[\overrightarrow{\mathbf{w}}(t) - \overrightarrow{\mathbf{w}}(0)\right]^{\mathrm{T}} \mathbf{P}(0) \left[\overrightarrow{\mathbf{w}}(t) - \overrightarrow{\mathbf{w}}(0)\right] \tag{3.166}$$

When $P(0) = \frac{1}{\varepsilon}\mathbf{I}$ and $\overrightarrow{\mathbf{w}}(0)$ is a small vector, the second term in (3.166) reduces to $\varepsilon\overrightarrow{\mathbf{w}}(t)^T\overrightarrow{\mathbf{w}}(t)$. Thus, the RLS method is implicitly a weight-decay technique whose weight-decay effect is governed by $\mathbf{P}(0)$. A smaller ε usually leads to a better training accuracy, while a larger ε results in a better generalization [689]. At iteration t, the Hessian for the above error function is related to the error covariance matrix $\mathbf{P}(t)$ by [689]

$$\mathbf{H}(t) \approx 2\mathbf{P}^{-1}(t) - 2\mathbf{P}^{-1}(0) \qquad (3.167)$$

A complex training problem can also be decomposed into separate, localized identification subproblems, each being solved by the RLS method [1055, 886]. In the local linearized LS (LLLS) method [1055], each subproblem has the objective function as the sum of the squares of the linearized back-propagated error signals for each neuron. In the block RLS (BRLS) algorithm [886], at a step of the algorithm, an M-layer FNN is divided into $M - 1$ subproblems, each of which is an overdetermined system of linear equations for each layer of the network. The total complexity is $O\left(\sum_{m=1}^{M-1}\left[N_{\mathrm{w}}^{(m)}\right]^2\right)$, where $N_{\mathrm{w}}^{(m)}$ is the number of weights at the mth layer. This is a considerable saving with respect to a global method whose complexity is $O\left(N_{\mathrm{w}}^2\right)$.

3.13 Miscellaneous Learning Algorithms

In addition to the aforementioned major MLP training methods, there is a vast literature concerning approaches to MLP learning. We describe some of these methods in this section.

The EM method [299] is the most popular optimization approach to the exact ML solution for the parameters given an incomplete data. The EM splits a complex learning problem into a group of separate small-scale subproblems and solves each of the subproblems using a simple method, and is thus computationally efficient. The method alternates between performing the expectation step (E-step) and the maximization step (M-step), which, respectively, compute the expected values of the latent variables and the ML estimates of the parameters. The EM method has been applied to obtain ML estimates of the network parameters for FNN learning [745]. In [745], the training of the three-layer MLP is first decomposed into a set of single neurons, and the individual neurons are then trained via a linear weighted regression algorithm.

The extreme learning machine (ELM) algorithm [513] is a learning method for three-layer FNNs. In this algorithm, the weights from the input to hidden layers and the bias of the hidden layer are randomly generated from a continuous distribution probability and kept fixed, and the weights from the hidden to output layers are then determined by using the generalized inverse. The ELM learns much faster than the BP without a loss of generalization performance. The ELM can also be effectively used for training neural networks with threshold units.

A generalization of the BP algorithm is developed for training FNNs in [1243]. The algorithm is proved to have a guaranteed convergence under certain sufficient conditions. The BP, Gauss–Newton, and LM algorithms are special cases of this general algorithm. The strength of the general algorithm lies in its ability to handle time-varying inputs.

As an alternative to the BP algorithm, a gradient-descent method without error backpropagation has been proposed for MLP learning [134]. Unlike the BP, the method feeds gradients forward rather than feeding errors backwards. The gradients of the final output are determined by feeding the gradients of the intermediate outputs forward at the same time that the outputs of the intermediate layers are fed forward. This method turns out to be equivalent to the BP for a three-layer MLP, but is much more readily extended to arbitrary number of layers without modification. This method has a great potential for concurrency.

The leap-frog [1045] is an optimization method based on the motion of an object of unit mass in a J_1-dimensional conservative force field. The total energy of the object, made up of kinetic and potential energy components, is conserved. The potential energy can be specified as the objective function to be optimized and the kinetic energy is derived from the fact that a force acts on the object. The method accelerates the kinetic energy to the point of its maximum, resulting in the minimum objective function. The leap-frog has been applied to the training of the MLP [497].

The LP method is also used for training FNNs [1026]. The LP method can be effective for training a small network and can converge rapidly and reliably to a better solution than the BP. However, the LP method may take too long a time for each iteration for very large networks. Some measures are considered in [1026] so as to extend the method for efficient implementations in large networks.

The fuzzy BP (FBP) algorithm [1057] is an extension to the BP. The FBP shows considerably greater convergence speed than the BP and can easily escape from local minima. For the aggregation of input values (forward propagation), the Sugeno fuzzy integral [1064] is employed. For weight learning, error backpropagation takes place. QuickFBP is a modification of the FBP algorithm, where the modified computation of the net function is significantly faster [848]. The FBP algorithm is proved to be of exponential complexity in the case of large-sized networks with a large number of inputs, while the QuickFBP algorithm is of polynomial complexity.

3.13.1 Layerwise Linear Learning

FNNs can be trained by iterative layerwise learning methods [56, 1000, 338, 517, 972]. Weight updating is performed layer by layer, and weight optimization at each layer is reduced to solving an SLE, $\mathbf{A}\overrightarrow{\mathbf{w}}^{(m)} = \mathbf{b}$, where $\overrightarrow{\mathbf{w}}^{(m)}$ is a weight vector associated with the layer, and $\mathbf{A}$ and $\mathbf{b}$ are a matrix and a vector of suitable dimensions, respectively. These algorithms are typically one to two orders of magnitude faster in computational time than the BP for a given accuracy.

The BP algorithm can be combined with a linear-algebra method [56] or Kalman filtering method [1000]. In [56], SLEs are formed based on the computation of target node values using inverse activation functions. The updated weights need to be transformed to ensure that target values are in the range of the activation functions.

An efficient method that combines the layerwise approach and the BP strategy is given in [972]. The objective function is constructed using the Taylor-series expansion of the nonlinear operators describing a neural network, and weights at each layer are updated by solving a linear problem using the iterative Kaczmarz method. The Kaczmarz method is a particular case of the BP. Thus, the combination of the Kaczmarz method and layerwise learning leads to a layerwise BP algorithm. The algorithm provides stable convergence for an arbitrary matrix $\mathbf{A}$. The method is more accurate and faster than the CG with Powell restarts and the Quickprop [346].

In [338], the layerwise learning algorithm introduces a linearization of the hidden node sigmoidal activation functions for the optimization of the hidden layer weights. A special penalty term is added to the cost function of the linearized network to eliminate the linearization error introduced. There are no user-adjustable optimization parameters.

In [517], the least squares error (LSE) problems are solved based on the QR decomposition, which is a recursive algorithm suitable for implementation on a systolic processor. A selective training strategy is applied to reduce the computational complexity and the training accuracy, which assess the effect of a particular training pattern on the weight estimates prior to its inclusion in any iteration. Data that do not significantly change the weights are not used in that iteration, obviating the computational expense of updating.

In [199], a fast algorithm for three-layer MLP, called *OWO-HWO*, a combination of hidden weight optimization (HWO) and output weight optimization (OWO) has been described. The HWO is a batch version of the Kalman filtering method given in [1000], restricted to hidden units. The HWO develops desired hidden unit net signals from delta functions. The resulting hidden unit error functions are minimized with respect to the hidden weights. The OWO solves a SLE to optimize the output weights. The OWO-HWO uses separate error functions for each hidden unit and solves multiple SLEs. The OWO-HWO is proved to be equivalent to a combination of linearly transforming the training data and performing the OWO-BP [758], which uses the OWO to update the output weights and the BP to update the hidden weights. OWO-HWO is superior to the OWO-BP in terms of convergence, and converges to the same training error as the LM method in the time that is an order of magnitude less [199].

3.13.2 Natural-gradient Method

When the parameter space is not Euclidiean but has a Riemannian metric structure, the ordinary gradient does not give the steepest direction of the cost function, while the natural gradient does [29]. Amari expresses the natural gradients explicitly in the case of the parameter space of perceptrons, and information geometry is used for calculating the natural gradients [29]. The online natural gradient learning gives the Fisher efficient estimator when the cost function is differentiable, implying that it is asymptotically equivalent to the optimal batch procedure. This suggests that the flat-spot problem that appears in the BP algorithm disappears when the natural gradient is used [29].

3.13.3 Binary Multilayer Perceptrons

The MLP primarily uses continuous activation functions such as the sigmoidal functions. This is necessary for using the first- or second-order optimization algorithms, which require a differentiable error function. The MLP with hard-limiting activation function, termed the *binary MLP*, has benefits such as simple internal representations, extremely simple operations for trained networks, easy hardware implementation as well as easy rule extraction from trained networks. The binary MLP is typically used for classification.

Methods for training the binary MLP can be based on fuzzy logic and the idea of error backpropagation [294, 174]. A set of fuzzy rules are designed for updating all the weights. For classification [1143], a key problem in its learning is to decide bigger linearly separable subsets. Multicore learning (MCL) and multicore expand-and-truncate learning (MCETL) are two algorithms for constructing binary MLPs, which are based on some lemmas about linear separability [1143]. MCL and MCETL simplify the equations for the computation of the weights and biases, and they result in the construction of a simpler hidden layer.

3.14 Escaping Local Minima

The problem of loading a set of training examples onto a neural network is NP-complete [120, 1036]. As a consequence, existing algorithms cannot be guaranteed to learn the optimal solution in polynomial time. Conventional first-order and second-order gradient-based methods cannot avoid local minima.

In the case of one neuron, the logistic function paired with the MSE function can lead to $\left(\frac{N}{J_1}\right)^{J_1}$ local minima, where N is the size of the training set and J_1 the input dimension [48], while with the entropic error function, the error function is convex and thus has only one minimum [86, 1048]. The use of the entropic error function considerably reduces the total number of local minima.

The error surface of an MLP has a stair-step appearance with many very flat and very steep regions [522]. For the case of a small number of training examples, there is often a one-to-one correspondence between the individual training examples and the steps on the surface. The surface becomes smoother as the number of training examples is increased. In all directions there are flat regions extending to infinity, which makes the line-search-related learning algorithms useless.

3.14.1 Some Heuristics for Escaping Local Minima

Many strategies have been explored to reduce the chances of getting trapped at a local minimum. One simple and effective technique to avoid local minima in incremental learning is to present examples to the network in a random order from the training set during each epoch. Another way is to run the learning algorithms using initial values in different regions of the weight space, and then to find the best solution. This is especially useful for fast convergent algorithms such as the CG algorithm.

The injection of noise into the learning process is an effective means for escaping from local minima. This also leads to a better generalization capability. Various annealing schemes actually use this strategy. Random noise can be added to the input [1034], to the desired output [1141], or to the weights [1134]. The level of the added noise should be decreased as the learning progresses. The three methods have the same effect, namely, the inclusion of an extra stochastic term in the weight vector adaptation.

Weight scaling [386, 950] is a technique used for escaping local minima and accelerating convergence. Using the weight scaling process, the weight vector to each neuron $\mathbf{w}_j^{(m)}$ is scaled by a factor $\beta_j^{(m)}$, where $\beta_j^{(m)} \in (0,1)$ is decided by a relation of the degree of saturation at each node $D_j^{(m)} = \left| o_j^{(m)} - 0.5 \right|$, the learning rate η, and the maximum error at the output nodes $e_{\max} = \max_{p,i} |\hat{y}_{p,i} - y_{p,i}|$. Weight scaling effectively reduces the degree of saturation of the activation function and thus maintains a relatively large derivative of the activation function. This enables relatively large weight updates, which may eventually lead the training algorithm out of the local minimum. In [722], weight scaling is applied whenever the training process gets stuck at a local minimum.

An effective way to escape local minima is realized by incorporating an annealing noise term into the gradient-descent algorithm [151]

$$w_{ij}^{(m)}(t+1) = w_{ij}^{(m)}(t) - \eta g_{ij}^{(m)}(t) + \alpha_0 r 2^{-\beta_0 t} \qquad (3.168)$$

where α_0 and β_0 are positive constants, and $r \in (-0.5, 0.5)$ is a random number. This heuristic has also been used in the SARprop algorithm [1098].

3.14.2 Global Optimization Techniques

Global optimization techniques based on stochastic or deterministic approaches can be used to find a global optimal solution to network training. Their search, however, covers the entire solution space, and thus they are generally much more computationally expensive than local methods. Stochastic search procedures such as EAs [420] and the SA [607], and deterministic optimization methods such as the tabu search [415] are popular global optimization techniques.

An overview of training neural networks using EAs is given in [1225]. The SA is also used for training the MLP [607, 333]. In [23], the particle swarm optimization (PSO) [598] is used for the training of the MLP.

In [83], a reactive tabu search technique has been proposed and applied for training neural networks, where the objective function $E(\overline{\mathbf{w}})$ associated with MLP learning is transformed into a combinatorial problem by choosing a discrete binary encoding of the weights.

3.14.3 Deterministic Global-descent Techniques

Deterministic global-descent methods usually use a tracing strategy decided by trajectory functions. These can be hybrid global/local minimization methods [68, 1021] or based on the concept of the terminal attractor [1251].

The TRUST technique [68] formulates the optimization problem in terms of the flow of a special deterministic dynamical system. This global optimization scheme possesses the capability of tunneling through higher values of the error function to reach the global minimum. The TRUST technique is applied in algorithms such as the BP with global descent [179, 180] and the BP with tunneling [970]. The global-descent method is merely a gradient descent on a specially defined criterion function.

The nonlinear optimization via external lead (NOVEL) method [1021] relies on an external traveling trace to pull a search trajectory out of a local optimum without having to restart the search from a new initial estimate. The search is performed in a continuous fashion. The NOVEL method has two stages: a global search to locate promising regions containing local minima and a local search to actually find them. The global search stage has a number of bootstrapping steps, each characterized by an ordinary differential equation of the type

$$\mathbf{x}^{(k)}(t) = A\left(\nabla_{\mathbf{x}}f\left(\mathbf{x}^{(k)}(t)\right)\right) + B\left(\mathbf{x}^{(k-1)}(t), \mathbf{x}^{(k)}(t)\right) \qquad (3.169)$$

for the steps $k = 1, 2, \cdots$, where $\mathbf{x}^{(0)} = T(t)$, $T(t)$ being an external trace function, and $A(\cdot)$ and $B(\cdot)$ are nonlinear functions. The first term on the right-hand side allows local minima to attract the trajectories, while the second term lets the trajectory escape from local minima. The trace function T is selected such that the entire space is finally traversed. The NOVEL selects one initial point for each promising region. These initial points are then used for searching local minima using a descent algorithm.

Nonlinear dynamic systems satisfying the Lipschitz condition have a unique solution for each initial condition, and the trajectory of the state can approach the solution asympototically, but never reach it. The concept of a terminal attractor was first introduced by Zak [1251]. Terminal attractors are fixed points in a dynamic system violating the Lipschitz condition. As a result, a terminal attractor is a singular solution that envelopes the family of regular solutions, while each regular solution approaches such an attractor in finite time. The terminal attractor-based BP (TABP) algorithm [1156, 110, 553] applies the concept of the terminal attractor to enable a finite time convergence to the global minimum. In contrast to the BP, η in the TABP is adapted by

$$\eta = \gamma \frac{\Omega(E)}{\|\mathbf{g}\|^2} \qquad (3.170)$$

where $\gamma > 0$, $\mathbf{g} = \nabla_{\overrightarrow{\mathbf{w}}} E$, and $\Omega(E)$ is a non-negative continuous function of E. This leads to an error function, which evolves by

$$\frac{\mathrm{d}E}{\mathrm{d}t} = -\gamma \Omega(E) \qquad (3.171)$$

When $\Omega(E)$ is selected as E^μ, with $\frac{1}{2} < \mu < 1$, E will stably reach zero in time [553]

$$T = \frac{E^{1-\mu}(0)}{\gamma(1-\mu)} \qquad (3.172)$$

According to (3.170), at local minima, $\eta \to \infty$ and the algorithm can escape from local minima. By selecting γ and μ, one can tune the time to exactly reach $E = 0$. The TABP method can be three orders of magnitude faster than the BP [1156]. When $\|\mathbf{g}\|$ is so large that η is less than γ, as a heuristic, one can force $\eta = \gamma$ temporarily to speed up the convergence, that is, switch to BP temporarily.

Based on the Lipschitzian property of the MSE function, the globally optimal training algorithm (GOTA) [1082] is derived by using the branch-and-bound based Lipschitz optimization method. The effectiveness of the GOTA is improved by using dynamically computed local Lipschitz constants over subsets of the weight space. Local-search procedures such as the BP algorithm can be incorporated into the GOTA, so as to improve the learning efficiency of the GOTA while retaining the globally convergent property.

3.14.4 Stochastic Learning Techniques

There are other stochastic learning algorithms for training neural networks [1126, 1127, 138, 1047]. Although their convergence is not as fast as the BP, they can generalize better and can effectively avoid local minima. Besides, they are flexible in network topology, error function, and activation function. Gradient information is not required.

Alopex is a biologically motivated stochastic parallel process designed to find the global minimum of error surfaces [459]. Alopex works by broadcasting an energy or cost function to all the weight processors in the network and changing each weight stochastically, based on a correlation between changes in its own output and in the energy. It is suitable for parallel implementation, since there is no interaction between the weight processors. Alopex has been applied for training the MLP [1126].

A random step-size strategy implemented in [1127] employs an annealing average step size. The large steps enable the algorithm to jump over local maxima/minima, while the small ones ensure convergence in a local area. In [1127], the random step-size algorithm has been compared with several other stochastic algorithms including the SA [607], Alopex [1126], the iterated adaptive memory stochastic search (IAMSS) [138], and the algorithm proposed in [1047].

3.15 Hardware Implementation of Perceptrons

Neural networks can be implemented using digital VLSI technology. Multiplication-free architectures are attractive, since digital multiplication operations in each neuron are very demanding in terms of time or chip area and create a bottleneck. In binary representation, multiplication between two integers can be substituted by a shift if one of the integers is a power of two.

When the logistic or the hyperbolic tangent function is used, an exponential function needs to be calculated. The value of an exponential function is usually computed by using a Taylor-series expansion, which requires many floating-point operations. In view of the piecewise-linear approximation of the sigmoidal and its derivative functions, we need to store two tables. The output of a unit can be approximated by linear interpolation of the points in the table. Since the activation function usually has output in the interval $(0, 1)$ or $(-1, 1)$, it would be possible to adopt a fixed-point representation[11]. Most neural chips integrate LUTs and fixed-point representation of the activation function in order to simplify the logic design and increase the processing speed. Unfortunately, the representation precision of numbers is directly proportional to the cost. In most applications, a representation precision with a

---

[11] A fixed-point number is a representation for a real number that has a fixed number of digits after the delimiting decimal, binary, or hexadecimal point, that is, $a = \pm d \cdots d.d_1 d_2 \cdots d_p$, where p is a fixed number specified by a user.

A fixed-point number can exactly represent all values up to its maximum value as long as the number of digits is suitably selected. Given the same number of digits in representation, a floating-point number cannot represent as many digits accurately as a fixed-point number due to the use of automatically managed exponential. However, the floating-point representation has a wider dynamic range.

16-bit coding for weights and an 8-bit coding of the outputs is enough for the convergence of a learning algorithm [44].

The computational power of the MLP using integer weights in a very restricted range has been analyzed in [317]. It is shown that if the weights are not restricted to proper range and precision, a solution may not exist. For classification problems, an existence result is derived for calculating a weight range able to guarantee the existence of a solution as a function of the minimum distance between patterns of different classes. A review of the MLP with various limited-precision weight representations is given in [317].

The parallel VLSI implementation of the MLP with the BP algorithm conducted in [763] uses only shift-and-add and rounding operations instead of multiplications. In the forward phase, weights are restricted to powers-of-two, and the activation function is computed through LUT. This avoids the multiplication operation. In the backward phase, the derivative of the activation function is computed through LUT, some internal terms are rounded to the nearest powers-of-two, and external terms like η are selected as powers-of-two terms. Decomposition of binary integers into power-of-two terms can be accomplished very quickly and with a limited amount of circuitry [705]. The gain with respect to multiplication is more than one order of magnitude both in speed and in chip area, and thus the overhead due to the decomposition operation is negligible. The rounding operations introduce randomness that helps the BP to escape from local minima. This randomness can be considered as an additive noise on the input data, and it helps to improve the generalization performance of the network [943].

The MLP with BP is implemented as a full digital system using pulse-mode neurons in [482]. A piecewise-linear activation function is used. The BP algorithm is simplified to make the hardware implementation easier. The derivative of the activation function is generated by a pulse differentiator. A random pulse sequence is injected to the pulse differentiator output to improve the learning capability.

The circuit complexity of the sigmoidal MLP can be examined in the framework of the classical Boolean and threshold gate circuit complexity by converting the sigmoidal MLP into an equivalent threshold gate circuit [92]. Sigmoidal MLPs can be implemented in polynomial size Boolean circuits with a small constant fan-in at the expense of an increase in the number of layers by a logarithmic factor. A recent survey of many results on the circuit complexity for networks of binary perceptrons is available in [93]. Nine different constructive solutions for the addition of two binary numbers using the threshold logic gate have been discussed and compared.

Perceptron learning can also be implemented using parallel digital hardware such as a systolic array of processors. Systolic arrays can be readily implemented in programmable logic devices. A systolic array implementation of the MLP with BP is given in [392], where the BP algorithm is implemented online by using a pipelined adaptation.

3.16 Backpropagation for Temporal Learning

The MLP is purely static and is incapable of processing time information. One can add a time window over the data to act as a memory for the past. Locally recurrent, globally feedforward networks [1103] and fully connected RNNs are two classes of models for dynamic system modeling. Both classes of models can be universal approximators for dynamical systems [698, 695].

In the applications of dynamical systems, we need to forecast an input at time $t + 1$ from the network state at time t. The resulting neural-network model for modeling a dynamical process is referred to as a *temporal association network*. Temporal association networks must have a recurrent architecture so as to handle the time-dependent nature of the association.

When the recurrent feature is integrated into the basic MLP architecture, this new model is capable of learning dynamic systems. The BP algorithm is required to be modified accordingly.

3.16.1 Recurrent Multilayer Perceptrons with Backpropagation

The time-delay neural network (TDNN) [1137] maps a finite-time sequence $\{x(t), x(t-1), \ldots, x(t-m)\}$ into a single output $y(t)$. It is an FNN equipped with time-delayed versions of a signal $x(t)$ as input, and becomes a partially recurrent network. The BP algorithm can be used to train the network. The architecture of a TDNN using a three-layer MLP is illustrated in Fig. 3.8. The input to the network is $m + 1$ continuous samples. If at time t, the input to the network is $\mathbf{x}_t = (x(t), x(t-1), \ldots, x(t-m))^T$, then at time $t + i$, it must be $\mathbf{x}_{t+i} = (x(t+i), x(t+i-1), \ldots, x(t+i-m))^T$. The TDNN has been successfully applied to speech recognition [1137, 718] and time-series prediction [1165]. The architecture can be generalized when $x(t)$ and $y(t)$ are vectors. This network practically functions as a finite impulse response (FIR) filter.

In Fig. 3.8, if the single output $y(t+1)$ is applied to a tapped-delayed-line memory of p units and the p delayed replicas of $y(t+1)$ are fed back to the input of the network, the input to this new RNN is then $\left(\mathbf{x}^T(t); \mathbf{y}^T(t)\right)^T = (x(t), x(t-1), \ldots, x(t-m); y(t), y(t-1), \ldots, y(t-p))^T$ and the output of the network is $y(t+1)$. The vector $\mathbf{x}(t)$ is an exogenous input originating from outside the network, and $\mathbf{y}(t)$ is regression of the model output $y(t+1)$. The model is called a *nonlinear autoregressive with exogenous inputs (NARX)* model.

By replacing each synapse weight with a linear, time-invariant filter, the MLP can be used for temporal processing [1020]. When the filter is an FIR filter, we get an FIR neural network. The FIR MLP can be implemented as a resistance-capacitance model [1020]. One can use a temporal extension of the BP algorithm to train the FIR MLP [1139, 1140]. Once the network is

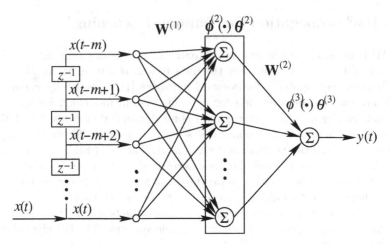

Fig. 3.8. Architecture of the TDNN. The TDNN is an FNN with a finite-time sequence as input.

trained, all the weights are fixed and the network can be used as an MLP. The FIR MLP is suitable for many applications of temporal processing.

An RNN can be unfolded, and this produces an equivalent layered FNN so that the BP algorithm can be adapted for training the unfolded network. This method is called *BP through time (BPTT)* [979, 1167]. For a long sequence, the unfolded network may be very large, and the algorithm will be inefficient. Truncated BPTT alleviates this problem by ignoring all the past contributions to the gradient beyond a certain time into the past [1188].

The recurrent MLP architecture proposed in [698] is proved to be a universal approximator for any dynamical system in which the output of the MLP is fed back to the input of the network. A BP-like learning algorithm with robust modification is used to train the recurrent MLP. This recurrent MLP generates a better performance with less nodes compared to the MLP.

3.16.2 Recurrent Neural Networks with Recurrent Backpropagation

RNNs with the sigmoidal activation function can be a universal approximator of any continuous or discrete differential trajectory on a compact time interval [695, 390].

The recurrent BP algorithm is used to train fully connected RNNs, where the units are assumed to have continuous states [918, 24]. After training on a set of N patterns $\{(\mathbf{x}_k, \mathbf{y}_k)\}$, the presentation of $\mathbf{x}_k$ will drive the network output to a fixed attractor state $\mathbf{y}_k$. Thus, the algorithm learns static mappings, and may be used as associative memories. The computational complexity of the recurrent BP is $O\left(J^2\right)$ for an RNN of J units. When the initial weights are selected as small values, the network almost always converges to a stable point.

The time-dependent recurrent learning (TDRL) algorithm is an extension of the recurrent BP to dynamic sequences that produce time-dependent trajectories [896]. It is a gradient-descent method searching the weights of a continuous RNN to minimize the error function of the temporal trajectory of the states.

The real-time recurrent learning (RTRL) algorithm is used for training fully connected RNNs with discrete-time states [1187]. It is a modified BP algorithm, and is an online algorithm without the need for allocating a memory proportional to the maximum sequence length. This RTRL is suitable for tasks that require retention of information over fixed or indefinite time length, and it is best suitable for real-time applications.

In the normalized RTRL algorithm [756], the learning rate of the RTRL is normalized at each step so that one has the optimal adaptive learning rate for every discrete time instant. The algorithm has the *a posteriori* learning in the RNNs. The normalized RTRL is both faster and more stable than the RTRL. In [596], the RTRL algorithm has been extended to its complex-valued form where the inputs, outputs, weights, and activation functions are complex-valued.

The delayed RNN is a continuous-time RNN having time-delayed feedbacks. Due to the presence of the time delay in the differential equation, the system has an infinite number of degrees of freedom. The TDRL [896] and RTRL [1187] algorithms are introduced into the delayed RNN [1093]. A comparative study of the ordinary RNN and the TDNN has been made in terms of the learning algorithms, learning capability, and robustness against noise in [1093].

Numerous methods for training RNNs are given in the literature. For an overview of various gradient-based learning algorithms for RNNs, see [897]. A comprehensive analysis and comparison of the BPTT [1167], the recurrent BP [918, 24], and the RTRL [1187] is given in [1188].

3.17 Complex-valued Multilayer Perceptrons and Their Learning

In the real domain, common nonlinear transfer functions are the hyperbolic tangent and logistic functions, which are bounded and analytic everywhere. According to Liouville's theorem, a complex transfer function, which is both bounded and analytic everywhere, has to be a constant. As a result, designing a neural network for processing complex-valued signals is a challenging task, since a complex nonlinear activation function cannot be both analytic and bounded everywhere in the complex plane $\mathcal{C}$. The Cauchy–Riemann equations are necessary and sufficient conditions for a complex function to be analytic at a point $z \in \mathcal{C}$.

The error function E for training complex MLPs is defined by

$$E = \frac{1}{2}\sum_{p=1}^{N} \mathbf{e}_p^{\mathrm{H}}\mathbf{e}_p \tag{3.173}$$

where $\mathbf{e}_p$ is defined by (3.20), but it is a complex-valued vector. E is not analytic since it is a real-valued function.

3.17.1 Split Complex Backpropagation

The conventional approach for learning complex-valued MLPs selects split complex activation functions. Each split complex function consists of a pair of real sigmoidal functions marginally processing the inphase and quadrature components. Based on this split strategy, the complex version of the BP algorithm is derived for complex-valued MLPs [687, 100, 850, 1110], and the complex-valued RTRL algorithm is derived for complex-valued RNNs [596]. This approach can avoid the unboundedness of fully complex activation functions, however, the split complex activation function cannot be analytic.

The split complex BP uses the split derivatives of the real and imaginary components instead of relying on well-defined fully complex derivatives. The derivatives cannot fully exploit the correlation between the real and imaginary components of the weighted sum of the input vectors. In the split approach, the activation function is split by

$$\phi(z) = \phi_{\mathrm{R}}\left(\mathrm{Re}(z)\right) + \mathrm{j}\phi_{\mathrm{I}}\left(\mathrm{Im}(z)\right) \tag{3.174}$$

where z is the net input to a neuron

$$z = \mathbf{w}^{\mathrm{T}}\mathbf{x} \tag{3.175}$$

Typically, both $\phi_{\mathrm{R}}(\cdot)$ and $\phi_{\mathrm{I}}(\cdot)$ are selected to be the same sigmoidal function. Here, $\mathbf{x} \in \mathcal{C}^J$ and $w \in \mathcal{C}^J$ are the J-dimensional complex input and weight vectors, respectively.

In [1238], $\phi_{\mathrm{R}}(\cdot)$ and $\phi_{\mathrm{I}}(\cdot)$ are selected as $\phi_{\mathrm{R}}(x) = \phi_{\mathrm{I}}(x) = x + a_0 \sin(\pi x)$, where $a_0 \in (0, 1/\pi)$ is a constant slope parameter. This slip complex function satisfies most properties of complex activation functions. This method can reduce the information redundancy among hidden neurons of a complex MLP, and results in a guaranteed weight update when the estimation error is not zero.

3.17.2 Fully Complex Backpropagation

The fully complex BP is derived based on a suitably selected complex activation function [406, 605]. In [406], an activation function of the type

$$\phi(z) = \frac{z}{c_0 + \frac{1}{r_0}|z|} \tag{3.176}$$

is used, where c_0 and r_0 are real positive constants. The function $\phi(\cdot)$ maps a point z on the complex plane to a unique point $\phi(z)$ on the open disc $\{z : |z| < r_0\}$, with the same phase angle, and c_0 controls the steepness of $|\phi(z)|$. This complex function satisfies most of the properties for activation function, and a circuit for such a complex neuron is designed in [406].

In [605], the fully complex BP algorithm [406] is simplified by using the Cauchy–Riemann equations. It is shown that the fully complex BP algorithm is the complex conjugate form of the BP algorithm and that the split complex BP is a special case of the fully complex BP. This generalization is possible by employing elementary transcendental functions (ETFs) that are almost everywhere bounded and analytic in $\mathcal{C}$. The complex ETFs provide well-defined derivatives for optimization of the fully complex BP algorithm. Nine complex ETFs, including $\sin z$, $\tan z$, $\sinh z$ and $\tanh z$, are listed as suitable nonlinear activation functions in [605]. Fully complex FNNs with these ETFs as the nonlinear activation functions are proved to be universal approximators in the complex domain [606].

The fully complex normalized BP algorithm [450] is an improvement on the complex BP [406] obtained by including an adaptive normalized learning rate. This is achieved by performing a minimization of the complex-valued instantaneous output error that has been expanded via a Taylor-series expansion. The method is valid for any complex activation function discussed in [605].

Many other algorithms for training complex-valued MLPs are available in the literature. They are typically complex versions of some algorithms used for training real-valued MLPs. The split complex EKF algorithm [935] has a faster convergence than the split complex BP [687]. The split complex RProp algorithm [577] outperforms the split complex BP [1238], and the fully complex BP [605] in terms of the computational complexity, convergence speed, and accuracy.

3.18 Applications and Computer Experiments

We are now in a position to end this chapter by considering some applications and examples.

3.18.1 Application 3.1: NETtalk — A Speech Synthesis System

One of the most well-known applications of the MLP is the speech synthesis system NETtalk [1014]. NETtalk is a three-layer classification network, as illustrated in Fig. 3.9 that translates English letters into phonemes. A string of characters forming English text is then converted into a string of phonemes, which can be further sent to an electronic speech synthesizer to produce speech.

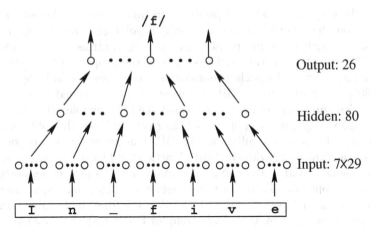

Fig. 3.9. Schematic drawing of the NETtalk architecture

The MLP-based TDNN model is applied. The input layer looks at a seven-character (character, space or punctuation) window of text. This seven-character window is enough to pronounce all but a few unusually difficult characters. Each character is fed to 29 input nodes, corresponding to the 29 possible letters. Thus, there is a total of 203 input nodes. Given a letter, only one of the 29 nodes is set to 1 and all the others are set to 0. The output layer uses 26 nodes, corresponding to 23 articulatory features and 3 stress features. The number of hidden units varies from simulation to simulation and 80 hidden nodes are used for continuous speech. The network has a total of 309 nodes and 18 629 weights. Both the BP and Boltzmann learning [9] algorithms are applied to train the network. The architecture of the original NETtalk is optimized in [460] by using the OBS procedure, which reduces the total number of free weight parameters by nine-tenths.

At the training phase, only the output unit corresponding to the correct phoneme is set to 1, while all the other output nodes are set to 0. At the performing stage, the network does not output exactly ones or zeros. Only the output node with the maximum value is determined as the pronouced phoneme. The hidden units play the same role as the rule extractor. The training set consists of a corpus from a dictionary, and phonetic transcriptions from informal, continuous speech of a child. NETtalk is suitable for fast implementation without any domain knowledge, while the development of conventional rule-based expert systems such as the DECtalk needs years of group work.

3.18.2 Application 3.2: Handwritten Digit Recognition

Handwritten character recognition is a classical problem in pattern recognition. It can be commercially used in postal services [655] or in banking services

[766] to recognize handwritten digits on envelopes or bank cheques. This is a real image-recognition problem. The input consists of black or white pixels. The mapping is from the two-dimensional image space to category space, which has considerable complexity. Handwritten digit recognition is a classification problem of ten output classes.

Le Cun *et al.* [655] have designed a large MLP trained with the BP to recognize handwritten digits. The architecture of the network is shown in Fig. 3.10. For a pattern of class i, the desired output for the ith output node is $+1$, while it is -1 for all the other output nodes. The network has four hidden layers named H1, H2, H3, and H4, where H1 and H3 are shared weight feature extractors while H2 and H4 are averaging/subsampling layers. This architecture is reminiscent of the neocognition architecture [388].

The training set is 9298 segmented numerals digitized from handwritten postal codes that appeared on real U.S. mail. Additional 3349 printed digits coming from 35 different fonts are also added to the training set. Around 79% of the training set is used for training and the remaining 21% for testing. The size of the characters is normalized to a 16×16 pixel image by using a linear transformation. Due to the linear transformation, the resulting image is not binary but has multiple gray levels. The gray-leveled image is further scaled

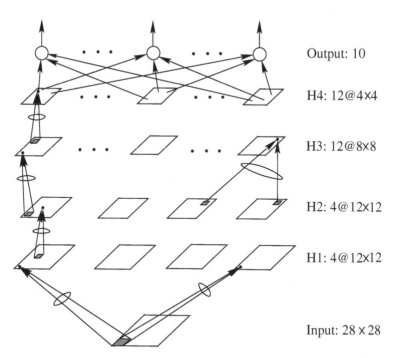

Fig. 3.10. Schematic drawing of the six-layer MLP architecture for handwritten digit recognition

Fig. 3.11. Samples of normalized digits from the testing set (From [655])

to the range of −1 to 1. Some samples of normalized digits from the testing set are shown in Fig. 3.11.

The input layer uses a 28 × 28 plane instead of a 16 × 16 plane to avoid problems arising from a kernel overlapping a boundary. H1 contains 4 independent 24 × 24 feature maps. Each unit in a feature map takes its input from a 5 × 5 neighborhood on the input plane. Weight sharing is applied to connections on each unit in a feature map, and all the 576 units in each feature map share a set of 26 weights. Layer H2 is composed of 4 planes of 12 × 12. Each unit in one of these planes has input from 4 units from its corresponding plane in H1. All the weights are forced to be equal. As a result, H2 performs a local averaging and a 2:1 subsampling in both the directions. H3 is comprised of 12 feature maps each arranged in an 8 × 8 plane. The connections between H2 and H3 are similar to that between the input layer and H1. Each unit in one of the 12 feature maps receives input of one or two 5 × 5 neighborhoods centered around units that are at identical positions within each of the H2 maps according to the scheme: H2.1→ H3.1, H2.1+H2.2→H3.2, H2.1+H2.2→ H3.3, H2.2→H3.4, H2.1+H2.2→ H3.5, H2.1+H2.2→H3.6, H2.3→H3.7, H2.3+H2.4→H3.8, H2.3+H2.4→ H3.9, H2.4→ H3.10, H2.3+H2.4→ H3.11, H2.3+H2.4→H3.12. Again, weight sharing is applied among all the units within each map. H4 is composed of 12 planes of size 4 × 4, and behaves exactly like H2. The output layer has 10 nodes, and is fully connected to H4.

The network has 4635 nodes, 98 442 connections, and 2578 independent parameters. This architecture was optimized from a previous architecture [654] by using the OBD technique [656], and has one fourth of the number of free parameters.

An incremental BP algorithm based on an approximate Newton's method was employed, and the training time for 30 epochs through the training set plus test required 3 days on a Sun SparcStation 1. The error rates on the training set and testing set were, respectively, 1.1% and 3.4%. When a rejection criterion was employed to the testing set, the result was 9% rejection rate for 1% error. These rejected samples may be due to fault segmentation, or ambiguous writing even to humans. The large network with weight sharing was shown to outperform a small network with the same number of free parameters. A similar result on using the MLP for handwritten digits scanned from real bank cheques has been reported in [766].

3.18.3 Example 3.1: Iris Classification

The well-known Iris classification problem is a benchmark for evaluating MLP learning algorithms. In the Iris data set, 150 patterns are classified into 3 classes. Each pattern has four numeric properties.

We use the MLP to learn this problem. The logistic sigmoidal function is selected, and two learning schemes are applied. Eighty per cent of the data set is used as training data, and the remaining 20% as testing data. We set the performance goal as 0.001, and the maximum number of epochs as 10^4. We simulate and compare ten popular MLP learning algorithms, namely, the BP, BPM, Rprop, BFGS, OSS, LM, scaled CG, CG with Powell–Beale restarts, Fletcher–Powell CG, and Polak–Ribiere CG algorithms. Simulation is conducted on a PC with a Pentium III 600 MHz CPU and 128M RAM based on the Matlab© Neural Networks Toolbox.

In the first scheme, we select a 4-3-1 network. To fully make use of the range of the activation function, we encode the attribute values of the three classes as 0.2, 0.5 and 0.8, respectively. During generalization, if the network output for an input pattern is closest to one of the attribute values, the pattern is identified as belonging to that class. For the BP method, $\eta = 0.9$. For the BPM, $\eta = 0.9$ and $\alpha = 0.3$. Table 3.1 lists the results based on an average of 50 random runs. The traces of the training error are illustrated in Fig. 3.12 for a random run.

The second strategy employs the conventional neural classifier architecture. The network has three outputs: Only the node corresponding to its class is set to 1, while all the other nodes are set to 0. When training the network, to avoid the saturation region, we use 0.9 and 0.1 to, respectively, replace 1 and 0. At the generalization stage, only the node with the largest output is treated as 1 and outputs at all the other nodes are treated as 0. We select a 4-4-3 network. All the other learning parameters are selected to be the same as those for the first strategy. The training results for 50 independent runs are listed in Table 3.2. The learning curves for a random run of these algorithms are shown in Fig. 3.13.

For this example, we see that the BFGS, RProp, and LM usually generate better classification accuracy. The performance of the BP as well as that of the BPM is highly dependent on suitably selected learning parameters, which are difficult to find for practical problems.

Table 3.1. Performance comparison of a 4-3-1 MLP trained with ten learning algorithms. Rp—Rprop, SCG—scaled CG, CGB—CG with Powell–Beale restarts, CGF—Fletcher–Powell CG, and CGP—Polak–Ribiere CG.

Algorithm	Number of epochs	Training MSE	Classification accuracy(%)	std	Mean training time (s)	std. (s)
BP	9961.8	0.0041	96.67	0.0539	165.8539	4.6304
BPM	10^4	0.0028	97.33	0.0286	180.6522	0.7288
RP	8788.3	0.0015	97.80	0.0248	160.3764	46.7865
BFGS	172.36	0.0052	100	0	8.988	10.5012
OSS	4873.8	0.0030	97.07	0.0424	274.5672	234.8215
LM	1996.8	0.0014	96.67	0	57.9253	104.9830
SCG	2918.3	0.0018	97.80	0.0229	105.2147	127.7863
CGB	156.56	0.0056	95.53	0.0813	9.4578	7.33
CGF	285.68	0.0043	96.27	0.0770	15.4889	22.0937
CGP	862.3	0.0042	96.93	0.0441	46.9216	128.098

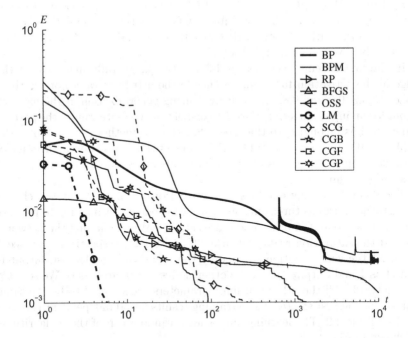

Fig. 3.12. IRIS classification using a 4-4-1 MLP trained with ten learning methods: the traces of the training error for a random run. t corresponds to the number of epochs.

Table 3.2. Performance comparison of a 4-4-3 MLP trained with ten learning algorithms

Algorithm	Number of epochs	Training MSE	Classification accuracy(%)	std	Mean training time (s)	std. (s)
BP	9999.0	0.0167	92.93	0.1409	166.9312	1.2553
BPM	10^4	0.0194	92.00	0.1421	181.4327	0.5029
RP	9889.0	0.0048	95.80	0.0342	181.6255	14.2879
BFGS	202.78	0.0306	96.67	0	11.2168	8.4474
OSS	6139.5	0.0125	90.47	0.1249	385.9305	258.8876
LM	876.78	0.0346	100	0	45.9425	96.0034
SCG	2428.0	0.0041	95.47	0.0321	88.3316	101.2338
CGB	438.12	0.0374	82.07	0.2692	25.0697	24.6787
CGF	696.54	0.0533	76.60	0.2906	37.1598	35.6401
CGP	1876.1	0.0290	90.73	0.1841	97.2245	130.6590

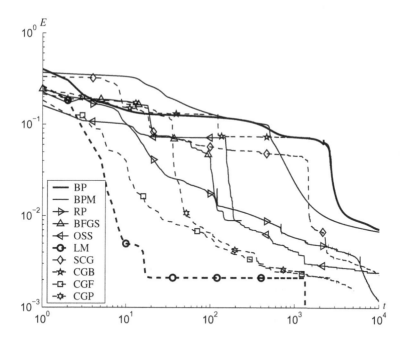

Fig. 3.13. IRIS classification using a 4-4-3 MLP trained with ten learning methods: the traces of the training error for a random run. t corresponds to the number of epochs.

3.18.4 Example 3.2: DoA Estimation

In this example, the MLP is trained so that it can perform as a direction-finding algorithm for antenna arrays. A uniform linear array of $L = 4$ elements, with a spacing of half the carrier wavelength between the neighboring elements, is used as the receiver. The carrier frequency is assumed to be $f_c = 1.0$ MHz. The impinging source is a BPSK signal with a power of 1 dB and a 100% cosine rolloff. The background noise of the array is assumed to be -10 dB and Gaussian. The input of the network is the array measurement and the output is the DoA of the signal. We perform three types of preprocessing to the array measurement $\mathbf{x}$ to remove the redundancy of the signal amplitude and initial phase. The result is used as the input to the network.

(a) $\mathbf{R} = [R_{ij}]$, the autocorrelation of vector $\mathbf{x}$, contains useful information of the sources. Since only the upper triangle of $\mathbf{R}$ contains useful information related to the DoAs, one can use the following vector as the network input [320]

$$\mathbf{u} = \left(R_{12}, R_{13}, \cdots, R_{1L}; R_{23}, \cdots, R_{2L}; \cdots; R_{(L-1)L}\right)^{\mathrm{T}}$$

This input vector is further normalized as $\mathbf{z} = \frac{\mathbf{u}}{\|\mathbf{u}\|}$ so as to remove the effect of the signal gain. In view of the real and imaginary parts of $\mathbf{z}$, we have a total of $2L(L-1) = 24$ input nodes.

(b) The phase differences between the array elements and the reference element can be used as the network input

$$\mathbf{z} = \left(\varphi\left(x_2\right) - \varphi\left(x_1\right), \cdots, \varphi\left(x_L\right) - \varphi\left(x_1\right)\right)^{\mathrm{T}} \in [0, 2\pi)^{L-1}$$

To compress the input space, the $\mathrm{mod}(2\pi)$ operator can be applied to each component due to its 2π periodic nature. This causes the 2π periodic discontinuity. A better way to deal with this difficulty is to customize the input vector so as to comprise the sines and cosines of all the phase differences. In this case, we have $2(L-1)$ nodes.

(c) The average of $\mathbf{x}$, i.e. $\overline{\mathbf{x}}$, $\mathbf{z} = \frac{\overline{\mathbf{x}}}{\|\overline{\mathbf{x}}\|}$ is used to get $2L$ input nodes.

We implement the simulations for the above three cases. For case (a), a 24-60-1 network is used. For case (b), we select a 6-30-1 network. The network architecture for case (c) is 8-60-1. For all the cases, the training set contains 500 samples uniformly distributed from $-90°$ to $90°$, and 100 random samples in this interval are selected for testing. Computation is conducted on a PC with a Pentium III 600 MHz CPU and 128M RAM based on the Matlab© Neural Networks Toolbox. The BFGS algorithm is used for training. The training goal is 10^{-5}, the maximum epoch is set to be 1000. The training results are shown in Table 3.3. The learning and testing errors are shown in Fig. 3.14.

The preprocessing schemes (a) and (b) employ correlation information between array elements, while scheme (c) performs simple averaging on the measurement of each element. According to the results, schemes (a) and (b) produce similar performance in terms of the estimation accuracy $\Delta\theta$, which are significantly better than that for scheme (c).

Table 3.3. Direction finding using MLP trained by the BFGS algorithm

Scheme	Number of epochs	Training MSE	Testing MSE	Training time (s)	Testing time /sample (s)
(a)	1000	0.1130	0.1143	2026.42	5.0×10^{-4}
(b)	1000	0.1613	0.1979	188.12	9.12×10^{-5}
(c)	1000	0.8942	1.0987	978.06	2.02×10^{-4}

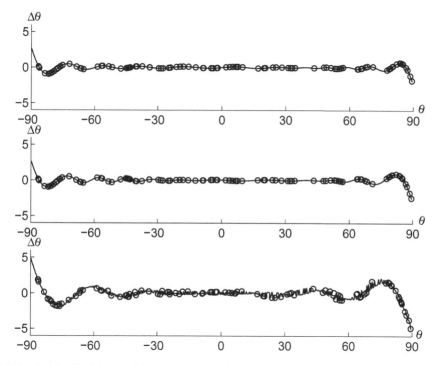

Fig. 3.14. Training and testing errors. θ and $\Delta\theta$ are, respectively, the DoA (in degrees) and the estimation error in the DoA (in degrees).

4

Hopfield Networks and Boltzmann Machines

Memory is important for transforming a static network into a dynamic one. Memories can be long term or short term. A long-term memory is used to store stable system information, while a short-term memory is useful for simulating a dynamic system with a temporal dimension. As discussed in Chapter 3, FNNs can become dynamic by embedding memory into the network using time delay.

The Hopfield model is a dynamic associative-memory model, and the Boltzmann machine as well as some other stochastic models are proposed as its generalization. By a combination of the concept of energy and the neural network topology, these models provide a method to deal with COPs, which are notorious NP-complete problems. In this chapter, we discuss these models, their learning algorithms as well as their analog implementations. Associative memories, SA, and chaotic neural networks are important topics treated in this chapter. The COP is also described in this chapter.

4.1 Recurrent Neural Networks

RNNs are inspired by ideas from statistical physics. There is at least one feedback connection in RNNs. RNNs are dynamical systems with temporal state representations. They are computationally powerful, and can be used in many temporal processing models and applications.

The Hopfield model can store information in a dynamically stable structure [502]. The Boltzmann machine [9], which relies on a stochastic algorithm of statistical thermodynamics called *simulated annealing (SA)* [607], is a generalization of the Hopfield model.

For an RNN of J units, we denote the input of unit i as x_i and the output or state of unit i as y_i. The architecture of a fully connected RNN is illustrated in Fig. 4.1. The dynamics of unit i with sound biological and electronic motivation are given as [461]

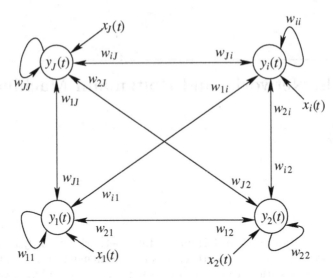

Fig. 4.1. Architecture of a fully connected RNN of J neurons. Each neuron has its input x_i and output y_i.

$$net_i(t) = \sum_{j=1}^{J} w_{ji} y_j(t) + x_i(t) \qquad (4.1)$$

$$\tau_i \frac{dy_i(t)}{dt} = -y_i(t) + \phi\left(net_i\right) + x_i(t) \qquad (4.2)$$

for $i = 1, \ldots, J$, where τ_i is a time constant, net_i is the net input to the ith unit, $\phi(\cdot)$ is a sigmoidal activation function, and the input $x_i(t)$ and the output $y_i(t)$ are continuous functions of time. In (4.2), $-y_i(t)$ denotes natural signal decay. It has been shown in [390] that any continuous state-space trajectory can be approximated to any desired degree of accuracy by the output of a sufficiently large continuous-time RNN described by (4.1) and (4.2). In other words, the RNN is a universal approximator of dynamical systems.

The universal approximation capability of RNNs has been investigated in a number of papers (see [695, 699], and the references given in [699]). In [695], a fully connected discrete-time RNN with the sigmoidal activation function has been proved to be a universal approximator of discrete- or continuous-time trajectories on compact time intervals. It is shown in [699] that a continuous-time RNN with the sigmoidal activation function and external input can approximate any finite-time trajectory of a dynamical time-variant system.

When used for the approximation of dynamic systems, due to its feedforward as well as feedback connections, an RNN is more powerful in representing complex dynamics than an FNN. Given the same approximation capability, an RNN has a compact network architecture whose size is considerably smaller than that of an FNN. RNNs can also be used as associative memories to build attractors $\mathbf{y}_p$ from input-output association $\{\mathbf{x}_p, \mathbf{y}_p\}$. The recurrent BP [918, 24] and the RTRL [1187] are usually used for RNN training.

4.2 Hopfield Model

The Hopfield model [504] is the most popular dynamic model. It is a fully interconnected RNN with J McCulloch–Pitts neurons [783], having a topology similar to that of Fig. 4.1. The Hopfield model is usually represented by using a J-J layered architecture, as illustrated in Fig. 4.2. The input layer only collects and distributes feedback signals from the output layer. The network has a symmetric architecture with a symmetric zero-diagonal real weight matrix, that is, $w_{ij} = w_{ji}$ and $w_{ii} = 0$. Each neuron in the second layer sums the weighted inputs from all the other neurons to calculate its current net activation net_i, then applies an activation function to net_i and broadcasts the result along the connections to all the other neurons. The Hopfield network is biologically plausible since it functions like the human retina [659].

4.2.1 Dynamics of the Hopfield Model

The Hopfield model operates in an unsupervised manner. The dynamics of the network are described by a system of nonlinear ordinary differential equations (ODE). The discrete form of the dynamics is defined by

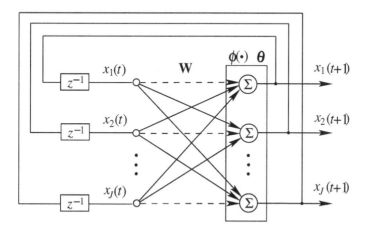

Fig. 4.2. Architecture of the Hopfield network. Note that $w_{ii} = 0$ is represented by a dashed-line connection. $\phi(\cdot)$ and $\boldsymbol{\theta}$ are, respectively, a vector comprising the activation functions for all the neurons and a vector comprising the biases for all the neurons.

$$net_i(t+1) = \sum_{j=1}^{J} w_{ji} x_j(t) + \theta_i \qquad (4.3)$$

$$x_i(t+1) = \phi(net_i(t+1)) \qquad (4.4)$$

where net_i is the weighted net input of the ith neuron, $x_i(t)$ is the output of the ith neuron, θ_i is a bias to the neuron, and $\phi(\cdot)$ is the sigmoidal function. The discrete time variable t in (4.3) and (4.4) takes values $0, 1, 2, \cdots$.

Correspondingly, the continuous Hopfield model is given by

$$\frac{d net_i(t)}{dt} = \sum_{j=1}^{J} w_{ji} x_j(t) + \theta_i \qquad (4.5)$$

$$x_i(t) = \phi(net_i(t)) \qquad (4.6)$$

Note that t in (4.5) and (4.6) denotes the continuous-time variable.

In order to characterize the performance of the network, the concept of energy is introduced and the following energy function defined [502]:

$$E = -\frac{1}{2} \sum_{i=1}^{J} \sum_{j=1}^{J} w_{ij} x_i x_j - \sum_{i=1}^{J} \theta_i x_i$$

$$= -\frac{1}{2} \mathbf{x}^{\mathrm{T}} \mathbf{W} \mathbf{x} - \mathbf{x}^{\mathrm{T}} \boldsymbol{\theta} \qquad (4.7)$$

where $\mathbf{x} = (x_1, x_2, \ldots, x_J)^{\mathrm{T}}$ is the input and state vector, and $\boldsymbol{\theta} = (\theta_1, \theta_2, \ldots, \theta_J)^{\mathrm{T}}$ is the bias vector.

4.2.2 Stability of the Hopfield Model

The stability of the Hopfield network is asserted by Lyapunov's second theorem. This is shown by the following derivations.

$$\frac{dE}{dt} = \sum_{i=1}^{J} \frac{dE}{dx_i} \frac{dx_i}{dt} = \sum_{i=1}^{J} \frac{dE}{dx_i} \frac{dx_i}{dnet_i} \frac{dnet_i}{dt} \qquad (4.8)$$

From (4.7),

$$\frac{dE}{dx_i} = -\left(\sum_{j=1}^{J} w_{ji} x_j + \theta_i \right) = -\frac{dnet_i}{dt} \qquad (4.9)$$

Thus

$$\frac{dE}{dt} = -\sum_{i=1}^{J} \left(\frac{dnet_i}{dt} \right)^2 \frac{dx_i}{dnet_i} \qquad (4.10)$$

Since the sigmoidal function $\phi(net_i)$ is monotonically increasing, $\frac{dx_i}{dnet_i}$ is always positive. Thus, we have $\frac{dE}{dt} \leq 0$. As a result, the Hopfield network is stable. The dynamic equations of the Hopfield network actually implement a gradient-descent algorithm based on the cost function E [503].

The Hopfield model is asymptotically stable when running in asynchronous or serial update mode. In asynchronous or serial mode, only one neuron at a time updates itself in the output layer, and the energy function either decreases or stays the same after each iteration. However, if the Hopfield memory is running in the synchronous or parallel update mode, that is, all neurons update themselves at the same time, it may not converge to a fixed point, but may instead become oscillatory between two states [719, 136, 228].

The Hopfield network with the signum activation has a smaller degree of freedom compared to that using the sigmoidal activation, since it is constrained to changing the states along the edges of a J-dimensional hypercube $\mathcal{O} = \{-1, 1\}^J$. The use of sigmoidal functions helps in smoothing out some of the local minima.

4.2.3 Applications of the Hopfield Model

Due to recurrence, the Hopfield network remembers cues from the past and does not complicate the training procedure. The dynamics of the network are described by a system of ODEs and by an associated energy function to be minimized. This network is a stable dynamic system, which converges to a stable fixed point when implemented in asynchronous update mode. The Hopfield network can be used for converting analog signals into the digital format, for associative memory, and for solving COPs.

The Hopfield network can be used as an effective interface between analog and digital devices, where the input signals to the network are analog and the output signals are discrete values. The neural interface has the capability of learning. The neural-based analog-to-digital (A/D) converter adapts to compensate for initial device mismatches or long-term drifts [1083]. By adjusting the parameters of the neuron circuit with a learning rule, an adaptive network is generated.

Associative memory is a major application of the Hopfield network. The fixed points in the network energy function are used to store feature patterns. When a noisy or incomplete pattern is presented to the trained network, a pattern in the memory is retrieved. This property is most useful for pattern recognition or pattern completion.

The energy minimization ability of the Hopfield network is used to solve optimization problems. The local minima of (4.7) correspond to the attractors in the phase space, which are nominal memories of the network. A large class of COPs can be expressed in this form of QP optimization problems, and thus can be solved using the Hopfield network. Signal estimation can also be formulated as QP problems. In this approach, the weights and biases are trained such that the global minimum state of the network energy corresponds to the optimum solution of the optimization problem.

4.3 Analog Implementation of Hopfield Networks

The high interconnection in physical topology makes the Hopfield network especially suitable for analog VLSI implementation. The analog implementation of the Hopfield model has found applications in many fields. The convergence time of the network dynamics is decided by a circuit time constant, which is of the order of a few nanoseconds [503]. The Hopfield network can be implemented by interconnecting an array of resistors, nonlinear operational amplifiers with symmetrical outputs, capacitors, and external bias current sources. Each neuron can be implemented by a capacitor, a resistor, and a nonlinear amplifier. A current source is necessary for representing the bias. The circuit structure of the neuron is shown in Fig. 4.3. A drawback of the Hopfield network is the necessity to update the complete set of network coefficients caused by the signal change, and this causes difficulties in its circuit implementation.

By applying Ohm's law and Kirchhoff's current law to the ith neuron, we obtain

$$C_i \frac{du_i}{dt} = -\frac{u_i}{R_i} + \sum_{j=1}^{J} \frac{v_j}{R_{ij}} + I_i \tag{4.11}$$

where $v_i = \phi(u_i)$, $\phi(u_i)$ is the sigmoidal function, and

$$\frac{1}{R_i} = \frac{1}{R_{i0}} + \sum_{j=1}^{J} \frac{1}{R_{ij}} = G_{i0} + \sum_{j=1}^{J} G_{ij} \tag{4.12}$$

and G_{ij} is the conductance. In the circuits shown in Fig. 4.3, the inverting output of each neuron is used to generate negative weights since the conductance G_{ij} is always positive.

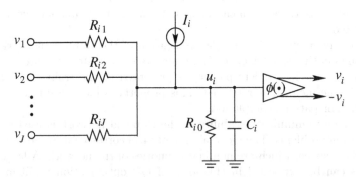

Fig. 4.3. A circuit for neuron i in the Hopfield model. v_i, $i = 1, \cdots, J$, is the output voltage of neuron i, I_i is the external bias current source for neuron i, u_i is the voltage at the interconnection point, C_i is a capacitor, and R_{ik}, $k = 0, 1, \cdots, J$, are resistors. The sigmoidal function $\phi(\cdot)$ is used as the transfer function of the amplifiers.

Equation (4.11) can be written as

$$\tau_i \frac{\mathrm{d}u_i}{\mathrm{d}t} = -\alpha_i u_i + \left(\sum_{j=1}^{J} w_{ji} x_j + \theta_i \right)$$

$$= -\alpha_i u_i + \mathbf{w}_i^{\mathrm{T}} \mathbf{x} + \theta_i \qquad (4.13)$$

where $x_i = v_i = \phi(u_i)$ is the input signal, $\tau_i = r_i C_i$ is the circuit time constant, r_i is a scaling resistance, $\alpha_i = \frac{r_i}{R_i}$ is a damping coefficient, $w_{ji} = \frac{r_i}{R_{ij}} = r_i G_{ij}$ is the synaptic weight, $\theta_i = r_i I_i$ is the external bias signal, and $\mathbf{w}_i = (w_{1i}, w_{1i}, \cdots, w_{Ji})^{\mathrm{T}}$ is the weight vector feeding neuron i.

The dynamics of the whole network can be written as

$$\tau \frac{\mathrm{d}\mathbf{u}}{\mathrm{d}t} = -\alpha \mathbf{u} + \mathbf{W}^{\mathrm{T}} \mathbf{x} + \boldsymbol{\theta} \qquad (4.14)$$

where the circuit time constant matrix $\tau = \mathrm{diag}(\tau_1, \tau_2, \ldots, \tau_J)$, the interconnect-point voltage vector $\mathbf{u} = (u_1, u_2, \ldots, u_J)^{\mathrm{T}}$, the damping coefficient matrix $\alpha = \mathrm{diag}(\alpha_1, \alpha_2, \ldots, \alpha_J)$, the input and state vector $\mathbf{x} = (x_1, x_2, \ldots, x_J)^{\mathrm{T}}$, the bias vector $\boldsymbol{\theta} = (\theta_1, \theta_2, \ldots, \theta_J)^{\mathrm{T}}$, and the $J \times J$ weight matrix $\mathbf{W} = [\mathbf{w}_1 \mathbf{w}_2 \cdots \mathbf{w}_J]$.

At the equilibrium of the system, $\frac{\mathrm{d}\mathbf{u}}{\mathrm{d}t} = \mathbf{0}$, thus

$$\alpha \mathbf{u} = \mathbf{W}^{\mathrm{T}} \mathbf{x} + \boldsymbol{\theta} \qquad (4.15)$$

The dynamics of the network are controlled by C_i and R_{ij}. A sufficient condition for the Hopfield network to be stable is that $\mathbf{W}$ is a symmetric matrix with diagonal elements being zero [503]. The stable states correspond to the local minima of the Lyapunov function [503]

$$E(\mathbf{x}) = -\frac{1}{2}\mathbf{x}^{\mathrm{T}}\mathbf{W}\mathbf{x} - \mathbf{x}^{\mathrm{T}}\boldsymbol{\theta} + \sum_{i=1}^{J} \alpha_i \int_0^{x_i} \phi^{-1}(\xi)\mathrm{d}\xi \qquad (4.16)$$

where $\phi^{-1}(\cdot)$ is the inverse of $\phi(\cdot)$. Equation (4.16) is a special case of the Cohen–Grossberg model [251].

An inspection of (4.13) and (4.5) shows that α is zero in the basic Hopfield model. This term corresponds to an integral related to $\phi^{-1}(\cdot)$ in the energy function. When the gain of the sigmoidal function $\beta \to \infty$, that is, when the sigmoidal function is selected as the hard-limiter function and the nonlinear amplifiers function as switches, the integral terms are insignificant and $E(\mathbf{x})$ in (4.16) approaches (4.7). In this case, the circuit model is exact for the basic Hopfield model. The stable states of the basic Hopfield network are the corners of the hypercube, namely, the local minima of (4.7) are in $\{-1, +1\}^J$ [503]. For large but finite gains, the sigmoidal function leads to a large positive contribution near the hypercube boundaries, but to a negligible contribution far from the boundaries. This leads to an energy surface that still has its maxima at the corners, but the minima slightly move inward from the corners of the hypercube. As β decreases, each minimum moves further inward and disappears one at a time. When β gets small enough, the energy minima start to disappear.

For the realization given in Fig. 4.3, it is not possible to independently adjust the network parameters, since the coefficient α_i is nonlinearly related to all the weights w_{ij}. Many other circuit implementations of the Hopfield model have been proposed, among which the work in [694] deserves mentioning.

In [694], α_i are removed by replacing the integrators and nonlinear amplifiers in the previous model by ideal integrators with saturation. The circuit of such a neuron is illustrated in Fig. 4.4. The dynamic equation of this neuron can be described by

$$\frac{\mathrm{d}v_i}{\mathrm{d}t} = \frac{1}{C_i} \left[\sum_{i=1}^{J} \frac{1}{R_{ij}} v_i + I_i \right] \tag{4.17}$$

where $|v_i| \leq 1$. Comparing (4.17) with (4.5), we have $w_{ji} = \frac{1}{C_i R_{ij}}$ and $\theta_i = \frac{I_i}{C_i}$. This model is referred to as a *linear system in a saturated mode (LSSM)*, which retains the basic structure of the Hopfield model and is easier to analyze, synthesize and implement than the Hopfield model. The energy function of the model is exactly the same as (4.7).

The Hopfield model is a network model that is most suitable for hardware implementation. Numerous circuits for the Hopfield model have been proposed, including analog VLSI and optical [345] implementations.

4.4 Associative-memory Models

Association is a salient feature of human memory. Associative memory models, known as *content-addressable memories (CAMs)*, are a class of the most extensively analyzed neural networks. A pattern can be stored in memory

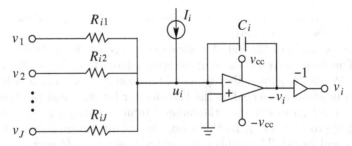

Fig. 4.4. A modified circuit for neuron i in the Hopfield model. The integrator and the nonlinear amplifier in Fig. 4.3 are replaced by an ideal integrator with saturation. v_{cc} is the power voltage.

through a learning process. For an imperfect input pattern, associative memory has the capability to recall the stored pattern correctly by performing a collective relaxation search. Associative memories can be either heteroassociative or autoassociative. For heteroassociation, the input and output vectors range over different vector spaces, while for autoassociation, both the input and output vectors range over the same vector space.

Many associative-memory models have been proposed, among which linear associative memories [28, 609, 34], the brain-states-in-a-box (BSB) [36], and bidirectional associative memories (BAMs) [621, 57] can be used as both autoassociative and heteroassociative memories, while the Hopfield model [502], the Hamming network [717], and the Boltzmann machine [9] can only be used as autoassociative models.

The Hopfield model is a continuous-time, continuous-state dynamic associative-memory model. It is a nonlinear dynamical system, where information retrieval is realized as an evolution of the system state. The binary Hopfield network is a well-known model for nonlinear associative memories. It can retrieve a pattern stored in memory in response to the presentation of a corrupted version of the pattern. This is done by mapping a fundamental memory **x** onto a stable point of a dynamical system. The states of the neurons can be considered as short-term memories (STMs) while the synaptic weights can be treated as long-term memories (LTMs). In [717], the Hopfield network has been trained as an associative memory to retrieve eight digits in 120 black-and-white pixels.

4.4.1 Hopfield Model: Storage and Retrieval

The operation of the Hopfield network as an associative memory includes two phases: storage and retrieval. Bipolar coding is often used for associative memory in that bipolar vectors have a greater probability of being orthogonal than binary vectors. We use bipolar coding in this chapter.

We now store in the network a set, $\{\mathbf{x}_p\}$, of N bipolar patterns, where $\mathbf{x}_p = (x_{p,1}, x_{p,2}, \ldots, x_{p,J})^{\mathrm{T}}$, $x_{p,i} = \pm 1$. These patterns are called *fundamental memories*. Storage is implemented by using a learning algorithm, while retrieval is based on the dynamics of the network.

Generalized Hebbian Rule

Conventional algorithms for associative storage are typically local algorithms based on the Hebbian rule [473]. The Hebbian rule [473] is known as the *outer product rule of storage* in connection with associative learning. Using this method, $\boldsymbol{\theta}$ is chosen as the zero vector. A generalized Hebbian rule for training the Hopfield network is defined by [502]

$$w_{ij} = \frac{1}{J} \sum_{p=1}^{N} x_{p,i} x_{p,j} \tag{4.18}$$

for all $i \neq j$, and $w_{ii} = 0$. In matrix form

$$\mathbf{W} = \frac{1}{J} \left(\sum_{p=1}^{N} \mathbf{x}_p \mathbf{x}_p^{\mathrm{T}} - N \mathbf{I}_J \right) \tag{4.19}$$

where $\mathbf{I}_J$ denotes the $J \times J$ identity matrix.

The generalized Hebbian rule can be written in an incremental form

$$w_{ij}(t) = w_{ij}(t-1) + \eta x_{t,i} x_{t,j} \tag{4.20}$$

for all $i \neq j$, where the step size $\eta = \frac{1}{J}$, $t = 1, \cdots, N$, and $w_{ij}(0) = 0$. As such, the learning is completed after each pattern $\mathbf{x}_t$ in the pattern set is presented exactly once.

The generalized Hebbian rule is both local and incremental. It has an absolute storage capability of $N_{\max} = \frac{J}{2 \ln J}$ [784], where the storage capability of an associative memory network is defined by the maximum number of fundamental memories, $N_{\max}$, that can be stored and retrieved reliably. For reliable retrieval, $N_{\max}$ is dropped to approximately $\frac{J}{4 \ln J}$ [784]. The generalized Hebbian rule, however, suffers severe degradation and $N_{\max}$ decreases significantly, if the training patterns are correlated. For example, time series usually include significant correlations in the measurements of adjacent samples. Some variants of the Hebbian rule, such as the weighted Hebbian rule [28] and the Hebbian rule with decay [611], can improve the storage capabilities.

When training associative memory networks using the classical Hebbian learning, an additional term called *crosstalk* may arise. When crosstalk becomes too large, spurious states other than the negative stored patterns appear [959]. The number of negative stored patterns is always equivalent to the number of stored patterns. Hebbian learning produces good results when the stored patterns are nearly orthogonal. This is the case when N bipolar vectors are randomly selected from R^J, and $N \ll J$. In practice, patterns are usually correlated and the incurred crosstalk may reduce the capacity of the network. The storage capability of the network is expected to decrease if the Hamming distance between the fundamental memories becomes smaller.

Pseudoinverse Rule

The pseudoinverse solution targets at minimizing the crosstalk between the stored patterns. The pseudoinverse rule uses the pseudoinverse of the pattern matrix, while the classical Hebbian learning uses the correlation matrix of the patterns [909, 576, 573, 959].

Denoting $\mathbf{X} = [\mathbf{x}_1, \mathbf{x}_2, \cdots, \mathbf{x}_N]$, the autoassociative memory is defined as

$$\mathbf{X}^{\mathrm{T}} \mathbf{W} = \mathbf{X}^{\mathrm{T}} \tag{4.21}$$

Using the pseudoinverse, we actually minimize $E = \left\| \mathbf{X}^{\mathrm{T}} \mathbf{W} - \mathbf{X}^{\mathrm{T}} \right\|_{\mathrm{F}}$, thus minimizing the crosstalk in the associative network. The pseudoinverse solution for the weight matrix is thus given as

$$\mathbf{W} = \left(\mathbf{X}^{\mathrm{T}}\right)^{\dagger} \mathbf{X}^{\mathrm{T}} \tag{4.22}$$

The pseudoinverse rule, also called the *projection learning rule*, is neither incremental nor local. It involves inverting an $N \times N$ matrix, thus the training is very slow and impractical. However, there are iterative procedures consisting of successive local corrections [308] and incremental procedures for the calculation of the pseudoinverse [432].

The pseudoinverse solution performs better than the Hebbian learning when the patterns are correlated. Both the Hebbian and pseudoinverse rules are general-purpose methods for training associative memory networks that can be represented as $\mathbf{X}^{\mathrm{T}}\mathbf{W} = \overline{\mathbf{X}}^{\mathrm{T}}$, where $\mathbf{X}$ and $\overline{\mathbf{X}}$ are, respectively, the stored and associated pattern matrices. For an autoassociated pattern $\mathbf{x}_i$, the weights generated from the Hebbian learning projects the whole input space into the linear subspace spanned by $\mathbf{x}_i$. The projection, however, is not orthogonal. Instead, the pseudoinverse solution provides orthogonal projection to the linear subspace spanned by the stored patterns [959]. Theoretically, for $N < J$ and uncorrelated patterns, the pseudoinverse solution has a zero error, and the storage capability in this case is $N_{\max} = J - 1$ [573, 959].

The pseudoinverse rule is also adapted to sorting sequences of prototypes, where an input $\mathbf{x}_i$ leads to an output $\mathbf{x}_{i+1}$. The MLP with BP learning can be used to compute the pseudoinverse solution when the dimension J is large, since direct methods to solve the pseudoinverse will use up the memory and the convergence time is intolerably large [573, 959].

Improved Hebbian Learning Rule

An alternative local and incremental learning rule based on the Hebbian rule is given by [1058, 1059]

$$w_{ij}(t) = w_{ij}(t-1) + \eta \left[x_{t,i} x_{t,j} - h_{ji}(t) x_{t,i} - h_{ij}(t) x_{t,j} \right] \tag{4.23}$$

$$h_{ij}(t) = \sum_{u=1, u\neq i,j}^{J} w_{iu}(t-1) x_{t,u} \tag{4.24}$$

where $\eta = \frac{1}{J}$, $t = 1, 2, \cdots, N$, $w_{ij}(0) = 0$ for all i and j, and h_{ij} is a form of local field at neuron i.

The improved Hebbian rule given by (4.23) and (4.24) has an absolute capacity of $\frac{J}{\sqrt{2 \ln J}}$ for uncorrelated patterns. It also performs better than the generalized Hebbian rule for correlated patterns [1058]. It does not suffer significant capacity loss when patterns with medium correlation are stored.

It is shown in [1059] that the Hebbian rule is the zeroth-order expansion of the pseudoinverse rule, and the improved Hebbian rule given by (4.23) and (4.24) is one form of the first-order expansion of the pseudoinverse rule.

Perceptron-type Learning Rule

The rules addressed above are *one-shot* methods, in which the network training is completed in a single epoch. A learning problem in a Hopfield network with J units can be transformed into a learning problem for a perceptron of dimension $\frac{J(J+1)}{2}$ [959]. This equivalence between Hopfield networks and perceptrons leads to the conclusion that every learning algorithm for perceptrons can be transformed into a learning algorithm for Hopfield networks.

Perceptron learning algorithms for storing bipolar patterns in Hopfield networks have been discussed in [573]. They are simple, online, local algorithms. Unlike Hebbian rule-based algorithms, the perceptron learning-based algorithms work over multiple epochs and often reduce the error nonmonotonically over the epochs. The perceptron-type learning rule is given by [538, 573]

$$w_{ij}(t) = w_{ij}(t-1) + \eta \left[x_{t,i} x_{t,j} - \frac{1}{2} \left(y_{t,i} x_{t,j} + y_{t,j} x_{t,i} \right) \right] \qquad (4.25)$$

where $\mathbf{y}_t = \text{sgn}(\mathbf{W}\mathbf{x}_t)$, $t = 1, \cdots, N$, the learning rate η is a small positive number, and the $w_{ji}(0)$s are small random numbers or zero[1]. If all the $w_{ji}(0)$ are selected as zero, η can be selected as any positive number; otherwise, η can be selected as a number of the same order of magnitude or larger than the weights. This accelerates the convergence process. Note that $w_{ji}(t) = w_{ij}(t)$.

However, when the signum vector is not realizable, the perceptron-type rule does not converge but oscillates indefinitely. The perceptron-type rule can be viewed as a supervised extension of the Hebbian rule by incorporating a term for correcting unstable bits. For an RNN, the storage capability of the perceptron-type algorithm can reach the upper bound $N_{\max} = J$, for uncorrelated patterns.

In [843], a complex perceptron learning algorithm has also been studied for associative memory by using complex weights and a decision circle in the complex plane for the output function.

Experimental Comparison: Perceptron-type Learning Rule vs. Generalized Hebbian Rule

An extensive experimental comparison between a perceptron-type learning rule [573] and the generalized Hebbian rule [502] has been made in [538] on a wide range of conditions on the library patterns: the number of patterns N, the pattern density p, and the amount of correlation of the bits in a pattern, decided by block size B.

During the experiment, the J bits of each pattern are divided into contiguous blocks, each of size B. All the bits in a block are given the same value: all 1 for a 1-block, or all -1 for a (-1)-block. p is the probability that an arbitrary block of an arbitrary generated pattern is a 1-block.

---

[1] In this chapter, the signum function $\text{sgn}(x)$ is defined as 1 for $x \geq 0$ and -1 for $x < 0$.

The results are evaluated on two criteria: stability of the library patterns and error-correction ability during the recall phase. The perceptron-type rule is found to be perfect in ensuring stability of the stored library patterns under all the evaluated conditions. The generalized Hebbian rule degrades rapidly as N is increased, or p is decreased, or B is increased. In many cases, the perceptron-type rule works much better than the generalized Hebbian rule in correcting pattern errors. The uniformly random case ($p = 0.5$, $B = 1$) is the main exception, when the generalized Hebbian rule systematically equals or outperforms the perceptron-type rule in the error-correction experiments.

Retrieval Stage

After the bipolar words have been stored, the network can be used for information retrieval. When a J-dimensional vector (bipolar word) $\mathbf{x}$, representing a corrupted or incomplete memory of the network, is presented to the network as its state, information retrieval is performed automatically according to the network dynamics given by (4.3) and (4.4), or (4.5) and (4.6). For hard-limiting activation function, the discrete form of the network dynamics can be written as

$$x_i(t+1) = \mathrm{sgn}\left(\sum_{j=1}^{J} w_{ji} x_j(t) + \theta_i \right) \tag{4.26}$$

for $i = 1, 2, \ldots, J$, or in matrix form

$$\mathbf{x}(t+1) = \mathrm{sgn}(\mathbf{W}\mathbf{x}(t) + \boldsymbol{\theta}) \tag{4.27}$$

where $\mathbf{x}(0)$ is the input corrupted memory, and $\mathbf{x}(t)$ represents the retrieved memory at time t. The retrieval process continues until the state vector $\mathbf{x}$ remains unchanged. The convergent $\mathbf{x}$ is a fixed point or the retrieved memory.

4.4.2 Storage Capability

In Subsect. 4.4.1, the storage capability for each of the four storage algorithms is given. In practice, there are some upper bounds on the storage capability of general RNNs.

Upper Bounds on Storage Capability

An upper bound on the storage capability of a class of RNNs with zero-diagonal weight matrix is derived deterministically in [8].

Theorem 4.1 (Upper Bound). *For any subset of N binary J-vectors, in order to find a corresponding zero-diagonal weight matrix $\mathbf{W}$ and a bias vector $\boldsymbol{\theta}$ such that these vectors are fixed points of the network*

$$\mathbf{x}_i = \mathrm{sgn}(\mathbf{W}\mathbf{x}_i + \boldsymbol{\theta}) \tag{4.28}$$

for $i = 1, 2, \cdots, N$, one needs to have $N \le J$.

Thus, the upper bound on the storage capability is $N_{\max} = J$. This bound is valid for any learning algorithm for RNNs with a zero-diagonal weight matrix. The Hopfield network, having a symmetric zero-diagonal weight matrix, is one such network, and as a result, the Hopfield network can at most stably store J patterns.

The upper bound introduced in Theorem 4.1 is too tight, since it requires that all the N-tuple subsets of bipolar J-vectors are retrievable. It is also noted in [1124] that any two datums differing in precisely one component cannot be jointly stored as stable states in the Hopfield network.

When permitting a small fraction ϵ of a set of N bipolar J-vectors irretrievable, the upper bound approximates $2J$ when $J \to \infty$. This is given by a theorem derived from the function counting theorem (Theorem 2.1) [267, 401, 1124, 573].

Theorem 4.2 (Asymptotical Upper Bound). *For N prototype vectors in general position, the storage capacity $N_{\max}$ can approach $2J$, in the sense that, for any $\epsilon > 0$ the probability of retrieving a fraction $(1 - \epsilon)$ of any set of $2J$ vectors tends to unity when $J \to \infty$.*

Here N prototype vectors in *general position* means that any subset of up to N vectors is linearly independent. Theorem 4.2 is more general than Theorem 4.1, since there is no constraint on **W**. This RNN is sometimes referred to as the *generalized Hopfield network*. Both the theorems hold true irrespective of the updating mode, be it synchronous or asynchronous.

The generalized Hopfield network with a general, zero-diagonal weight matrix has stable states in randomly asynchronous mode [741]. The asymptotic storage capacity of such a network using the perceptron learning scheme [963] has been analyzed in [742]. The perceptron learning rule with zero bias is used to compute the columns of **W** for each neuron independently, and as such the entire **W** is constructed. A lower and an upper bound of the asymptotic storage capacity are obtained as $J - 1$ and $2J$, respectively.

In a special case of the generalized Hopfield network with zero bias vector, some spectral strategies are used for constructing **W** [1124]. All the spectral storage algorithms have a storage capacity of J for uncorrelated patterns [1124]. A recursive implementation of the pseudoinverse spectral storage algorithm has also been given.

Other Developments

When the Hopfield network is used as associative memory, there are cases where the fundamental memories are not stable. In addition, spurious states, which are other stable states different from the fundamental memories and their negative counterparts, may arise [16, 959]. The Hopfield network trained with the generalized Hebbian rule can have a large number of spurious states, depending exponentially on N, the number of fundamental memories, even in the case when these vectors are orthogonal [137]. These spurious states are the corners of the unit hypercube that lie on or near the subspace spanned by the N fundamental memories. The presence of spurious states and limited storage capacity are the two major restrictions for the Hopfield network being used as associative memory. It has been proved that as long as N, the number of fundamental memories, is small compared to J, the number of neurons in the network, the fundamental memories are stable in a probabilistic sense [16].

The Gardner conditions [401] are often used as a measure of the stability of the patterns. Associative learning can be designed to enhance the basin of attraction for every pattern to be stored by optimizing these conditions. The Gardner algorithm [401] combines maximal storage with a predefined level of stability for the patterns. Based on the Gardner conditions, the inverse Hebbian rule [260, 259] is given by

$$w_{ij} = - \left(\mathbf{C}^{-1} \right)_{ij} \tag{4.29}$$

$$\mathbf{C} = \frac{1}{N} \sum_{p=1}^{N} \mathbf{x}_p \mathbf{x}_p^{\mathrm{T}} \tag{4.30}$$

Unlike the generalized Hebbian rule, which can only store unbiased patterns, the inverse-Hebbian rule is capable of storing N patterns, biased or unbiased, in a Hopfield network of N neurons. The patterns have zero basins of attraction, and the correlation matrix $\mathbf{C}$ must be nonsingular. Matrix inversion can be implemented using a local learning algorithm [540]. The inverse-Hebbian rule provides ideal initial conditions for any algorithm capable of increasing the pattern stability. The quadratic Oja algorithm, as a generalization of Oja's algorithm [864], can be applied to adapt the weights so as to enhance the size of the basin of attraction of a subset of the stored patterns until the storage of the pattern subset is compromised. Unfortunately the quadratic Oja algorithm is nonlocal.

In [562], by using the Gardner algorithm for training the weights [401] and using a nonmonotonic activation function

$$\phi(net) = \begin{cases} +1, & net < -b_1 \text{ or } 0 < net < b_1 \\ -1, & net > b_1 \text{ or } -b_1 < net < 0 \end{cases} \tag{4.31}$$

the storage capacity of the network can be made to be always larger than $2J$ and reach its maximum value of $10.5J$ for $b_1 = 1.22$. In [822, 1237], a continuous nonmonotonic activation function is used to improve the performance of the Hopfield network. The exact form of the nonmonotonic activation and its parameters are not very critical. The storage capacity of the Hopfield network can be greatly improved to approximately $0.4J$, and spurious states can be totally eliminated. When it fails to recall a memory, a chaotic behavior will occur.

The eigenstructure learning rule [694] is developed for continuous-time Hopfield models in linear saturated mode. The design method allows linear combinations of the prototype vectors to be stored as asymptotically stable equilibrium points as well. The storage capacity is better than those of the pseudoinverse solution [909] and the generalized Hebbian rule [502]. All the desired patterns are guaranteed to be stored as asymptotically stable equilibrium points. Guidelines for reducing the number of spurious states and for estimating the extent of the domains of attraction for the stored patterns are also provided. The method has been extended to discrete-time neural networks in [803].

A quantum computational learning algorithm, which is a combination of quantum computation with the Hopfield network, has been developed in [1125]. The quantum associative memory has a capacity that is exponential in the number of neurons, namely, offering a storage capacity of $O\left(2^J\right)$. It employs simple spin-1/2 (two-state) quantum systems and represents patterns as quantum operators.

4.4.3 Multilayer Perceptrons as Associative Memories

Most RNN-based associative memories have low storage capacity as well as poor retrieval ability. RNNs exhibit asymptotic behavior and as such are difficult to analyze. MLP-based autoassociative memories with equal numbers of input and output nodes have been introduced to overcome these limitations [228, 1196].

Recurrent Correlation Associative Memory

The recurrent correlation associative memory (RCAM) uses a J-N-J MLP-based recurrent architecture [228], as shown in Fig. 4.5. At each time instant, the hidden layer computes an intermediate mapping, while the output layer completes an association of the input pattern to an approximate prototype pattern. The approximated pattern is fed back to the network and the process continues until convergence to a prototype is achieved. The activation function for the ith neuron in the hidden layer is $\phi_i(\cdot)$, and the activation function at the output layer is the signum function.

The matrix $\mathbf{W}^{(1)}$, a $J \times N$ matrix, is made up of the N J-bit bipolar memory patterns $\mathbf{x}_i, i = 1, 2, \cdots, N$, that is, $\mathbf{W}^{(1)} = [\mathbf{x}_1, \mathbf{x}_2, \cdots, \mathbf{x}_N]$. The matrix $\mathbf{W}^{(2)}$, an $N \times J$ matrix, is the transpose of $\mathbf{W}^{(1)}$, namely, $\mathbf{W}^{(2)} = \left[\mathbf{W}^{(1)}\right]^T$. At the presentation of pattern $\mathbf{x}$, the net input to neuron j in the hidden layer is $net_j^{(2)} = \mathbf{x}_j^T \mathbf{x}$. The network evolution is

$$\mathbf{x}(t+1) = \mathrm{sgn}\left(\sum_{j=1}^{N} \phi_j\left(\mathbf{x}_j^T \mathbf{x}(t)\right) \cdot \mathbf{x}_j\right) \qquad (4.32)$$

The correlation of two patterns, $\mathbf{x}_1^T \mathbf{x}_2 = J - 2d_{\mathrm{H}}\left(\mathbf{x}_1, \mathbf{x}_2\right)$, where $d_{\mathrm{H}}(\cdot)$ is the Hamming distance between two binary vectors within. The Hamming distance is the number of bits in the two vectors that do not match each other.

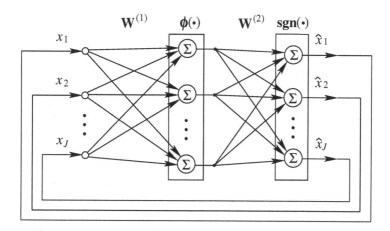

Fig. 4.5. Architecture of the J-N-J MLP-based RCAM. The ith entry of $\phi(\cdot)$, $\phi_i(\cdot)$, corresponds to the activation function at the ith neuron in the hidden layer, and $\mathbf{sgn}(\cdot)$ represents that the activation function of each neuron in the output layer is the signum function. Note that the number of hidden units is taken as N, the number of stored patterns.

In the case when all the $\phi_i(net) = \phi(net)$, $\phi(net)$ being any continuous, monotonic nondecreasing weighting function over $[-J, J]$, the RCAM (4.32) is proved to be asymptotically stable in both the synchronous and asynchronous update modes [228]. Based on this, a family of RCAMs have been proposed that possess the asympototical stability. This property is especially suitable for hardware implementation, since there are faults in the manufacture of any physical device.

When all the $\phi_i(net) = net$, the RCAM model is equivalent to the correlation-matrix associative memory [609, 34], that is, the connection corresponding to the case of the Hopfield network can be written as $\mathbf{W} = \sum_{p=1}^{N} \mathbf{x}_p \mathbf{x}_p^{\mathrm{T}}$. By suitably selecting $\phi_i(\cdot)$, the model is reduced to some existing associative memories, which have a storage capacity that grows polynomially or exponentially with J [228].

In particular, when all the $\phi_i(net) = a^{net}$ with radix $a > 1$, an exponential correlation associative memory model (ECAM) [228] is obtained. The exponential activation function stretches the ratios among the weights and makes the largest weight more overwhelming. This significantly increases the storage capacity. The ECAM model exhibits an asymptotic storage capacity that scales exponentially with J [228]. Under the noise-free condition, this storage capacity is 2^J patterns [228]. A VLSI chip for the ECAM model has been fabricated and tested [228]. The multivalued RCAM [229] can increase the error-correction capability with large storage capability and less interconnection complexity.

Discussion

The local identical index (LII) model [1196] is an autoassociative memory model that uses the J-N-J MLP architecture. The weight matrices $\mathbf{W}^{(1)}$ and $\mathbf{W}^{(2)}$ are the same as those defined in the RCAM model. It utilizes the signum activation function and biases in both the hidden and output layers. The LII model utilizes the local characteristics of the fundamental memories through two metrics, namely, the global identical index (GII) and the LII. Based on the minimum Hamming distance as the underlying association principle, the scheme can be viewed as an approximate Hamming decoder. The LII model exhibits low structural as well as operational complexity. It is a one-shot associative memory, and can accommodate up to 2^J prototype patterns. The LII model outperforms the LSSM [694] and its discrete version [803] in recognition accuracy at the presentation of the corrupted patterns, controlled by using the Hamming distance. It can successfully associate input patterns that are even loosely correlated with the corresponding prototype pattern.

For a J-J_2-J MLP-based autoassociative memory, the hidden layer is a bottleneck layer with fewer nodes, $J_2 < J$. This bottleneck layer is used to discover a limited set of unique prototypes that cluster the training set. When a linear activation function is employed, the MLP-based autoassociative memory has a serious limitation, namely, it does not allow the user to control the granularity of the clusters formed. Due to the lack of cluster-competition mechanism, different clusters that are close to one another in the input space may merge. This problem can be overcome by using the sigmoidal activation function.

4.4.4 The Hamming Network

The Hamming network [717] is a straightforward associative memory. It calculates the Hamming distance between the input pattern and each memory pattern, and selects the memory with the smallest Hamming distance. The network output is the index of a prototype pattern and thus the network can be used as a pattern classifier. The Hamming network is used as the classical Hamming decoder or Hamming associative memory. It provides the minimum-Hamming-distance solution.

The Hamming network has a J-N-N layered architecture, as illustrated in Fig. 4.6. The third layer is called the *memory layer*, each of whose neurons corresponds to a prototype pattern. The input and hidden layers are feedforward, fully connected, while each hidden node has a feedforward connection to its corresponding node in the memory layer. Neurons in the memory layer are fully interconnected, and form a competitive subnetwork known as the *MAXNET*. The MAXNET responds to an input pattern by generating a winner neuron through iterative competitions. The Hamming network is implicitly recurrent due to the interconnections in the memory layer.

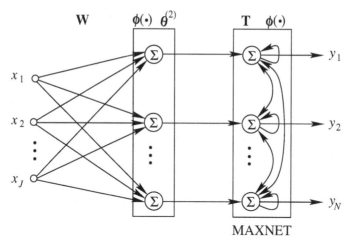

MAXNET

Fig. 4.6. Architecture of the J-N-N Hamming network. The activation functions at all the units in each layer, denoted by vector ϕ, are the signum function, and $\boldsymbol{\theta}^{(2)}$ is a vector comprising the biases for all the neurons at the hidden layer. The number of neurons in the memory layer, N, corresponds to the number of stored patterns. $\mathbf{T} = [t_{kl}]$, $k, l = 1, \cdots, N$.

The second layer generates matching scores that are equal to J minus the Hamming distances to the stored patterns, that is, $J - d_{\mathrm{H}}(\mathbf{x}, \mathbf{x}_i)$, $i = 1, \cdots, N$, for pattern $\mathbf{x}$. These matching scores range from 0 to J. The unit with the highest matching score corresponds to the stored pattern that best matches the input. The weights between the input and hidden layers and the biases of the hidden layer are, respectively, set as

$$w_{ij} = \frac{x_{j,i}}{2}, \qquad \theta_j^{(2)} = \frac{J}{2} \qquad (4.33)$$

for $j = 1, \cdots, N$, $i = 1, \cdots, J$.

All the thresholds and the weights t_{kl} in the MAXNET are fixed. The thresholds are set as zero. The weights from each node to itself are set as unity and weights between nodes are inhibitory, that is

$$t_{kl} = \begin{cases} 1, & k = l \\ -\varepsilon, & k \neq l \end{cases} \qquad (4.34)$$

where $\varepsilon < \frac{1}{N}$.

When a binary pattern is presented to the network, the network first generates an initial input for the MAXNET

$$y_j(0) = \phi \left(\sum_{i=1}^{N} w_{ij} x_i - \theta_j^{(2)} \right) \qquad (4.35)$$

for $j = 1, \cdots, N$, where $\phi(\cdot)$ is the threshold-logic nonlinear function.

The input pattern is then removed and the MAXNET continues the iteration

$$y_j(t+1) = \phi\left(\sum_{k=1}^{N} t_{kj}y_k(t)\right) = \phi\left(y_j(t) - \varepsilon \sum_{k=1,k\neq j}^{N} y_k(t)\right) \qquad (4.36)$$

for $j = 1, \cdots, N$, until the output of only one node is positive. This node corresponds to the selected class.

The Hamming network implements the minimum error classifier, when the bit errors are random and independent. For the J-N-N Hamming network, there are $J \times N + N^2$ connections, while for the Hopfield network the number of connections is J^2. When $J \gg N$, the number of connections in the Hamming network is significantly less than that in the Hopfield network. In addition, the Hamming network offers a storage capacity that is exponential in the input dimension [463], and it does not have any spurious state that corresponds to a no-match result. Under the noise-free condition, the Hamming network has a storage capacity of 2^J patterns [463]. For a sufficiently large but finite radix a, the ECAM operates as a Hamming associative memory [463].

However, the Hamming network suffers from difficulties in hardware implementation and low retrieval speed. Based on the correspondence between the Hamming network and the ECAM [463], the ECAM can be used to compute the minimum Hamming distance, in a distributed fashion by analog exponentiation and thresholding devices. The two-level Hamming network [531] generalizes the Hamming memory by providing for local Hamming distance computations in the first level and a voting mechanism in the second level. It allows for a much more practical hardware implementation and a faster retrieval.

4.5 Simulated Annealing

Annealing is referred to as tempering certain alloys of metal by heating and then gradually cooling them. Thus, the simulation of this process is SA. A metal is first heated above its melting point and then cooled slowly until it solidifies into a perfect crystalline structure. The defect-free crystal state corresponds to the global minimum energy configuration. The Metropolis algorithm is a simple method for simulating the evolution to the thermal equilibrium of a solid for a given temperature [797]. The SA algorithm [607] is a variant of the Metropolis algorithm, where the temperature is changing from high to low. SA is a successful global optimization method for many important classes of problems such as in the layout of integrated circuits [984].

4.5.1 Classic Simulated Annealing

SA is a general, serial algorithm for finding a global minimum for a continuous function [607]. The solutions by this technique are close to the global minimum within a polynomial upper bound for the computational time and are independent of the initial conditions. Some parallel algorithms for SA have been proposed aiming to improve the accuracy of the solutions by applying parallelism [274].

According to statistical thermodynamics, P_α, the probability of a physical system being in state α with energy E_α at temperature T satisfies the Boltzmann distribution[2]

$$P_\alpha = \frac{1}{Z} e^{\frac{-E_\alpha}{k_B T}} \qquad (4.37)$$

where k_B is the Boltzmann's constant, T is the absolute temperature, and Z is the partition function, defined by

$$Z = \sum_\beta e^{-\frac{E_\beta}{k_B T}} \qquad (4.38)$$

the summation being taken over all states β with energy E_β at temperature T. At high T, the Boltzmann distribution exhibits uniform preference for all the states, regardless of the energy. When T approaches zero, only the states with minimum energy have nonzero probability of occurrence.

In SA, we omit the constant k_B. At high T, the system ignores small changes in the energy and approaches thermal equilibrium rapidly, that is, it performs a coarse search of the space of global states and finds a good minimum. As T is lowered, the system responds to small changes in the energy, and performs a fine search in the neighborhood of the already determined minimum and finds a better minimum. At $T = 0$, any change in the system states does not lead to an increase in the energy, and thus, the system must reach equilibrium if $T = 0$.

When performing the SA, theoretically a global minimum is guaranteed to be reached with a high probability. The artificial thermal noise is gradually decreased in time. T is a control parameter called the *computational temperature*, which controls the magnitude of the perturbations of the energy function $E(\mathbf{x})$. The probability of a state change is determined by the Boltzmann distributions of the energy difference of the two states

$$P = e^{-\frac{\Delta E}{T}} \qquad (4.39)$$

The probability of uphill moves in the energy function ($\Delta E > 0$) is large at a high T, and is low at a low T. SA allows uphill moves in a controlled fashion: It attempts to improve on greedy local search by occasionally taking a risk and accepting a worse solution. SA can be performed as given by Algorithm 4.5.1 [607, 247].

The classical SA procedure is known as *Boltzmann annealing*. The cooling schedule for T is critical to the efficiency of SA. If T is reduced too rapidly, a premature convergence to a local minimum may occur. In contrast, if it is

---

[2] Also known as the Boltzmann–Gibbs distribution.

Algorithm 4.5.1 (SA)

1. Initialize the system configuration. Randomize $\mathbf{x}(0)$.
2. Initialize T with a large value.
3. Repeat:
   a) Repeat:
      i. Apply random perturbations to the output state
         of neurons $\mathbf{x} = \mathbf{x} + \Delta\mathbf{x}$.
      ii. Evaluate $\Delta E(\mathbf{x}) = E(\mathbf{x} + \Delta\mathbf{x}) - E(\mathbf{x})$:
         A. If $\Delta E(\mathbf{x}) < 0$, keep the new state;
         B. Otherwise, accept the new state with the
            probability $P = e^{-\frac{\Delta E}{T}}$.
      until the number of accepted transitions becomes below a threshold
      level.
   b) Set $T = T - \Delta T$.
   until T is small enough.

---

too slow, the algorithm is very slow to converge. Based on a Markov-chain analysis on the SA process, Geman and Geman [404] have proved that T must be decreased according to

$$T(t) \geq \frac{T_0}{\ln(1+t)}, \qquad t = 1, 2, \ldots \qquad (4.40)$$

to ensure convergence to the global minimum with probability one, where T_0 is the initial temperature. In other words, in order to guarantee the Boltzmann annealing to converge to the global minimum with probability one, $T(t)$ is needed to decrease logarithmically with time. This is practically too slow. In practice, one usually applies, in Step 3b, a fast schedule $T(t) = \alpha T(t-1)$ with $0.85 \leq \alpha \leq 0.96$, to achieve a suboptimal solution.

4.5.2 Variants of Simulated Annealing

The classical SA procedure is a stochastic search method, and the convergence to the global optimum is too slow for a reliable cooling schedule. Many methods, such as Cauchy annealing [1071], simulated reannealing [532], generalized SA [1101], and the SA with known global value [732], have been proposed to accelerate the SA search. There are also global optimization methods that make use of the idea of annealing [948, 960]. Some VLSI designs of SA are also available [659].

In Cauchy annealing [1071], the Cauchy distribution[3] is used to replace the Boltzmann distribution. The infinite variance provides a better ability to

---

[3] Also known as the Cauchy–Lorentz distribution.

escape from local minima and allows for the use of faster schedules, such as T decreasing according to

$$T(t) = \frac{T_0}{t} \tag{4.41}$$

A stochastic neural network (SNN) trained with Cauchy annealing is also called the *Cauchy machine*.

In simulated reannealing [532], T decreases exponentially with t, that is,

$$T = T_0 \mathrm{e}^{-\frac{c_1 t}{J}} \tag{4.42}$$

where $c_1 > 0$ is a constant, and J is the dimension of the input space. The introduction of reannealing also permits adaptation to changing insensitivities in the multidimensional parameter space. The generalized SA [1101] generalizes both Cauchy annealing [1071] and Boltzmann annealing [607] within a unified framework inspired by the generalized thermostatistics.

An SA algorithm under the simplifying assumption of known global value has been investigated in [732]. The algorithm is the same as Algorithm 4.5.1 except that at each iteration a uniform random point is generated over a sphere, whose radius depends on the difference between the current function value $E(\mathbf{x}(t))$ and the optimal value E^*, namely, $E(\mathbf{x}(t)) - E^*$, and T is also decided by this difference. The algorithm has guaranteed convergence and an upper bound for the expected first hitting time, *i.e.* the expected number of iterations before reaching the global optimum value within a given accuracy ε, is established.

The idea of annealing is a general optimization principle, which can be extended by using fuzzy logic. In the fuzzy annealing scheme [948], fuzzification is performed by adding an entropy term. The fuzziness at the beginning of the entire procedure is used to prevent the optimization process getting stuck at an inferior local optimum. The fuzziness is reduced step by step. The fuzzy annealing scheme results in an increase in the computation speed by a factor of one hundred or more compared to the SA [948].

SA makes a random search on the energy surface. Deterministic annealing [961, 960] is a method where randomness is incorporated into the energy or cost function, which is then deterministically optimized at a sequence of decreasing temperature. The approach is derived within the framework of information theory and probability theory. Deterministic annealing has been used for nonconvex optimization problems such as clustering, MLP training, and RBFN training [961, 960].

4.6 Combinatorial Optimization Problems

Any problem that has a large set of discrete solutions and a cost function for rating those solutions relative to one another is a COP. COPs are known to be NP-complete[4] [1069, 1087]. In COPs, the number of solutions grows exponentially with n, the size of the problem, at $O(n!)$ or $O(e^n)$ so that no algorithm can find the global minimum solution in a polynomial computational time. The goal for COPs is to find an optimal solution or sometimes a nearly optimal solution. An exhaustive search of all the possible solutions for the optimum is impractical.

The TSP is perhaps the most famous COP [502]: Given a set of points, either nodes on a graph or cities on a map, find the shortest possible tour that visits every point exactly once and then returns to its starting point. There are $(n-1)!/2$ possible tours for an n-city TSP. The Hopfield network was the first neural network used for the TSP, and it achieves a near-optimum solution [504]. Routing of wires on a printed circuit board (PCB) is a typical TSP.

The location-allocation problem is another example of COPs: Given a set of facilities, each of which serves a certain number of nodes on a graph, the objective is to place the facilities on the graph so that the average distance between each node and its serving facility is minimized.

4.6.1 Formulation of Combinatorial Optimization Problems

The Hopfield network can be effectively used to deal with COPs with the objective functions of the linear or quadratic form, linear equalities and/or inequalities as the constraints, and binary variable values so that the constructed energy function can be of quadratic form. For example, a class of COPs including the location-allocation problem can be formulated as [777]

$$\min \sum_i \sum_j c_{ij} v_{ij} \tag{4.43}$$

subject to

$$\sum_j a_{ij} v_{ij} = b_i \tag{4.44}$$

$$\sum_i v_{ij} = 1 \tag{4.45}$$

$$v_{ij} \in \{0, 1\} \tag{4.46}$$

for $i = 1, \cdots, m$, $j = 1, \cdots, n$, where $\mathbf{V} = [v_{ij}] \in \{0,1\}^{m \times n}$ is a variable matrix, $c_{ij} \geq 1$, $a_{ij} \geq 1$, and $b_i \geq 1$, are constant integers.

To make use of the Hopfield network, one needs first to convert the COP into a constrained real optimization problem and solve the latter using the penalty method. The COP defined by (4.43) through (4.46) can be transformed into the minimization of the following total cost

---

[4] Namely, nondeterministic polynomial-time complete.

$$E = \frac{A_1}{2} \sum_i \left(\sum_j a_{ij} v_{ij} - b_i \right)^2 + \frac{A_2}{2} \sum_j \left(\sum_i v_{ij} - 1 \right)^2$$

$$+ \frac{B}{2} \sum_i \sum_j v_{ij} \left(1 - v_{ij} \right) + \frac{C}{2} \sum_i \sum_j c_{ij} v_{ij} \qquad (4.47)$$

When the first three terms are all zeros, the solution is a feasible one. The constants A_1, A_2, B, and C are weights of individual constraints, which can be tuned for an optimal or good solution. The cost E given by (4.47) has the same form as that of the energy function of the Hopfield network, and thus can be solved by using the Hopfield network.

For the penalty method, there is always a compromise between good-quality solution and convergence. For a feasible solution, the weighting factors for the penalty terms should be sufficiently large, which however causes the constraints on the original problem to become relatively weaker, resulting in a deterioration of the quality of the solution. A trial-and-error process for choosing some of the penalty parameters is inevitable in order to obtain feasible solutions. Moreover, the gradient-descent method often leads to a local minimum of the energy landscape.

By minimizing the square of the cost function (4.43), the network distinguishes optimal solutions more sharply than with (4.47) and this greatly overcomes many of the weaknesses of the network with (4.47) [777]. When the constraint (4.44) is an inequality constraint, it can be expressed as an equality constraint by introducing some slack variables (slack neurons).

4.6.2 Escaping Local Minima for Combinatorial Optimization Problems

The SA, as a general-purpose method for searching the global optimum [607], is a popular method for any optimization problem including COPs. However, due to its Monte Carlo nature, the SA would require even more iterations than complete enumeration, for some problems, in order to guarantee convergence to an exact solution. For example, for an n-city TSP, the SA using the logarithmic cooling schedule needs a computational complexity of $O\left(n^{n^{2n-1}} \right)$, which is far more than $O((n-1)!)$ for complete enumeration and $O\left(n^2 2^n \right)$ for dynamic programming [201, 1]. Thus, one has to apply heuristic fast cooling schedules to improve the convergence speed.

The Hopfield network is more desirable for solving COPs that can be formulated into quadratic functions. The Hopfield network converges very fast, and it can also be easily implemented using RC circuits. However, due to its gradient-descent nature, it always gets trapped at the nearest local minimum of the initial random state. Some strategies are necessary for escaping from the local minima.

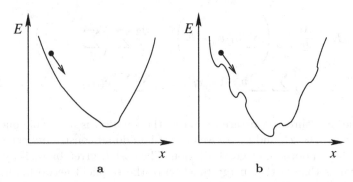

Fig. 4.7. Schematics of the landscapes of the energy function with one-dimensional variable x and different values of gain β. (**a**) Low β smoothes the surface. (**b**) High β reveals more details in the surface.

Gain Annealing

To help the Hopfield network escape from the local minima, a popular strategy is to change the sigmoidal gain β, by starting from a low gain and gradually increasing it. When β is low, the energy landscape is smooth, and the algorithm can easily find a good local minimum. As β increases, more details of the energy landscape are revealed, and the algorithm can find a better solution. This is illustrated in Fig. 4.7. This is analogous to the cooling process of the SA [607], and this process is usually called *gain annealing*. In the limit, when $\beta \to \infty$, the hypobolic tangent function becomes the signum function.

Balancing Objective and Constraint Terms

In order to use the Hopfield network for solving optimization problems, the cost function can be constructed as

$$E(\mathbf{x}) = \lambda_o E_o(\mathbf{x}) + \lambda_c E_c(\mathbf{x}) \tag{4.48}$$

where E_o and E_c represent the objective and constraint terms, respectively, and λ_o, $\lambda_c > 0$ are their corresponding weights.

By adaptively adjusting the balance between the constraint and objective terms, the network can avoid falling into a local minimum and continue to update in a gradient-descent direction of energy [1154].

At a local minimum of E, the following relations are always satisfied

$$\frac{\partial E}{\partial x_i} \Delta x_i = \left(\lambda_o \frac{\partial E_o}{\partial x_i} + \lambda_c \frac{\partial E_c}{\partial x_i} \right) \Delta x_i \geq 0 \tag{4.49}$$

for $i = 1, \cdots, J$. To escape from the local minimum, one can adjust λ_o and/or λ_c so that (4.49) is not satisfied for at least one neuron. A learning rule is given by Algorithm 4.6.2 [1154].

Algorithm 4.6.2 (Escaping Local Minima)

1. Select δ as a small positive constant.
2. At the local minimum, select a neuron that satisfies

$$\frac{\partial E_o}{\partial x_i} \frac{\partial E_c}{\partial x_i} < 0 \qquad (4.50)$$

3. Adjust λ_o and λ_c until $\frac{\partial E}{\partial x_i} \Delta x_i < 0$.
   The procedure is given as
   a) if

$$\frac{\partial E_o}{\partial x_i} \Delta x_i < 0 \quad \text{and} \quad \frac{\partial E_c}{\partial x_i} \Delta x_i > 0 \qquad (4.51)$$

   modify λ_o according to

$$\lambda_o = -\lambda_c \frac{\frac{\partial E_c}{\partial x_i}}{\frac{\partial E_o}{\partial x_i}} + \delta \qquad (4.52)$$

   b) else if

$$\frac{\partial E_o}{\partial x_i} \Delta x_i > 0 \quad \text{and} \quad \frac{\partial E_c}{\partial x_i} \Delta x_i < 0 \qquad (4.53)$$

   modify λ_c according to

$$\lambda_c = -\lambda_o \frac{\frac{\partial E_o}{\partial x_i}}{\frac{\partial E_c}{\partial x_i}} + \delta \qquad (4.54)$$

In Algorithm 4.6.2, δ is used to control the learning speed, and the updated λ_o and λ_c are always positive. The energy of the network decreases with the state change of the ith neuron. Thus, the learning eliminates the local minimum that the network would fall into. The minimum found is not only a minimum of the total energy function, but also the minima of both the constraint and objective terms. This minimum is always a global or a near-global one. The method is capable of finding an optimal or near-optimal solution in a short time for the TSP.

4.6.3 Combinatorial Optimization Problems with Equality and Inequality Constraints

The Hopfield network can be used to solve COPs under equality as well as inequality constraints, as long as the constructed energy function is of the form of (4.7). Constraints are treated by introducing in the objective function some additional energy terms that penalize any infeasible state. Some extensions to the Hopfield model are necessary in order to handle both equality and inequality constraints [1083, 2].

Assume that we have both linear equality and inequality constraints

$$\mathbf{r}_i^{\mathrm{T}} \mathbf{x} = s_i \qquad (4.55)$$

$$\mathbf{q}_j^{\mathrm{T}} \mathbf{x} \leq h_j \quad \text{or} \quad \mathbf{q}_j^{\mathrm{T}} \mathbf{x} \geq h_j \qquad (4.56)$$

for $i = 1, \cdots, l, j = 1, \cdots, k$, where $\mathbf{r}_i = (r_{i,1}, \cdots, r_{i,J})^{\mathrm{T}}$, $\mathbf{q}_j = (q_{j,1}, \cdots, q_{j,J})^{\mathrm{T}}$, s_i is a constant, and $h_j > 0$.

In the extended Hopfield model [2], each inequality constraint is converted to an equality constraint by introducing an additional variable managed by a new neuron, known as the *slack neuron*. Each slack neuron is connected to the initial neurons, where their corresponding variables occur in its linear combination. The extended Hopfield model has the drawback of being frequently stabilized in neuron states far from the suitable ones, *i.e.* zero and one. To deal with this drawback, a new penalty energy term is derived to significantly reduce the number of neurons with unsuitable states [674]. h_j is extended to $h_j \geq 0$. The derived rules introduce competitions between the variables involved into the same constraint. The competitive mechanism also deals with the upper bounded inequality constraints. This mechanism has the capacity to distribute the neurons into the two states.

The k-out-of-n design rule [1074] is used to facilitate the construction of the network energy functions for multiple k-out-of-n equality constraints, $\sum_{i=1}^{n} x_i = k$, and inequality constraints, $\sum_{i=1}^{n} x_i \leq k$. For k-out-of-n inequality constraints, slack neurons are used. A generalized architecture for the Hopfield network with k-out-of-n design is achieved by adding to the original J neurons one adjustable neuron associated with each given constraint [701] so as to improve the quality of the solutions. This architecture also applies when slack neurons are used for inequality constraints.

4.7 Chaotic Neural Networks

An RNN such as the Hopfield network, when introduced with chaotic dynamics, is sometimes called a *chaotic neural network*. The chaotic dynamics are temporarily generated for searching and self-organizing, and eventually vanish with the autonomous decrease of a bifurcation parameter corresponding to the temperature in the SA process. Thus, the chaotic neural network gradually approaches to a dynamical structure of the RNN.

Since the operation of the chaotic neural network is similar to that of the SA, not in a stochastic way but in a deterministically chaotic way, the operation is known as *chaotic SA (CSA)*. More specifically, the transiently chaotic dynamics are used for searching a basin containing the global optimum, followed by a stable and convergent phase when the chaotic noise decreases to zero. As a result, the chaotic neural network has a high ability for searching globally optimal or near-optimal solutions [201]. The CSA is a good alternative approach to the SA.

The SA, employing the Monte Carlo scheme, searches all the possible states by temporally changing the probability distributions. In contrast, the

CSA searches a possible fractal subspace with continuous states by temporally changing invariant measures that are determined by its dynamics. Thus, the search region in the CSA is very small compared with the state space, and the CSA can perform an efficient search.

A small amount of chaotic noise can be injected into the output of the neurons and/or to the weights during the operation of the Hopfield network. In [472], a chaotic neural network is obtained by adding chaotic noise to each neuron of the discrete-time continuous-output Hopfield network and gradually reducing the noise so that it is initially chaotic, but eventually convergent.

The chaotic neural network introduced in [201, 14] is obtained by adding a negative self-coupling to the Hopfield network. By gradually removing the self-coupling, the transient chaos is used for searching and self-organizing. The updating rule for the chaotic neural network is given by [201]

$$net_i(t+1) = \left(1 - \frac{\alpha_i}{\tau_i}\right) net_i(t) + \frac{1}{\tau_i}\left(\mathbf{w}_i^{\mathrm{T}}\mathbf{x} + \theta_i\right) - c(t)(x_i - \gamma) \qquad (4.57)$$

$$x_i(t) = \phi(net_i(t)) \qquad (4.58)$$

where $c(t+1) = \beta c(t)$, $\beta \in [0,1]$, the bias $\gamma > 0$, and other parameters are the same as for (4.13). A large initial value of $c(t)$ is used so that the self-coupling is strong enough to generate chaotic dynamics for searching the global minima. The damping of $c(t)$ produces successive bifurcations so that the neurodynamics eventually converge from strange attractors to a stable equilibrium point.

The CSA approach is derived by varying the time step Δt of an Euler discretization of the Hopfield network described by (4.13) [1148]

$$net_i(t+\Delta t) = net_i(t)\left(1 - \frac{\alpha_i \Delta t}{\tau_i}\right) + \frac{\Delta t}{\tau_i}\left(\mathbf{w}_i^{\mathrm{T}}\mathbf{x} + \theta_i\right) \qquad (4.59)$$

The time step is analogous to the temperature parameter in the SA, and the method starts with a large Δt, where the dynamics are chaotic, and gradually decreases it. When $\Delta t \to 0$, the system approaches the Hopfield model (4.13) and minimizes its energy function. When $\Delta t = 1$, the Euler-discretized Hopfield network is identical to the chaotic neural network given in [201]. The simulation results for COPs are comparable to that of the method proposed in [201].

Many chaotic approaches [201, 1148, 472] can be unified and compared under the framework of adding an extra energy term E_{CSA} into the original computational energy (4.16) of the Hopfield model [646]. The extra energy term modifies the original Hopfield energy landscape to accommodate transient chaos. For example, for the CSA proposed in [201], E_{CSA} can be selected as

$$E_{\mathrm{CSA}} = \frac{c(t)}{2}\sum_i x_i(x_i - 1) \qquad (4.60)$$

This results in many logistic maps being added to the Hopfield energy function. E_{CSA} is convex, and hence drives $\mathbf{x}$ toward the interior of the hypercube. This driving force is diminished as $E_{\mathrm{CSA}} \rightarrow 0$ when $\lambda(t) \rightarrow 0$.

The CSA has a better search ability for solving COPs compared to the SA. However, a number of network parameters must be subtly adjusted so as to guarantee the convergence of the chaotic network. Unlike the SA, the CSA may not find a globally optimal solution no matter how slowly the annealing is carried out, because the chaotic dynamics are completely deterministic. Stochastic CSA [1147] has been proposed as a combination of the SA and the CSA [201] by using a noisy chaotic neural network, which is obtained by adding decaying stochastic noise into the chaotic neural network proposed in [201]. Stochastic CSA restricts the random search to a subspace of chaotic attracting sets, and this subspace is much smaller than the entire state space searched by the SA. Simulation results show that the stochastic CSA performs more efficiently than the SA and CSA [201] for the TSP and the channel-assignment problem.

4.8 Hopfield Networks for Other Optimization and Signal-processing Problems

The LS problem is a typical method for optimization and in signal processing. Matrix inversion can be performed using the Hopfield network [540]. Given a nonsingular $n \times n$ matrix $\mathbf{A}$, the energy function can be defined by $\|\mathbf{AV} - \mathbf{I}\|_{\mathrm{F}}^2$, where $\mathbf{V}$ denotes the inverse of $\mathbf{A}$ and the subscript F denotes the Frobenius norm. This energy function can be decomposed into n energy functions, and n similar networks are required, each optimizing an energy function. This method can be used to solve a system of n linear equations with n variables, $\mathbf{Ax} = \mathbf{b}$, where $\mathbf{A} \in R^{n \times n}$ and $\mathbf{x}, \mathbf{b} \in R^n$, if the SLE has a unique solution, that is, $\mathbf{A}$ is nonsingular. In [183], this SLE is solved by using a continuous Hopfield network with n nodes. The Hopfield network is designed to minimize the energy function $E = \frac{1}{2}\|\mathbf{Ax} - \mathbf{b}\|^2$, and the activation function is selected as a linear transfer function. This method is also applicable when there exists infinitely many solutions and $\mathbf{A}$ is singular. Another neural LS estimator that uses continuous Hopfield network and a nonlinear activation function has been proposed in [397].

A Hopfield network with linear transfer functions augmented by an additional feedforward layer can be used to solve an SLE [1144] and to compute the pseudoinverse[5] of a matrix [682]. The resultant augmented linear Hopfield network can be used to solve constrained LS optimization problems.

The LP network [1083] is designed based on the Hopfield model for solving LP problems

$$\min \mathbf{a}^{\mathrm{T}} \mathbf{x} \tag{4.61}$$

---

[5] Also known as the Moore–Penrose generalized inverse.

subject to

$$\mathbf{d}_j^{\mathrm{T}} \mathbf{x} \geq h_j \qquad (4.62)$$

for $j = 1, \cdots, M$, where $\mathbf{d}_j = (d_{j,1}, d_{j,2}, \cdots, d_{j,J})^{\mathrm{T}}$, $\mathbf{D} = [d_{j,i}]$ is a $M \times J$ matrix, and h_j is a constant. Each inequality constraint is modeled by a slack neuron. The network contains a signal plane with J neurons and a constraint plane with M neurons. The energy function decreases until the net reaches a state where all time derivatives are zero.

With some modifications, the LP network [1083] can be used to solve LSE problems [1216]. In [270], a circuit based on a modification of the LP network [1083] is designed for computing the discrete Hartley transform (DHT). A circuit for computing the discrete Fourier transform (DFT) is obtained by simply adding a few adders to the DHT circuit. The circuits can compute the DHT and DFT within circuit time constants of the order of nanoseconds.

The stability, computational speed, and computational accuracy of the LP network depends substantially on the location of the eigenvalues of the matrix product $\mathbf{D}_g^{\mathrm{T}} \mathbf{D}_f$ (or $\mathbf{D}_f \mathbf{D}_g^{\mathrm{T}}$), where $\mathbf{D}_g$ and $\mathbf{D}_f$ are, respectively, approximations to $\mathbf{D}$ at the signal and constraint planes of the network in the nonideal case. The case of purely real eigenvalues has been treated in [1216], and the result has been extended to the general case of complex eigenvalues in [449].

4.9 Multistate Hopfield Networks

The multilevel Hopfield network [361] and the complex-valued multistate Hopfield network [546, 827] are two direct generalizations of the Hopfield network. The multilevel Hopfield network uses neurons with an increasing multistep function as the activation function, while the complex-valued multistate Hopfield network uses a multivalued complex-signum function as the activation function. The complex-valued Hopfield-like network [489], like the complex-valued multistate Hopfield network [546, 827], uniformly quantizes the phase of the net input of each neuron and disregards the corresponding amplitude, but uses different dynamic equations.

The use of multistate neurons leads to a network architecture that is significantly smaller than that of the conventional Hopfield network, and hence, a simple hardware implementation. The reduction in the network size is highly desirable in large-scale applications such as image restoration and the TSP. In addition, the complex-valued multistate Hopfield network is also more efficient and convenient than the Hopfield network in the manipulation of complex-valued signals.

4.9.1 Multilevel Hopfield Networks

The multilevel Hopfield network provides a compact architecture for associative memory. In [361], a multilevel Hopfield network is obtained by replacing the threshold activation function with an increasing multistep function and modifying the generalized Hebbian rule. The storage capability of the multilevel Hopfield network is proved to be $O\left(J^3\right)$ bits for a network of J neurons, which is of the same order as that of the Hopfield network [8]. Given a network of J neurons, the number of patterns that the multilevel network can reliably store and retrieve may be considerably less than that for the Hopfield network, since each codeword in the multilevel Hopfield network typically contains more bits.

The multilevel sigmoidal function has typically been used as the activation function in the multilevel Hopfield network [1246, 1032, 1269]. In [1032], a storage procedure for the multilevel Hopfield network in the synchronous mode has been developed based on the LS solution, and also examined by using an image restoration example. A retrieval procedure for the multilevel Hopfield network has been proposed and applied to COPs such as the TSP in [337]. In [1246], a multilevel Hopfield-like network is obtained by using a new neuron with self-feedback and the multilevel sigmoidal activation function. The multilevel model has been applied for A/D conversion, and a circuit implementation for the neural A/D converter has been fabricated [1246].

4.9.2 Complex-valued Multistate Hopfield Networks

The complex-valued multistate Hopfield network [546, 827] employs the multivalued complex-signum activation function that is defined as an L-stage phase quantizer for complex numbers

$$\text{csign}(u) \widehat{=} \begin{cases} z^0, & \arg(u) \in [0, \varphi_0) \\ z^1, & \arg(u) \in [\varphi_0, 2\varphi_0) \\ \vdots \\ z^{L-1}, & \arg(u) \in [(L-1)\varphi_0, L\varphi_0) \end{cases} \tag{4.63}$$

where $z = e^{j\varphi_0}$ with $\varphi_0 = \frac{2\pi}{L}$ is the Lth root of unity. Each state takes one of the equally spaced L points on the unit circle of the complex plane.

Similar to the Hopfield network, the system dynamics are defined by

$$net_i(t) = \sum_{k=1}^{J} w_{ki} x_k(t) \tag{4.64}$$

$$x_i(t+1) = \text{csign}\left(net_i(t) \cdot z^{\frac{1}{2}}\right) \tag{4.65}$$

for $i = 1, \cdots, J$, where J is the number of neurons, and the factor $z^{\frac{1}{2}} = e^{j\frac{\varphi_0}{2}}$ places the resulting states in the angular centers of each sector. A sufficient condition for the stability of the dynamics is that the weight matrix is Hermitian with non-negative diagonal entries, that is, $\mathbf{W} = \mathbf{W}^{\text{H}}$, $w_{ii} \geq 0$ [546]. The energy can be defined as

$$E(\mathbf{x}) = -\frac{1}{2}\mathbf{x}^{\mathrm{H}}\mathbf{W}\mathbf{x} \tag{4.66}$$

In order to store a set of N patterns, $\{\mathbf{x}_i\} \subset \{0, 1, \cdots, L-1\}^J$, $\mathbf{x}_i$ is first encoded to its complex memory state $\epsilon_i = (\epsilon_{i,1}, \cdots, \epsilon_{i,J})^{\mathrm{T}}$ with

$$\epsilon_{i,j} = z^{x_{i,j}} \tag{4.67}$$

The decoding of a memory state to a pattern is the inverse of (4.67). The complex-valued pattern set $\{\epsilon_i\}$ can be stored in weights by the generalized Hebbian rule [502]

$$w_{ji} = \frac{1}{J}\sum_{p=1}^{N}\epsilon_{p,i}\epsilon_{p,j}^* \tag{4.68}$$

for $i, j = 1, 2, \ldots, J$, where the superscript $*$ denotes the conjugate operation. Thus, $\mathbf{W}$ is Hermitian.

The storage capability of the memory, N_{max}, is dependent upon the resolution L for an acceptable level of the error probability P_{max}. As L is increased, N_{max} decreases, but each pattern contains more information.

Due to the use of the generalized Hebbian rule, the storage capacity of the network is very low and the problem of spurious memories is very pronounced. In [664], a gradient descent-based learning rule has been proposed to enhance the storage capacity and also reduce the number of spurious memories. In [827], an LP method has been proposed for storing into the network each pattern in an integral set $M \subset \{0, 1, 2, \cdots, L-1\}^J$ as a fixed point. A set of inequalities are employed to render each memory pattern as a strict local minimum of a quadratic energy landscape, and the LP method is employed to obtain the weight matrix and the threshold vector. Compared to the generalized Hebbian rule, the LP method significantly reduces the number of spurious memories, and provides better results in the case of noisy gray-level image reconstruction.

Since gray-scale images can be represented by integral vectors, reconstruction of such images from their distorted versions constitutes a straightforward application of multistate associative memory. The complex-valued Hopfield network is particularly suitable for interpreting images transformed by two-dimensional Fourier transform and two-dimensional autocorrelation functions [546].

Complex Hopfield-like Networks

The complex-valued Hopfield-like network [489] processes input vectors fully in the complex space using complex weights. Real and imaginary parts of the data are treated with equal significance in nondegenerate complex space.

Given vector $\mathbf{x}$, whose elements are normalized complex numbers, *i.e.* $|x_k| = 1$, $k = 1, \cdots, J$. The dynamics of the network are given as

$$net_i = \sum_{k=1, k \neq i}^{J} w_{ki} x_k(t) \qquad (4.69)$$

$$r_i(t) = |net_i|, \quad \alpha_i(t) = [\arg(net_i)]_{\varphi_0} \qquad (4.70)$$

$$\frac{d\beta_i}{dt} = K [\alpha_i(t) - \beta_i(t)] \qquad (4.71)$$

$$x_i(t) = e^{j\beta_i(t)} \qquad (4.72)$$

where $\varphi_0 = \frac{2\pi}{L}$ represents the resolution of the phase quantizer, L is a positive integer, the operator $[\cdot]_{\varphi_0}$ finds in the interval $(-\pi, \pi]$ the quantization of the variable, which can be a discrete argument $m\varphi_0$ with the integer m in $(-\frac{L}{2}, \frac{L}{2})$, $K > 0$ is a real gain, and the amplitude of output signal $x_i(t)$ is fixed at unity. Note that the amplitudes r_i are not used in this model.

The pseudostationary attractor is analyzed by defining the network energy E as the real part of the energy calculated in the complex space

$$E = \frac{1}{2} \mathrm{Re} \left(-\mathbf{x}^H \mathbf{W} \mathbf{x} \right) \qquad (4.73)$$

and since $\frac{dE}{dt} < 0$ [489], the system converges [489].

The complex-valued Hopfield-like network [489] can be used as complex autoassociative memory. In order to store the attractors $\mathbf{x}_i$, $i = 1, \cdots, N$, whose elements are normalized complex numbers, that is, $|x_{i,k}| = 1$, $k = 1, \cdots, J$, one can specify $\mathbf{W}$ as a complex autocorrelation matrix

$$w_{ki} = \sum_{p=1}^{N} x_{p,i} x_{p,k}^* \qquad (4.74)$$

In this case, $\mathbf{W}$ is Hermitian, $\mathbf{W} = \mathbf{W}^H$, that is, $w_{ij} = w_{ji}^*$, and the energy E in (4.73) can be defined by $E = -\frac{1}{2} \mathbf{x}^H \mathbf{W} \mathbf{x}$ since it is real-valued.

A modification of this network has been successfully applied to the DoA estimation of antenna arrays [1224].

4.10 Boltzmann Machines and Learning

Boltzmann machines are a class of stochastic RNNs based on physical systems [9, 487]. The Boltzmann machine has the same network architecture as that of the Hopfield model, that is, it is highly recurrent with $w_{ij} = w_{ji}$ and $w_{ii} = 0$, $i, j = 1, \cdots, J$. In contrast to the Hopfield network, the Boltzmann machine can have hidden units. The Hopfield network operates in an unsupervised manner, while the Boltzmann machine can also be trained in a supervised manner.

4.10.1 The Boltzmann Machine

Unlike the Hopfield model, neurons of a Boltzmann machine are divided into visible and hidden units, as illustrated in Fig. 4.8. In Fig. 4.8a, the visible units are clamped onto specific states determined by the environment, while the hidden units always operate freely. By capturing high-order statistical correlations in the clamping vector, the hidden units simulate the underlying constraints contained in the input vectors. This type of Boltzmann machine uses unsupervised learning, and can perform pattern completion. When the visible units are further divided into input and output neurons, as shown in Fig. 4.8b, this type of Boltzmann machine can be trained in a supervised manner. The recurrence eliminates the difference in input and output cells. The Boltzmann machine, operated in sequential or synchronous mode, is a universal approximator for arbitrary functions defined on finite sets [1239].

Instead of using a sigmoidal function in the Hopfield network, the activation at each neuron takes the value of 0 or 1, which is dependent on the probability of a temperature variable T

$$net_i = \sum_{j=1, j \neq i}^{J} w_{ji} x_j = \mathbf{w}_i \mathbf{x} \tag{4.75}$$

$$P_i = \frac{1}{1 + e^{-\frac{net_i}{T}}} \tag{4.76}$$

$$x_i = \begin{cases} 1, & \text{with probability } P_i \\ 0, & \text{with probability } 1 - P_i \end{cases} \tag{4.77}$$

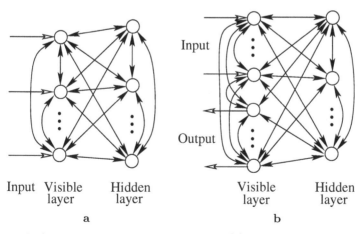

Input Visible   Hidden   Visible   Hidden
       layer   layer   layer   layer
            a   b

Fig. 4.8. Architectures of Boltzmann machines. (**a**) An architecture with visible and hidden neurons. (**b**) An architecture with input, output, and hidden neurons. Hollow arrows denote information exchange with the environment.

When $net_i = 0$ or T is very large, x_i is either 1 or 0 with equal probability. For very small T, x_i is deterministically 1. The input and output states can be fixed or variable.

The search for all the possible states is performed at a temperature T in the Boltzmann machine. At the steady state, the relative probability of two states in the Boltzmann machine is determined by the Boltzmann distribution of the energy difference between the two states

$$\frac{P_\alpha}{P_\beta} = e^{-\frac{E_\alpha - E_\beta}{T}} \tag{4.78}$$

where E_α and E_β are the corresponding energy levels of the two states. The energy can be computed by the same formula as for the Hopfield model

$$E(t) = -\frac{1}{2}\mathbf{x}^T(t)\mathbf{W}\mathbf{x}(t) = -\sum_{i<j} x_i(t)x_j(t)w_{i,j} \tag{4.79}$$

Like the multistate Hopfield model [546], a multivalued Boltzmann machine has been proposed as an extension of the two-valued Boltzmann machine [706]. Each neuron of the multivalued Boltzmann machine can only take L discrete stable states, and the angle between two adjacent directions is given by $\varphi_0 = \frac{2\pi}{L}$. The probability of state change is according to the Boltzmann distribution of the energy difference between the two states. In [50], a synchronous Boltzmann machines as well as its learning algorithm has been introduced to facilitate parallel implementations.

4.10.2 The Boltzmann Learning Algorithm

When using the Boltzmann machine with hidden units, the generalized Hebbian rule cannot be used as in an unsupervised manner. For the supervised learning of the Boltzmann machine, the BP is not applicable due to the different network architecture. SA is used by Boltzmann machines to learn weights corresponding to the global optimum. The learning process in the Boltzmann machine is computationally very expensive. The computational complexity is exponential in the number of neurons [559], which is much higher than that of the BP.

For constraint-satisfaction problems, some of the neurons are externally clamped to some input patterns, and we then find the global minimum for these particular input patterns. The integration of the SA into the Boltzmann learning rule makes the Boltzmann machine especially suitable for constraint-satisfaction tasks involving a large number of weak constraints [487].

The original Boltzmann learning algorithm [487, 9] is based on counting occurrences. The Boltzmann learning algorithm based on correlations [912] provides a better performance than the original algorithm. The correlation-based learning procedure for the Boltzmann machine is given in Algorithm 4.10.2 [912, 470].

Algorithm 4.10.2 (Boltzmann Learning)

1. Initialization.

   a) Initialize the weights w_{ji}.
      Set w_{ji} as uniform random values in $[-a_0, a_0]$, where a_0 is typically selected as 0.5 or 1.
   b) Set the initial temperature T_0 and the final temperature T_{f}.

2. Clamping phase.

   Present the patterns for network training. For unsupervised learning, all the visible nodes are clamped to the patterns. For supervised learning, all the input and output nodes are clamped to the pattern pairs.

   a) For each example, perform the SA procedure until T_{f} is reached.

      i. At each temperature T, let the network relax according to the Boltzmann distribution for a length of time. Relaxation is performed by updating the states of the unclamped (hidden) units

      $$x_i = \begin{cases} +1, & \text{with probability } P_i \\ -1, & \text{with probability } 1 - P_i \end{cases} \tag{4.80}$$

      where P_i is calculated according to (4.75) and (4.76).
      ii. Update T according to the annealing schedule.
   b) At T_{f}, estimate the correlation in the clamped condition

      $$\rho_{ij}^+ = \mathrm{E}\left[x_i x_j\right], \quad i, j = 1, 2, \ldots, J, i \neq j \tag{4.81}$$

3. Free-running phase.

   a) Repeat Step 2a. For the unsupervised learning, all the visible neurons are now free running. For the supervised learning, only the input neurons are clamped and the output neurons are free running.
   b) At T_{f}, estimate the correlation in the free-running condition

      $$\rho_{ij}^- = \mathrm{E}\left[x_i x_j\right], \quad i, j = 1, 2, \ldots, J, i \neq j \tag{4.82}$$

4. Weight update.

   The weight update is performed as

   $$\Delta w_{ij} = \eta \left(\rho_{ij}^+ - \rho_{ij}^-\right), \quad i, j = 1, 2, \ldots, J, i \neq j \tag{4.83}$$

   where the learning rate $\eta = \frac{\varepsilon}{T}$, and ε is a small positive constant. Equation (4.83) is called the *Boltzmann learning rule*.

5. Repeat Steps 2 through 4 until there is no change in w_{ij}, for all i, j.

4.10.3 The Mean-field-theory Machine

Mean-field approximation is a well-known method in statistical physics [414]. The mean-field annealing algorithm was proposed to accelerate the convergence of the Boltzmann machine [912, 911]. The Boltzmann machine with such an algorithm is also termed the *mean-field-theory (MFT) machine* or *deterministic Boltzmann machine*.

The mean-field annealing algorithm is a deterministic method. Instead of the stochastic binary neuron output for the Boltzmann machine, continuous neuron outputs, which are calculated as the average of the probability of the binary neuron variables at temperature T, are used. The average of state x_i is calculated for a specific value of activation net_i according to (4.80), (4.75) and (4.76)

$$\mathrm{E}\left[x_i\right] = (+1)P_i + (-1)\left(1 - P_i\right) = 2P_i - 1$$

$$= \tanh\left(\frac{net_i}{T}\right) \tag{4.84}$$

The correlation in the Boltzmann learning rule is replaced by the mean-field approximation

$$\mathrm{E}\left[x_i x_j\right] \simeq \mathrm{E}\left[x_i\right]\mathrm{E}\left[x_j\right] \tag{4.85}$$

The above approximation method is usually termed the *naive or zero-order mean-field approximation*.

The mean-field annealing, which replaces all the states in the Boltzmann machine by their averages, can be treated as a deterministic form of Boltzmann learning. The MFT machine provides a substantial speedup over the Boltzmann machine, and is one to two orders of magnitude faster than the Boltzmann machine [458, 912].

In [644], the equivalence between the asynchronous MFT machine and the continuous Hopfield model is established in terms of the same fixed points for networks using the same Hopfield topology and energy function. The naive MFT machine is shown to perform the steepest descent on an appropriately defined cost function under certain circumstances, and has been empirically used to solve a variety of supervised learning problems [485].

An approximate mean-field algorithm for the Boltzmann machine is presented in [559], which has a computational complexity of $O\left(J^3\right)$, J being the number of neurons. In the absence of hidden unit, the weights can be directly computed from the fixed-point equation of the learning rules, and thus a gradient descent procedure is avoided. The solutions are close to the optimal ones and thus, the method gives a significant improvement over the naive mean-field algorithm.

The mean-field annealing algorithm can be derived following the optimization of the Kullback–Leibler divergence between the factorial approximating distribution and the ideal joint distribution of the binary neural variables in terms of the mean activations. In [1194], two interactive mean-field algorithms are derived by extending the internal representations to include both the mean activations and the mean correlations. The two algorithms, respectively, estimate the mean activations subject to the mean correlations, and the mean correlations subject to the mean activations by optimizing the objective quantified by a combination of the Kullback–Leibler divergence and the correlation strength between any two distinct variables. The interactive mean-field algorithms improve the mean-field approximation in both the performance and the relaxation efficiency.

The mean-field annealing dynamics are isomorphic to the steady-state equations of an RC circuit. The mean-field annealing algorithm can be simulated by RC circuits, coupled with the local nature of the Boltzmann machine, which makes the MFT machine suitable for massively parallel VLSI implementation [26, 660, 1005].

However, the validity of the naive mean-field algorithm is challenged in [393, 559]. Naive mean-field learning does not converge in general. By applying naive mean-field approximation to a finite system with non-random interactions, the true stochastic system is not faithfully represented in many situations. This suggests that the naive mean-field approximation is inadequate for learning. The independence assumption is shown to be unacceptably inaccurate in multiple-hidden-layer configgurations, thus accounting for the empirically observed failure of MFT learning in such networks. As a result, the MFT machine only works in the supervised mode with only a single hidden layer [393, 470]. The mean state is not a sufficient representation for the free-running probability distribution and thus, the mean-field method is ineffective for unsupervised learning.

4.11 Discussion

As generalizations of the Hopfield network, both the Boltzmann machine and the MFT machine can be used as associative memory. The Boltzmann and MFT machines that use hidden units have a far higher capacity for storage and error-correcting retrieval of random patterns and improved basins of attraction than the Hopfield network [458, 26].

When the Boltzmann machine is trained as associative memory using an adaptive association rule [569], it does not suffer from spurious states. The association rule, which creates a sphere of influence around each stored pattern, is a generalization of the generalized Hebbian rule. Spurious fixed points, whose regions of attraction are not recognized by the rule, are skipped, due to the finite probability to escape from any state. The upper and lower bounds on retrieval probabilities of each stored pattern from an initial state at a given Hamming distance from it are also given.

Due to the existence of the hidden units, neither the Boltzmann machine nor the MFT machine can be trained and retrieved in the same way as in the case of the Hopfield model. The training for the two models is given in the preceding sections. The retrieval process is as follows [458]. The visible neurons are clamped to a corrupted pattern, the whole network is annealed to a lower temperature, where the state of the hidden neurons approximates the learned internal representation of the stored pattern, and then the visible neurons are released. The annealing process continues until the whole network is settled.

The Boltzmann machine is important historically and theoretically, since it sparked the reviving interests in neural networks. They are suitable for modeling biological phenomena, since biological neurons are stochastic systems. However, a large number of iterations at each temperature and a very small temperature step are required to reach the global minimum, since this process is necessary for estimating the probabilities. As a result, this process is too slow although it can find the global optimum. For practical problems, we have to increase the step size of the temperature and thus, the algorithm is no longer guaranteed to find the global minimum with probability one. In so doing, the algorithm is capable of finding near-optimum solutions for many practical applications.

In addition to the Boltzmann and MFT machines, there are some other implementations of stochastic Hopfield networks. The Gaussian machine [20] is a general framework that includes the Hopfield network, the Boltzmann machine and also other stochastic networks. Stochastic distribution is realized by adding thermal noise, a stochastic external input ε, to each unit, and the network dynamics are the same as that of the Hopfield network. The stochastic term ε obeys a Gaussian distribution with zero mean and variance σ^2, where the deviation $\sigma = kT$, and T is the temperature. The stochastic term ε can occasionally bring the network to states with a higher energy. When $k = \sqrt{\frac{8}{\pi}}$, the distribution of the outputs has the same behavior as a Boltzmann machine. When employing noise obeying a logistic distribution rather than a Gaussian distribution in the original definition, we can obtain a Gaussian machine identical to a Boltzmann machine. When the noise in the Gaussian machine takes a Cauchy distribution with zero as the peak location and the half-width at the maximum $\sigma = T\sqrt{\frac{8}{\pi}}$, we get a Cauchy machine [1071]. The Gaussian machine may be more suitable than the Boltzmann machine for some tasks. A similar idea was embodied in the stochastic network given in [692], where in addition to a cooling schedule for temperature T, gain annealing is also applied. The gain $\frac{1}{\beta}$ has to be decreased more slowly than T, and kept bounded away from zero.

4.12 Computer Experiments

We conclude this chapter by two examples. In Example 4.1, the Hopfield network is used as associative memory, and three learning algorithms for storing patterns into the Hopfield network are simulated and compared. In Example 4.2, the optimization capability of the Hopfield network is applied for DoA estimation.

4.12.1 Example 4.1: A Comparison of Three Learning Algorithms

This example is designed to check the storage capabilities of the Hopfield network trained with three local algorithms, namely, the generalized Hebbian, improved Hebbian, and perceptron-type learning rules. After the Hopfeild network is trained with a pattern set, we present the same pattern set and examine the average retrieval bit error rates (BERs) and the average storage error rates (SERs) for a number of random runs.

A set of bipolar patterns each having a bit-length of $J = 40$ is given. The pattern set $\{\mathbf{x}_i\}$ is generated randomly. Theoretically, the storage capacities for the generalized Hebbian, improved Hebbian, and perceptron-type learning rules are $\frac{J}{2\ln J} = 5.42$, $\frac{J}{\sqrt{2\ln J}} = 14.73$, and $J = 40$, respectively. These capacities have been verified during our experiments. After the Hopfield network is trained, the maximum number of iterations at the performing stage is set as 30. The BERs and SERs are calculated based on 50 random runs.

Simulation is conducted for the case of N uncorrelated patterns as well as N slightly correlated patterns. In the case of N uncorrelated patterns, the matrix composed of the randomly generated patterns are of full rank, that is, having a rank of N. In the case of N slightly correlated patterns, $N - 1$ patterns are randomly generated and are uncorrelated; the remaining one pattern is generated by linearly combining any three of the $N - 1$ patterns and then applying the signum function, until the corresponding matrix has a rank of $N - 1$.

For the perceptron-type learning, we select $\mathbf{W}(0)$ according to two schemes: (a) a symmetrical, random, zero-diagonal matrix with each entry in the range of $(-0.05, 0.05)$ and (b) the zero matrix. η is selected as 0.2. The maximum number of epochs is set as 50. Training terminates when the relative energy change between two epochs is below 10^{-4}.

To begin with, we store a set of $N = 20$ patterns. The training and performing results are shown in Table 4.1, and the system energy evolution traces during the training process for a random run are illustrated in Fig. 4.9.

During the retrieval stage, if the fundamental memories are presented to the network, the desired patterns can usually be produced by the network after one iteration if N is not over the capacity of the network. The network trained by the generalized Hebbian rule cannot correctly retrieve any of the patterns, since the number of patterns is much greater than its storage capacity. The network trained with the improved Hebbian rule can, on average, correctly retrieve 14.90 patterns, which is close to its theoretical capacity. The perceptron-type rule can almost correctly retrieve all the patterns. This accuracy can be further improved by training with more epochs. It is noted that the results for the uncorrelated and slightly correlated cases are very close to each other for all these algorithms.

Table 4.1. Comparison of three associative memory algorithms ($J = 40$, $N = 20$): GH—generalized Hebbian rule, IH—improved Hebbian rule, PT(a)—perceptron-type rule with nonzero $\mathbf{W}(0)$, PT(b)—perceptron-type rule with zero $\mathbf{W}(0)$. Note that the words *epochs* and *iterations* stand for *the number of epochs for training*, and *the number of iterations for retrieval*, respectively.

Algorithm	Uncorrelated				Correlated			
	Epochs	Iterations	BER	SER	Epochs	Iterations	BER	SER
GH	1	28.90	0.2379	19.72	1	24.18	0.2930	17.72
IH	1	18.44	0.02118	5.10	1	15.22	0.0128	3.82
PT(a)	5.86	1.76	5.25×10^{-4}	0.1	5.2	1.6	3.45×10^{-4}	0.12
PT(b)	5.80	1.58	2.25×10^{-4}	0.04	5.74	1.14	2.50×10^{-4}	0.04

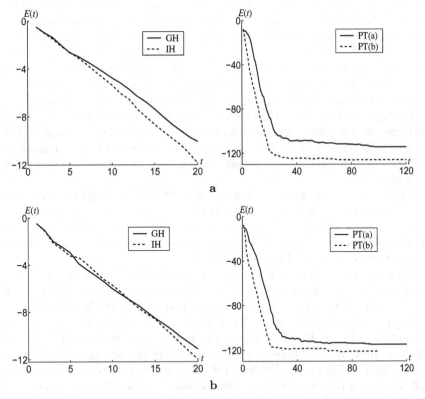

Fig. 4.9. Energy evolutions: $J = 40$, $N = 20$. (**a**) Uncorrelated patterns. (**b**) Slightly correlated patterns. t is the number of iterations. Note that one epoch corresponds to 20 iterations.

Table 4.2. Comparison of three associative memory algorithms ($J = 40$, $N = 40$)

Algorithm	Uncorrelated				Correlated			
	Epochs	Iterations	BER	SER	Epochs	Iterations	BER	SER
GH	1	30	0.2868	39.98	1	30	0.2966	38.88
IH	1	30	0.2988	38.08	1	30	0.2839	37.44
PT(a)	33.90	11.44	0.01974	2.04	32.66	9.2	0.02244	2.28
PT(b)	34.44	11.48	0.01875	1.90	33.44	10.9	0.01329	1.34

Now, let us increase N to 40; the corresponding results are listed in Table 4.2 and shown in Fig. 4.10. The capacity of the perceptron-like learning is shown to be 38. However, by increasing the number of epochs to 100, the storage capability of the perceptron-like learning can be further improved, the average iteration for the retrieval stage is 1, and the SER is close to 0. Thus, the storage capability can reach 40 for both the uncorrelated and slightly correlated cases. The evolution traces of energy for 100 epochs are shown in Fig. 4.11. From the simulation results, the perceptron-like learning initialized with the zero weight matrix is superior to that initialized with a nonzero weight matrix.

4.12.2 Example 4.2: Using the Hopfield Network for DoA Estimation

There are many applications of the Hopfield network for ASP [321]. To make use of the Hopfield network, the cost function for a DoA estimation or beam-forming problem is first mapped onto the quadratic energy function of the Hopfield network in a minimum MSE sense. For Gaussian noise and uniform linear arrays, the Hopfield network using such a cost function is equivalent to the ML estimator. We give one as an example here.

In the p^3-neuron Hopfield model [937, 550], each neuron corresponds to a unique combination of the possible values of amplitude, frequency and phase, where each of the three parameters is discretized to its respective desired accuracy, typically p levels. A hypothetical signal vector $\mathbf{s}_{u,v,q}$ can be defined as

$$\mathbf{s}_{u,v,q} = a_u e^{j\theta_v} \left(1, e^{-j\tau_q}, \cdots, e^{-j(n-1)\tau_q} \right)^{\mathrm{T}}$$

where $1 \leq u, v, q \leq p$. $\mathbf{s}_{u,v,q}$ can be permuted as $\mathbf{s}_i$, $i = 1, \cdots, p^3$. The DoA problem is transformed into minimizing the error E_{DoA}

$$E_{\mathrm{DoA}} = \|\mathbf{y} - \mathbf{Sx}\|^2$$

where $\mathbf{y}$ is the measured signal vector on the antenna array, $\mathbf{S} = [\mathbf{s}_1, \mathbf{s}_2, \cdots, \mathbf{s}_{p^3}]$ and $\mathbf{x}$ is the output vector of the network. Expanding E_{DoA} and eliminating terms irrelevant to $\mathbf{x}$, and also introducing a penalty term to force the neurons to converge to digital values 0 or 1, we get the redefined objective function

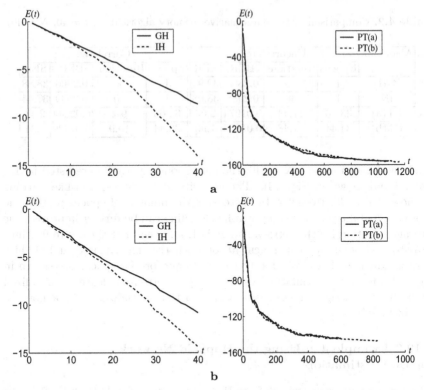

Fig. 4.10. Energy evolutions: $J = 40$, $N = 40$. (**a**) Uncorrelated patterns. (**b**) Slightly correlated patterns. t is the number of iterations. Note that one epoch corresponds to 40 iterations.

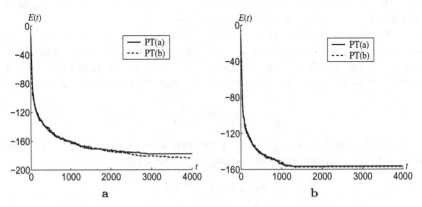

Fig. 4.11. Energy evolutions for 100 epochs: $J = 40$, $N = 40$. (**a**) Uncorrelated patterns. (**b**) Slightly correlated patterns. Note that one epoch corresponds to 40 iterations.

$$E_{\text{DoA}} = \mathbf{x}^{\text{T}} \mathbf{S}^{\text{H}} \mathbf{S} \mathbf{x} - 2\text{Re}\left(\mathbf{y}^{\text{H}} \mathbf{S} \mathbf{v}\right) - \sum_{i=1}^{p^3} \left(\mathbf{s}_i^{\text{H}} \mathbf{s}_i\right) x_i \left(x_i - 1\right)$$

Expanding $\mathbf{S}$ by block matrix operations leads to

$$E_{\text{DoA}} = \text{Re}\left(\sum_{i=1}^{p^3} \sum_{j=1, j \neq i}^{p^3} \left(\mathbf{s}_i^{\text{H}} \mathbf{s}_j\right) x_i x_j - \sum_{i=1}^{p^3} \left(2\mathbf{y}^{\text{H}} \mathbf{s}_i + \mathbf{s}_i^{\text{H}} \mathbf{s}_i\right) x_i \right)$$

Based on a comparison of the cost function with the energy function of the Hopfield network, the weights and external input to the ith neuron can be defined by

$$w_{ij} = -\text{Re}\left(\mathbf{s}_i^{\text{H}} \mathbf{s}_j\right), \quad i \neq j$$

$$\theta_i = \text{Re}\left(\mathbf{y}^{\text{H}} \mathbf{s}_i - \frac{1}{2} \mathbf{s}_i^{\text{H}} \mathbf{s}_i\right)$$

The number of neurons p^3 may be huge for implementation. For narrow-band point sources, one can eliminate the frequency and amplitude variables and only discretize the phase into p levels and get a network of p neurons [937]. Accordingly, the cost function can be defined by

$$E_{\text{DoA}} = \left\| \mathbf{y} - [\mathbf{v}_1, \mathbf{v}_2, \cdots, \mathbf{v}_p] \mathbf{x} \right\|^2$$

where

$$\mathbf{v}_i = \mathbf{P}_{\mathbf{d}_i} \mathbf{y}$$

$$\mathbf{P}_{\mathbf{d}_i} = \mathbf{d}_i \left(\mathbf{d}_i^{\text{H}} \mathbf{d}_i\right)^{-1} \mathbf{d}_i^{\text{H}}$$

and $\mathbf{d}_i$ is the hypothetical steering vector, which is used to represent the signals $\mathbf{s}_i$. A comparison with the p^3-neuron case leads to

$$w_{ij} = -\text{Re}\left(\mathbf{y}^{\text{H}} \mathbf{P}_{\mathbf{d}_i}^{\text{H}} \mathbf{P}_{\mathbf{d}_j} \mathbf{y}\right), \quad i \neq j$$

$$\theta_i = \frac{1}{2}\text{Re}\left(\mathbf{y}^{\text{H}} \mathbf{P}_{\mathbf{d}_i} \mathbf{y}\right)$$

The definition for E_{DoA} can be extended to L snapshots by applying $E_{\text{DoA}} = \frac{1}{L} \sum_{l=1}^{L} E_{\text{DoA},l}$, where $E_{\text{DoA},l}$ is the cost function for the lth snapshot. In this case, w_{ij} and θ_i are their respective averages for L snapshots.

For a ULA of 10 elements with an intersensor separation of half a wavelength, the normalized frequency is 0.2. A network of 100 neurons was used to discretize the bearing space between 0 and $45°$. Three sources with SNR 5 dB are impinging from $12.0°$, $20.0°$, and $32.0°$. Twenty-five (25) snapshots of data were collected for averaging.

By using the Hopfield network with a gain annealing schedule [550], the evolution of the network and the DoA estimates are shown in Fig. 4.12. The hypothetical 28th, 46th, and 72th neurons represent the DoAs of the sources. The network settled to the digital values in 14 iterations.

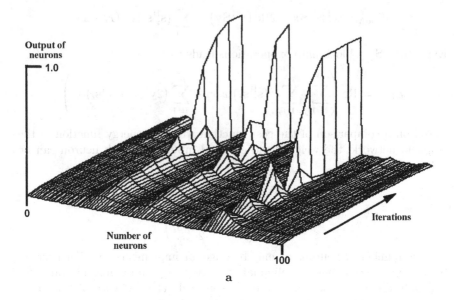

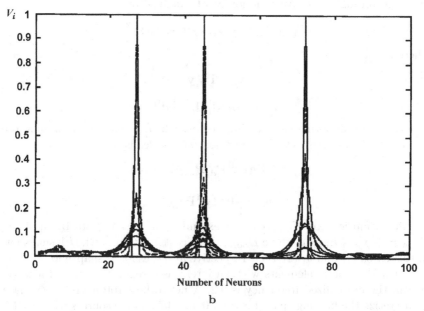

Fig. 4.12. DoA estimation using the Hopfield network with a gain annealing scheme, the evolution of the network: (**a**) In three-dimensional representation. (**b**) In two-dimensional representation. (From [550].)

5

Competitive Learning and Clustering

Clustering is a fundamental data-analysis method that has been most extensively studied and also widely used in many engineering and scientific fields. Clustering is an unsupervised classification technique that identifies some inherent structure present in a set of objects based on a similarity measure. Clustering methods can be classified into two main categories, namely, statistical model identification-based clustering [786] and competitive learning-based clustering.

This chapter is dedicated to clustering. We describe a number of competitive learning-based neural networks and clustering algorithms. We place emphasis on clustering models such as the SOM [610], the learning vector quantization (LVQ) [611], and the ART model [159, 160], as well as clustering algorithms such as the C-means [746, 714] and fuzzy C-means (FCM) [104] algorithms. Many associated topics such as the underutilization problem, robust clustering, hierarchical clustering, and cluster validity are also discussed. Clustering can be used for the selection of prototypes in the RBFN and rule extraction for neurofuzzy systems, which will be dealt with in Chapters 6 and 7, respectively. Clustering is also widely used for pattern recognition, feature extraction, VQ, image segmentation, function approximation, and data mining.

5.1 Vector Quantization

VQ is a classical method that produces an approximation to a continuous PDF $p(\mathbf{x})$ of the vector variable $\mathbf{x} \in R^n$ using a finite number of prototypes. That is, VQ represents a set of feature vectors $\mathbf{x}$ by a finite set of prototypes $\{\mathbf{c}_1, \cdots, \mathbf{c}_K\} \subset R^n$. The finite set of prototypes is referred to as the *codebook*. Codebook design can be performed by using clustering algorithms. Once the codebook is specified, the approximation of $\mathbf{x}$ involves finding the reference vector $\mathbf{c}$ closest to $\mathbf{x}$ so that [611, 614]

$$\|\mathbf{x} - \mathbf{c}\| = \min_i \|\mathbf{x} - \mathbf{c}_i\| \tag{5.1}$$

This is the nearest-neighbor paradigm, and the procedure is actually the simple competitive learning (SCL).

The codebook can be designed by minimizing the expected squared quantization error

$$E = \int \|\mathbf{x} - \mathbf{c}\|^2 p(\mathbf{x}) d\mathbf{x} \tag{5.2}$$

where $\mathbf{c}$ satisfies (5.1), that is, $\mathbf{c}$ is a function of $\mathbf{x}$ and $\mathbf{c}_i$.

An iterative approximation scheme for finding the codebook is derived from the criterion (5.2) [614]

$$\mathbf{c}_i(t+1) = \mathbf{c}_i(t) + \eta(t)\delta_{wi}\left[\mathbf{x}(t) - \mathbf{c}_i(t)\right] \tag{5.3}$$

where the subscript w corresponds to the prototype closest to $\mathbf{x}(t)$, termed the *winning prototype*, δ_{wi} is the Kronecker delta[1], and η is a small positive learning rate, satisfying the classical Robbins–Monro conditions

$$\sum \eta(t) = \infty, \quad \text{and} \quad \sum \eta^2(t) < \infty \tag{5.4}$$

Typically, η is selected to be decreasing monotonically in time. For example, one can select $\eta(t) = \eta_0 \left(1 - \frac{t}{T}\right)$, where $\eta_0 \in (0, 1]$ and T is the iteration bound. This is the SCL-based VQ.

Voronoi tessellation, also called a *Voronoi diagram*, is useful for the illustration of VQ results. The space is partitioned into a finite number of regions bordered by hyperplanes. Each region is represented by a codebook vector, which is the nearest neighbor to any point within the same region. An illustration of Voronoi tessellation in the two-dimensional space is shown in Fig. 5.1. All vectors in one of the regions constitute a *Voronoi set*. For a smooth underlying probability density $p(\mathbf{x})$ and large K, all regions in an optimal Voronoi partition have the same within-region variance σ^2 [407].

5.2 Competitive Learning

Competitive learning can be implemented using a J-K neural network whose output layer is termed the *competition layer*. The neurons in the competition layer are fully connected to the input nodes. In the competition layer, lateral connections are used to perform lateral inhibition. The architecture of the competitive learning network is shown in Fig. 5.2.

The basic principle underlying competitive learning is the mathematical statistics problem called *cluster analysis*. Competitive learning is usually based on the minimization of a functional such as [1105]

---

[1] $\delta_{wi} = 1$ for $w = i$, and 0 otherwise.

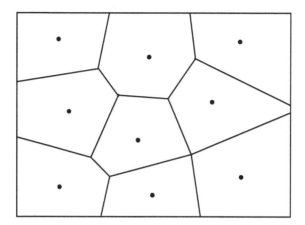

Fig. 5.1. A schematic of Voronoi tessellation in the two-dimensional space. Code-book vectors are depicted as points.

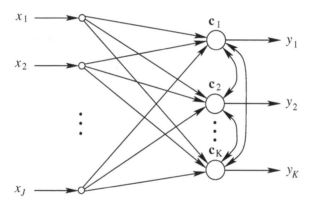

Fig. 5.2. Architecture of the J-K competitive learning network. For input $\mathbf{x}$, the network selects one of the K prototypes $\mathbf{c}_i$ by setting $y_i = 1$ and $y_j = 0, j = 1, \cdots, K$, $j \neq i$.

$$E = \frac{1}{N} \sum_{p=1}^{N} \sum_{k=1}^{K} \mu_{kp} \left\| \mathbf{x}_p - \mathbf{c}_k \right\|^2 \tag{5.5}$$

where N is the size of the pattern set, and μ_{kp} is the connection weight assigned to prototype $\mathbf{c}_k$ with respect to $\mathbf{x}_p$, denoting the membership of pattern p into cluster k.

Although minimization of (5.5) can lead to batch algorithms, it is difficult to use the gradient-descent method, since the winning prototypes must be determined with respect to each input pattern $\mathbf{x}_p$. By using the following functional, the gradient-descent method leads to sequential updating of the prototypes with respect to the input pattern $\mathbf{x}_p$

$$E_p = \sum_{k=1}^{K} \mu_{kp} \|\mathbf{x}_p - \mathbf{c}_k\|^2 \tag{5.6}$$

When $\mathbf{c}_k$ is the closest prototype to $\mathbf{x}_p$ in the Euclidean metric, that is, when it is the winning prototype of $\mathbf{x}_p$, the weight μ_{kp} is equal to unity; otherwise, μ_{kp} is less than unity.

The SCL is derived by minimizing (5.5) under the assumption that the weights are obtained according to the nearest prototype condition

$$\mu_{kp} = \begin{cases} 1, & k = \arg_k \min \|\mathbf{x}_p - \mathbf{c}_k\| \\ 0, & \text{otherwise} \end{cases} \tag{5.7}$$

Thus (5.5) becomes

$$E = \frac{1}{N} \sum_{p=1}^{N} \left\{ \min_{1 \le k \le K} \|\mathbf{x}_p - \mathbf{c}_k\|^2 \right\} \tag{5.8}$$

This is the average of the squared Euclidean distances between the inputs $\mathbf{x}_p$ and their closest prototypes $\mathbf{c}_k$. The minimization of (5.8) implies that each input attracts only its winning prototype and has no effect on its nonwinning prototypes at all.

Based on the squared error criterion (5.6) and the gradient-descent method, assuming $\mathbf{c}_w = \mathbf{c}_w(t)$ to be the winning prototype of $\mathbf{x} = \mathbf{x}_t$, we get the SCL as

$$\mathbf{c}_w(t+1) = \mathbf{c}_w(t) + \eta(t) [\mathbf{x}_t - \mathbf{c}_w(t)] \tag{5.9}$$
$$\mathbf{c}_i(t+1) = \mathbf{c}_i(t), \quad i \ne w \tag{5.10}$$

where $\eta(t)$ can be selected according to (5.4). The process is known as *winner-take-all (WTA)*. The WTA mechanism plays an important role in the design of most unsupervised learning neural networks. If each cluster has its own learning rate as $\eta_i = \frac{1}{N_i}$, where N_i is the number of samples assigned to the ith cluster, the algorithm achieves the minimum output variance [1210].

Many WTA models can be achieved based on the continuous-time Hopfield network topology [752, 298, 1077, 1065], or based on the CNN [239] model with linear circuit complexity [1013, 38]. There are also some circuits for realizing the WTA function such as a series of compact CMOS integrated circuits [653] and a simple analog circuit with its dynamic equation being governed by just one parameter [1077].

k-winners-take-all (k-WTA) is a process of selecting the k largest components from an N-vector. k-WTA is a key task in decision making, pattern recognition, associative memories, or competitive learning networks. k-WTA networks are usually based on the continuous-time Hopfield network

[752, 1230, 152]. The k-WTA circuit devised in [653] has infinite resolution, and is implemented using the Hopfield network based on the penalty method. In [1111], a k-WTA circuit with $O(n)$ interconnect complexity has been developed by extending the WTA circuits given in [653]. The k-WTA is formulated as a mathematical programming problem solved by the direct analog implementation of the Lagrange multiplier method, where n is the number of neurons. The circuit has merits of real-time responses and short wire length, although it has a finite resolution.

5.3 The Kohonen Network

Von der Malsburg's model [1133] and Kohonen's self-organization map (SOM)[2] [610, 611] are two well-known topology-preserving competitive learning models. Both models have their biological plausibility in the cortex of mammals. The SOM have now become a popular model for VQ, clustering analysis, feature extraction, and data visualization.

The Kohonen network is a J-K feedforward structure with fully interconnected processing units that compete for signals. The output layer is called the *Kohonen layer*. Input nodes are fully connected to output neurons with their associated weights. Lateral connections between neurons are used as a form of feedback whose magnitude is dependent on the lateral distance from a specific neuron, which is characterized by a neighborhood parameter.

The Kohonen network defined on R^n is a one-, two-, or higher-dimensional grid A of neurons characterized by prototypes $\mathbf{c}_k \in R^n$ [612, 611][3]. The architecture of the network in the two-dimensional space is illustrated in Fig. 5.3.

The Kohonen network uses competitive learning. Input patterns are presented sequentially in time through the input layer, without specifying the desired output. The Kohonen network is extended to the SOM when the lateral feedback is more sophisticated than the WTA rule. For example, the lateral feedback used in the SOM can be selected as the so-called *Mexican hat function*, which is found in the visual cortex. The SOM is more successful in classification and pattern-recognition problems.

5.3.1 Self-organizing Maps

The SOM employs the Kohonen network topology. For each neuron k, compute the Euclidean distance to the input pattern $\mathbf{x}$, and find the neuron whose prototype is closest to $\mathbf{x}$

$$\|\mathbf{x}_t - \mathbf{c}_w\| = \min_{k \in A} \|\mathbf{x}_t - \mathbf{c}_k\| \tag{5.11}$$

---

[2] In the literature, the self-organizing map is also written as the self-organizing feature map (SOFM).

[3] $\mathbf{c}_k$ is also viewed as the weight vector to neuron k, namely, $\mathbf{w}_k = (w_{1k}, w_{2k}, \cdots, w_{Jk})^{\mathrm{T}}$.

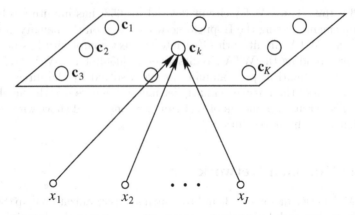

Fig. 5.3. Architecture of the two-dimensional J-K Kohonen network. Each neuron is characterized by its prototype c_k. The neurons in the Kohonen layer compete with each other.

Neuron w is the winning neuron, called the *excitation center*, which becomes the center of a group of input vectors that lie closest to c_w.

For all the input vectors closest to c_w, update all the propotype vectors by

$$c_k(t+1) = c_k(t) + \eta(t)h_{kw}(t)[x_t - c_k(t)] \tag{5.12}$$

for $k = 1, \cdots, K$, where $\eta(t)$ is selected according to (5.4), and $h_{kw}(t)$ is the so-called *excitation response, neighborhood function* or *kernel function*, which defines the response of neuron k when c_w is the excitation center. Equation (5.12) is known as the *Kohonen learning rule* [612].

If $h_{kw}(t)$ takes the value unity for $k = w$ and zero otherwise, (5.12) reduces to the SCL. $h_{kw}(t)$ can be selected as a function that decreases with the increasing distance between c_k and c_w, and is typically selected as the Gaussian function

$$h_{wk}(t) = h_0 e^{-\frac{\|c_k - c_w\|^2}{\sigma^2(t)}} \tag{5.13}$$

where h_0 is a positive constant. In the SOM, the size of the topological neighborhood shrinks with time, thus $\sigma(t)$ is a decreasing function of t, and a popular choice is the exponential decay with time [859]

$$\sigma(t) = \sigma_0 e^{-\frac{t}{\tau}} \tag{5.14}$$

where σ_0 is a positive constant and τ is a time constant.

The Gaussian topological neighborhood is more biologically reasonable than a rectangular one. The SOM using the Gaussian neighborhood converges more quickly than the SOM using a rectangular one [731]. The Gaussian neighborhood function enforces ordering in the neighborhood sets at every training iteration. We now give the SOM algorithm in Algorithm 5.3.1.

Algorithm 5.3.1 (SOM)

1. Set $t = 0$.
2. Initialize all $\mathbf{c}_k(0)$ and all learning parameters such as $\eta(0)$, h_0, and σ_0.
3. Repeat:
   a) Present input pattern $\mathbf{x}_t$ at time t.
   b) Select the winning neuron whose prototype is closest to $\mathbf{x}_t$ by (5.11).
   c) Update the prototypes for all neurons by (5.12).
   d) Set $t = t + 1$.
   until a criterion is satisfied.

The initialization of the codebook vectors $\mathbf{c}_k(0)$ can be selected as random values, or from available samples, or any ordered initial state. The algorithm can be stopped when the map achieves an equilibrium with a given accuracy or when a specified number of iterations is reached.

In the convergence phase, h_{wk} can be selected as time invariant, and each prototype is recommended to be updated by using an individual learning rate η_k [614]

$$\eta_k(t + 1) = \frac{\eta_k(t)}{1 + h_{wk}\eta_k(t)} \tag{5.15}$$

Normalization of $\mathbf{x}$ is sometimes suggested since the resulting reference vectors tend to have the same dynamic range. This may improve the numerical accuracy [612].

The SOM is deemed to converge to an organized configuration in one- or higher-dimensional SOM with probability one. In the literature, there are some proofs for the convergence of the one-dimensional SOM based on the Markov-chain analysis [360]. However, no general proof of convergence for multidimensional SOM is available [360, 614].

The SOM [611] is a clustering network with a set of heuristic procedures that suffers from several major problems, such as forced termination, unguaranteed convergence, nonoptimized procedures, and the output being often dependent on the sequence of data. The SOM is not based on the minimization of any known objective function, termination is not based on optimizing any model of the process or its data. It is well known that the Kohonen network is closely related to the C-means clustering [717].

After the learning phase is completed, the network is ready for generalization. When a new input pattern $\mathbf{x}$ is presented to the map, the corresponding output $\mathbf{c}$ is determined according to the mapping: $\mathbf{x} \to \mathbf{c}$ such that $\|\mathbf{x} - \mathbf{c}\| = \min_{r \in A} \|\mathbf{x} - \mathbf{c}_r\|$. The mapping performs VQ of the input space into the map A.

5.3.2 Applications of Self-organizing Maps

The SOM is well known for its ability to perform clustering while preserving topology. It compresses information while preserving the most important topological and metric relationships of the primary data elements. The SOM is useful for VQ, clustering, feature extraction, and data visualization. The Kohonen learning rule is a major development of the competitive learning.

Like the classical VQ method, the SOM was originally intended to approximate input signals or their PDFs, by quantified codebook vectors that are localized in the input space to minimize a quantization error functional [612]. By assigning each input vector to a neuron with the nearest reference vector, the SOM is able to divide the input space into regions with common nearest reference vectors.

The SOM is related to adaptive C-means, but performs a topological feature map that is more complex than just cluster analysis. After training, the input vectors are spatially ordered in the array, $i.e.$ the neighboring input vectors in the map are more similar than the more remote ones. The Kohonen learning rule provides a codebook in which the distortion effects are automatically taken into account. The SOM is especially powerful for the visualization of high-dimensional data. It converts complex, nonlinear statistical relations between high-dimensional data into simple geometric relations at a low-dimensional display.

The topology preservation property makes the SOM a popular choice in data analysis. However, the SOM is not a good choice in terms of clustering performance compared to other popular clustering algorithms such as the C-means, the neural gas, and the ART 2A [768, 471]. Besides, for large output dimensions, the number of nodes in the adaptive SOM grid increases exponentially with the number of function parameters. The prespecified standard grid topology may not be able to match the structure of the distribution, and can thus lead to poor topological mappings.

The SOM can be used to decompose complex information-processing systems into a set of simple subsystems [398]. A comprehensive survey of engineering applications of the SOM is given in [613]. A fully analog integrated circuit of the SOM has been designed in [757].

5.3.3 Extensions of Self-organizing Maps

Adaptive-subspace SOM (ASSOM) [613, 616, 614] is a modular neural network model comprising an array of topologically ordered SOM submodels. ASSOM creates a set of local subspace representations by competitive selection and cooperative learning. Each submodel is responsible for describing a specific region of the input space by its local principal subspace, and represents a manifold such as a linear subspace with a small dimensionality, whose basis vectors are determined adaptively. The ASSOM model not only inherits the topological representation property of the SOM, but provides learning results that reasonably describe the kernels of various transformation groups like the PCA. The ASSOM is used to learn a number of invariant features, usually pieces of elementary one- or two-dimensional waveforms with different frequencies called *wavelets*, independent of their phases.

The hyperbolic SOM (HSOM) [954] implements its lattice by a regular triangulation of the hyperbolic plane. The number of neighbors of a neuron in a hyperbolic grid can increase exponentially as a function of the distance between the neurons on the hyperbolic lattice, whereas the growth in the number of neighbors of regular lattices in the Euclidean space follows a power law. Therefore, the hyperbolic lattice provides more freedom to map a complex information space such as language into spatial relations.

The extraction of knowledge from large databases is an essential task of data analysis and data mining. The multidimensional data may involve quantitative variables and qualitative (nominal, ordinal) variables such as categorical data, which is the case in survey data. The PCA is a popular technique that projects quantitative data on significant axes, while correspondence analysis is a technique for analyzing the relations existing between the modalities of all the qualitative variables by completing a simultaneous projection of the modalities. The SOM can be viewed as an extension of the PCA due to its topology-preserving property. For qualitative variables, the SOM has been generalized for multiple correspondence analysis [266].

The SOM is designed for real-valued vectorial data analysis, and it is not suitable for nonvectorial data analysis such as the structured data analysis. Examples of structured data are temporal sequences such as the time series, language, and words, spatial sequences like the DNA chains, and tree- or graph-structured data arising from natural language parsing and from chemistry. Various unsupervised models for nonvectorial data have been proposed recently. Prominent unsupervised self-organizing methods are the temporal Kohonen map (TKM), the recurrent SOM (RSOM), the recursive SOM (RecSOM), the SOM for structured data (SOMSD), and the merge SOM (MSOM). All these models introduce recurrence into the SOM. These models have been reviewed and compared in [447, 1061].

5.4 Learning Vector Quantization

The k-NN algorithm [327] is a widely applicable classification technique. It is also used for outlier detection. All training patterns are used as prototypes and an input pattern is assigned to the class with the closest prototype. It generalizes well for large training sets, and the training set can be extended at any time. The theoretical asymptotic classification error is upper-bounded by twice the Bayes error. However, it uses a large storage space, and has a computational complexity of $O\left(N^2\right)$, where N is the size of the pattern set. It also takes a long time for recall. Thus, the k-NN is impractical for large training sets.

The LVQ [612] employs exactly the same network architecture as the Kohonen network with the exception that each output neuron is specified with a classmembership and no assumption is made concerning the topological structure. There are two families of the LVQ-style models, supervised models such as the LVQ1, LVQ2, and LVQ3 [611] as well as unsupervised models such as the LVQ[4] [611] and the incremental C-means [746]. The LVQ network is associated with the two-layer competitive learning network shown in Fig. 5.2.

The LVQ in the supervised mode is based on the known classification of feature vectors, and can be treated as a supervised version of the SOM. The LVQ is used for VQ and classification, as well as for the fine tuning of the SOM [611, 612]. LVQ algorithms define near-optimal decision borders between classes, even in the sense of classical Bayesian decision theory.

The LVQ minimizes the functional (5.5), where $\mu_{kp} = 1$ if neuron k is the winner and zero otherwise, when a pattern pair p is presented. The LVQ works on a set of N pattern pairs $(\mathbf{x}_p, \mathbf{y}_p)$, where $\mathbf{x}_p \in R^J$ is the input vector and $\mathbf{y}_p \in R^K$ is the binary target vector coding the classmembership, that is, only one entry of $\mathbf{y}_p$ takes the value unity, while all its other entries are zero. Kohonen proposed a family of LVQ algorithms including the LVQ1, LVQ2, and LVQ3 [612]. Assuming that the pth pattern is presented at time t, the LVQ1 is given as [612]

$$
\begin{aligned}
\mathbf{c}_w(t+1) &= \mathbf{c}_w(t) + \eta(k)\left[\mathbf{x}_t - \mathbf{c}_w(t)\right], & y_{p,w} = 1 \\
\mathbf{c}_w(t+1) &= \mathbf{c}_w(t) - \eta(t)\left[\mathbf{x}_t - \mathbf{c}_w(k)\right], & y_{p,w} = 0 \\
\mathbf{c}_i(t+1) &= \mathbf{c}_i(t), & i \neq w
\end{aligned}
\tag{5.16}
$$

where w is the index of the winning neuron, $\mathbf{x}_t = \mathbf{x}_p$, $y_{p,w} = 1$ and 0, respectively, represent correct and incorrect classifications of $\mathbf{x}_p$, and $\eta(t)$ is defined as in earlier formulations. When it is used to fine tune the SOM, one should start with a small $\eta(0)$, usually less than 0.1, such as 0.01. This algorithm tends to reduce the point density of $\mathbf{c}_i$ around the Bayesian decision surfaces.

The OLVQ1 is an optimized version of the LVQ1 [615]. In the OLVQ1, each codebook vector $\mathbf{c}_i$ is assigned an individual adaptive learning rate η_i, which is adjusted by [615]

$$
\eta_i(t) = \frac{\eta_i(t-1)}{1 + s(t)\eta_i(t-1)}
\tag{5.17}
$$

where $s(t) = +1$ if the classification is correct and $s(t) = -1$ if the classification is wrong. Since $\eta_i(t)$ can increase, it should be limited to be less than 1. One can limit $\eta_i(t)$ to be less than $\eta_i(0)$, and select $\eta_i(0)$ as 0.3. The convergence of the OLVQ1 may be up to one order of magnitude faster than that of the LVQ1.

---

[4] The unsupervised LVQ is essentially the SCL-based VQ, which is discussed in Sect. 5.1.

LVQ2 and LVQ3 comply better with the Bayesian decision surface. In LVQ1, only one codebook vector $\mathbf{c}_i$ is updated at each step, while LVQ2 and LVQ3 change two codebook vectors simultaneously. Different LVQ algorithms can be combined in the clustering process[5]. However, both LVQ2 and LVQ3 have the problem of reference vector divergence [999]. In a generalization of the LVQ2 [999], this problem is eliminated by applying gradient descent on a nonlinear cost function. Some applications of the LVQ in engineering have been reviewed in [616].

The addition of training counters to individual neurons of the LVQ can effectively record the training statistics of the LVQ [861]. During the course of training, this allows for dynamic self-allocation of the neurons to classes. At the generalization stage, these counters provide an estimate of the reliability of classification of the individual neurons. The method turns out to be especially valuable in handling strongly overlapping class distributions in the pattern space.

5.5 *C*-means Clustering

The most well-known data-clustering technique is the statistical C-means[6] clustering [1096, 817], which is due to MacQueen [746]. The C-means algorithm approximates the ML solution for determining the location of the means of a mixture density of component densities. The C-means clustering is closely related to simple competitive learning, and is a special case of the SOM. The algorithm partitions a set of N input patterns, $\mathcal{X}$, into K separated subsets $\mathcal{C}_k$, each containing N_k input patterns by minimizing the MSE function

$$E\left(\mathbf{c}_1, \cdots, \mathbf{c}_K\right) = \frac{1}{N} \sum_{k=1}^{K} \sum_{\mathbf{x}_n \in \mathcal{C}_k} \|\mathbf{x}_n - \mathbf{c}_k\|^2 \qquad (5.18)$$

where $\mathbf{c}_k$ is the prototype or *center* of the cluster $\mathcal{C}_k$. To improve the similarity of samples in each cluster, one can minimize E with respect to $\mathbf{c}_k$ by setting $\frac{\partial E}{\partial \mathbf{c}_k} = 0$; thus, the optimal location of $\mathbf{c}_k$ is the mean of the samples in the cluster

$$\mathbf{c}_k = \frac{1}{N_k} \sum_{\mathbf{x}_i \in \mathcal{C}_k} \mathbf{x}_i \qquad (5.19)$$

The C-means clustering can be implemented in either the batch mode [714, 817] or the incremental mode [746]. The batch C-means [714], frequently called the *Linde–Buzo–Gray, LBG* or *generalized Lloyd algorithm*, is applied when the whole training set is available. When the training set is obtained online, the incremental C-means is commonly applied.

---

[5] A program package for LVQ algorithms called LVQPAK is introduced in [615].
[6] In the literature, C-means is also known as k-means.

In the batch C-means, the initial partition is arbitrarily defined by placing each input pattern into a randomly selected cluster. The prototypes are defined to be the average of the patterns in the individual clusters. When the C-means is performed, at each step the patterns keep changing from one cluster to the closest cluster $\mathbf{c}_k$ according to the SCL rule

$$\|\mathbf{x}_i - \mathbf{c}_k\| = \min_j \|\mathbf{x}_i - \mathbf{c}_j\| \tag{5.20}$$

and the prototypes are then recalculated according to (5.19).

In the incremental C-means, each cluster is initialized with a random pattern as its prototype. The C-means continues to update the prototypes upon the presentation of each new pattern. If at time t the kth prototype is $\mathbf{c}_k(t)$ and the input pattern is $\mathbf{x}_t$, then at time $t+1$ the incremental C-means gives the new prototype as

$$\mathbf{c}_k(t+1) = \begin{cases} \mathbf{c}_k(t) + \eta(t)\,(\mathbf{x}_t - \mathbf{c}_k(t)), & k = \arg_j \min \|\mathbf{x} - \mathbf{c}_j\| \\ \mathbf{c}_k(t), & \text{otherwise} \end{cases} \tag{5.21}$$

where $\eta(t)$ is a learning rate. Usually $\eta < 1$ and should slowly decrease to zero. Typically

$$\eta(t) = \frac{\eta(0)}{\sqrt{1 + \text{int}(\frac{t}{M})}} \tag{5.22}$$

where $\text{int}(\cdot)$ extracts the integer part of a number and M is a positive integer.

The general procedure for the C-means clustering is given by Algorithm 5.5.

Algorithm 5.5 (C-means)

1. Set K.
2. Arbitrarily select an initial cluster partition.
3. Repeat:
   a) Decide K cluster prototypes $\mathbf{c}_k$.
   b) Redistribute patterns among the clusters using criterion (5.18).
   until the change in all $\mathbf{c}_k$ is sufficiently small.

After the algorithm converges, we can calculate the variance vector, $\boldsymbol{\sigma}_k = (\sigma_{k,1}, \cdots, \sigma_{k,J})^{\mathrm{T}}$, for each cluster

$$\sigma_{k,i} = \sqrt{\frac{\sum_{\mathbf{x}_j \in C_k} (x_{j,i} - c_{k,i})^2}{N_k - 1}} \tag{5.23}$$

for $k = 1, \cdots, K$, $i = 1, \cdots, J$. As a gradient-descent-based algorithm, the C-means clustering achieves a local optimum solution that depends on the initial selection of the cluster prototypes. The number of clusters must also be prespecified.

5.5.1 Improvements on the *C*-means

When an initial prototype is in a region with few training patterns, this results in a large cluster. This disadvantage can be remedied by a modified *C*-means [1190]. The clustering starts from one cluster, $k = 1$. It splits the cluster with the largest intracluster distance into two clusters. After each splitting, the *C*-means is applied until the existing k clusters are convergent. This procedure is continued until K clusters are obtained.

In [225], the incremental *C*-means is improved by adding two mechanisms, one for biasing the clustering towards an optimal Voronoi partition [407] by using a cluster variance-weighted MSE as the objective function and the other for adjusting the learning rate dynamically according to the current variances in all partitions. The method always converges to an optimal or near-optimum configuration.

The local minimum problem of the *C*-means clustering can be eliminated by using global optimization methods such as the GA [630, 62], the SA [61], and a hybrid SA/EA system [295]. In the genetic *C*-means [630], the chromosomes encode the classmembership matrix $\mathbf{U} = [\mu_{kp}]$. The *C*-means operator, defined as one step of the batch *C*-means, is a local-search operator, and the distance-based mutation is a biased mutation operator specific to clustering. The crossover operator is not used in this technique. In the GA-based *C*-means [62], the chromosomes encode the K prototypes of the clusters using floating-point coding. The algorithm tries to evolve the appropriate cluster prototypes while optimizing a given clustering metric. The SA-based *C*-means clustering [61] integrates the power of the SA for obtaining the minimum energy configuration and the searching capability of the *C*-means. The SA-based *C*-means is used to search for appropriate clusters in the multidimensional feature space such that a similarity metric of the resulting clusters is optimized. Data points are redistributed among the clusters probabilistically, so that points that are farther away from the cluster prototype have higher probabilities of migrating to other clusters.

The enhanced LBG [889] is derived directly from the LBG with a negligible overhead. The concept of utility of a codeword is a powerful instrument to overcome the problem of bad local minima arising from a bad choice of the initial codebook. The utility allows the identification of those badly positioned codewords, and guides their movement from the proximity of a local minimum in the error function. The enhanced LBG outperforms the LBG with utility (LBG-U) [383] both in terms of the accuracy and the number of required iterations. The LBG-U is also based on the LBG algorithm and the concept of utility.

The relation between the PCA and the *C*-means has been established in [309]. Principal components have been proved to be the continuous solutions to the discrete cluster membership indicators for the *C*-means clustering, with a clear simplex cluster structure [309]. PCA-based dimensionality reductions are particularly effective for the *C*-means clustering. Lower bounds for the

C-means objective function (5.18) are derived as the total variance minus the eigenvalues of the data covariance matrix [309].

In the two-stage clustering procedure [1129], the SOM is first used to cluster the data set, and the prototypes produced are further clustered using an agglomerative clustering algorithm or the C-means. The clustering results using the SOM as an intermediate step are comparable with direct clustering of the data, but with a significantly reduced computation time.

5.6 Mountain and Subtractive Clustering

The mountain clustering [1208, 1209] is a simple and effective method for estimating the number of clusters and the initial locations of the cluster centers, which are the difficulties faced by most conventional methods. The method grids the data space and computes a potential value for each grid point based on its distance to the actual data points. Each grid point is treated as a potential cluster center depending on its potential value. A measure of the potential for each grid is calculated based on the density of the surrounding data points. The grid with the highest potential is selected as the first cluster center and then the potential values of all the other grids are reduced according to their distance to the first cluster center. Grid points closer to the first cluster center have greater reduction in potential. The next cluster center is located at the grid point with the highest remaining potential. This process is repeated until the remaining potential values of all the grids fall below a threshold. However, the grid structure causes the method to grow exponentially with the dimension of the problem.

The subtractive clustering method [226] is a modified form of the mountain clustering method. The idea is to use all the data points to replace all the grid points as potential cluster centers. By this means, the effective number of grid points is reduced to the size of the pattern set, which is independent of the dimensionality of the problem. The problems of gridding resolution and high computational complexity are thus avoided. The subtractive clustering is a fast method for estimating clusters in the data [226].

The subtractive clustering assumes each of the N data points in the pattern set, $\mathbf{x}_i$, to be a potential cluster center, and the potential measure is defined as a function of the Euclidean distances to all the other input data points

$$P(i) = \sum_{j=1}^{N} e^{-\alpha \|\mathbf{x}_i - \mathbf{x}_j\|^2} \tag{5.24}$$

for $i = 1, \cdots, N$, where $\alpha = \frac{4}{r_a^2}$, the constant $r_a > 0$ being effectively a normalized radius defining the neighborhood. Data points outside this radius have insignificant influence on the potentials. A data point surrounded by many neighboring data points has a high potential value. Thus, the mountain and subtractive clustering techniques are less sensitive to noise than other clustering algorithms, such as the C-means [746, 714] and the FCM [104].

After the data point with the highest potential, $\mathbf{x}_u$, where $u = \arg_i \max P(i)$, is selected as the kth cluster center, that is, $\mathbf{c}_k = \mathbf{x}_u$ with $\overline{P}(k) = P(u)$ as its potential value, the potential of each data point $\mathbf{x}_i$ is revised by subtracting a term associated with $\mathbf{c}_k$

$$P(i) = P(i) - \overline{P}(k)e^{-\beta\|\mathbf{x}_i - \mathbf{c}_k\|^2} \tag{5.25}$$

where $\beta = \frac{4}{r_b^2}$, and the constant $r_b > 0$ is a normalized radius defining the neighborhood. In order to avoid closely located cluster centers, r_b is set greater than r_a, typically $r_b = 1.25 r_a$.

The algorithm continues until the remaining potential of all the data points is below some fraction of the potential of the first cluster center, that is,

$$\overline{P}(k) = \max_i P(i) < \varepsilon \overline{P}(1) \tag{5.26}$$

where ε is selected within $(0, 1)$. When ε is close to 0, a large number of hidden nodes will be generated. On the contrary, a value of ε close to 1 will lead to a small network structure. Typically, ε is selected as 0.15.

The subtractive clustering algorithm is described by Algorithm 5.6 [226, 227].

Algorithm 5.6 (Subtractive Clustering)

1. Set r_a, r_b, and ε.
2. Calculate the potential values $P(i)$, $i = 1, \cdots, N$.
3. Set $k = 1$.
4. Repeat:
   a) Find the data point $\mathbf{x}_u$ with $u = \arg_i \max P(i)$.
   b) Set the kth cluster center as $\mathbf{x}_u$, that is, $\mathbf{c}_k = \mathbf{x}_u$ and $\overline{P}(k) = P(u)$.
   c) Revise the potential of each data point $\mathbf{x}_i$ by (5.25).
   d) Set $k = k + 1$.
   until $\overline{P}(k) < \varepsilon \overline{P}(1)$.

The training data $\mathbf{x}_i$ is recommended to be scaled before applying the method. This helps in selecting proper values for α and β. Since it is difficult to select a suitable ε for all data patterns, additional criteria for accepting/rejecting cluster centers can be used. One method is to select two thresholds [226, 227], namely, $\overline{\varepsilon}$ and $\underline{\varepsilon}$. Above $\overline{\varepsilon}$, $\mathbf{c}_k$ is definitely accepted as a cluster center, while below $\underline{\varepsilon}$ it is definitely rejected. If $\overline{P}(k)$ falls between the two thresholds, a tradeoff between a reasonable potential and its distance to the existing cluster centers must been examined

$$R = \frac{d_{\min}}{r_a} + \frac{\overline{P}(k)}{\overline{P}(1)} \tag{5.27}$$

where $d_{\min}$ is the shortest of the distances between $\mathbf{c}_k$ and $\mathbf{c}_i$, $i = 1, \cdots, k-1$. If $R \geq 1$, accept $\mathbf{c}_k$ and continue the algorithm. If $R < 1$, reject $\mathbf{c}_k$ and set $\overline{P}(k) = P(u) = 0$, and select the data point with the next highest potential as $\mathbf{c}_k$ and retest.

Unlike the C-means and FCM algorithms, which require iterations of many epochs, the subtractive clustering requires only one pass of the training data. Besides, the number of clusters does not need to be specified *a priori*. The subtractive clustering is a deterministic method: For the same neural network structure, the same network parameters are always obtained.

The C-means and FCM algorithms require $O(KNT)$ computations, where T is the total number of epochs and each computation requires the calculation of the distance and the memberships. The computational load for the subtractive clustering is $O\left(N^2 + KN\right)$, each computation involving the calculation of the exponential function. Thus, for small- or medium-size training sets, the subtractive clustering is relatively fast. However, when $N \gg KT$, the subtractive clustering requires more training time [279]. The subtractive clustering is suitable for problems with moderate size of data set.

The subtractive clustering provides only rough estimates of the cluster centers, since the cluster centers obtained are situated at some data points. Moreover, since α and β are not determined from the data set and no cluster validity is used, the clusters produced may not appropriately represent the clusters. For small data sets, one can try a number of values for α, β and ε and select a proper network structure. The results by the subtractive clustering can be used to determine the number of clusters and their initial values for initializing iterative optimization-based clustering algorithms such as the C-means and the FCM.

The subtractive clustering can be improved by performing a search over α and β, which makes it essentially equivalent to the least-biased fuzzy clustering algorithm [98]. The least-biased fuzzy clustering, based on the deterministic annealing approach [961, 960], tries to minimize the clustering entropy of each cluster, namely, the entropy of the centroid with respect to the clustering membership distribution of data points, under the assumption of unbiased centroids. The subtractive clustering can be realized by replacing the Gaussian potential function with a Cauchy-type function of first order, and an online clustering method has been implemented based on this new potential function in [39].

In [881], the mountain clustering and the subtractive clustering methods are improved by tuning the prototypes obtained using the gradient-descent method so as to maximize the potential function. By modifying the potential function, the mountain method can also be used to detect other types of clusters like circular shells [881].

A kernel-induced distance is used to replace the Euclidean distance in the potential function [602]. This enables clustering of the data that is linearly inseparable in the original space into homogeneous groups in the transformed high-dimensional space, where the data separability is increased; the proposed method is characterized by higher clustering accuracy than the subtractive method.

5.7 Neural Gas

The neural gas (NG) [768] is a VQ model that minimizes a known cost function and converges to the C-means quantization error via a soft-to-hard competitive model transition. The soft-to-hard annealing process helps the algorithm to escape from local minima. The NG is a topology-preserving network, and can be treated as an extension to the C-means. It has a fixed number of processing units, K, with no lateral connection.

A data optimal topological ordering is achieved by using neighborhood ranking within the input space at each training step. To find its neighborhood rank, each neuron compares its distance to the input vector with the distances of all the other neurons to the input vector. Neighborhood ranking provides the training strategy with mechanisms related to robust statistics, and the NG does not suffer from the prototype underutilization problem (see Sect. 5.11) [980]. At each step t, the Euclidean distances between an input vector $\mathbf{x}_t$ and all the prototype vectors $\mathbf{c}_k(t)$, $k = 1, \cdots, K$, are calculated by

$$d_k\left(\mathbf{x}_t\right) = \|\mathbf{x}_t - \mathbf{c}_k(t)\| \tag{5.28}$$

and $\mathbf{d}((t)) = (d_1\left(\mathbf{x}_t\right), \cdots, d_K\left(\mathbf{x}_t\right))^{\mathrm{T}}$. Each prototype $\mathbf{c}_k(t)$ is assigned a rank $r_k(t)$, which takes an integer value from $0, \cdots, K - 1$, with 0 for the smallest and $K - 1$ for the largest $d_k\left(\mathbf{x}_t\right)$.

The prototypes are updated by

$$\mathbf{c}_k(t + 1) = \mathbf{c}_k(t) + \eta h\left(r_k(t)\right)\left(\mathbf{x}_t - \mathbf{c}_k(t)\right) \tag{5.29}$$

where $h(r) = \mathrm{e}^{-\frac{r}{\rho(t)}}$ realizes a soft competition, and $\rho(t)$ is the neighborhood width. When $\rho(t) \to 0$, (5.29) reduces to the C-means update rule (5.21). During the iterations, both $\rho(t)$ and $\eta(t)$ decrease exponentially from their initial positive values

$$\eta(t) = \eta_0 \left(\frac{\eta_\mathrm{f}}{\eta_0}\right)^{\frac{t}{T_\mathrm{f}}} \tag{5.30}$$

$$\rho(t) = \rho_0 \left(\frac{\rho_\mathrm{f}}{\rho_0}\right)^{\frac{t}{T_\mathrm{f}}} \tag{5.31}$$

where η_0 and ρ_0 are the initial decay parameters, η_f and ρ_f are the final decay parameters, and T_f is the maximum number of iterations.

Algorithm 5.7 (Neural Gas)

1. Initialize K, c_k, $k = 1, \cdots, K$, ρ_0, η_0, ρ_f, η_f, and T_f.
2. Set $t = 1$.
3. Repeat:
    a) Calculate distances $d_k(x_t)$, $k = 1, \cdots, K$, by (5.28).
    b) Sort the components of $d(t)$ and assign each prototype with a rank $r_k(t)$, $k = 1, \cdots, K$, which is a unique value from $0, 1, \cdots, K - 1$.
    c) Calculate $\eta(t)$, $\rho(t)$ by (5.30) and (5.31), respectively.
    d) Update c_k, $k = 1, \cdots, K$, by (5.29).
    e) Set $t = t + 1$.
    until a stopping criterion is satisfied.

The prototypes c_k are initialized by randomly assigning vectors from the training set. The NG algorithm is given by Algorithm 5.7.

Unlike the SOM, which uses predefined static neighborhood relations, the NG determines a dynamical neighborhood relation as learning proceeds. The NG can be derived from an explicit cost function by using the gradient-descent method. The NG is an efficient and reliable clustering algorithm that is not sensitive to the neuron initialization.

The NG converges faster to a smaller error E than the C-means, the maximum-entropy clustering [961], and the SOM. This advantage is achieved at the price of a higher computational effort. In a serial implementation, the complexity for the NG is $O(K \log K)$ while the other three methods all have a complexity of $O(K)$, where K is the number of prototypes. Nevertheless, in parallel implementation all the four algorithms have the same complexity, $O(\log K)$ [768]. The NG can be derived from a gradient-descent procedure on a potential function associated with the framework of fuzzy clustering [104].

Some efforts have been made to accelerate the sequential NG. In the fast implementation given in [238], a truncated exponential function is used as the neighborhood function and the neighborhood ranking is implemented without evaluating and sorting all the distances. Given the same quality of the resulting codebook, this fast realization gains a speedup of five times over the original NG for codebook design in image vector quantization. In [968], an improved NG and its analog VLSI subcircuitry have been developed based on partial sorting. The VLSI architecture includes two chips, one for the Euclidean distance computation and the other for the programmable sorting of code vectors. The latter is based on the WTA structure [653]. The approach is empirically shown to reduce the training time by up to two orders of magnitude, without reducing the performance quality.

5.7.1 Competitive Hebbian Learning

In the Voronoi tessellation, when the prototype of each Voronoi region is connected to all the prototypes of its bordering Voronoi regions, a Delaunay triangulation is obtained. Competitive Hebbian learning [767, 769] is a method that generates a subgraph of the Delaunay triangulation, called the *induced Delaunay triangulation* by masking the Delaunay triangulation with a data distribution $P(\mathbf{x})$. The induced Delaunay triangulation has been proved to be optimally topology-preserving in a general sense [767].

Given a number of prototypes in R^J, competitive Hebbian learning successively adds connections among them by evaluating input data drawn from a distribution $P(\mathbf{x})$. The method does not change the prototypes, but only generates topology according to these prototypes. For each input vector $\mathbf{x}$, the two closest prototypes are connected by an edge. This leads to the induced Delaunay triangulation, which is limited to those regions of the input space R^J, where $P(\mathbf{x}) > 0$. The Delaunay triangulation and the induced Delaunay triangulation are illustrated in Fig. 5.4.

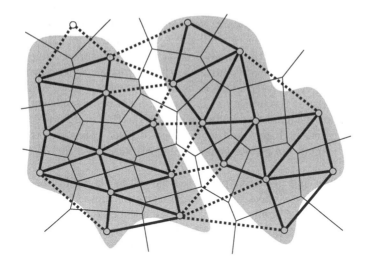

Fig. 5.4. An illustration of the Delaunay triangulation and the induced Delaunay triangulation. The Delaunay triangulation is represented by a mix of thick and thick dashed lines, the induced Delaunay triangulation by thick lines, Voronoi tessellation by thin lines, prototypes by circles, and a data distribution $P(\mathbf{x})$ by shaded regions. To generate the induced Delaunay triangulation, two prototypes are connected only if at least a part of the common border of their Voronoi polygons lies in a region where $P(\mathbf{x}) > 0$.

5.7.2 The Topology-representing Network

The topology-representing network [769] is obtained by alternating the learning steps of the NG and the competitive Hebbian learning, where the NG is used to distribute a certain number of prototypes and the competitive Hebbian learning is then used to generate the topology. An edge aging scheme is used to remove obsolete edges. Competitive Hebbian learning avoids the topological defects observed for the SOM.

5.8 ART Networks

Adaptive resonance theory (ART) [435] is biologically motivated and is a major development of the competitive learning paradigm. The theory leads to an evolving series of real-time unsupervised network models for clustering, pattern recognition, and associative memory [159–164, 166]. These models are capable of stable category recognition in response to arbitrary input sequences with either fast or slow learning. ART models are characterized by systems of differential equations, which formulate stable self-organizing learning methods. Instar and outstar learning rules are the two learning rules used. The ART has the ability to adapt, yet not forget the past training, and this is what Grossberg refers to as the *stability–plasticity dilemma* [435, 159].

At the training stage, the stored prototype of a category is adapted when an input pattern is sufficiently similar to the prototype. When novelty is detected, the ART adaptively and autonomously creates a new category with the input pattern as the prototype. The meaning of being sufficiently similar is dependent on a vigilance parameter $\rho \in (0, 1]$. If ρ is large, the similarity condition becomes stringent and many finely divided categories are formed. In contrast, a smaller ρ gives a coarser categorization, resulting in fewer categories.

The stability and plasticity properties as well as the ability to efficiently process dynamic data make the ART attractive for clustering large, rapidly changing sequences of input patterns, such as in the case of data mining [772]. However, the ART approach does not correspond to the C-means algorithm for cluster analysis and the VQ in the global optimization sense [717].

5.8.1 ART Models

The ART model family includes a series of unsupervised learning models. ART networks employ a J-K recurrent architecture. The input layer F1, called the *comparing layer*, has J neurons, while the output layer F2, called the *recognizing layer*, has K neurons. Layers F1 and F2 are fully interconnected in both the directions. Layer F2 acts as a WTA network. The feedforward weights connecting to the F2 neuron j are represented by the vector $\mathbf{w}_j$, while the feedback weights from the same neuron are represented by the vector $\mathbf{c}_j$.

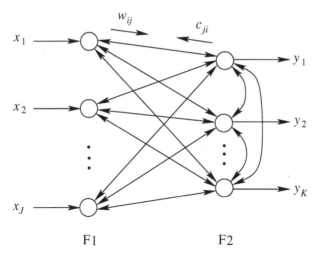

Fig. 5.5. Architecture of the ART model. w_{ij} and c_{ji} are, respectively, feedforward and feedback weights between layers F1 and F2. The feedforward weights connecting to the F2 neuron j are represented by the vector $\mathbf{w}_j = (w_{1j}, \cdots, w_{Jj})^{\mathrm{T}}$, while the feedback weights from the same neuron are represented by the vector $\mathbf{c}_j = (c_{j1}, c_{j2}, \cdots, c_{jJ})^{\mathrm{T}}$. The vector $\mathbf{c}_j$ stores the prototype of cluster j. Layer F2 is a WTA network. The output selects one of the K prototypes $\mathbf{c}_i$ by setting $y_i = 1$ and $y_j = 0$, $j = 1, \cdots, K$, $j \neq i$

The vector $\mathbf{c}_j$ stores the prototype of cluster j. J is the number of features used to represent a pattern, and K, the number of clusters, varies with the size of the problem. The architecture of the ART model is shown in Fig. 5.5.

The ART models are characterized by a set of STM and LTM time-domain nonlinear differential equations. The STM equations describe the evolution of the neurons and the interactions between them, while the LTM equations describe the change of the interconnection weights with time as a function of the system state. Layer F1 stores the STM for the current input pattern, while F2 stores the prototypes of clusters as the LTM.

Three types of ART implementations can be distinguished, namely, full mode, STM steady-state mode, and fast-learning mode [160, 1015]. In the full-mode implementation, both the STM and LTM differential equations are realized. The STM steady-state mode only implements the LTM differential equations, while the STM behavior is governed by nonlinear algebraic equations. In the fast-learning mode, both the STM and LTM are implemented by their steady-state nonlinear algebraic equations, and thus proper sequencing of STM and LTM events is required. The fast-learning mode is inexpensive and is most popular.

Like the incremental C-means, the ART model family is sensitive to the order of presentation of the input patterns. ART models tend to build clusters of the same size, independently of the distribution of the data.

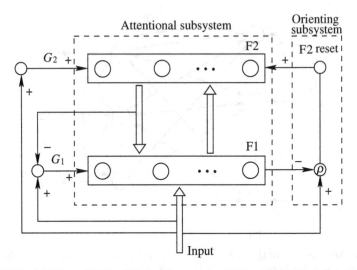

Fig. 5.6. Architecture of the ART 1 with supplemental units. G_1 and G_2 are outputs of gain control units. The F2 reset unit controls the vigilance matching.

ART 1

The simplest and most popular ART model is the ART 1 [159] for learning to categorize arbitrarily many, complex binary input patterns presented in an arbitrary order.

The main elements of a minimum ART 1 model are shown in Fig. 5.6. The two fields of neurons, F1 and F2, are linked both bottom-up and top-down by adaptive filters. The unsupervised two-layer feedforward (bottom-up) pattern recognition network is termed an *attentional subsystem*. There is also an auxiliary subsystem, called the *orienting subsystem*, that becomes active during search.

Since ART 1 in fast learning mode is most widely used, we only discuss this mode here. To begin with, all cluster categories are set as *uncommitted*. When a new pattern is presented at time t, the net input or activation to F2 neuron j is given by

$$net_j(t) = \mathbf{x}_t^{\mathrm{T}} \mathbf{w}_j(t) \tag{5.32}$$

for $j = 1, \cdots, K$, where $\mathbf{w}_j = (w_{1j}, \cdots, w_{Jj})^{\mathrm{T}}$.

Competition between F2 neurons is performed to select the most similar category represented by the winning neuron w such that

$$net_w(t) = \max_{j=1,\cdots,K} net_j(t) \tag{5.33}$$

Neuron w then undergoes a vigilance test so as to determine whether it is close enough to $\mathbf{x}_t$:

$$\frac{\|\mathbf{x}_t \wedge \mathbf{c}_w(t-1)\|}{\|\mathbf{x}_t\|} \geq \rho \qquad (5.34)$$

where $\wedge$ denotes the logical AND operation. For binary values of x_i, the Euclidean norm $\|\mathbf{x}\| = \sum_i x_i$.

If the winning neuron passes the vigilance test, the system enters the resonance mode, and the weights for the winning neuron are updated by

$$\mathbf{c}_w(t) = \mathbf{c}_w(t-1) \wedge \mathbf{x}_t \qquad (5.35)$$

$$\mathbf{w}_w(t) = \frac{L\left[\mathbf{c}_w(t-1) \wedge \mathbf{x}_t\right]}{L - 1 + \|\mathbf{c}_w(t-1) \wedge \mathbf{x}_t\|} \qquad (5.36)$$

where $L > 1$ is a constant parameter.

Otherwise, the F2 neuron reset mechanism is applied to remove the neuron w from the current search by setting $net_w = -1$ and the system enters the search mode. If all the stored categories cannot pass the vigilance test, one of the uncommitted categories of the K categories is assigned to this pattern.

A popular fast-learning implementation is given by Algorithm 5.8.1.1 [820, 772, 1015][7].

Algorithm 5.8.1.1 (ART 1)

1. Initialization.
   a) Select $0 < \rho \leq 1$ and $L > 1$.
   b) Set $\mathbf{w}_j(0) = \frac{1}{1+J}\mathbf{1}$ and $\mathbf{c}_j(0) = \mathbf{1}$, $j = 1, \cdots, K$, where $\mathbf{1}$ denotes a J-dimensional vector, all of whose entries are unity.
   c) Set $t = 1$.
2. Set F2 neuron activations $net_j(t) = 0$, $j = 1, \cdots, K$, and present pattern $\mathbf{x}_t$.
3. Compute F2 neuron activations $net_j(t)$, $j = 1, \cdots, K$, by (5.32).
4. Competition between F2 neurons.
   Select the winning neuron w according to (5.33).
5. Vigilance test.
   Determine if $\mathbf{c}_w(t)$ is close enough to $\mathbf{x}_t$ by (5.34).
   a) If true, go to Step 6 (resonance mode).
   b) Otherwise, go to Step 8 (search mode).
6. Update weights for winning node by (5.35) and (5.36).
7. Set $t = t + 1$ and return to Step 2 until all patterns are presented.
8. Set $net_w(t) = -1$ and return to Step 4.

In the algorithm, ρ determines the level of abstraction at which the ART discovers clusters. The minimal number of clusters present in the data can be determined by $\rho_{\min} < \frac{1}{J}$ [772]. Initial bottom-up weights are usually selected

---

[7] In [1015], $\mathbf{w}_j(0) = \frac{L}{L-1+J}\mathbf{1}$.

as $0 < w_{jk}(0) \leq \frac{L}{L-1+J}$. Larger values of $w_{jk}(0)$ favor the creation of new nodes, while smaller values attempt to put a pattern into an existing cluster. The ART 1 is stable for a finite training set. However, the order of the training patterns may influence the final prototypes and clusters.

Unlike many alternative methods such as the SOM [610], the Hopfield network [502], the neocognitron [388], the ART 1 can deal with arbitrary combinations of binary input patterns. In addition, the ART 1 has no restriction on memory capacity since its memory matrices are not square.

Other popular ART 1-based clustering algorithms include the improved ART 1 (IART 1) [1031], the adaptive Hamming net (AHN) [516], the fuzzy ART [163, 166, 168], the fuzzy AHN [516], and the projective ART (PART) [155]. The IART 1 was proposed to deal with pattern classification and image enhancement in the presence of noise without prior knowledge [1031]. The AHN has the same architecture as the Hamming network [717], but uses a weight learning and thresholding technique based on the ART 1 technique. The AHN is functionally equivalent to the ART 1 and obtains the same recognition categories as the ART 1 without any searching. It optimizes the ART 1 in terms of easy implementation, training time and memory storage [516]. The fuzzy ART [163] simply extends the logic AND concept in the ART 1 to the fuzzy AND concept. The fuzzy AHN is an extension of the AHN obtained by integrating ideas used in the fuzzy ART. Both the fuzzy ART and the fuzzy AHN have an analog architecture, and function like the ART 1 but for analog input patterns. The PART [155] is especially designed for data mining by considering the sparsity of information. The principal difference between the PART and the ART 1 is in the F1 layer. In the PART, the F1 layer selectively sends signals to nodes in the F2 layer.

The ART models are typically governed by differential equations, which result in a high computational complexity for numerical implementations. Implementations using analog or optical hardware are more desirable. A modified ART 1 algorithm in fast-learning mode has been derived for easy hardware implementation in [1015]. The method has also been extended for the full mode and the STM steady-state mode. A number of hardware implementations of the ART 1 in different modes are also surveyed in [1015].

ART 2

The ART 2 [160] is designed to categorize analog or binary random input sequences. It is similar to the ART 1, but has a more complex F1 field that allows the ART 2 to stably categorize sequences of analog inputs that can be arbitrarily close to one another. The F1 field includes a combination of normalization and noise suppression, and also the comparison of the bottom-up and top-down signals needed for the reset mechanism. By characterizing the clustering behavior of the ART 2, Burke [149] has found similarity between ART-based clustering and the C-means clustering.

The ART 2 employs a computationally expensive architecture that presents difficulties in parameter selection. The ART 2A [164], which accurately reproduces the behavior of the ART 2 in the fast-learn limit, is two to three orders of magnitude faster than the ART 2, and also suggests efficient parallel implementations. The ART 2A is also fast at intermediate learning rates, which can capture many of the desirable properties of slow learning of the ART 2 including noise tolerance.

The ART 2A employs the same architecture as the ART 2. Only feedforward connection between F1 and F2 is used in the ART 2A learning, and dot product is used as the similarity measure. In the ART 2A, an input pattern is first normalized as

$$\bar{\bar{\mathbf{x}}}_t = \frac{\mathbf{x}_t}{\|\mathbf{x}_t\|} \tag{5.37}$$

For each F2 neuron j, the net input or activation $net_j(t)$ is given by

$$net_j(t) = \bar{\bar{\mathbf{x}}}_t^{\mathrm{T}} \mathbf{w}_j(t) \tag{5.38}$$

Competition is performed to find the winning F2 neuron w such that

$$net_w(t) = \max_j net_j(t) \tag{5.39}$$

If more than one node is maximal, choose one at random.

As in Algorithm 5.8.1, a vigilance test is conducted on the neuron w to select between resonance and reset

$$net_w \geq \rho \tag{5.40}$$

If the vigilance test is passed, the system enters the resonance mode and the weight vector is updated by

$$\mathbf{w}_w(t) = \frac{\eta \mathbf{x}_t + (1 - \eta)\mathbf{w}_w(t - 1)}{\|\eta \mathbf{x}_t + (1 - \eta)\mathbf{w}_w(t - 1)\|} \tag{5.41}$$

where $\eta \in [0, 1]$ is the learning rate. Otherwise, the system enters the search mode that repeatedly looks for the next winning category that can pass the vigilance test. If none of the categories can pass the vigilance test, mismatch reset takes place and a new category j is created by setting

$$\mathbf{w}_j = \bar{\bar{\mathbf{x}}}_t \tag{5.42}$$

In [160], the F2 layer initially contains a number of uncommitted nodes, which get committed one by one upon the input presentation. F2 can also be initialized as the null set and dynamically grown during learning. An implementation of ART 2A using the latter strategy is given in Algorithm 5.8.1.2 [471].

The ART 2A with intermediate learning rates η copes better with noisy inputs than it does with fast learning rates, and the emergent category structure is also less dependent on the order of the input presentation [164].

Algorithm 5.8.1.2 (ART 2A)

1. Initialization.
   a) Set $0 < \rho < 1$ and $0 \leq \eta \leq 1$.
   b) Initialize F2 as the null set.
   c) Set $t = 1$.
2. Present a pattern $\mathbf{x}_t$.
3. Input normalization.
   Get $\bar{\bar{\mathbf{x}}}_t$ by (5.37).
4. Category choice.
   a) For each F2 neuron j, the net input or activation, $net_j(t)$, is calculated by (5.38).
   b) The winning neuron indexed w is obtained according to (5.39).
5. Vigilance test.
   a) If the vigilance test (5.40) is passed, the network is in resonance. Go to Step 6.
   b) Otherwise, enter search mode.
      The system repeatedly looks for the next winning category that can pass the vigilance test:
      i. If successful, enter resonance mode and go to Step 6.
      ii. Otherwise, mismatch reset takes place and a new category j is created by (5.42)
      iii. Resonance mode.
      The weight vector is updated by (5.41).
6. Set $t = t + 1$. Return to Step 2 with a new pattern.

The ART-C 2A [471] applies a constraint-reset mechanism on the ART 2A to allow a direct control on the number of output clusters generated during the self-organizing process. The constraint reset mechanism adaptively adjusts the value of ρ, while in the ART 2A prior knowledge is required to estimate an appropriate ρ. The ART 2A and the ART-C 2A have a clustering quality comparable to that of the C-means and the SOM, but with an advantage in the computational time [471].

Other ART Models

The ART 3 [162] carries out parallel searches by testing hypotheses about distributed recognition codes in a multilevel network hierarchy. The ART 3 introduces a search process for ART architectures that can robustly cope with sequences of asynchronous analog input patterns in real time.

The distributed ART (dART) [157] combines the stable fast-learning capabilities of ART systems with the noise-tolerance and code-compression capabilities of the MLP. Distributed activation helps to achieve memory compression and generalization. With a WTA code, the unsupervised dART model reduces to the fuzzy ART [163].

Other ART-based algorithms include the efficient ART (EART) family [67], the simplified ART (SART) family [67], the symmetric fuzzy ART (S-Fuzzy ART) [67], the Gaussian ART [1189] as an instance of the SART family, and the fully self-organizing SART (FOSART) [65].

5.8.2 ARTMAP Models

The ARTMAP model family, which is self-organizing and goal oriented, is a class of supervised learning methods. The ARTMAP, also termed *predictive ART*, autonomously learns to classify arbitrarily many, arbitrarily ordered vectors into recognition categories based on predictive success [165]. Although both the ARTMAP and the BP learning methods are supervised learning, the ARTMAP has a number of advantages such as being self-organizing, self-stabilizing, match learning, and real time. The ARTMAP learns orders of magnitude more quickly and also is more accurate than the BP. These are achieved by using an internal controller that conjointly maximizes predictive generalization and minimizes predictive error by linking predictive success to category size on a trial-by-trial basis, using only local operations. However, the ARTMAP is very sensitive to the order of presentation of the training patterns compared to learning using RBFN.

The architecture of the ARTMAP [165, 166, 168] is illustrated in Fig. 5.7. The ARTMAP learns predetermined categories of binary input patterns in a supervised manner. The ARTMAP is based on a pair of ART modules, namely, ART_a and ART_b. ART_a and ART_b can be fast-learning ART 1 modules coding binary input vectors. These modules are connected by an inter-ART module that resembles ART 1. The inter-ART module includes a map field that controls the learning of an associative map from ART_a recognition categories to ART_b recognition categories. The map field also controls match tracking of the ART_a vigilance parameter. The inter-ART vigilance resetting signal is a form of backpropagation of information.

Given a stream of input-output pairs $\{(\mathbf{x}_p, \mathbf{y}_p)\}$, where $\mathbf{y}_p$ is the correct prediction given $\mathbf{x}_p$. During the course of training, ART_a receives a stream $\{\mathbf{x}_p\}$ and ART_b receives a stream $\{\mathbf{y}_p\}$. These ART modules are linked by an associative learning network and an internal controller that ensures autonomous system operation in real time. During generalization, when a pattern $\mathbf{x}$ is presented to ART_a, its prediction at ART_b is produced.

The fuzzy ARTMAP [166, 168] can be taught to supervisedly learn predetermined categories of binary or analog input patterns. The fuzzy ARTMAP incorporates two fuzzy ART modules. Fuzzy versions of the ART and the ARTMAP minimize the predictive error and improve the generalization by designing a neural-network structure, which realizes a new min-max learning rule. The fuzzy ARTMAP is capable of fast, but stable, online recognition learning, hypothesis testing, and adaptive naming in response to an arbitrary

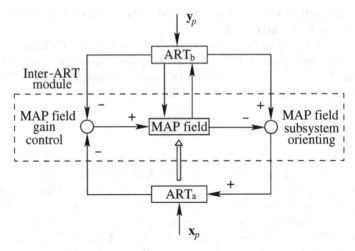

Fig. 5.7. Block diagram of the ARTMAP network. ART_a and ART_b self-organize categories for $\{x_p\}$ and $\{y_p\}$. The inter-ART module consists of the MAP field, the MAP field gain control, and the MAP field orienting subsystem. Positive and negative signs denote the excitatory and the inhibitory paths, respectively.

stream of analog or binary input patterns. The fuzzy ARTMAP is also shown to be a universal approximator [1128].

Other members of the ARTMAP family include the ART-EMAP [168], the ARTMAP-IC [167], the Gaussian ARTMAP [1189], the distributed ARTMAP (dARTMAP) [157], the default ARTMAP [158], and the simplified fuzzy ARTMAP [594, 1113]. The ART-EMAP [168] extends the fuzzy ARTMAP to perform classification, based on spatial and temporal evidence accumulated during the presentation of a sequence of patterns from each target to be identified. The ART-EMAP uses distributed category representation to improve the fuzzy ARTMAP performance in noisy or ambiguous situations. The ARTMAP-IC [167] modifies the ARTMAP match-tracking algorithm to allow the network to encode inconsistent cases, and combines instance counting during training with distributed category representation. The ARTMAP-IC adds an instance-counting layer F3 to the ARTMAP. Training is the same as for the ARTMAP, except that a counting weight enumerates the number of instances placed in each category. The ARTMAP-IC consistently improves both the predictive accuracy and the code compression, compared to the ARTMAP and the ART-EMAP. The Gaussian ARTMAP [1189] is based on the synthesis of a Gaussian classifier and the ARTMAP model. Compared with the fuzzy ARTMAP, the Gaussian ARTMAP has more complex learning rules as well as choice and match functions, but generalizes better. The Gaussian ARTMAP retains the fast learning and parallel computing properties of the fuzzy ARTMAP. The dARTMAP [157] seeks to combine the computational advantages of the MLP and the ART in a real-time neural network for supervised learning.

The distributed vs. the WTA-coding representation is a primary factor differentiating the various ARTMAP networks. With a WTA coding during training but distributed during testing, the dARTMAP reduces to the ARTMAP-IC, and further reduces to the fuzzy ARTMAP with WTA coding during both the testing and the training. The default ARTMAP is the same as the fuzzy ARTMAP during the training, but uses a distributed code representation during the testing. The ARTMAP-IC equals the default ARTMAP plus instance counting, which biases the output of a category node for the test set by the size of the training set coded by that node. The dARTMAP employs a distributed code and instance counting during both the training and the testing. The relations of some of the ARTMAP variants are given by[8] [158]

$$\text{fuzzy ARTMAP} \subset \text{default ARTMAP} \subset \text{ARTMAP-IC} \subset \text{dARTMAP}$$

More precisely, the dARTMAP reduces to the ARTMAP-IC when coding is set to WTA during the training; the ARTMAP-IC reduces to the default ARTMAP when counting weights are set to unity; the default ARTMAP reduces to the fuzzy ARTMAP when coding is set to WTA during the testing as well as the training.

5.9 Fuzzy Clustering

Fuzzy clustering is an important class of clustering algorithms. Fuzzy clustering helps to find natural vague boundaries in data. We introduce some fuzzy algorithms in this section. Preliminaries of fuzzy sets and fuzzy logic are given in Chapter 8.

5.9.1 Fuzzy C-means Clustering

The discreteness of each cluster endows the C-means algorithm with analytical and algorithmic intractabilities. Partitioning the data set in a fuzzy manner helps to circumvent such difficulties. The unsupervised FCM clustering [103, 104], also known as the *fuzzy ISODATA* [328], considers each cluster as a fuzzy set, and each feature vector may be assigned to multiple clusters with some degree of certainty measured by the membership function taking values in the interval $[0, 1]$.

The FCM optimizes the following objective function [103, 104]

---

[8] Many ART and ARTMAP variants are due to G. A. Carpenter and S. Grossberg, and their group at Boston University. Interested readers are referred to the website *http://cns.bu.edu/~gail/* for details of these models and algorithms.

$$E = \sum_{j=1}^{K} \sum_{i=1}^{N} \mu_{ji}^{m} \|\mathbf{x}_i - \mathbf{c}_j\|^2 \tag{5.43}$$

where N is the size of the input pattern set, $\mathbf{U} = \{\mu_{ji}\}$ denotes the membership matrix whose element μ_{ji} denotes the membership of $\mathbf{x}_i$ into cluster j and $\mu_{ji} \in [0,1]$. For better interpretation, the condition

$$\sum_{j=1}^{K} \mu_{ji} = 1 \tag{5.44}$$

must be valid for $i = 1, \cdots, N$.

The parameter $m \in (1, \infty)$ is a weighting factor called the *fuzzifier*. By minimizing (5.43) subject to (5.44), the optimal membership function μ_{ji} and cluster centers are derived as

$$\mu_{ji} = \frac{\left(\frac{1}{\|\mathbf{x}_i - \mathbf{c}_j\|^2}\right)^{\frac{1}{m-1}}}{\sum_{l=1}^{K} \left(\frac{1}{\|\mathbf{x}_i - \mathbf{c}_l\|^2}\right)^{\frac{1}{m-1}}} \tag{5.45}$$

$$\mathbf{c}_j = \frac{\sum_{i=1}^{N} (\mu_{ji})^m \mathbf{x}_i}{\sum_{i=1}^{N} (\mu_{ji})^m} \tag{5.46}$$

for $i = 1, \cdots, N$, $j = 1, \cdots, K$. Equation (5.45) corresponds to a soft-max rule and (5.46) is similar to the mean of the data points in a cluster. The two equations are dependent on each other.

The iteration process terminates when the change in the prototypes

$$e(t) = \sum_{j=1}^{K} \|\mathbf{c}_j(t) - \mathbf{c}_j(t-1)\|^2 \tag{5.47}$$

is sufficiently small. The FCM algorithm is summarized in Algorithm 5.9.1 [104, 584][9].

In the FCM algorithm, the fuzzifier m determines the fuzziness of the partition produced, and the value of m reduces the influence of small membership values. If $m \to 1+$, the resulting partition asymptotically approaches a hard or crisp partition. On the other hand, the partition becomes a maximally fuzzy partition if $m \to \infty$. The FCM clustering with a high degree of fuzziness diminishes the probability of getting stuck at local minima [104]. A typical value for m is 1.5 or 2.0.

The FCM algorithm needs to store the membership matrix $\mathbf{U}$ and all the prototypes $\mathbf{c}_i$. The alternating estimation of $\mathbf{U}$ and $\mathbf{c}_i$s causes a computational and storage burden for large-scale data sets. The computation can be

---

[9] An iterative optimization algorithm that is derived from an objective function using necessary conditions for local extrema is known as an *alternating optimization* algorithm.

Algorithm 5.9.1 (Fuzzy C-means)

1. Set $t = 0$.
2. Initialize K, ε, and m.
3. Randomize and normalize $\mathbf{U}(0)$ according to (5.44), and then calculate $\mathbf{c}_j(0)$, $j = 1, \cdots, K$, by (5.46).
   Or alternatively, set $\mathbf{c}_j(0)$, $j = 1, \cdots, K$, and then calculate $\mathbf{U}(0)$ by (5.45).
4. Repeat:
   a) Set $t = t + 1$.
   b) Calculate $\mu_{ji}(t)$ and $\mathbf{c}_j(t)$ according to (5.45) and (5.46).
   c) Calculate $e(t)$ by (5.47).
   until $e(t) \leq \varepsilon$.

---

accelerated by combining their updates [563], and consequently the storage of $\mathbf{U}$ is avoided. The single iteration timing of the accelerated method grows linearly with K, while that of the FCM grows quadratically with K since the norm calculation introduces another nested summation [563].

The C-means is a special case of the FCM, when u_{ji} is unity for only one class and zero for all the other classes. Like the C-means, the FCM may find a local optimum for a specified number of centers. The result is dependent on the initial membership matrix $\mathbf{U}(0)$ or cluster centers $\mathbf{c}_j(0)$, $j = 1, \cdots, K$.

Variants of the Fuzzy C-means

The penalized FCM [1222] is a convergent generalized FCM obtained by adding a penalty term associated with μ_{ji}. The compensated FCM [710] is achieved by modifying the penalty term of the penalized FCM to converge faster. These generalizations are more meaningful and effective than the FCM. A weighted FCM clustering [1102] has been used for fuzzy modeling towards developing a Takagi–Sugeno–Kang (TSK) fuzzy model of optimal structure. All these and many other existing generalizations of the FCM can be analyzed under a unified framework, termed the *generalized FCM (GFCM)* algorithm [1242]. The GFCM is derived by using the Lagrange multiplier method from an objective function that comprises a generalization of the FCM criterion and a regularization term representing the constraints. New algorithms can be designed by selecting different parameters and functions in the GFCM.

The multistage random sampling FCM algorithm [219] can reduce the clustering time normally by a factor of 2 to 3. A series of subsets of the full data set are used to create initial cluster centers in order to provide an approximation to the final cluster centers. The quality of the final partitions is equivalent to that created by the FCM. The algorithm is divided into two phases, where the first phase is a multistage iterative process of a modified

FCM and the second phase is the standard FCM with the cluster centers initialized by the first phase. The FCM has been generalized by introducing the generalized Boltzmann distribution to escape local minima [948]. Existing global optimization techniques can also be incorporated into the FCM so as to provide globally optimum solutions.

The ε-insensitive FCM (εFCM) clustering algorithm is an extension to the FCM obtained by introducing the robust statistics using Vapnik's ε-insensitive estimator as the loss function to reduce the effect of outliers [685]. The εFCM is based on L_1-norm clustering [601]. Other robust extensions to the FCM includes the L_p-norm clustering ($0 < p < 1$) [465] and the L_1-norm clustering [601].

When the data set is a blend of unlabeled and labeled patterns, the FCM with partial supervision [903] can be applied. The classification information is added to the objective function used in the FCM, and the FCM with partial supervision is derived following the same procedure as that of the FCM. A weighting factor balances the supervised and unsupervised components within the objective function [903]. The FCM has also been extended for clustering other data types, such as symbolic data [331].

5.9.2 Conditional Fuzzy C-means Clustering

The conditional FCM clustering [899] develops clusters preserving the homogeneity of the clustered patterns with regard to their similarity in the input space, as well as their respective values assumed in the output space. It is a supervised clustering.

The conditional FCM clustering is based on the FCM algorithm, but requires the output variable of a cluster to satisfy a particular condition. This condition can be treated as a fuzzy set, defined via the corresponding membership. Corresponding to the FCM clustering, we have

$$\mu_{ji} = \overline{y}_{i,j} \frac{\left(\frac{1}{\|\mathbf{x}_i - \mathbf{c}_j\|^2}\right)^{\frac{1}{m-1}}}{\sum_{l=1}^{K} \left(\frac{1}{\|\mathbf{x}_i - \mathbf{c}_l\|^2}\right)^{\frac{1}{m-1}}} \tag{5.48}$$

where $\overline{y}_{i,j}$ describes a level of involvement of $\mathbf{x}_i$ in the constructed cluster j, and can be obtained by normalizing the output patterns of the training set. The computation of the prototypes is performed using (5.46). The constraint $\overline{y}_{i,j}$ maintains the requirement that if the ith pattern of the data set is considered not to belong to the cluster j and thus its membership degree is null, then this element is not considered in the calculation of the prototype of the cluster [899]. This results in a reduction in the computational complexity for classification problems by splitting the original problem into a series of condition-driven clustering problems. A family of generalized weighted conditional FCM clustering algorithms have been derived in [686].

5.9.3 Other Fuzzy Clustering Algorithms

There are numerous other clustering algorithms based on the concept of fuzzy membership. Two early fuzzy clustering algorithms are the Gustafson–Kessel [443] and the adaptive fuzzy clustering (AFC) algorithms [35]. The Gustafson–Kessel algorithm extends the FCM by using the Mahalanobis distance, and is suited for hyperellipsoidal clusters of equal volume. This algorithm takes typically five times as long as the FCM to complete cluster formation [586]. The AFC also employs the Mahalanobis distance, and is suitable for ellipsoidal or linear clusters. The Gath–Geva algorithm [402] is derived from a combination of the FCM and the fuzzy ML estimation. The method incorporates the hypervolume and density criteria as cluster-validity measures and performs well in situations of large variability of cluster shapes, densities, and number of data points in each cluster.

The C-means and FCM algorithms are based on the minimization of the trace of the (fuzzy) within-cluster scatter matrix. The minimum scatter volume (MSV) and minimum cluster volume (MCV) algorithms are two iterative clustering algorithms based on determinant (volume) criteria [633]. The MSV algorithm minimizes the determinant of the sum of the scatter matrices of the clusters, while the MCV minimizes the sum of the volumes of the individual clusters. The behavior of the MSV is shown to be similar to that of the C-means, whereas the MCV is more versatile. The MCV in general gives better results than the C-means, MSV, and Gustafson–Kessel algorithms, and is less sensitive to initialization than the EM algorithm [299].

Volume prototypes extend the cluster prototypes from points to regions in the clustering space [595]. A cluster represented by a volume prototype implies that all the data points close to a cluster center belong fully to that cluster. In [595], the Gustafson–Kessel algorithm and the FCM have been extended by using the volume prototypes and similarity-driven merging of clusters.

Some categories of fuzzy clustering algorithms are described below.

Fuzzy Clustering Based on the Kohonen Network and Learning Vector Quantization

There are various fuzzy clustering methods that have been developed based on the Kohonen network. The fuzzy SOM [519] incorporates fuzziness into the learning process of the SOM by replacing the learning rate with fuzzy membership of the nodes in each class.

The fuzzy LVQ (FLVQ) [105], originally named the *fuzzy Kohonen clustering network (FKCN)* [107], combines the ideas of fuzzy membership values for learning rates, the parallelism of the FCM, and the structure and self-organizing update rules of the Kohonen network. The FLVQ is a batch clustering algorithm whose learning rates are derived from fuzzy memberships, and usually terminates in such a way that the FCM objective function is approximately minimized. The FLVQ employs the same update equations as that of the FCM, but $m = m(t)$ is a monotonic decreasing function of time; usually, the maximum and minimum values of m are heuristically selected in the interval $(1.1, 7)$. The FLVQ yields an improved convergence as well as reduced labeling errors when compared with the FCM.

Soft competitive learning in clustering algorithms has the same function as fuzzy clustering [66]. The softcompetition scheme (SCS) [1210] is another soft version of the LVQ. The SCS asympotically evolves into the Kohonen learning algorithm. The SCS is a sequential, deterministic VQ algorithm, which is realized by modifying the neighborhood mechanism of the Kohonen learning algorithm and incorporating the stochastic relaxation principles. The SCS consistently provides better codebooks than the incremental C-means [714], even for the same computation time. The SCS is relatively insensitive to the choice of the initial codebook. The learning rates of the SCS are partially based on posterior probabilities. The learning rates of the FLVQ and the SCS algorithms have opposite tendencies [105]. The SCS has difficulty in choosing good algorithmic parameters [105].

An extended family of the batch FLVQ algorithm is derived in [582] based on the minimization of a functional defined as the average generalized distance between the feature vectors and the prototypes. The FLVQ [105] and the FCM can be seen as special cases of the extended FLVQ family learning schemes [582]. In [586], the non-Euclidean FLVQ (NEFLVQ) and the non-Euclidean FCM (NEFCM) are derived by minimizing a reformulation function that employs distinct weighted norms to measure the distance between each of the prototypes and the feature vectors under a set of equality constraints imposed on the weight matrices.

The generalized LVQ (GLVQ) algorithm [880] introduces softcompetition into the LVQ by updating every prototype for each input vector. If there is a perfect match between the incoming input and the winner node, then the GLVQ reduces to the LVQ. On the other hand, the greater the mismatch to the winner, the larger the impact of an input vector on the update of the nonwinner nodes. The GLVQ is very sensitive to simple scaling of the input data, since its learning rates are reciprocally dependent on the sum of the squares of the distances from an input vector to the node weight vectors.

The generalized LVQ family (GLVQ-F) algorithm[10] [583] is a fuzzy modification to the GLVQ, which overcomes the problem of scaling that the GLVQ suffers from. The GLVQ-F reduces to the LVQ as its weighting exponent approaches unity from above.

The GLVQ-F and the family of fuzzy algorithms for LVQ (FALVQ) [585, 579] are online fuzzy clustering algorithms. The design of specific FALVQ algorithms reduces to the selection of the membership function assigned to the prototypes of an LVQ network [579].

---

[10] It is actually one algorithm wherein each value of the fuzzifier m corresponds to one family member.

The reformulation of the LVQ or clustering algorithms [581] provides a framework for the development of specific clustering algorithms by appropriate selection of a generator function. Linear generator functions lead to the FCM [104] and FLVQ [105] algorithms, while exponential generator functions lead to entropy-constrained fuzzy clustering (ECFC) algorithms and entropy-constrained LVQ (ECLVQ) algorithms [581].

Fuzzy Clustering Based on ART Networks

In Sect. 5.8.1, we have mentioned some fuzzy ART models such as the fuzzy ART, S-fuzzy ART, and the fuzzy AHN, as well as some fuzzy ARTMAP models such as the fuzzy ARTMAP, ART-EMAP, default ARTMAP, ARTMAP-IC, and dARTMAP.

The supervised fuzzy min-max classification network [1038] as well as the unsupervised fuzzy min-max clustering network [1039] is a kind of combination of fuzzy logic and the ART 1 model [159]. The operations in these models are very simple, requiring only complements, additions and comparisons that are most suitable for parallel hardware execution. The fuzzy min-max classification network [1038] has a three-layer feedforward architecture, and utilizes min-max hyperboxes as fuzzy sets, which are aggregated into fuzzy set classes. The fuzzy min-max clustering network employs a two-layer architecture similar to the fuzzy ART model.

Some clustering and fuzzy clustering algorithms including the SOM [611], the FLVQ [105], the fuzzy ART [163], the growing neural gas (GNG) [380], and the FOSART [65] have been surveyed and compared in [66].

Fuzzy Clustering Based on the Hopfield Network

The clustering problem can be cast as a problem of minimization of the criterion defined as the MSE between the training patterns and the cluster centers. This optimization problem can then be solved using the Hopfield network [711, 710].

In the fuzzy Hopfield network (FHN) [711] and the compensated fuzzy Hopfield network (CFHN) [710], the training patterns are mapped to a Hopfield network of a two-dimensional neuron array, where each column represents a cluster and each row a training pattern. The state of each neuron corresponds to a fuzzy membership function. A fuzzy clustering strategy is included in the Hopfield network to eliminate the need for finding the weighting factors in the energy function. This energy function is called the *scatter energy function*, and is formulated based on the within-class scatter matrix in pattern recognition. These models have inherent parallel structures.

In the FHN [711], an FCM clustering strategy is imposed for updating the neuron states. The CFHN [710] integrates the compensated FCM model into the learning scheme and updating strategies of the Hopfield network to avoid the NP-hard problem and to speed up the convergence rate for the clustering procedure. The training scheme of the CFHN enables the network to learn more rapidly and more effectively than clustering using the Hopfield network, the FCM [104], and the penalized FCM [1222]. The CFHN is trained to classify the input patterns into feasible clusters when the defined energy function converges to near-global minimum.

The CFHN has been used for VQ in image compression [726], so that the parallel implementation for codebook design is feasible. The CFHN is trained to classify the divided vectors on a real image into feasible class to generate an available codebook when the defined energy function converges to a near-global minimum.

5.10 Supervised Clustering

Most of the previously discussed clustering methods are unsupervised clustering, where unlabeled patterns are involved. When output patterns are used in clustering, this yields supervised clustering methods. The locations of the cluster centers are influenced not only by the input pattern spread, but also by the output pattern deviations.

For classification problems, the classmembership of each pattern in the training set is available and can be used for clustering. Examples of such supervised clustering methods include the LVQ family [612], the ARTMAP family [165], the conditional FCM [899], and the C-means plus k-NN based clustering [140]. For classification problems, supervised clustering significantly improves the decision accuracy.

In the case of supervised learning using kernel-based neural networks such as the RBFN, the structure (kernels) is usually determined by using unsupervised clustering. This method, however, is not effective for finding a parsimonious network. Supervised clustering can be implemented by augmenting the input pattern with its output pattern, $\widetilde{\mathbf{x}}_i = \left[\mathbf{x}_i^{\mathrm{T}}, \mathbf{y}_i^{\mathrm{T}}\right]^{\mathrm{T}}$, so as to obtain an improved distribution of the cluster centers by an unsupervised clustering [899, 194, 981, 1112]. When the range of $\mathbf{y}_i$ is smaller than that of $\mathbf{x}_i$, the role of the dependent variable $\mathbf{y}_i$ will be de-emphasized. A scaling factor β is introduced to balance between the underlying similarity in the input space and the similarity in the output space [899]

$$\widetilde{\mathbf{x}} = \left[\mathbf{x}_i^{\mathrm{T}}, \beta\left(\mathbf{y}_i\right)^{\mathrm{T}}\right]^{\mathrm{T}} \tag{5.49}$$

The resulting extended objective function in the case of the FCM is given by

$$E = \sum_{j=1}^{K}\sum_{i=1}^{N} \mu_{ji}^m \left\|\mathbf{x}_i - \mathbf{c}_{x,j}\right\|^2 + \sum_{j=1}^{K}\sum_{i=1}^{N} \mu_{ji}^m \left\|\beta\mathbf{y}_i - \mathbf{c}_{y,j}\right\|^2 \tag{5.50}$$

where the new cluster center $\mathbf{c}_j = \left[\mathbf{c}_{x,j}^{\mathrm{T}}, \mathbf{c}_{y,j}^{\mathrm{T}}\right]^{\mathrm{T}}$. The first term in (5.50) corresponds to the conventional FCM, and the second term applies to supervised learning. The resulting cluster codebook vectors are rescaled and projected into the input space to obtain the centers.

Based on the enhanced LBG algorithm [889], the clustering for function approximation (CFA) [426] algorithm is specially designed for function approximation problems. The CFA increases the density of the prototypes in the input areas where the target function presents a more variable response, rather than just in the zones with more input examples [426]. The CFA minimizes the variance of the output response of the training examples belonging to the same cluster. The value of the jth output node for the prototype of cluster j is calculated as a weighted average of the output responses of the training data belonging to cluster j. In [1054], a prototype regression function is built as a linear combination of local linear regression models, one for each cluster, and is then inserted into the FCM functional. In this way, the prototypes can be adjusted according to both the input distribution and the regression function in the output space.

5.11 The Underutilization Problem

Conventional competitive learning-based clustering algorithms like the C-means and the LVQ are plagued by a severe initialization problem [980, 436]. If the initial values of the prototypes are not in the convex hull formed by the input data, the clustering algorithms may not produce meaningful results. This is the so-called *prototype underutilization* or *dead-unit problem* since some prototypes, called *dead units*, may never win the competition. This problem is caused by the fact that these algorithms update only the winning prototype for every input. The underutilization problem is illustrated in Fig. 5.8.

In order to alleviate the sensitivity of competitive learning to the initialization of the clustering centers, many efforts have been made to solve the underutilization problem. Initializing the prototypes with random input vectors can reduce the probability of the underutilization problem, but does not eliminate it. In the leaky learning strategy [980, 436], all the prototypes are updated. The winning prototype is updated by employing a fast learning rate, while all the losing prototypes move towards the input vector with a much slower learning rate, for example, one-tenth of that of the winning prototype.

5.11.1 Competitive Learning with Conscience

To avoid the underutilization problem, one can assign each processing unit with a threshold, and then increase the threshold if the unit wins, and decrease it otherwise [980]. A similar idea is embodied in the conscience strategy, which reduces the winning rate of the frequent winners [303]. The frequent winner receives a bad conscience by adding a penalty term to its distance from the input signal. This leads to an entropy maximization, that is, each unit wins

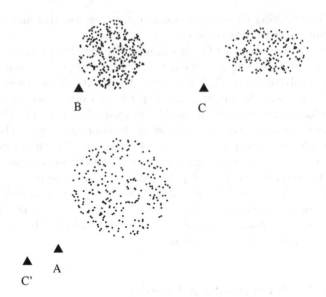

Fig. 5.8. An illustration of the underutilization problem for competitive learning-based clustering. There are three clusters in the data set. If the three prototypes c_1, c_2, and c_3 are initialized at A, B, and C, respectively, they will correctly move to the centers of the three clusters. However, if they are initialized at A, B, and C', respectively, C' will never become a winner and thus becomes a dead unit. In the latter case, the system divides the three data clusters into two newly formed clusters, and the prototypes c_1 and c_2 will, respectively, move to the centroids of the two clusters.

at an approximately equal probability. Thus, the probability of under-utilized neurons being selected as winners is increased.

The popular frequency-sensitive competitive learning (FSCL) [12] employs this conscience strategy. The FSCL scheme reduces the underutilization problem by introducing a distortion measure that ensures all codewords in the codebook to be updated with a similar probability. The codebooks obtained by the FSCL algorithm have sufficient entropy so that Huffman coding of the VQ indices would not provide significant additional compression.

In the FSCL, each prototype incorporates a count of the number of times it has been the winner, u_i, $i = 1, \cdots, K$. The distance measure is modified to give prototypes with a lower count value a chance to win the competition. The algorithm is similar to the VQ algorithm, and the only difference is that the winning neuron is found according to [12]

$$\mathbf{c}_w(t) = \arg_{\mathbf{c}_i} \min_{i=1,\cdots,K} \left\{ u_i(t-1) \left\| \mathbf{x}_t - \mathbf{c}_i(t-1) \right\| \right\} \qquad (5.51)$$

where w is the index of the winning node, u_i is updated by

$$u_i(t) = \begin{cases} u_i(t-1) + 1, & i = w \\ u_i(t-1), & \text{otherwise} \end{cases} \qquad (5.52)$$

and $u_i(0) = 0$, $i = 1, \cdots, K$. In (5.51), $u_i \|\mathbf{x}_t - \mathbf{c}_i\|$ can be generalized as $F(u_i) \|\mathbf{x}_t - \mathbf{c}_j\|$. When selecting the fairness function as $F(u_i) = u_i^{\beta_0 e^{-t/T_0}}$, where β_0 and T_0 are constants, the FSCL emphasizes the winning uniformity of codewords initially and gradually turns into competitive learning as training proceeds to minimize the MSE objective function.

The multiplicatively biased competitive learning (MBCL) model [237] is a class of competitive learning models in which the competition among the neurons is biased by a multiplicative term. The MBCL avoids neuron underutilization with probability one, as time goes to infinity. The FSCL [12, 631] is a member of the MBCL family. In the MBCL, only one weight vector is updated per learning step. This is of practical interest, since its instances have computational complexities among the lowest in existing competitive learning models.

The fuzzy FSCL (FFSCL) [243] combines the frequency sensitivity with fuzzy competitive learning. Since both the FSCL and the FFSCL use non-Euclidean distance to determine the winner, they may lead to the problem of shared clusters in the sense that a number of prototypes may be updated into the same cluster during the learning process.

5.11.2 Rival Penalized Competitive Learning

This problem of shared clusters for the FSCL and the FFSCL has been considered in the rival penalized competitive learning (RPCL) algorithm [1204]. The RPCL adds a new mechanism to the FSCL by creating a rival penalizing force. For each input, not only is the winning unit modified to adapt to the input but also the second-place winner called the *rival* is updated by a smaller learning rate along the opposite direction, all the other prototypes being unchanged:

$$\mathbf{c}_i(t+1) = \begin{cases} \mathbf{c}_i(t) + \eta_w (\mathbf{x}_t - \mathbf{c}_i(t)), & i = w \\ \mathbf{c}_i(t) - \eta_r (\mathbf{x}_t - \mathbf{c}_i(t)), & i = r \\ \mathbf{c}_i(t), & \text{otherwise} \end{cases} \qquad (5.53)$$

where w and r are the indices of the winning and rival prototypes, which are decided by (5.51), and η_w and η_r are their respective learning rates. In practice, $\eta_w(t) \gg \eta_r$. The learning rate η_r is also called the *delearning rate* for the rival.

This actually pushes the rival away from the sample pattern so as to prevent its interference in the competition. The RPCL automatically allocates an appropriate number of prototypes for an input data set, and all the extra candidate prototypes will finally be pushed to infinity. It provides a better performance than the FSCL. The RPCL can be regarded as an unsupervised extension of the supervised LVQ2 [612]. It simultaneously modifies the weight vectors of both the winner and its rival, when the winner is in a wrong class but the rival is in a correct class for an input vector [1204]. The lotto-type competitive learning (LTCL) [736] can be treated as a generalization of the RPCL, where instead of just penalizing the nearest rival, all the losers are penalized equally. The generalized LTCL [737] modifies the LTCL by allowing more than one winner, which are divided into tiers, with each tier being rewarded differently. However, the RPCL may encounter the overpenalization and underpenalization problems due to an inappropriate delearning rate [1263].

The stepwise automatic rival-penalized (STAR) C-means [223] is a generalization of the C-means. The STAR C-means consists of two separate phases. The first phase implements the FSCL [12], which assigns each cluster with at least a prototype. The second phase is derived from a Kullback–Leibler divergence-based criterion, and adjusts the units adaptively by a learning rule that automatically penalizes the winning chance of all rival prototypes in the subsequent competitions while tuning the winning one to adapt to the input. The STAR C-means has a mechanism similar to the RPCL, but penalizes the rivals in an implicit way, whereby circumventing the determination of the rival delearning rate of the RPCL. The STAR C-means is applicable to ellipse-shaped data clusters as well as sphere-shaped ones without the underutilization problem and without having to predetermine the correct cluster number.

5.11.3 Softcompetitive Learning

By relaxing the WTA criterion, clustering methods can treat more than a single neuron as winners to a certain degree and update their prototypes accordingly, resulting in the winner-take-most paradigm, namely, the softcompetitive learning. Examples are the SCS [1210], the SOM [611], the NG [768], maximum-entropy clustering [961], the GLVQ [880], the FCM [104] and the fuzzy competitive learning (FCL) [243].

The FCL algorithms [243] are a class of sequential algorithms obtained by fuzzifying competitive learning algorithms, such as the SCL, the unsupervised LVQ, and the FSCL. The concept of winning is formulated as a fuzzy set and the network outputs become the winning memberships of the competing neurons. The enhanced sequential fuzzy clustering (ESFC) [1255] is a modification to the FCL, obtained by introducing a nonunity weighting on the winning centroid and an excitation-inhibition mechanism so as to better overcome the underutilization problem.

The SOM [611] is a learning process, which takes the winner-take-most strategy at the early stages and becomes a WTA approach, while its neighborhood size reduces to unity as a function of time in a predetermined manner. Due to the softcompetitive strategy, the SOM, the GNG [380] and fuzzy clustering algorithms such as the FLVQ, the fuzzy ART, and the FOSART, are less likely to be trapped in local minima and to generate dead units than hard competitive alternatives [66].

The maximum-entropy clustering [961] avoids the underutilization problem and local minima in the error function by using softcompetitive learning and deterministic annealing. The prototypes are updated by

$$\mathbf{c}_i(t+1) = \mathbf{c}_i(t) + \eta(t) \left[\frac{e^{-\beta \|\mathbf{x}_t - \mathbf{c}_i(t)\|^2}}{\sum_{j=1}^{K} e^{-\beta \|\mathbf{x}_t - \mathbf{c}_j(t)\|^2}} \right] (\mathbf{x}_t - \mathbf{c}_i(t)) \qquad (5.54)$$

for $i = 1, \cdots, K$, where η is the learning rate and the parameter $\frac{1}{\beta}$ anneals from a large parameter to zero, and the term within the bracket turns out to be the Boltzmann distribution. The SCS [1210] employs a similar softcompetitive strategy, but β is fixed as unity.

The winner-take-most criterion, however, detracts some prototypes from their corresponding clusters, and consequently becomes biased toward the global mean of the clusters, since all the prototypes are attracted to each input pattern [729].

5.12 Robust Clustering

Data sets usually contain noisy points or outliers. This will affect the results of clustering, and the concept of robust clustering needs to be introduced. As described in Chapter 2, the influence of outliers can be substantially eliminated by using the loss function in the robust statistics approach [514]. This idea has also been integrated into robust clustering methods [132, 601, 279, 376, 465, 685]. The C-median clustering [132] is derived by solving a bilinear programming problem that utilizes the L_1-norm distance. The fuzzy C-median clustering [601] is a robust FCM method that uses the L_1-norm with the exemplar estimation based on the fuzzy median.

Robust clustering algorithms can be derived by optimizing a specially designed objective function. The objective function usually has two terms:

$$E_{\mathrm{T}} = E + E_{\mathrm{c}} \qquad (5.55)$$

where E is the cost for the conventional algorithms such as (5.43), and the constraint term E_{c} is used for describing the noise cluster.

5.12.1 Noise Clustering

In the noise-clustering approach [278], noise and outliers are collected into a separate, amorphous noise cluster, whose prototype has the same distance, δ, from all the data points. The other points are collected into K clusters. The threshold δ is a relatively high value compared to the distances of the *good* points to the cluster prototypes. If a noisy point is far away from all the K

clusters, it is attracted to the noise cluster. In the noise-clustering approach [278], the second term is given by

$$E_c = \sum_{i=1}^{N} \delta^2 \left(1 - \sum_{j=1}^{K} \mu_{ji} \right)^m \tag{5.56}$$

Following the procedure for the derivation of the FCM, we have

$$\mu_{ji} = \frac{\left(\frac{1}{\|\mathbf{x}_i - \mathbf{c}_j\|^2} \right)^{\frac{1}{m-1}}}{\sum_{k=1}^{K} \left(\frac{1}{\|\mathbf{x}_i - \mathbf{c}_k\|^2} \right)^{\frac{1}{m-1}} + \left(\frac{1}{\delta^2} \right)^{\frac{1}{m-1}}} \tag{5.57}$$

The second term in the denominator of (5.57) being large for outliers, leads to small μ_{ji}. The formula for the prototypes is the same as that for the FCM method, given by (5.46).

The noise-clustering algorithm can be treated as a robustified FCM algorithm. When all the K clusters have about the same size, the noise clustering is very effective. However, a single threshold is too restrictive if the cluster size varies widely in the data set.

5.12.2 Possibilistic C-means

The possibilistic C-means (PCM) algorithm [634], as opposed to the FCM, does not require that the memberships of a data point across the clusters sum to unity. This allows the membership functions to represent a possibility of belonging rather than a relative degree of membership between clusters. As a result, the derived degree of membership does not decrease as the number of clusters increases. Due to the elimination of this constraint, the modified objective function is decomposed into many individual objective functions, one for each cluster, which can be optimized separately.

The constraint term for the PCM is given by a sum associated with the fuzzy complements of all the K clusters

$$E_c = \sum_{j=1}^{K} \beta_j \sum_{i=1}^{N} (1 - \mu_{ji})^m \tag{5.58}$$

where β_j are suitable positive numbers. The individual objective functions are given as

$$E_T^j = \sum_{i=1}^{N} \mu_{ji}^m \|\mathbf{x}_i - \mathbf{c}_j\|^2 + \beta_j \sum_{i=1}^{N} (1 - \mu_{ji})^m \tag{5.59}$$

for $j = 1, \cdots, K$. Differentiating (5.59) with respect to μ_{ji} and setting it to zero leads to the solution

$$\mu_{ji} = \frac{1}{1 + \left(\frac{\|\mathbf{x}_i - \mathbf{c}_j\|^2}{\beta_j}\right)^{\frac{1}{m-1}}} \tag{5.60}$$

where the second term in the denominator is large for outliers, leading to small μ_{ji}. Some heuristics for selecting β_j have also been given in [634].

Given a number of clusters K, the FCM will arbitrarily split or merge real clusters in the data set to produce exactly the specified number of clusters. The PCM, in contrast to the FCM, can find those natural clusters in the data set. When K is smaller than the number of actual clusters, only K good clusters are found, and the other data points are treated as outliers. When K is larger than the number of actual clusters, all the actual clusters can be found and some clusters will coincide. In other words, K can be specified somewhat arbitrarily.

In the noise-clustering algorithm, there is only one noise cluster, while in the PCM there are K noise clusters. The PCM algorithm functions as a collection of K independent noise-clustering algorithms, each looking for a single cluster. The performance of the PCM, however, relies heavily on good initialization of cluster prototypes and estimation of β_j, and the PCM tends to converge to coincidental clusters [279].

5.12.3 A Unified Framework for Robust Clustering

By extending the idea of treating outliers as the fuzzy complement, a family of robust clustering algorithms has been obtained [1223]. Assume that a noise cluster exists outside each data cluster. The fuzzy complement of μ_{ji}, denoted as $f(\mu_{ji})$, may be interpreted as the degree to which $\mathbf{x}_i$ does not belong to the ith data cluster. Thus, the fuzzy complement can be viewed as the membership of $\mathbf{x}_i$ in the noise cluster with a distance β_j. Based on this, one can propose many different implementations of the probabilistic approach [279, 1223]. For robust fuzzy clustering, a general form of E_c is given as a generalization of that for the PCM [1223]

$$E_c = \sum_{i=1}^{N} \sum_{j=1}^{K} \beta_j \left[f(\mu_{ji})\right]^m \tag{5.61}$$

Note that the PCM [634] uses the standard fuzzy complement $f(\mu_{ji}) = 1 - \mu_{ji}$.

By setting to zero the derivatives of E_T with respect to the variables, a fuzzy clustering algorithm is obtained. For example, by setting $m = 1$ and $f(\mu_{ji}) = \mu_{ji} \ln(\mu_{ji}) - \mu_{ji}$ [279] or $f(\mu_{ji}) = 1 + \mu_{ji} \ln(\mu_{ji}) - \mu_{ji}$ [1223], we can obtain

$$\mu_{ji} = e^{-\frac{\|\mathbf{x}_i - \mathbf{c}_j\|^2}{\beta_j}} \tag{5.62}$$

and $\mathbf{c}_j$ has the same form as that for the FCM. The alternating cluster estimation method [981] is a simple extension of the general method given in [279] and [1223]. β_j can be adjusted by [981]

$$\beta_j = \min_k \|\mathbf{c}_k - \mathbf{c}_j\|^2, \quad k \neq j \qquad (5.63)$$

The fuzzy robust C-spherical shells algorithm [1223] searches the clusters belonging to the spherical shells by combining the concept of the fuzzy complement and the fuzzy C-spherical shells algorithm [632]. The hard robust clustering algorithm [1223] is an extension of the GLVQ-F algorithm [583] and is obtained by setting $\beta_j = \infty$ if neuron j is the winner and setting β_j by (5.63) if it is a loser. In these robust algorithms, the initial values and adjustment of β_j are very important.

5.12.4 Other Robust Clustering Problems

The clustering of a vectorial data set with missing entries belongs to the category of robust clustering. In [466], four strategies, namely, the whole data, partial distance, optimal completion, and nearest prototype strategies, have been discussed for implementing the FCM clustering for incomplete data.

The robust competitive agglomeration (RCA) [376] is an algorithm that combines the advantages of hierarchical and partitional clustering techniques. The objective function also contains a constraint term given in the form of (5.61). An optimum number of clusters is determined via a process of competitive agglomeration, while the knowledge of the global shape of the clusters is incorporated via the use of prototypes. Robust statistics like the M-estimator is incorporated to combat the outliers. Overlapping clusters are handled by the use of fuzzy memberships.

Relational data can be clustered by using the non-Euclidean relational FCM (NERFCM) [464, 465]. A number of fuzzy clustering algorithms for relational data have been reviewed in [280]. The introduction of the concept of noise clustering into these relational clustering techniques leads to their robust versions [280]. For a review of robust clustering methods, the reader is referred to [279].

5.13 Clustering Using Non-Euclidean Distance Measures

Conventional clustering methods are based on the Euclidean distance, which favors hyperspherically shaped clusters of equal size. The Euclidean distance measure results in the undesirable property of splitting big and elongated clusters [327]. Other distance measures can be defined to search for clusters of specific shapes in the feature space.

The Mahalanobis distance can be used to look for hyperellipsoid-shaped clusters. However, the C-means algorithm with the Mahalanobis distance tends to produce unusually large or unusually small clusters [761]. The hyperellipsoidal clustering (HEC) network [761] integrates the PCA and clustering into one network, and can adaptively estimate the hyperellipsoidal shape of each cluster. The HEC implements a clustering algorithm using a regularized

Mahalanobis distance, which is a linear combination of the Mahalanobis distance and the Euclidean distance. The regularized distance achieves a tradeoff between the hyperspherical and hyperellipsoidal cluster shapes so as to prevent the HEC network from producing unusually large or unusually small clusters. The Mahalanobis distance is used in the Gustafson–Kessel algorithm [443] and the AFC [35].

The symmetry-based C-means [1062] employs the point-symmetry distance as the dissimilarity measure. The point-symmetry distance is defined as

$$d_{j,i} = d\left(\mathbf{x}_j, \mathbf{c}_i\right) = \min_{p=1,\cdots,N,\, p \neq j} \frac{\|(\mathbf{x}_j - \mathbf{c}_i) + (\mathbf{x}_p - \mathbf{c}_i)\|}{\|\mathbf{x}_j - \mathbf{c}_i\| + \|\mathbf{x}_p - \mathbf{c}_i\|} \qquad (5.64)$$

where $\mathbf{c}_i$ is a prototype vector, and the pattern set $\{\mathbf{x}_i\}$ is of size N. Note that $d_{j,i} = 0$ only when $\mathbf{x}_p = 2\mathbf{c}_i - \mathbf{x}_j$. The symmetry-based C-means uses the C-means as a coarse search for the K cluster centroid. A fine-tuning procedure is then performed based on the point-symmetry distance using the nearest-neighbor paradigm. The symmetry-based C-means can effectively find clusters with symmetric shapes, such as the human face.

A number of algorithms for detecting circles and hyperspherical shells have been proposed as extensions of the C-means and FCM algorithms. These include the fuzzy C-shells [277], fuzzy C-ring [754], hard C-spherical shells [632], unsupervised C-spherical shells [632], fuzzy C-spherical shells [632], and possibilistic C-spherical shells [634] algorithms. These algorithms are based on iterative optimization of objective functions similar to that for the FCM, but defines the distance from a prototype $\boldsymbol{\lambda}_i = (\mathbf{c}_i, r_i)$ to the point $\mathbf{x}_j$ as

$$d_{j,i}^2 = d^2\left(\mathbf{x}_j, \boldsymbol{\lambda}_i\right) = \left(\|\mathbf{x}_j - \mathbf{c}_i\| - r_i\right)^2 \qquad (5.65)$$

where $\mathbf{c}_i$ is the center of the hypersphere and r_i is the radius. These algorithms can effectively estimate the optimal number of substructures in the data set by using some validity criteria such as spherical-shell thickness [632], fuzzy hypervolume and fuzzy density [402, 754]. These criteria will be introduced in Subsect. 5.17.2.

By using different distance measures, many clustering algorithms can be derived for detecting clusters of various shapes such as lines and planes [104, 279, 376, 1263, 595], circles and spherical shells [632, 881, 1263], ellipses [403, 376], curves, curved surfaces, ellipsoids [104, 402, 761, 376, 595], rectangles, rectangular shells and polygons [494].

5.14 Hierarchical Clustering

Existing clustering algorithms are broadly classified into partitional clustering, hierarchical clustering, and density-based clustering. Clustering methods discussed so far belong to partitional clustering. In this section, we deal with hierarchical clustering.

5.14.1 Partitional, Hierarchical, and Density-based Clustering

Partitional clustering can be either hard clustering or fuzzy clustering. Fuzzy clustering can deal with overlapping cluster boundaries. Partitional clustering is dynamic, where points can move from one cluster to another. Knowledge of the shape or size of the clusters can be incorporated by using appropriate prototypes and distance measures. Due to the optimization of a certain criterion function, partitional clustering is sensitive to initialization and susceptible to local minima. Partitional clustering has difficulty in determining the suitable number of clusters K, which is usually prespecified. In addition, it is also sensitive to noise and outliers. Typical partitional clustering algorithms have a computational complexity of $O(N)$, where N is the number of the input data.

Hierarchical clustering consists of a sequence of partitions in a hierarchical structure, which can be represented graphically as a clustering tree, called a *dendrogram*. Hierarchical clustering techniques can be classified into agglomerative and divisive techniques. New clusters are formed by reallocating the membership degree of one point at a time, based on some measure of similarity or distance. It is suitable for data with dendritic substructure. Divisive clustering performs in a way opposite to that of agglomerative clustering, but is computationally more costly. Hierarchical clustering usually takes the form of agglomerative clustering [1207].

Hierarchical clustering has a number of advantages over the partitional clustering. In hierarchical clustering, outliers can be easily identified, since they merge with other points less often due to their larger distances from the other points. Consequently, the number of points in a collection of outliers is typically much less than the number in a cluster. In addition, the number of clusters K does not need to be specified, and the local minimum problem arising from initialization is no longer a problem any more.

However, prior knowledge of the shape or size of the clusters cannot be incorporated, and consequently overlapping clusters cannot always be separated. Moreover, hierarchical clustering is static, and points committed to a given cluster in the early stages cannot move to a different cluster. Also, hierarchical clustering typically has a computational complexity of at least $O\left(N^2\right)$. This makes it impractical for larger data sets. Classical hierarchical clustering algorithms are sensitive to noise and outliers.

Density-based clustering groups neighboring objects of a data set into clusters based on density conditions. Clusters are dense regions of objects in the data space and are separated by regions of low density. Density-based clustering is robust against outliers since an outlier affects the clustering only in the neighborhood of this data point. It can handle outliers and discover clusters of arbitrary shape. The computational complexity of density-based clustering is in the same order of magnitude as that of the hierarchical algorithms. The DBSCAN (density-based spatial clustering of applications with noise) [342] is a widely known density-based clustering algorithm. In DBSCAN, a region is defined as the set of points that lie in the ϵ-neighborhood of some point p. Cluster label propagation from p to the other points in a region $\mathcal{R}$ happens if $|\mathcal{R}|$, the cardinality of $\mathcal{R}$, exceeds a given threshold for the minimal number of points.

5.14.2 Distance Measures, Cluster Representations, and Dendrograms

The simplest and most popular methods for calculating the intercluster distance are the single linkage and complete linkage techniques. The single linkage technique, also called the *nearest-neighbor paradigm*, calculates the intercluster distance using the two closest data points in different clusters

$$d\left(\mathcal{C}_1, \mathcal{C}_2\right) = \min_{\mathbf{x} \in \mathcal{C}_1, \mathbf{y} \in \mathcal{C}_2} d(\mathbf{x}, \mathbf{y}) \qquad (5.66)$$

where $d\left(\mathcal{C}_1, \mathcal{C}_2\right)$ denotes the distance between two clusters $\mathcal{C}_1$ and $\mathcal{C}_2$, and $d(\mathbf{x}, \mathbf{y})$ the distance between two data points $\mathbf{x}$ and $\mathbf{y}$. The single linkage technique is more suitable for finding well-separated stringy clusters.

In contrast, the complete linkage method uses the farthest distance between any two data points in different clusters to define the intercluster distance. Other more complicated methods are group average linkage, median linkage, and centroid linkage methods.

The representation of clusters is also necessary in hierarchical clustering. The shape and extent of a cluster are conventionally represented by its centroid or prototype. This is desirable only for spherically shaped clusters, but causes cluster splitting for a large or arbitrary shaped cluster, since the centroids of its subclusters can be far apart. At the other extreme, all the data points in a cluster are used as its representatives, and this makes the clustering algorithm extremely sensitive to noise in the data points and the outliers. This all-points representation can cluster arbitrary shapes. The scatter-points representation [439], as a tradeoff between the two extremes, represents each cluster by a certain fixed number of points that are generated by selecting well-scattered points from the cluster and then shrinking them toward the center of the cluster by a specified fraction. This reduces the adverse effects of the outliers since the outliers are typically farther away from the mean and are thus shifted by a larger distance due to the shrinking. The scatter-points representation achieves robustness to outliers, and identifies clusters having nonspherical shape and wide variations in size.

Dendrograms for Agglomerative Clustering

Agglomerative clustering starts from N clusters, each containing exactly one data point. A series of nested merging is performed until finally all the data points are grouped into one cluster. The agglomerative clustering processes

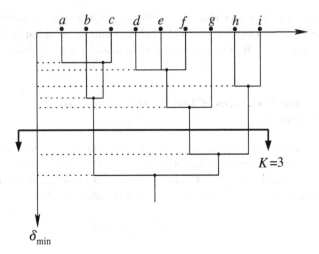

Fig. 5.9. A single-linkage dendrogram. The process of successive merging of the clusters is guided by the set distance δ_{min}. At a cross section with δ_{min}, the number of clusters can be decided. At the cross section shown in the figure, there are three clusters: $\{a, b, c\}$, $\{d, e, f, g\}$, and $\{h, i\}$.

a set of N^2 numerical relationships between the N data points, and agglomerates according to their similarity, usually described by a distance. Agglomerative clustering is based on a local connectivity criterion. The run time is $O\left(N^2\right)$. The process of agglomerative clustering can be easily illustrated by using a dendrogram, as shown in Fig. 5.9.

5.14.3 Agglomerative Clustering Methods

Agglomerative clustering methods can be based on the centroid [1261], all-points [1250], and scatter-points [439] representations. For large data sets, storage or multiple input/output scans of the data points is a bottleneck for the existing clustering algorithms. Some strategies can be applied to combat this problem [1261, 439, 1129, 1145].

The minimum spanning tree (MST) algorithm [1250] is a conventional agglomerative clustering technique. The MST method is a graph-theoretical technique [1069, 1087]. It uses the all-points representation. The method first finds an MST for the input data. Then, by removing the longest $K - 1$ edges, K clusters are obtained. Initially each point is a separate cluster. An agglomerative algorithm starts with the disjoint set of clusters. Pairs of clusters with minimum distance are then successively merged until a criterion is satisfied. The MST algorithm is good at clustering arbitrary shapes. However, the method is very sensitive to the outliers, and it may merge two clusters due to the existence of a chain of outliers connecting them.

The BIRCH (balanced iterative reducing and clustering using hierarchies) method [1261] first performs an incremental and approximate preclustering phase in which dense regions of points are represented by compact summaries, and then a centroid-based hierarchical algorithm is used to cluster the set of summaries. During preclustering, the entire database is scanned, and cluster summaries are stored in a data structure called the *clustering feature (CF) tree*. Robustness is achieved by eliminating the outliers from the summaries via the identification of the sparsely distributed data points in the feature space. After the CF tree is built, agglomerative clustering is applied to the set of summaries to perform global clustering. The BIRCH needs only a little more than one scan of the data. However, the BIRCH fails to identify clusters with nonspherical shapes or wide variation in size by splitting larger clusters and merging smaller clusters.

The CURE (clustering using representation) method [439] is a robust clustering algorithm based on the scatter-points representation. To handle large databases, the CURE employs a combination of random sampling and partitioning to reduce the computational complexity. Random samples drawn from the data set are first partitioned and each partition is partially clustered. The partial clusters are then clustered in a second pass to yield the desired clusters. The CURE method uses the k-d tree and heap data structures. The computational complexity of the CURE is $O\left(N^2\right)$ for low-dimensional data, which is no worse than that of the centroid-based hierarchical algorithm. The CURE provides better clustering performance with less execution time compared to the BIRCH [439]. The CURE can discover clusters with interesting shapes and is less sensitive than the MST to the outliers.

The CHAMELEON [593] first creates a graph, where each node represents a pattern and the edges between each node and all the other nodes exist according to the k-NN paradigm. A graph-partitioning algorithm is used to recursively partition the graph into many small unconnected subgraphs, each partitioning yielding two subgraphs of roughly equal size. Agglomerative clustering is applied, each subcluster being used as an initial subcluster. The CHAMELEON merges two subclusters only when the interconnectivity as well as the closeness of the individual clusters is very similar to that of the merged cluster. It automatically adapts to the characteristics of the clusters being merged. The CHAMELEON is more effective than the CURE in discovering clusters of arbitrary shapes and varying densities. It has a computational complexity of $O\left(N^2\right)$ [593].

5.14.4 Combinations of Hierarchical and Partitional Clustering

Some methods make use of the advantages of both the hierarchical and partitional clustering techniques [1129, 1145, 408, 376].

The VQ-clustering and VQ-agglomeration [1145] methods involve a VQ process followed, respectively, by a clustering algorithm and an agglomerative clustering algorithm that treat codewords as initial prototypes. Each codeword is associated with a gravisphere that has a well-defined attraction radius. The agglomeration algorithm requires that each codeword be moved directly to the centroid of its neighboring codewords. The movements of codewords in the feature space are synchronous, and converge quickly to certain sets of concentric circles for which the centroids identify the resulting clusters. A similar two-stage clustering procedure that uses the SOM for VQ and an agglomerative clustering or the C-means algorithm for further clustering is given in [1129]. The performance results of these two-stage methods are comparable to those of direct methods, with a significantly reduced computational time [1145, 1129].

A clustering algorithm that clusters data with arbitrary shapes without knowing the number of clusters in advance is given in [1063]. It is a two-stage algorithm: An ART-like algorithm is first used to partition data into a set of small multidimensional hyperellipsoids, and a dendrogram is then built to sequentially merge those hyperellipsoids. Dendrograms and the so-called *tables of relative frequency counts* are then used so as to pick some trustable clustering results from many different clustering results.

The hierarchical unsupervised fuzzy clustering (HUFC) [408] has the advantages of both the hierarchical clustering and the fuzzy clustering. The PCA is applied to each cluster for optimal feature extraction. This method is effective for data sets with a wide dynamic range in both the covariance and the number of members in each class. The RCA [376] employs competitive agglomeration to find the optimum number of clusters, uses prototypes to represent the global shape of the clusters, and integrates robust statistics to achieve noise immunity.

5.15 Constructive Clustering Techniques

Conventional partitional clustering algorithms assume a network with a fixed number of clusters (nodes) K, which needs to be prespecified. However, selecting the appropriate value of K is a difficult task without prior knowledge of the input data. This difficulty can be solved by using constructive clustering.

A simple strategy for determining the optimal K is to perform clustering for a range of K, and select the value of K that minimizes a cluster-validity measure. The scatter-based FSCL clustering [1046] is an FSCL algorithm that uses a scatter matrices-based validity measure called *sphericity*. The algorithm gradually increases K until there is no significant decrease in sphericity. This procedure is, however, computationally too expensive when the actual number of clusters is large. In [146], the optimal number of prototypes is determined by optimizing a cost function that includes, in addition to the distortion errors, a codebook-complexity term.

Like the C-means [746, 714], the ISODATA method [60] is another popular early statistical method for unlabeled data clustering. They are both nearest-centroid clustering methods. The ISODATA can be treated as a variant of the incremental C-means [746] by incorporating some heuristics for merging and splitting clusters, and for handling outliers. Thus, the ISODATA has a variable number of clusters K. For implementation of the ISODATA, one needs to specify an initial value of K, the initial cluster means, the split threshold, and the lump threshold, and the minimum number of data for a cluster.

The self-creating mechanism in the competitive learning process can adaptively determine the natural number of clusters. In self-creating competitive learning networks, each node is associated with a local statistical variable, which is used to control the growing and pruning of the network architecture.

The self-creating and organizing neural network (SCONN) [230] employs adaptively modified node thresholds to control its self-growth. At the presentation of a new input, if the winning node is active, the winning node is updated; otherwise, a new node is recruited from the winning node. Activation levels of all the nodes decrease with time, so that the weight vectors are distributed at the final stage according to the input distribution. Nonuniform VQ is also realized by decreasing the activation levels of the active nodes and increasing those of the other nodes to estimate the asymptotic point density automatically. The SCONN avoids the underutilization problem, and has VQ accuracy and speed advantage over the SOM and the batch C-means [714].

The growing cell structures (GCS) network [377] can be regarded as a modification of the SOM by integrating the node-recruiting and pruning functions. The GCS assigns each node with a local accumulated statistical variable called a *signal counter* u_i, $i = 1, \cdots, K$. At each pattern presentation, only the winning node increases its signal counter u_w by 1, and then all the signal counters u_i decay with a forgetting factor. After a fixed number of learning iterations, the node with the largest signal counter gets the right to insert a new node between itself and its farthest neighbor. The network occasionally prunes a node whose signal counter is less than a specified threshold during a complete epoch. The growing grid network [381] is strongly related to the GCS. In contrast to the GCS, the growing grid has a strictly rectangular topology. By inserting complete rows or columns of units, the grid may adapt its height/width ratio to the given pattern distribution. The branching competitive learning (BCL) network [1199] adopts the same method for recruiting and pruning nodes as the GCS except that a new geometrical criterion is applied to the winning node before updating its signal counter u_w. The BCL network is efficient for capturing the spatial distribution of the input data.

The GNG model [380, 382] is based on the GCS [377] and NG [768] models. The GNG is capable of generating and removing neurons and lateral connections dynamically. In the GNG, lateral connections are generated according to the competitive Hebbian learning rule. The GNG achieves robustness against noise and performs perfect topology-preserving mapping. By integrating an online criterion so as to identify and delete useless neurons, the GNG with utility criterion (GNG-U) [382] is also able to track nonstationary data input.

A similar online clustering method, which uses the concept of utility as well as the node insertion and pruning mechanisms, is given in [385].

The dynamic cell structures (DCS) model [139] uses a modified Kohonen learning rule to adjust the prototypes and the competitive Hebbian rule to establish a dynamic lateral connection structure. Applying the principle of DCS to the GCS yields the DCS-GCS algorithm, which has a behavior similar to that of the GNG. The life-long learning cell structures (LLCS) algorithm [445] is an online clustering and topology-representation method. It employs a strategy similar to that of the ART, which can effectively deal with the stability–plasticity dilemma [159]. A similarity-based unit pruning and an aging-based edge pruning procedures are incorporated.

The self-splitting competitive learning (SSCL) [1263] can find the natural number of clusters based on the one-prototype-take-one-cluster (OPTOC) paradigm and a validity measure for self-splitting. The OPTOC enables each prototype to situate at the centroid of one natural cluster when the number of clusters is greater than that of the prototypes. The SSCL starts with a single prototype and splits adaptively during the learning process until all the clusters are found. During the learning process, one prototype is chosen to split into two prototypes according to the validity measure. This splitting process terminates only when the SSCL achieves an appropriate number of clusters. After learning, each cluster is labeled by a prototype located at its center.

In [438], an online agglomerative clustering algorithm called *AddC* is designed for nonstationary data. For each new data point, the method updates the closest centroid, merges the two nearest centrids, and then treats the data point as a new centroid. Each centroid has a counter to record the number of points it represents. After all the data is presented, those centroids with negligible counters are merged with their nearest centroids. The algorithm clusters the data in a single pass without a knowledge of the number of clusters, and the performance is comparable to that of the exisiting clustering algorithms, such as the incremental C-means [746] and the maximum-entropy clustering [961]. The algorithm provides an efficient framework to determine the natural number of clusters, given the scale of the problem. The algorithm implicitly minimizes the local distortion, a measure used to represent clusters with relatively small mass.

5.16 Miscellaneous Clustering Methods

There are also numerous density-based and graph-theory-based clustering algorithms. In the following, we briefly mention some algorithms that are associated with competitive learning and neural networks.

A simple and efficient implementation of the LBG is given in [578], where the data points are stored by a k-d tree. The algorithm is typically one order of magnitude faster than the LBG.

The EM clustering [131] represents each cluster using a probability distribution, typically a Gaussian distribution. Each clustering is represented by a mean and $J_1 \times J_1$ covariance matrix. Each pattern belongs to all the clusters with the probabilities of membership determined by the distributions of the corresponding clusters. Thus, the EM clustering can be treated as a fuzzy clustering technique. The EM is derived by maximizing the log likelihood of the mixture model probability density function. The C-means algorithm has been shown to be equivalent to the classification EM (CEM) algorithm corresponding to the uniform spherical Gaussian model [178, 1207].

Kernel-based clustering first nonlinearly maps the patterns into an arbitrarily high-dimensional feature space. Clustering is then performed in the feature space. Some examples are the kernel C-means [1008], kernel subtractive clustering [602], variants of kernel C-means based on the SOM and ART [252], a kernel-based algorithm that minimizes the trace of the within-class scatter matrix [412], and support vector clustering (SVC) [97, 224, 153]. The SVC can effectively deal with the outliers.

5.17 Cluster Validity

So far, we have described many clustering algorithms. One problem arises as to how to select and evaluate a clustering algorithm. The number of clusters is application specific, and is usually specified by the user. An optimal number of clusters or a good clustering algorithm is only in the sense of a certain cluster validity criterion. Many cluster validity measures are defined for this purpose.

5.17.1 Measures Based on Maximal Compactness and Maximal Separation of Clusters

A good clustering algorithm should generate clusters with small intracluster deviations and large intercluster separations. Cluster compactness and cluster separation are two measures for describing the performance of clustering. Given the clustered result of a data set $\mathcal{X}$, the cluster compactness, cluster separation, and overall cluster quality measures are, respectively, defined by [471]

$$E_{\mathrm{CMP}} = \frac{1}{K} \frac{\sum_{i=1}^{K} \sigma(\mathbf{c}_i)}{\sigma(\mathcal{X})} \tag{5.67}$$

$$E_{\mathrm{SEP}} = \frac{1}{K(K-1)} \sum_{i=1}^{K} \sum_{j=1, j\neq i}^{K} e^{-\frac{d^2(\mathbf{c}_i, \mathbf{c}_j)}{2\sigma_0^2}} \tag{5.68}$$

$$E_{\mathrm{OCQ}}(\gamma) = \gamma \cdot E_{\mathrm{CMP}} + (1-\gamma) \cdot E_{\mathrm{SEP}} \tag{5.69}$$

where K is the number of clusters, $\mathbf{c}_i$ and $\mathbf{c}_j$ are, respectively, the centers of clusters i and j, $\sigma(\mathbf{c}_i)$ denotes the standard deviation of cluster i, $\sigma(\mathcal{X})$ is the standard deviation of data set $\mathcal{X}$, σ_0 is a Gaussian constant, $d(\mathbf{c}_i, \mathbf{c}_j)$ is the distance between $\mathbf{c}_i$ and $\mathbf{c}_j$, and the constant $\gamma \in [0, 1]$. A small E_{CMP} means that all the clusters have small deviations, and a small E_{SEP} value corresponds to a better separation performance. The overall performance E_{OCQ} is a linear combination of E_{CMP} and E_{SEP}.

Another popular cluster validity measure is defined as a function of the ratio of the sum of the within-cluster scatters to the between-cluster separation [282]

$$E_{\mathrm{WBR}} = \frac{1}{K} \sum_{k=1}^{K} \max_{l \neq k} \left\{ \frac{d_{\mathrm{WCS}}(\mathbf{c}_k) + d_{\mathrm{WCS}}(\mathbf{c}_l)}{d_{\mathrm{BCS}}(\mathbf{c}_k, \mathbf{c}_l)} \right\} \tag{5.70}$$

where the within-cluster scatter for cluster k, denoted as $d_{\mathrm{WCS}}(\mathbf{c}_k)$, and the between-cluster separation for cluster k and cluster l, denoted as $d_{\mathrm{BCS}}(\mathbf{c}_k, \mathbf{c}_l)$, are, respectively, calculated by

$$d_{\mathrm{WCS}}(\mathbf{c}_k) = \frac{\sum_i \|\mathbf{x}_i - \mathbf{c}_k\|}{N_k} \tag{5.71}$$

$$d_{\mathrm{BCS}}(\mathbf{c}_k, \mathbf{c}_l) = \|\mathbf{c}_k - \mathbf{c}_l\| \tag{5.72}$$

where N_k is the number of data points in cluster k. The best clustering minimizes E_{WBR}. This index indicates good clustering results for spherical clusters [1129].

A cluster-validity criterion has been defined in [1198] for evaluating fuzzy clustering; it minimizes the ratio of compactness E_{CMP1} and separation E_{SEP1} defined by

$$E_{\mathrm{CMP1}} = \frac{1}{N} \sum_{i=1}^{K} \sum_{p=1}^{N} \mu_{ip}^m \|\mathbf{x}_p - \mathbf{c}_i\|_{\mathbf{A}}^2 \tag{5.73}$$

$$E_{\mathrm{SEP1}} = \min_{i \neq j} \|\mathbf{c}_i - \mathbf{c}_j\|_{\mathbf{A}}^2 \tag{5.74}$$

where $\|\cdot\|_{\mathbf{A}}$ denotes a weighted norm, and $\mathbf{A}$ is a positive-definite symmetric matrix. Note that E_{CMP1} is equal to the criterion function for the FCM given by (5.43) when $m = 2$ and $\mathbf{A}$ is the identity matrix.

Entropy cluster validity measures based on class conformity have been given in [122, 471]. Some cluster validity measures are described and compared in [106].

5.17.2 Measures Based on Minimal Hypervolume and Maximal Density of Clusters

A good partitioning of the data usually leads to a small total hypervolume and a large average density of the clusters. Cluster-validity measures can be thus selected as the hypervolume and average density of the clusters. The fuzzy hypervolume criterion is defined by [402, 632]

$$E_{\text{FHV}} = \sum_{i=1}^{K} V_i \tag{5.75}$$

where V_i is the volume of the ith cluster

$$V_i = [\det(\mathbf{F}_i)]^{\frac{1}{2}} \tag{5.76}$$

and $\mathbf{F}_i$, the fuzzy covariance matrix of the ith cluster, is defined by [443]

$$\mathbf{F}_i = \frac{1}{\sum_{j=1}^{N} \mu_{ij}^m} \sum_{j=1}^{N} \mu_{ij}^m (\mathbf{x}_j - \mathbf{c}_i)(\mathbf{x}_j - \mathbf{c}_i)^{\mathrm{T}} \tag{5.77}$$

The average fuzzy density criterion is defined by

$$E_{\text{AFD}} = \frac{1}{K} \sum_{i=1}^{K} \frac{S_i}{V_i} \tag{5.78}$$

where S_i sums the membership degrees of only those members within the hyperellipsoid

$$S_i = \sum_{j=1}^{N} \mu_{ij}, \quad \forall \mathbf{x}_j \in \left\{ \mathbf{x}_j \mid (\mathbf{x}_j - \mathbf{c}_i)^{\mathrm{T}} \mathbf{F}_i^{-1} (\mathbf{x}_j - \mathbf{c}_i) < 1 \right\} \tag{5.79}$$

The fuzzy hypervolume criterion typically has a clear extremum; the average fuzzy density criterion is not desirable when there is a substantial cluster overlapping and a large variability in the compactness of the clusters [402]. The average fuzzy density criterion averages the fuzzy densities of individual clusters, and a partitioning that results in both dense and loose clusters may lead to a large average fuzzy density.

Measures for Shell Clustering

For shell clustering, the hypervolume and average density measures are still applicable. However, the distance vector between a pattern and a prototype needs to be redefined. In the case of spherical shell clustering, the distance vector between a pattern $\mathbf{x}_j$ and a prototype $\boldsymbol{\lambda}_i = (\mathbf{c}_i, r_i)$ is defined by

$$\mathbf{d}_{j,i} = (\mathbf{x}_j - \mathbf{c}_i) - r_i \frac{\mathbf{x}_j - \mathbf{c}_i}{\|\mathbf{x}_j - \mathbf{c}_i\|} \tag{5.80}$$

The fuzzy hypervolume and average fuzzy density measures for spherical-shell clustering are obtained by replacing $(\mathbf{x}_j - \mathbf{c}_i)$ in (5.77) and (5.79) by $\mathbf{d}_{j,i}$.

For shell clustering, the shell thickness measure can be used to describe the compactness of a shell. In the case of fuzzy spherical shell clustering, the fuzzy shell thickness of a cluster can be defined by [632]

$$T_j = \frac{\sum_{i=1}^{N} (\mu_{ji})^m \left(\|\mathbf{x}_i - \mathbf{c}_j\| - r_j \right)^2}{r_j \sum_{i=1}^{N} (\mu_{ji})^m} \tag{5.81}$$

The average shell thickness of all clusters

$$E_{\text{THK}} = \frac{1}{K} \sum_{j=1}^{K} T_j \tag{5.82}$$

can be used as a cluster-validity measure for shell clustering.

5.18 Computer Experiments

We have addressed various aspects of clustering techniques. Interested readers are referred to [539, 1207] for recent surveys on the data-clustering techniques and their applications. Other topics like global search-based clustering are also reviewed in [1207]. We now conclude this chapter by providing some simulation examples.

5.18.1 Example 5.1: Vector Quantization Using the Self-organizing Map

Given a data set of 2000 random data points in the two-dimensional space: 1000 uniformly random points in each of two opposite quarters of a unit square, as shown in Fig. 5.10. We use the SOM to realize VQ by producing a grid of cells[11].

The link distance[12] is employed. All prototypes of the cells are initialized at the center of the range of the data set, namely, $(0.5, 0.5)$. The ordering phase starts from a learning rate of 0.9 and decreases to the tuning-phase learning rate 0.02 in 1000 epochs, and then the tuning phase lasts a much longer time with a slowly decreasing learning rate. In the tuning phase, the neighborhood distance is set as 1. When the training is completed, two points, $\mathbf{p}_1 = (0.8, 0.6)$ and $\mathbf{p}_2 = (0.2, 0.4)$, are used as test points.

In the first group of simulations, the output cells are arranged in a 10×10 grid. The hexagonal neighborhood topology is employed. The training results for 10, 100, 1000 and 5000 epochs are shown in Fig. 5.11. At 5000 epochs, we tested $\mathbf{p}_1$ and $\mathbf{p}_2$, and found that they, respectively, belong to the 88th and 53rd clusters.

In the second group of simulations, the output cells are arranged in a one-dimensional grid of 100 nodes. The corresponding results are shown in Fig. 5.12. In this case, $\mathbf{p}_1$ and $\mathbf{p}_2$, respectively, belong to the 24th and 94th clusters.

---

[11] Simulations are based on the Matlab© Neural Network Toolbox.

[12] The link distance between two points A and B inside a polygon P is defined to be the minimum number of edges required to connect A and B inside P.

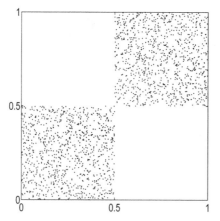

Fig. 5.10. Random data points in the two-dimensional space. In each of the two quarters, there are 1000 uniformly random points.

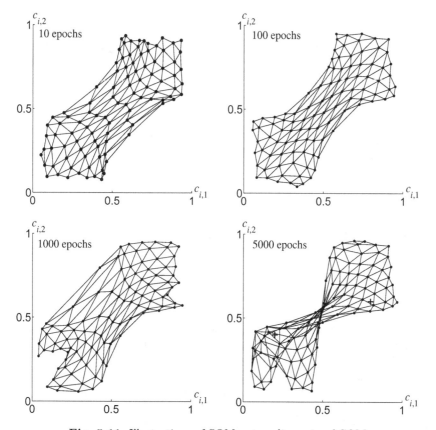

Fig. 5.11. Illustrations of SOM: a two-dimensional SOM

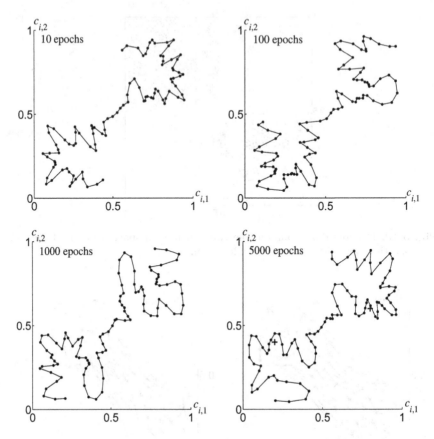

Fig. 5.12. Illustrations of SOM: a one-dimensional SOM. The data points p_1 and p_2 are denoted by plus ($+$) signs.

5.18.2 Example 5.2: Solving the TSP Using the Self-organizing Map

The SOM can be applied to solve the TSP. Assume that 30 cities are randomly located in a unit square. The objective is to find the shortest route that passes through all the cities, each city being visited exactly once. No constraint is applied to the Kohonen network since the topology of the solution is contained in the network topology. A one-dimensional grid of 60 units is used by the SOM. The desired solution is that all the cities are covered by nodes, and all the additional nodes are along the lines between cities. The Euclidean distance is employed. Other parameters are the same as for Example 5.1. The search results at the 10th, 100th, 1 000th, and 10 000th epochs are illustrated in Fig. 5.13, and the total map length at the 10 000th epoch is 4.0020. The

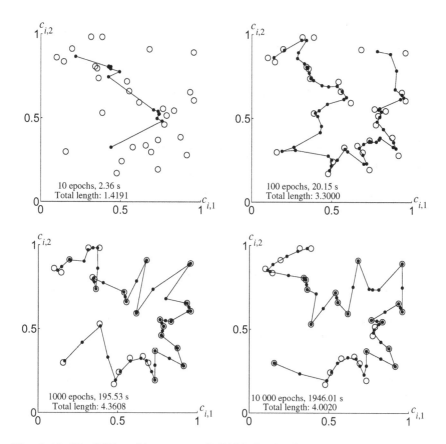

Fig. 5.13. The TSP problem using 1-D SOM. Circles denote positions of the cities.

computation time for the 10 000 epochs is 1946.01 seconds on a PC with a
Pentium III 600 MHz CPU and 128M RAM using the Matlab©.

It is seen that the results from the SOM are not satisfactory, and the routes
are not feasible solutions since some cities are not covered by nodes. Nevertheless, the SOM can be used to find a preliminary search for a suboptimal
route, which can be modified manually to obtain a feasible solution. The SOM
solution can be used as an initialization of other TSP solvers. In this case, we
do not need to run the SOM for many epochs.

The classical SOM is not efficient for searching suboptimal solution for the
TSP. Many practical TSP solvers have been developed using self-organizing
neural-network models based on the SOM, among which some solvers can find
a suboptimal solution for a TSP of hundred cities within dozens of epochs.
Most of them are based on the concept of *elastic ring* [515]. Some Kohonen
network-based models for the TSP have been discussed and simulated in [43].

The Lin–Kernighan algorithm [712] is one of the methods that achieves the best results for large-scale TSPs. A large-scale TSP can be rapidly solved by a divide-and-conquer technique, where clustering methods, such as the ART, are first used to group the cities and a local optimization algorithm, such as the Lin–Kernighan algorithm, is used to find the minimum in each group [830]. This speedup is offset by a slight loss in tour quality, but the structure is suitable for parallel implementation.

5.18.3 Example 5.3: Three Clustering Algorithms — A Comparison

In this example, we illustrate three popular clustering algorithms: the C-means, the FCM, and the subtractive clustering. The input data represents three clusters centered at $(2, 2)$, $(2, -2)$, and $(-2, -2)$, each having 200 data with a Gaussian distribution $N(0, 1)$ in both x and y directions. The initial cluster centers for the C-means is randomly sampled from the data set. The fuzzifier of the FCM is selected as $m = 2$. For the C-means and the FCM, the termination criterion is that the error in the objective function for two adjacent iteration is less than 10^{-5} and the maximum number of epochs is 100. We specify the number of clusters as 3. For the subtractive clustering, the parameter ε is selected as 0.4.

Simulations are performed based on averaging of 100 random runs. The simulation results are listed in Table 5.1. As far as this artificial data set is concerned, the C-means and the FCM have almost the same performance, which is considerably superior to that of the subtractive clustering. However, the subtractive clustering is a deterministic method, and it can automatically detect a suitable number of clusters for a wide range of ε.

The clustering results for a random run are illustrated in Fig. 5.14. There are minor differences in the cluster boundaries and the cluster centers between the algorithms.

Table 5.1. Comparison of the C-means, the FCM, and the subtractive clustering for an artificial data set. $\overline{d}_{wc}$ denotes the mean within-cluster distance, and $\overline{d}_{bc}$ denotes the mean between-cluster distance. A smaller value of $\overline{d}_{wc}/\overline{d}_{bc}$ corresponds to a better performance.

	C-means	FCM	Subtractive
$\overline{d}_{wc}$	1.2274	1.2275	1.2542
$\overline{d}_{bc}$	4.6077	4.6480	4.5963
$\overline{d}_{wc}/\overline{d}_{bc}$	0.2664	0.2641	0.2729
time	0.06384	0.06802	0.4978

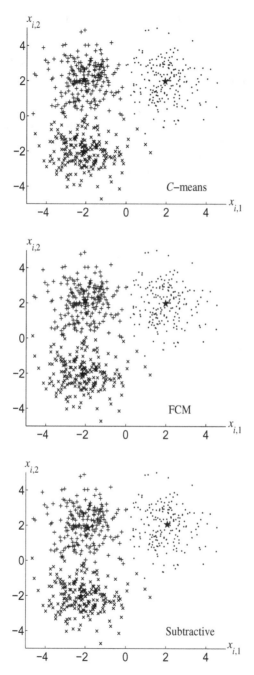

Fig. 5.14. Clustering using three methods: the C-means, the FCM, and the subtractive clustering. The five-point stars denote the cluster centers.

5.18.4 Example 5.4: Clustering Analog Signals Using ART 2A

We take an application of the ART 2A from [164]. The ART 2A was used to group 50 analog input patterns. When η in (5.41) is selected as 1, the ART 2A gives the same clustering results as that of a fast-learn ART 2 system with comparable parameters [160]. The inputs, listed in the left column of Fig. 5.15, are cyclically presented until the learning is stabilized into 23 categories. The ART 2A fast-learn simulation took only 4 s of Sun4/110 CPU time to cyclically present the 50 patterns three times. For the same objective, the computation time of the ART 2 was around 100 times that of the ART 2A. Thus, the ART 2A is an efficient algorithm for clustering analog signals.

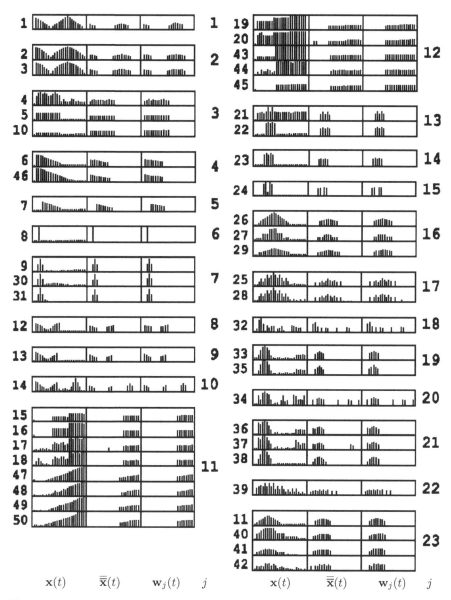

Fig. 5.15. Analog signals grouping using ART 2A. 50 signals are grouped into 23 clusters; input dimension $J = 25$, $\eta = 1$, and $\rho = 0.92058$. In each column, the three sections represent the input signal $\mathbf{x}(t)$, the normalized signal $\overline{\overline{\mathbf{x}}}(t)$, and the resulting LTM vector $\mathbf{w}_j(t)$, respectively. The indices of the inputs to the left of each column gives the order of presentation, and the indices to the right of each column correspond to the winning nodes j. All $\mathbf{x}(t)$ have the same, arbitrary scale. All $\overline{\overline{\mathbf{x}}}(t)$ and $\mathbf{w}_j(t)$ are scaled so that their norms are unity. (From [164].)

6

Radial Basis Function Networks

The RBFN is a universal approximator, with a solid foundation in the conventional approximation theory. The RBFN is a popular alternative to the MLP, since it has a simpler structure and a much faster training process.

This chapter is dedicated to the RBFN and its learning. In addition to the various RBFN learning methods, we also describe the applications of the RBFN to dynamic system modeling, VLSI implementations of the RBFN, normalized RBFNs, and complex RBFNs. A comparison of the RBFN with the MLP is then made. Finally, we use the RBFN to solve two ASP problems.

6.1 Introduction

Learning is an approximation problem, which is closely related to the conventional approximation techniques, such as generalized splines and regularization techniques. The RBFN has its origin in performing exact interpolation of a set of data points in a multidimensional space [926]. The RBFN has a network architecture similar to the classical regularization network [923], where the basis functions are the Green's functions of the Gram operator associated with the stabilizer. If the stabilizer exhibits radial symmetry, the basis functions are radially symmetric as well and an RBFN is obtained. From the viewpoint of approximation theory, the regularization network has three desirable properties [923, 413] given below.

1. It can approximate any multivariate continuous function on a compact domain to an arbitrary accuracy, given a sufficient number of units.
2. The approximation has the best-approximation property since the unknown coefficients are linear.
3. The solution is optimal in the sense that it minimizes a functional that measures how much it oscillates.

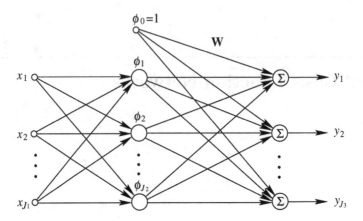

Fig. 6.1. Architecture of the RBFN. The input layer has J_1 nodes, the hidden and output layers have J_2 and J_3 neurons, respectively. $\phi_0(\mathbf{x}) = 1$ corresponds to the bias in the output layer, while $\phi_i(\mathbf{x}) = \phi(\mathbf{x} - \mathbf{c}_i)$, $\mathbf{c}_i$ being the center of the ith node and $\phi(\mathbf{x})$ an RBF.

6.1.1 Architecture of the Radial Basis Function Network

The RBFN is a J_1-J_2-J_3 FNN, and is shown in Fig. 6.1. Each node in the hidden layer uses an RBF, denoted by $\phi(r)$, as its nonlinear activation function. The hidden layer performs a nonlinear transform of the input, and the output layer is a linear combiner mapping the nonlinearity into a new space. The biases of the output layer neurons can be modeled by an additional neuron in the hidden layer, which has a constant activation function $\phi_0(r) = 1$. The RBFN can achieve a global optimal solution to the adjustable weights in the minimum MSE sense by using the linear optimization method.

For an input pattern $\mathbf{x}$, the output of the network is given by

$$y_i(\mathbf{x}) = \sum_{k=1}^{J_2} w_{ki}\phi\left(\|\mathbf{x} - \mathbf{c}_k\|\right) \tag{6.1}$$

for $i = 1, \cdots, J_3$, where $y_i(\mathbf{x})$ is the ith output of the RBFN, w_{ki} is the connection weight from the kth hidden unit to the ith output unit, $\mathbf{c}_k$ is the prototype or center of the kth hidden unit, and $\|\cdot\|$ denotes the Euclidean norm. The RBF $\phi(\cdot)$ is typically selected as the Gaussian function.

For a set of N pattern pairs $\{(\mathbf{x}_p, \mathbf{y}_p)\}$, (6.1) can be expressed in the matrix form

$$\mathbf{Y} = \mathbf{W}^{\mathrm{T}}\boldsymbol{\Phi} \tag{6.2}$$

where $\mathbf{W} = [\mathbf{w}_1, \cdots, \mathbf{w}_{J_3}]$ is a $J_2 \times J_3$ weight matrix, $\mathbf{w}_i = (w_{1i}, \cdots, w_{J_2i})^{\mathrm{T}}$, $\boldsymbol{\Phi} = [\boldsymbol{\phi}_1, \cdots, \boldsymbol{\phi}_N]$ is a $J_2 \times N$ matrix, $\boldsymbol{\phi}_p = (\phi_{p,1}, \cdots, \phi_{p,J_2})^{\mathrm{T}}$ is the output of the hidden layer for the pth sample, $\phi_{p,k} = \phi(\|\mathbf{x}_p - \mathbf{c}_k\|)$, $\mathbf{Y} = [\mathbf{y}_1\ \mathbf{y}_2\ \cdots\ \mathbf{y}_N]$ is a $J_3 \times N$ matrix, and $\mathbf{y}_p = (y_{p,1}, \cdots, y_{p,J_3})^{\mathrm{T}}$.

6.1.2 Universal Approximation of Radial Basis Function Networks

The RBFN has universal approximation and regularization capabilities. Theoretically, the RBFN can approximate any continuous function arbitrarily well, if the RBF is suitably chosen [923, 887, 704].

Micchelli considered the solution of the interpolation problem $s(\mathbf{x}_k) = y_k, k = 1, \cdots, J_2$, by functions of the form $s(\mathbf{x}) = \sum_{k=1}^{J_2} w_k \phi \left(\|\mathbf{x} - \mathbf{x}_k\|^2 \right)$, and proposed Micchelli's interpolation theorem [800]. A condition for suitable $\phi(\cdot)$ is that $\phi(\cdot)$ is completely monotonic on $(0, \infty)$, that is, it is continuous on $(0, \infty)$ and its lth-order derivative $\phi^{(l)}(x)$ satisfies $(-1)^l \phi^{(l)}(x) \geq 0$, $\forall x \in (0, \infty)$ and $l = 0, 1, 2, \cdots$. A less restrictive condition has been given in [580], where $\phi(\cdot)$ is continuous on $(0, \infty)$ and its derivatives satisfy $(-1)^l \phi^{(l)}(x) > 0$, $\forall x \in (0, \infty)$ and $l = 0, 1, 2$.

RBFs possess excellent mathematical properties. In the context of the exact interpolation problem, many properties of the interpolating function are relatively insensitive to the precise form of the nonlinear function $\phi(\cdot)$ [926]. It has been reported that the choice of RBF is not crucial to the performance of the RBFN [206].

The Gaussian RBFN can approximate, to any degree of accuracy, any continuous function by a sufficient number of centers $\mathbf{c}_i$, $i = 1, \cdots, J_2$, and a common standard deviation $\sigma > 0$ in the L_p-norm, $p \in [1, \infty]$ [887]. A class of RBFNs can achieve universal approximation when the RBF is continuous and integrable [887].

The requirement of the integrability of the RBF is relaxed in [704]. According to [704], the RBFN can approximate any continuous function in $L_p(\mu)$ with respect to the L_p-norm, $p \in [1, \infty)$, and μ is any finite measure, if the RBF is essentially bounded and not a polynomial. That is, for an RBF that is continuous almost everywhere, locally essentially bounded, and not a polynomial, the RBFN can approximate any continuous function with respect to the uniform norm [704]. Based on this result, such RBFs as $\phi(r) = \mathrm{e}^{-\frac{r}{\sigma^2}}$ and $\phi(r) = \mathrm{e}^{\frac{r}{\sigma^2}}$ also lead to universal approximation capability [704].

6.1.3 Learning for Radial Basis Function Networks

Like the MLP learning, the learning of the RBFN is formulated as the minimization of the MSE function[1]

$$E = \frac{1}{N} \sum_{i=1}^{N} \|\mathbf{y}_p - \mathbf{W}^{\mathrm{T}} \phi_p\|^2 = \frac{1}{N} \|\mathbf{Y} - \mathbf{W}^{\mathrm{T}} \mathbf{\Phi}\|_{\mathrm{F}}^2 \qquad (6.3)$$

where $\mathbf{Y} = [\mathbf{y}_1, \mathbf{y}_2, \cdots, \mathbf{y}_N]$, $\mathbf{y}_i$ is the target output for the ith sample in the training set, and $\| \cdot \|_{\mathrm{F}}$ is the Frobenius norm defined as $\|\mathbf{A}\|_{\mathrm{F}}^2 = \mathrm{tr} \left(\mathbf{A}^{\mathrm{T}} \mathbf{A} \right)$.

---

[1] When the coefficient $\frac{1}{N}$ is dropped, the objective function is called the *sum-of-squares error (SSE)* function.

The learning of the RBFN requires the determination of the RBF centers and the weights. The selection of the RBF center vectors is most critical to a successful RBFN implementation. The centers can be placed on a random subset or all of the training examples, or determined by clustering or via a learning procedure. One can also use all the data points as centers in the beginning, and then selectively remove centers using the k-NN classification scheme [304]. For some RBFs such as the Gaussian, it is also necessary to determine the smoothness parameter σ. The RBFN using the Gaussian RBF is usually termed the *Gaussian RBFN*. Existing RBFN learning algorithms are mainly developed for the Gaussian RBFN, and can be modified accordingly when other RBFs are used.

The Gaussian RBFN can be regarded as an improved alternative to the four-layer probabilistic neural network (PNN) [1050], which is based on the Parzen classifier from the pattern-recognition literature [327, 304, 627]. In a PNN, a Gaussian RBF node is placed at the position of each training pattern so that the unknown density can be well interpolated and approximated. This technique yields optimal decision surfaces in the Bayes' sense. Training is to associate each node with its target class. This approach, however, severely suffers from the *curse of dimensionality* and results in a poor generalization.

6.2 Radial Basis Functions

A number of functions can be used as the RBF [923, 800, 704]

$$\phi(r) = e^{-\frac{r^2}{2\sigma^2}}, \qquad \text{Gaussian} \qquad (6.4)$$

$$\phi(r) = \frac{1}{(\sigma^2 + r^2)^\alpha}, \qquad \alpha > 0 \qquad (6.5)$$

$$\phi(r) = (\sigma^2 + r^2)^\beta, \qquad 0 < \beta < 1 \qquad (6.6)$$

$$\phi(r) = r, \qquad \text{linear} \qquad (6.7)$$

$$\phi(r) = r^2 \ln(r), \qquad \text{thin-plate spline} \qquad (6.8)$$

$$\phi(r) = r^3, \qquad \text{cubic} \qquad (6.9)$$

$$\phi(r) = \frac{1}{1 + e^{\frac{r}{\sigma^2} - \theta}}, \qquad \text{logistic function} \qquad (6.10)$$

where $r > 0$ denotes the distance from a data point $\mathbf{x}$ to a center $\mathbf{c}$, the parameter σ in (6.4) through (6.6) and (6.10) is used to control the smoothness of the interpolating function, and θ in (6.10) is an adjustable bias. These RBFs are illustrated in Fig. 6.2.

When the RBF is selected as the logistic function (6.10), the RBFN can be derived from the three-layer MLP by replacing the weighted sum of the ith hidden neuron, namely, $net_i = \mathbf{x}^T \mathbf{w}_i^{(1)}$, with $\frac{\|\mathbf{x} - \mathbf{c}_i\|}{\sigma_i^2}$, where $\mathbf{c}_i$ is treated as $\mathbf{w}_i^{(1)}$ in the MLP [461]. When β in (6.6) takes the value of $\frac{1}{2}$, the RBF

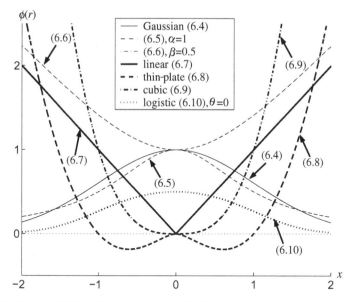

Fig. 6.2. Plots of RBFs: (6.4) through (6.10). The distance $r = |x - x_0|$ with $x_0 = 0$, and $\sigma = 1$.

becomes Hardy's multiquadric function, which is extensively used in surface interpolation with very good results [923]. When α in (6.5) is unity, $\phi(r)$ is suitable for DSP implementation [320].

Among these RBFs, the Gaussian (6.4), function (6.5), and the logistic function (6.10) are localized RBFs with the property that $\phi(r) \to 0$ as $r \to \infty$. Physiologically, there exist Gaussian-like receptive fields in cortical cells [923]. As a result, the RBF $\phi(r)$ is typically selected as the Gaussian or other localized RBFs.

The RBFN conventionally uses the Gaussian function (6.4) as the RBF. The Gaussian is compact and positive. It is motivated from the point of view of kernel regression and kernel density estimation. In fitting data in which there is normally distributed noise with the inputs, the Gaussian is the optimal basis function in the LS sense [1161]. The Gaussian is the only factorizable RBF, and this property is desirable for hardware implementation of the RBFN.

The thin-plate spline function (6.8) is another popular RBF for universal approximation. Unlike the Gaussian, the use of the thin-plate spline is motivated from a curve-fitting perspective [734]. The thin-plate spline is the solution when fitting a surface through a set of points and by using a roughness penalty [792]. It diverges at infinity and is negative over the region of $r \in (0, 1)$. However, for training purposes, the approximated function needs to be defined only over a specified range. There is some limited empirical evidence to suggest that the thin-plate spline better fits the data in high-dimensional settings [734].

A pseudo-Gaussian function in the one-dimensional space is introduced by selecting the standard deviation σ in the Gaussian (6.4) as two different positive values, namely, σ_- for $x < 0$ and σ_+ for $x > 0$ [957]. In the n-dimensional space, the pseudo-Gaussian function can be defined by

$$\phi(\mathbf{x}) = \prod_{i=1}^{n} \varphi_i(x_i) \tag{6.11}$$

$$\varphi_i(x_i) = \begin{cases} e^{-\frac{(x_i - c_i)^2}{\sigma_-^2}}, & x_i < c_i \\ e^{-\frac{(x_i - c_i)^2}{\sigma_+^2}}, & x_i \geq c_i \end{cases} \tag{6.12}$$

where $\mathbf{c} = (c_1, \cdots, c_n)^{\mathrm{T}}$ is the center vector, and index i runs over the dimension of the input space n. The pseudo-Gaussian function is not strictly an RBF due to its radial asymmetry, and this, however, eliminates the symmetry restriction and provides the hidden units with a greater flexibility with respect to function approximation.

Radial Basis Functions for Approximating Constant Values

Approximating functions with constant-valued segments using localized RBFs such as the Gaussian is most difficult. If a function has nearly constant values in some intervals, the RBFN with the Gaussian RBF is inefficient in approximating these values unless its variance is very large approaching infinity. The sigmoidal RBF, as a composite of a set of sigmoidal functions, can be used to deal with this problem [661]

$$\phi(x) = \frac{1}{1 + e^{-\beta[(x-c)+\theta]}} - \frac{1}{1 + e^{-\beta[(x-c)-\theta]}} \tag{6.13}$$

where the bias $\theta > 0$, and the gain $\beta > 0$. $\phi(x)$ is radially symmetric with the maximum at c. β controls the steepness and θ controls the width of the function. The shape of $\phi(x)$ is approximately rectangular if $\beta \times \theta$ is large. For large β and θ, it has a soft trapezoidal shape, while for small β and θ it is bell-shaped. $\phi(x)$ can be extended for an n-dimensional approximation

$$\phi(\mathbf{x}) = \prod_{i=1}^{n} \varphi_i(x_i) \tag{6.14}$$

$$\varphi_i(x_i) = \frac{1}{1 + e^{-\beta_i[(x_i-c_i)+\theta_i]}} - \frac{1}{1 + e^{-\beta_i[(x_i-c_i)-\theta_i]}} \tag{6.15}$$

where $\mathbf{x} = (x_1, \cdots, x_n)^{\mathrm{T}}$, $\mathbf{c} = (c_1, \cdots, c_n)^{\mathrm{T}}$ is the center vector, $\boldsymbol{\theta} = (\theta_1, \cdots, \theta_n)^{\mathrm{T}}$, and $\boldsymbol{\beta} = (\beta_1, \cdots, \beta_n)^{\mathrm{T}}$.

When β_i and θ_i are small, the sigmoidal RBF $\phi(\mathbf{x})$ will be close to zero and the corresponding node will have little contribution to the approximation task regardless of the tuning of the other parameters thereafter. To accommodate constant values of the desired output and to avoid diminishing the kernel functions, $\phi(\mathbf{x})$ can be modified by adding an additional term to the product term $\varphi_i(x_i)$ [671]

$$\phi(\mathbf{x}) = \prod_{i=1}^{n} (\varphi_i(x_i) + \widetilde{\varphi}_i(x_i)) \tag{6.16}$$

where

$$\widetilde{\varphi}_i(x_i) = [1 - \varphi_i(x_i)]\, e^{-a_i(x_i - c_i)^2} \tag{6.17}$$

with $a_i \geq 0$, and $\widetilde{\varphi}_i(x_i)$ being used as a compensating function to keep the product term from decreasing to zero when $\varphi_i(x_i)$ is small. β_i and a_i are, respectively, associated with the steepness and sharpness of the product term and θ_i controls the width of the product term. The parameters are adjusted by the gradient-descent method.

An alternative approach is to use the following raised-cosine function as a one-dimensional RBF [1004]

$$\phi(x) = \begin{cases} \cos^2(\frac{\pi x}{2}) & |x| \leq 1 \\ 0 & |x| > 1 \end{cases} \tag{6.18}$$

$\phi(x)$ is a zero-centered function with compact support since $\phi(0) = 1$ and $\phi(x) = 0$ for $|x| \geq 1$. The raised-cosine RBF can represent a constant function exactly using two terms. This RBF can be generalized to n dimensions [1004]

$$\phi(\mathbf{x}) = \prod_{i=1}^{n} \phi(x_i - c_i) \tag{6.19}$$

Apparently, $\phi(\mathbf{x})$ is nonzero only when $\mathbf{x}$ is in the $(-1, 1)^n$ vicinity of $\mathbf{c}$.

Figure 6.3 illustrates the sigmoidal RBF, the compensated sigmoidal RBF, and the raised-cosine RBF with different selections of β and θ. In Chapter 8, we will introduce some popular fuzzy membership functions, which can serve the same purpose by suitably constraining some parameters.

6.3 Learning RBF Centers

RBFN learning is usually implemented using a two-phase strategy. The first phase specifies and fixes suitable centers $\mathbf{c}_i$ and their respective standard deviations, also known as *widths* or *radii*[2], σ_i, and then the second phase adjusts the network weights $\mathbf{W}$. In this section, we describe the first phase.

---

[2] In this chapter, we use width as the terminology.

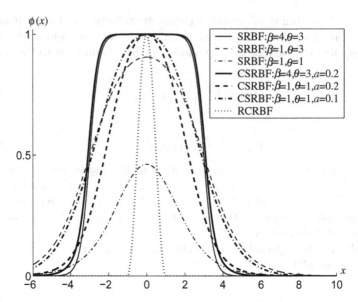

Fig. 6.3. RBFs for approximating constant-valued segments in the one-dimensional space. SRBF—the sigmoidal RBF (6.13), CSRBF—the compensated sigmoidal RBF (6.16), and RCRBF—the raised-cosine RBF (6.18). The center is selected as $c_i = c = 0$, while the shape parameters $\beta_i = \beta$ and $a_i = a$ are given in the figure.

6.3.1 Selecting RBF Centers Randomly from Training Sets

A simple method to specify the RBF centers is to randomly select a subset of the input patterns from the training set. Each RBF center is exactly situated at an input pattern. If the training set is representative of the learning problem, this method is regarded as appropriate. The training method based on a random selection of centers from a large training set of fixed size is found to be relatively insensitive to the use of the pseudoinverse, hence the method may itself be a regularization method [470].

However, if the training set is not sufficiently large or the training set is not representative of the learning problem, the randomly selected RBF centers may not be suitable for characterizing the data set and learning based on this may lead to undesirable performance. If the subsequent learning using a selection of random centers is not satisfactory, the process has to be repeated with another set of random centers until a desired performance is achieved.

For function approximation, one heuristic is to place the RBF centers at the extrema of the second-order derivative of a function with the highest absolute values and to place the RBF centers more densely in areas of higher absolute second-order derivative than in areas of lower absolute second-order derivative [991]. The second-order derivative of a function is usually associated with its curvature. This achieves a better function approximation than uniformly distributed center placement.

The Gaussian RBFN using the same σ for all RBF centers has universal approximation capability [887]. This global width can be selected as the average of all the Euclidian distances between the ith RBF center and its nearest neighbor

$$\sigma = \left\langle \|\mathbf{c}_i - \mathbf{c}_j\| \right\rangle \tag{6.20}$$

where

$$\|\mathbf{c}_i - \mathbf{c}_j\| = \min_{k=1,\cdots,J_2;k\neq i} \|\mathbf{c}_i - \mathbf{c}_k\| \tag{6.21}$$

Another simple and fast method for selecting σ is given by [133]

$$\sigma = \frac{d_{\max}}{\sqrt{2J_2}} \tag{6.22}$$

where $d_{\max}$ is the maximum distance between the selected centers,

$$d_{\max} = \max_{i,k=1,\cdots,J_2;k>i} \|\mathbf{c}_i - \mathbf{c}_k\| \tag{6.23}$$

This choice makes the Gaussian RBF neither too steep nor too flat.

In practice, the width of each RBF σ_i, $i = 1, \cdots, J_2$, can be determined according to the data distribution in the region of the corresponding RBF center. A heuristics for selecting σ_i is to average the distances between the ith RBF center and its L nearest neighbors

$$\sigma_i = \frac{1}{L} \sum_{l=1}^{L} \|\mathbf{c}_i - \mathbf{c}_l\| \tag{6.24}$$

where $\mathbf{c}_l, l = 1, \cdots, L$, are L nearest neighbors of $\mathbf{c}_i$. Alternatively, σ_i is selected according to the distance of unit i to its nearest neighbor unit j

$$\sigma_i = a \|\mathbf{c}_i - \mathbf{c}_j\| \tag{6.25}$$

where the scaling factor a is chosen between 1.0 and 1.5.

6.3.2 Selecting RBF Centers by Clustering Training Sets

Clustering is a data-analysis tool for characterizing the distribution of a data set, and is usually used for determining the RBF centers. The training set is grouped into appropriate clusters, whose prototypes are then used as RBF centers. The number of clusters can be specified or determined automatically depending on the clustering algorithm. All the unsupervised and supervised clustering algorithms described in Chapter 5 are applicable for clustering the RBF centers. The performance of the clustering algorithm is important to the efficiency of RBFN learning.

Unsupervised Clustering of RBF Centers

Unsupervised clustering such as the C-means is popular for clustering RBF centers [817]. Other examples of applying unsupervised clustering algorithms to the RBF center selection include the enhanced C-means [204], subtractive clustering [997], ART 2 [665], ART-like clustering [671], GCS [377], probabilistic SOM [40], FSCL [12], RPCL [1204], scatter-based FSCL clustering [1046], fuzzy clustering [22], hybrid RPCL and soft competitive learning clustering [771], mean-tracking clustering [1068], marginal median LVQ clustering [127], and iterative agglomerative clustering [833]. These clustering algorithms are discussed in Chapter 5.

Supervised Clustering of RBF Centers

RBF centers determined by supervised clustering are usually more efficient for RBFN learning than those determined by unsupervised clustering [194], since the distribution of the output patterns in the training set is also considered.

When the RBFN is trained for classification, the LVQ1 [611] algorithm is a popular method for clustering the RBF centers. Other examples for RBF center selection using supervised clustering include the conditional FCM [899], LVQ algorithms [1012], C-means plus k-NN based clustering [140], augmented unsupervised clustering by concatenating the scaled output vector to the input vector [194, 1112], CFA [426], fuzzy ARTMAP [1095], and supervised fuzzy clustering [1054].

The relationship between the augmented unsupervised clustering process and the MSE of RBFN learning has been investigated in [1112]. In the case of the Gaussian RBF and any Lipschitz continuous RBF, a weighted MSE for supervised quantization yields an upper bound to the MSE of RBFN learning. This upper bound and consequently the output error can be made arbitrarily small by decreasing the quantization error, which can be accomplished by increasing the number of hidden units.

Determining Covariance Matrices of RBF Centers

After the RBF centers are determined, the covariance matrices of the RBFs are set to the covariances of the input patterns in each cluster. In this case, the Gaussian RBFN is extended to the generalized RBFN using the Mahalanobis distance, defined by the weighted norm [923]

$$\phi\left(\|\mathbf{x} - \mathbf{c}_k\|_\mathbf{A}\right) = e^{-\frac{1}{2}(\mathbf{x} - \mathbf{c}_k)^\mathrm{T} \mathbf{\Sigma}^{-1}(\mathbf{x} - \mathbf{c}_k)} \tag{6.26}$$

where the squared weighted norm $\|\mathbf{x}\|_\mathbf{A}^2 = (\mathbf{A}\mathbf{x})^\mathrm{T}(\mathbf{A}\mathbf{x}) = \mathbf{x}^\mathrm{T}\mathbf{A}^\mathrm{T}\mathbf{A}\mathbf{x}$ and $\mathbf{\Sigma}^{-1} = 2\mathbf{A}^\mathrm{T}\mathbf{A}$.

When the Euclidean distance is employed, one can also select the width of the Gaussian RBFN according to Subsect. 6.3.1.

6.4 Learning the Weights

After RBF centers and their widths or covariance matrices are determined, learning of the weights $\mathbf{W}$ is reduced to a linear optimization problem, which can be solved using the LS method or the gradient-descent method.

6.4.1 Least Squares Methods for Weight Learning

After the parameters related to the RBF centers are determined, the weight matrix $\mathbf{W}$ is then trained to minimize the MSE (6.3), which is the LS problem. This LS problem requires a computational complexity of $O\left(NJ_2^2\right)$ flops for $N > J_2$ when the popular orthogonalization techniques such as the SVD and the QR desomposition are applied [424]. A simple representation of the solution is given explicitly by [133]

$$\mathbf{W} = \left(\mathbf{\Phi}^{\mathrm{T}}\right)^{\dagger}\mathbf{Y}^{\mathrm{T}} = \left(\mathbf{\Phi}\mathbf{\Phi}^{\mathrm{T}}\right)^{-1}\mathbf{\Phi}\mathbf{Y}^{\mathrm{T}} \tag{6.27}$$

where $[\cdot]^{\dagger}$ is the pseudoinverse of the matrix within. This is the batch LS method. The over- or underdetermined linear LS system is an ill-conditioned problem. The SVD is an efficient and numerically robust technique for dealing with such an ill-conditioned problem and is preferred. For regularly sampled inputs and exact interpolation, $\mathbf{W}$ can be computed by using the Fourier transform of the RBFN and the FFT technique [4], which reduces the computational complexity to $O(N\ln N)$.

When the full data set is not available and samples are obtained online, the RLS method can be used to train the weights online [45]

$$\mathbf{w}_i(t) = \mathbf{w}_i(t-1) + \mathbf{K}(t)e_i(t) \tag{6.28}$$

$$\mathbf{K}(t) = \frac{\mathbf{P}(t-1)\phi_t}{\phi_t^{\mathrm{T}}\mathbf{P}(t-1)\phi_t + \mu} \tag{6.29}$$

$$e_i(t) = y_{t,i} - \phi_t^{\mathrm{T}}\mathbf{w}_i(t-1) \tag{6.30}$$

$$\mathbf{P}(t) = \frac{1}{\mu}\left[\mathbf{P}(t-1) - \mathbf{K}(t)\phi_t^{\mathrm{T}}\mathbf{P}(t-1)\right] \tag{6.31}$$

for $i = 1, \cdots, J_3$, where $0 < \mu \leq 1$ is the forgetting factor. Typically, $\mathbf{P}(0) = a_0\mathbf{I}_{J_2}$, where a_0 is a sufficiently large number and $\mathbf{I}_{J_2}$ is the $J_2 \times J_2$ identity matrix, and $\mathbf{w}_i(0)$ is selected as a small random matrix.

6.4.2 Kernel Orthonormalization-based Weight Learning

According to (6.27), if $\mathbf{\Phi}^{\mathrm{T}}\mathbf{\Phi} = \mathbf{I}$, that is, the RBFs are orthonormalized with

$$\langle \phi_i(\mathbf{x}), \phi_j(\mathbf{x}) \rangle = \delta_{ij} \tag{6.32}$$

where δ_{ij} is the Kronecker delta, and $\langle \phi_i(\mathbf{x}), \phi_j(\mathbf{x}) \rangle$ is the inner product of the two RBFs, defined by

$$\langle \phi_i(\mathbf{x}), \phi_j(\mathbf{x}) \rangle = \sum_{\mathbf{x} \in \{\mathbf{x}_p\}} \phi_i(\mathbf{x}) \phi_j(\mathbf{x}) \tag{6.33}$$

the inversion operation is unnecessary. The optimum weight can be computed by

$$\mathbf{w}_k = \mathbf{\Phi} \overline{\mathbf{y}}_k \tag{6.34}$$

for $k = 1, \cdots, J_3$, where $\overline{\mathbf{y}}_k = (y_{1,k}, \cdots, y_{N,k})^{\mathrm{T}}$ corresponds to the kth row of $\mathbf{Y}$.

Based on this observation, an efficient, noniterative weight learning technique has been introduced by applying the GSO of RBFs [572]. The RBFs are first transformed into a set of orthonormal RBFs for which the optimum weights are computed. These weights are then recomputed in such a way that their values can be fitted back into the original RBFN structure, $i.e.$ with kernel functions unchanged.

By using the GSO procedure, a set of J_2 orthonormal RBFs $u_k(\mathbf{x})$ are obtained as

$$u_1(\mathbf{x}) = \frac{\phi_1(\mathbf{x})}{\|\phi_1(\mathbf{x})\|} \tag{6.35}$$

$$u_k(\mathbf{x}) = \frac{\phi_k(\mathbf{x}) - \sum_{i=1}^{k-1} \langle \phi_k(\mathbf{x}), u_i(\mathbf{x}) \rangle u_i(\mathbf{x})}{\left\| \phi_k(\mathbf{x}) - \sum_{i=1}^{k-1} \langle \phi_2(\mathbf{x}), u_i(\mathbf{x}) \rangle u_i(\mathbf{x}) \right\|} \tag{6.36}$$

for $k = 2, \cdots, J_2$. Note that $u_k(\mathbf{x})$, $k = 1, \cdots, J_2$, is a linear combination of $\phi_j(\mathbf{x})$, $j = 1, \cdots, k$

$$u_k(\mathbf{x}) = \sum_{j=1}^{k} c_{jk} \phi_j(\mathbf{x}) \tag{6.37}$$

where c_{jk} can be calculated from (6.35) and (6.36). Using $u_k(\mathbf{x})$ as RBFs, the corresponding weights are given by

$$\widetilde{\mathbf{w}}_k = \mathbf{U} \overline{\mathbf{y}}_k \tag{6.38}$$

for $k = 1, \cdots, J_3$, where $\mathbf{U} = [u_{ij}]$ is a $J_2 \times N$ matrix, and $u_{ij} = u_i(\mathbf{x}_j)$.

The original RBF weights can be solved by the transform

$$\mathbf{W} = \mathbf{C} \widetilde{\mathbf{W}} \tag{6.39}$$

where $\mathbf{C} = [c_{ki}]$, and $\widetilde{\mathbf{W}} = [\widetilde{w}_{ij}]$. More specifically

$$w_{kj} = \sum_{i=k}^{J_2} c_{ki} \widetilde{w}_{ij} \tag{6.40}$$

The coefficients c_{ki} can be calculated recursively by [572]

$$c_{ii} = \left(\sum_{p=1}^{N} \left[-\sum_{j=1}^{i-1} \left(\sum_{l=j}^{i-1} \langle \phi_i(\mathbf{x}), u_l(\mathbf{x}) \rangle c_{jl} \right) \phi_j(\mathbf{x}_p) + \phi_i(\mathbf{x}_p) \right]^2 \right)^{-\frac{1}{2}} \quad (6.41)$$

for $i = 1, \cdots, J_2$, and

$$c_{ki} = -c_{ii} \sum_{j=k}^{i-1} \left(\sum_{l=1}^{j} c_{lj} \langle \phi_i(\mathbf{x}), \phi_l(\mathbf{x}) \rangle \right) c_{kj}, \quad (6.42)$$

for $k = 1, \cdots, i-1$. The weights of the transformed RBFN, $\tilde{w}_{ij}$, can be calculated by

$$\tilde{w}_{ij} = \sum_{l=1}^{i} c_{lj} \left((\bar{\mathbf{y}}_i)^{\mathrm{T}} \overline{\phi}_l \right) \quad (6.43)$$

for $i = 1, \cdots, J_3$, where $\overline{\phi}_l = (\phi_{1,l}, \cdots, \phi_{N,l})^{\mathrm{T}}$ corresponds to the lth row of $\mathbf{\Phi}$.

The requirement for computing the offdiagonal terms in the solution of the linear set of weight equations is eliminated. In addition, the method has low storage requirements since the weights can be computed recursively, and the computation procedure can be organized in a parallel manner. This significantly reduces the computation time in comparison with other methods for the calculation of the weights. Incorporation of new hidden nodes aimed at improving the network performance does not require recomputation of the network weights already calculated. This allows for a very efficient network training procedure, where network hidden nodes are added one at a time until an adequate error goal is reached. The contribution of each RBF to the overall network output can be evaluated.

6.5 RBFN Learning Using Orthogonal Least Squares

Optimal subset selection techniques are computationally prohibitive. The OLS method [206, 208, 210] is an efficient way for subset model selection. The approach chooses and adds RBF centers one by one until an adequate network is constructed. All the training examples are considered as candidates for the centers, and the one that reduces the MSE the most is selected as a new hidden unit. The GSO is first used to construct a set of orthogonal vectors in the space spanned by the vectors of the hidden unit activation ϕ_p, and a new RBF center is then selected by minimizing the residual MSE. Model-selection criteria are used to determine the size of the network.

6.5.1 Batch Orthogonal Least Squares

The batch OLS method can not only determine the weights, but also choose the number and the positions of the RBF centers. The batch OLS can employ the forward [208, 210, 217] and the backward [501] center-selection approaches.

When the RBF centers are distinct, $\mathbf{\Phi}^T$ is of full rank. The orthogonal decomposition of $\mathbf{\Phi}^T$ is performed using the QR decomposition

$$
\mathbf{\Phi}^T = \mathbf{Q}
\begin{bmatrix}
\mathbf{R} \\
\cdots \\
\mathbf{0}
\end{bmatrix}
\tag{6.44}
$$

where $\mathbf{Q} = [\mathbf{q}_1 \cdots \mathbf{q}_N]$ is an $N \times N$ orthogonal matrix and $\mathbf{R}$ is a $J_2 \times J_2$ upper triangular matrix. By minimizing the MSE given by (6.3), one can make use of the invariant property of the Frobenius norm

$$
E = \frac{1}{N} \left\| \mathbf{YQ} - \mathbf{W}^T \mathbf{\Phi Q} \right\|_F^2 = \frac{1}{N} \left\| \mathbf{Q}^T \mathbf{Y}^T - \mathbf{Q}^T \mathbf{\Phi}^T \mathbf{W} \right\|_F^2
\tag{6.45}
$$

Let

$$
\mathbf{Q}^T \mathbf{Y}^T =
\begin{bmatrix}
\widetilde{\mathbf{B}} \\
\cdots \\
\overline{\mathbf{B}}
\end{bmatrix}
\tag{6.46}
$$

where $\widetilde{\mathbf{B}} = \left[\tilde{b}_{ij} \right]$ and $\overline{\mathbf{B}} = \left[\overline{b}_{ij} \right]$ are, respectively, a $J_2 \times J_3$ and an $(N - J_2) \times J_3$ matrices. We then have

$$
E = \frac{1}{N} \left\|
\begin{bmatrix}
\widetilde{\mathbf{B}} \\
\cdots \\
\overline{\mathbf{B}}
\end{bmatrix}
-
\begin{bmatrix}
\mathbf{R} \\
\cdots \\
\mathbf{0}
\end{bmatrix}
\mathbf{W}
\right\|_F^2
= \frac{1}{N} \left\|
\begin{bmatrix}
\widetilde{\mathbf{B}} - \mathbf{RW} \\
\cdots\cdots\cdots \\
\overline{\mathbf{B}}
\end{bmatrix}
\right\|_F^2
\tag{6.47}
$$

Thus, the optimal $\mathbf{W}$ that minimizes E is derived from

$$
\mathbf{RW} = \widetilde{\mathbf{B}}
\tag{6.48}
$$

In this case, the residual

$$
E = \frac{1}{N} \left\| \overline{\mathbf{B}} \right\|_F^2
\tag{6.49}
$$

This is the batch OLS.

Due to the orthogonalization procedure, it is very convenient to implement the forward and backward center-selection approaches. The forward selection approach is to build up a network by adding, one at a time, centers at the data points that result in the largest decrease in the network output error at each stage. The backward selection algorithm is an alternative approach that sequentially removes from the network, one at a time, those centers that cause the smallest increase in the residual.

The error reduction ratio (ERR) due to the kth RBF neuron is defined by [210]

$$
ERR_k = \frac{\left(\sum_{i=1}^{J_3} \tilde{b}_{ki}^2 \right) \mathbf{q}_k^T \mathbf{q}_k}{\text{tr} \left(\mathbf{YY}^T \right)}
\tag{6.50}
$$

for $k = 1, \cdots, N$. RBFN training can be in a constructive way and the centers with the largest ERR values are recruited until

$$1 - \sum_{k=1}^{J_2} ERR_k < \rho \tag{6.51}$$

where $\rho \in (0,1)$ is a tolerance. ERR is a performance-oriented criterion. An alternative terminating criterion can be based on the AIC [210], which balances between the performance and the complexity. The weights are determined at the same time. The criterion used to stop center selection is a simple threshold on the ERR. If the threshold chosen results in very large variances for Gaussian functions, poor generalization performance may occur. To improve generalization, regularized forward OLS methods can be implemented by penalizing large weights [873, 207].

The computation complexity of the orthogonal decomposition of an information matrix $\boldsymbol{\Phi}^T$ is $O\left(NJ_2^2\right)$. When the size of a training data set N is large, the batch OLS is computationally demanding and also needs a large amount of computer memory.

The RBF center clustering method based on the Fisher ratio class separability measure [762] is similar to the forward selection OLS algorithm [208, 210]. Both methods employ the the QR decomposition-based orthogonal transform to decorrelate the responses of the prototype neurons as well as the forward center-selection procedure. The OLS evaluates candidate centers based on the approximation error reduction in the context of nonlinear approximation, while the Fisher ratio-based forward selection algorithm evaluates candidate centers using the Fisher ratio class-separability measure for the purpose of classification. The two algorithms have similar computational cost.

6.5.2 Recursive Orthogonal Least Squares

Recursive OLS (ROLS) algorithms have been proposed for updating the weights of single-input single-output (SISO) [121] and multi-input multi-output (MIMO) systems [1241, 425]. In [1241], the ROLS algorithm determines the increment of the weight matrix. In [425], the full weight matrix is determined at each iteration, and this reduces the accumulated error in the weight matrix, and the ROLS has been extended for the selection of the RBF centers.

At iteration $t-1$, following a procedure similar to that for the batch OLS, and applying QR decomposition, we have [425]

$$\boldsymbol{\Phi}^T(t-1) = \mathbf{Q}(t-1) \begin{bmatrix} \mathbf{R}(t-1) \\ \cdots\cdots\cdots \\ \mathbf{0} \end{bmatrix} \tag{6.52}$$

$$\mathbf{Q}^T(t-1)\left(\mathbf{Y}(t-1)\right)^T = \begin{bmatrix} \widetilde{\mathbf{B}}(t-1) \\ \cdots\cdots\cdots \\ \overline{\mathbf{B}}(t-1) \end{bmatrix} \tag{6.53}$$

The sizes of the matrices $\boldsymbol{\Phi}$ and $\mathbf{Y}$ increase with a new data for each iteration.

Then at iteration t, the update for $\mathbf{R}(t-1)$ and $\widetilde{\mathbf{B}}(t-1)$ are calculated using another QR decomposition

$$
\begin{bmatrix} \mathbf{R}(t-1) \\ \cdots\cdots\cdots \\ \boldsymbol{\phi}^{\mathrm{T}}(t) \end{bmatrix} = \mathbf{Q}_1(t) \begin{bmatrix} \mathbf{R}(t) \\ \cdots\cdots \\ 0 \end{bmatrix} \tag{6.54}
$$

$$
\begin{bmatrix} \widetilde{\mathbf{B}}(t) \\ \cdots\cdots \\ \overline{\mathbf{y}}^{\mathrm{T}}(t) \end{bmatrix} = \mathbf{Q}_1^{\mathrm{T}}(t) \begin{bmatrix} \widetilde{\mathbf{B}}(t-1) \\ \cdots\cdots\cdots \\ (\mathbf{y}_t)^{\mathrm{T}} \end{bmatrix} \tag{6.55}
$$

The minimization of $E(t)$ leads to the optimal $\mathbf{W}(t)$, which is solved by

$$
\mathbf{R}(t)\mathbf{W}(t) = \widetilde{\mathbf{B}}(t) \tag{6.56}
$$

Since $\mathbf{R}(t)$ is an upper triangular matrix, $\mathbf{W}(t)$ can be easily solved by backward substitution. Update the residual at iteration t using the recursive equation

$$
\begin{aligned}
E(t) &= \frac{1}{t} \left\| \overline{\mathbf{B}} \right\|_{\mathrm{F}}^2 = \frac{1}{t} \left(\left\| \overline{\mathbf{y}}^{\mathrm{T}}(t) \right\|_{\mathrm{F}}^2 + \left\| \overline{\mathbf{B}}(t-1) \right\|_{\mathrm{F}}^2 \right) \\
&= \frac{t-1}{t} E(t-1) + \frac{1}{t} \left\| \overline{\mathbf{y}}^{\mathrm{T}}(t) \right\|_{\mathrm{F}}^2
\end{aligned} \tag{6.57}
$$

Initial values can be selected as $\mathbf{R}(0) = \alpha \mathbf{I}$, where α is a small positive number such as 0.01, $\widetilde{\mathbf{B}}(0) = 0$, and $\|\widetilde{\mathbf{B}}(0)\|_{\mathrm{F}}^2 = 0$. In offline training, the weights need only be computed once at the end of the training, since their values do not affect the recursive updates in (6.54) and (6.55).

After training with the ROLS, the final triangular system of (6.56), with $t = N$, contains important information about the learned network, and can be used to sequentially select the centers to minimize the network output error. Forward and backward center selection methods are developed from this information, and Akaike's FPE criterion [17] is used in the model selection [425]. Both of the ROLS selection algorithms sort the selected centers in the order of their significance in reducing the MSE [1241, 425].

6.6 Supervised Learning of All Parameters

The preceding methods for selecting the network parameters are practical, but by no means optimal as far as the supervised learning procedure is concerned. Although all the conventional unconstrained optimization methods, such as those introduced in Chapter 3, can be used for training the RBFN, the gradient-descent method is the simplest method for finding the minimum value of E. In this section, we apply the gradient-descent based supervised methods to RBFN learning.

6.6.1 Supervised Learning for General Radial Basis Function Networks

To derive the supervised learning algorithm for the RBFN with any useful RBF, we rewrite the error function (6.3) as

$$E = \frac{1}{N} \sum_{n=1}^{N} \sum_{i=1}^{J_3} (e_{n,i})^2 \tag{6.58}$$

where $e_{n,i}$ is the approximation error at the ith output node for the nth example

$$e_{n,i} = y_{n,i} - \sum_{m=1}^{J_2} w_{mi} \phi \left(\|\mathbf{x}_n - \mathbf{c}_m\| \right) = y_{n,i} - \mathbf{w}_i^{\mathrm{T}} \boldsymbol{\phi}_n \tag{6.59}$$

Taking the first-order derivative of E with respect to w_{mi} and $\mathbf{c}_m$, respectively, we have

$$\frac{\partial E}{\partial w_{mi}} = -\frac{2}{N} \sum_{n=1}^{N} e_{n,i} \phi \left(\|\mathbf{x}_n - \mathbf{c}_m\| \right) \tag{6.60}$$

$$\frac{\partial E}{\partial \mathbf{c}_m} = \frac{2}{N} \sum_{i=1}^{J_3} w_{mi} \sum_{n=1}^{N} e_{n,i} \dot{\phi} \left(\|\mathbf{x}_n - \mathbf{c}_m\| \right) \frac{\mathbf{x}_n - \mathbf{c}_m}{\|\mathbf{x}_n - \mathbf{c}_m\|} \tag{6.61}$$

for $m = 1, \cdots, J_2$, $i = 1, \cdots, J_3$, where $\dot{\phi}(\cdot)$ is the first derivative of $\phi(\cdot)$.
The gradient-descent method is defined by the update equations

$$\Delta w_{mi} = -\eta_1 \frac{\partial E}{\partial w_{mi}} \tag{6.62}$$

$$\Delta \mathbf{c}_m = -\eta_2 \frac{\partial E}{\partial \mathbf{c}_m} \tag{6.63}$$

where η_1 and η_2 are learning rates.

To prevent the situation that two or more centers are too close or coincide with one another during the learning process, one can add a term such as $\sum_{\alpha \neq \beta} \psi \left(\|\mathbf{c}_\alpha - \mathbf{c}_\beta\| \right)$ to E, where $\psi(\cdot)$ is an appropriate repulsive potential. The gradient-descent method given by (6.62) and (6.63) needs to be modified accordingly.

A simple strategy for initialization is to select the RBF centers based on a random subset of the examples and the weights $\mathbf{W}$ as a matrix with small random components. A more delicate initialization can usually improve the problem of local minima and lead to a considerable acceleration of the search process. Thus, one can use clustering to find the initial RBF centers and the LS to find the initial weights, and the gradient-descent procedure is then applied to refine the learning result.

When the gradients given above are set to zero, the optimal solutions to the weights and centers can be derived. The gradient-descent procedure is the iterative approximation to the optimal solutions. For each sample n, if we set $e_{n,i} = 0$, then the right-hand side of (6.60) is zero, we then achieve the global optimum and accordingly get the equation

$$\mathbf{y}_n = \mathbf{W}^{\mathrm{T}} \boldsymbol{\phi}_n \tag{6.64}$$

For all samples

$$\mathbf{Y} = \mathbf{W}^{\mathrm{T}} \boldsymbol{\Phi} \tag{6.65}$$

This is exactly the same SLE as (6.2). The optimum solution to weights is given by (6.27). Equating (6.61) to zero leads to

$$\mathbf{c}_m = \frac{\sum_{i=1}^{J_3} w_{mi} \sum_{n=1}^{N} \frac{e_{n,i} \dot{\phi}_{mn}}{\|\mathbf{x}_n - \mathbf{c}_m\|} \mathbf{x}_n}{\sum_{i=1}^{J_3} w_{mi} \sum_{n=1}^{N} \frac{e_{n,i} \dot{\phi}_{mn}}{\|\mathbf{x}_n - \mathbf{c}_m\|}} \tag{6.66}$$

where $\dot{\phi}_{mn} = \dot{\phi}(\|\mathbf{x}_n - \mathbf{c}_m\|)$. Thus, the optimal centers are weighted sums of the data points, corresponding to a task-dependent clustering problem.

6.6.2 Supervised Learning for Gaussian Radial Basis Function Networks

For the Guassian RBFN, the RBF at each center can be assigned a different width σ_i

$$\phi_i(\mathbf{x}) = e^{-\frac{\|\mathbf{x} - \mathbf{c}_i\|^2}{2\sigma_i^2}} \tag{6.67}$$

The RBFs can be further generalized to allow for arbitrary covariance matrices $\boldsymbol{\Sigma}_i$

$$\phi_i(\mathbf{x}) = e^{-\frac{1}{2}(\mathbf{x} - \mathbf{c}_i)^{\mathrm{T}} \boldsymbol{\Sigma}_i^{-1} (\mathbf{x} - \mathbf{c}_i)} \tag{6.68}$$

where $\boldsymbol{\Sigma}_i \in R^{J_1 \times J_1}$ is a positive-definite, symmetric covariance matrix. When $\boldsymbol{\Sigma}_i^{-1}$ is in general form, the shape and orientation of the axes of the hyperellipsoid are arbitrary in the feature space.

If $\boldsymbol{\Sigma}_i^{-1}$ is a diagonal matrix with nonconstant diagonal elements, $\boldsymbol{\Sigma}_i^{-1}$ is completely defined by a vector $\boldsymbol{\sigma}_i \in R^{J_1}$, and each ϕ_i is a hyperellipsoid whose axes are along the axes of the feature space

$$\boldsymbol{\Sigma}_i^{-1} = \mathrm{diag}\left(\frac{1}{\sigma_{i,1}^2}, \cdots, \frac{1}{\sigma_{i,J_1}^2}\right) \tag{6.69}$$

For the J_1-dimensional input space, each RBF of the form (6.68) has a total of $\frac{J_1(J_1+3)}{2}$ independent adjustable parameters, while each RBF of the form (6.67) and each RBF of the form (6.69) have only $J_1 + 1$ and $2J_1$ independent parameters, respectively. There is a tradeoff between using a small network with many adjustable parameters and using a large network with fewer adjustable parameters.

When using the RBF of the form (6.67), we get the gradients as

$$\frac{\partial E}{\partial \mathbf{c}_m} = -\frac{2}{N} \sum_{n=1}^{N} \phi_m \left(\mathbf{x}_n \right) \frac{\mathbf{x}_n - \mathbf{c}_m}{\sigma_m^2} \sum_{i=1}^{J_3} e_{n,i} w_{i,m} \qquad (6.70)$$

$$\frac{\partial E}{\partial \sigma_m} = -\frac{2}{N} \sum_{n=1}^{N} \phi_m \left(\mathbf{x}_n \right) \frac{\| \mathbf{x}_n - \mathbf{c}_m \|^2}{\sigma_m^3} \sum_{i=1}^{J_3} e_{n,i} w_{i,m} \qquad (6.71)$$

Similarly, for the RBF of the form (6.69), the gradients are given by

$$\frac{\partial E}{\partial c_{m,j}} = -\frac{2}{N} \sum_{n=1}^{N} \phi_m \left(\mathbf{x}_n \right) \frac{x_{n,j} - c_{m,j}}{\sigma_{m,j}^2} \sum_{i=1}^{J_3} e_{n,i} w_{i,m} \qquad (6.72)$$

$$\frac{\partial E}{\partial \sigma_{m,j}} = -\frac{2}{N} \sum_{n=1}^{N} \phi_m \left(\mathbf{x}_n \right) \frac{(x_{n,j} - c_{m,j})^2}{\sigma_{m,j}^3} \sum_{i=1}^{J_3} e_{n,i} w_{i,m} \qquad (6.73)$$

Adaptations for $\mathbf{c}_i$ and $\boldsymbol{\Sigma}_i$ are along the negative gradient directions. The weights $\mathbf{W}$ are updated by (6.60) and (6.62). To prevent unreasonable radii, the updating algorithms can also be derived by adding to the MSE E a constraint term that penalizes small radii, $E_c = \sum_i \frac{1}{\sigma_i}$ or $E_c = \sum_{i,j} \frac{1}{\sigma_{i,j}}$.

6.6.3 Implementations of Supervised Learning

The gradient-descent algorithms introduced heretofore are batch learning algorithms, since parameter update is carried out after all the samples are presented. As discussed in Chapter 3, by dropping $\frac{1}{N} \sum_{p=1}^{N}$ in the error function E and accordingly in the algorithms, one can update the parameters at each example $(\mathbf{x}_p, \mathbf{y}_p)$. This yields incremental learning algorithms, which are typically much faster than their batch counterparts for suitably selected learning parameters.

Although the RBFN trained by the gradient-descent method is capable of providing equivalent or better performance compared to that of the MLP trained with the BP, the training time for the two methods are comparable [1171]. The gradient-descent method is slow in convergence since it cannot efficiently use the locally tuned representation of the hidden-layer units. When the hidden-unit receptive fields, controlled by the widths σ_i, are narrow, for a given input only a few of the total number of hidden units will be activated and hence only these units need to be updated. However, in the gradient-descent method, there is no limitation on σ_i, thus there is no guarantee that the RBFN remains localized after the supervised learning [817]. As a result, the computational advantage of locality is not utilized.

The gradient-descent method is prone to finding local minima of the error function. For reasonably well-localized RBF, an input will generate a significant activation in a small region, and the opportunity of getting stuck at a local minimum is small. Unsupervised methods can be used to determine σ_i. Unsupervised learning is used to initialize the network parameters, and supervised learning is usually used for the fine tuning of the network parameters. The ultimate RBFN learning algorithm is typically a blend of unsupervised and supervised algorithms. Usually, the centers are selected by using a random subset of the training set or obtained by using clustering, the variances are selected using a heuristic, and the weights are solved by using a linear LS method or the gradient-descent method. This combination may yield a fast learning procedure with a sufficient accuracy.

6.7 Evolving Radial Basis Function Networks

EAs can be used to optimize network parameters of the RBFN. The fitness can be defined as

$$f(\mathbf{s}) = \frac{1}{1 + E} \tag{6.74}$$

where $\mathbf{s}$ is the genetic coding of the network parameters, and E is the objective function for neural-network learning. The fitness function can also be selected as a complexity criterion such as the AIC [18], which gives a tradeoff between the learning error and the network complexity [499].

The RBF centers can be very conveniently clustered using EAs such as the GA [13]. In order to partition a set of input patterns $\{\mathbf{x}_p\}$ into J_2 clusters, the cluster centers can be coded as $(c_{1,1}, \cdots, c_{1,J_1}; c_{2,1}, \cdots, c_{2,J_1}; \cdots; c_{J_2,1}, \cdots, c_{J_2,J_1})$, where $c_{i,j}$ is coded as an l-bit binary string. Accordingly, E in the fitness function (6.74) corresponds to the quantization error defined in Chapter 5. The width σ_i can then be calculated from the variance of each cluster, and the weights are trained using the RLS method.

In [499], the RBFN is trained by using the ES. Only the RBF centers and their corresponding widths are encoded into the variable-length chromosomes. For each chromosome, the RBF weights are computed using the RLS method. The fitness function is selected as the AIC. In [1027], all the RBFN parameters are simultaneously optimized by the GA. Usually, the parameters associated with each node are coded together in the chromosomes [13, 499, 1027].

In [1174], the GA is used to evolve the RBF centers $\mathbf{c}_i$ and their widths σ_i. To decrease the search space, the locations of $\mathbf{c}_i$ are not directly encoded into the genetic string, but are governed by a small set of space-filling curves whose parameters are in turn evolved. The values of σ_i are directly encoded into the genetic string. New centers can be inserted into positions for more details by evolving space-filling curves to distribute the RBF centers over the input space. The space-filling curves appear to yield better coverage of the relevant portions of the input space than simply generating the RBF centers randomly. The weights are trained by a conventional method that minimizes the MSE over the training set. The training procedure employed within the GA implicitly rewards the use of the minimal number of RBFs to achieve a given level of network performance by fixing the product of the number of RBFs and the number of epochs allowed through the data set. Thus, a smaller network has more epochs to decrease the MSE, thus improving the fitness. A scaled fitness at each generation is also defined based on the MSE.

The selection of the network structure and parameters can be performed simultaneously by employing GA-based optimization methods [112]. The network configuration is to set hidden nodes as an optimal subset from the training set. Each network is coded as a chromosome of a variable length string with distinct integers, and genetic operators including crossover, mutation, deletion and addition are defined to evolve a population of individuals. Single objective and multiobjective functions are used for evaluating the fitness of individual networks. The error function for approximating the training set is selected as the optimization criterion dependent on the network structure, namely the number of centers J_2, and weights $\mathbf{W}$ and centers $\mathbf{c}_i, i = 1, \cdots, J_2$. AIC [18], which provides a compromise between network complexity and network performance, is another objective to minimize.

The GA is suitable for the selection of the RBF centers in the case of RBFN classifiers [639]. The bit-length of a chromosome is the size of the training set. The 1 in the ith position of a chromosome representing the ith sample is selected as an RBF center. The fitness function is defined as the sum of the number of correct classifications, and a penalty term that is related to the difference between the size of the training set and a predefined maximum number of centers. After the RBF centers are determined, the SA [607] is then used to train the weights of the classifier with maximum correct classification as the optimization objective.

RBF centers and their widths can be evolved by using a cooperative-competitive GA strategy so that the resulting RBFs are mutually orthogonal activation sequences [1175]. The genetic approach operates on a population of competing RBFs ϕ_i, where the entire population corresponds to a single RBFN. Individual RBF centers in the population are competitive as well as cooperative in modeling the function over the entire domain of interest. Each chromosome contains the center and the width of an RBF prototype. A niche-based fitness-selection mechanism is used, and the size of population, J_2, is fixed. For a training set of size N, the normalized versions of all the RBFs are used

$$\bar{\bar{\phi}}_i(\mathbf{x}) = \frac{\phi_i(\mathbf{x})}{\sqrt{\sum_{j=1}^{N} \phi_i^2(\mathbf{x}_j)}}, \quad i = 1, \cdots, J_2 \tag{6.75}$$

Denote $\bar{\bar{\phi}}_i = \left(\bar{\bar{\phi}}_i(\mathbf{x}_1), \cdots, \bar{\bar{\phi}}_i(\mathbf{x}_N) \right)^{\mathrm{T}}$, and naturally, $\left\| \bar{\bar{\phi}}_i \right\| = 1$. Consider the inner product $\bar{\bar{\phi}}_i \cdot \bar{\bar{\phi}}_j$. If $\bar{\bar{\phi}}_i \cdot \bar{\bar{\phi}}_j$ is close to unity, the two nodes will compete and they will share the fitness. In contrast, if $\bar{\bar{\phi}}_i \cdot \bar{\bar{\phi}}_j$ is close to zero, the two nodes independently contribute to the overall prediction of the network and are more cooperative. The fitness measure for each RBF i is selected as

$$f(i) = \frac{|w_i|^\beta}{\frac{1}{J_2} \sum_{j=1}^{J_2} |w_j|^\beta} \tag{6.76}$$

where $1 < \beta < 2$, and w_i is the weight of the ith RBF to the output node. At each generation, the weights can be first solved using the LS method. In addition to the crossover and mutation operators, a creep operator is defined. The creep operator first decodes the genetic bit string into real-valued RBF parameters, then perturbs the parameters values, and finally encodes the perturbed values back into the bit string.

The PSO can also be used for architecture learning of the RBFN [728]. In [427], RBFN learning has been performed by using a multiobjective EA. A brief overview of the current state-of-the-art of research in applying the GA to RBFN training is given in [457].

6.8 Robust Learning of Radial Basis Function Networks

When a training set contains outliers, conventional training methods suffer from substantial performance deterioration. Robust statistics can be applied for robust learning of the RBFN [514]. Robust learning algorithms are usually derived from the M-estimator method. In Chapters 1 and 3, we have, respectively, described the theory of robust learning and robust learning algorithms for MLP learning. The derivation of robust RBFN learning algorithms follows the same procedure [992, 242].

The robust RBFN learning algorithm given in [992] is based on Hampel's tanh-estimator function. The RBFN architecture is initialized by using the conventional SVD-based learning method, which is outlier sensitive. This sensitivity is removed using the robust learning method. The robust part of the learning method is implemented iteratively using the CG method.

The annealing robust RBFN (ARRBFN) [242] improves the robustness of the RBFN against outliers for function approximation by using the M-estimator and the annealing robust learning algorithm (ARLA) [241]. The support vector regression (SVR) approach with the ε-insensitive loss function [1120] is first applied to determine an initial structure of the ARRBFN, and the ARLA [241] overcomes the problems of initialization and the cutoff points in the M-estimator method.

The median RBF algorithm [127] is based on robust parameter estimation of the RBF centers, and employs the Mahalanobis distance. It employs the marginal median estimator for evaluating the RBF centers and the MAD for estimating their dispersion parameters. A histogram-based fast implementation is provided for the median RBF algorithm. Compared to the conventional RBFN, the median RBF-based training is less biased by the presence of the outliers in the training set and provides an accurate estimation of the implied probabilities.

6.9 Various Learning Methods

In addition to the aforementioned RBFN learning methods, all general-purpose unconstrained optimization methods, including those introduced in Chapter 3, are applicable for RBFN learning, with no or very little modification, since the RBFN is a special FNN and RBFN learning is an unconstrained optimization problem. These include popular second-order approaches like the LM, CG, BFGS, and EKF, and global optimization methods like EAs, SA, and Tabu search. We also mention some other RBFN learning approaches in this section.

The LM method [764] is used for RBFN learning [429, 787, 904]. In [787, 904], the LM method is used for estimating nonlinear parameters, and the LS method using SVD is used for linear weight estimation at each iteration. All model parameters are optimized simultaneously. In [904], at each iteration the weights are updated many times during the process of looking for the search direction to update the nonlinear parameters. This further accelerates the convergence of the optimization search process.

RBFN learning can be viewed as a system-identification problem. As such, the EKF can be used for the learning procedure [1037]. After the number of centers is chosen, the EKF simultaneously solves for the prototype vectors and the weight matrix. A decoupled EKF is derived, and this further decreases the computational complexity of the training algorithm [1037]. The EKF training provides almost the same performance as gradient-descent training, but with only a fraction of the computational cost. In [249], a pair of parallel running extended Kalman filters are used to sequentially update both the output weights and the RBF centers.

In [1114], the BP with selective training [1113] has been applied to RBFN learning. The method improves the performance of the RBFN substantially compared with that of the gradient-descent method, in terms of convergence speed and accuracy. The method is quite effective when the dataset is error free and nonoverlapping. In [580], the RBFN is reformulated by using RBFs formed in terms of admissible generator functions, and provides a fully supervised gradient-descent training method. LP models with polynomial time complexity are also employed to train the RBFN [969]. In [475], a multiplication-free Gaussian RBFN with a gradient-based nonlinear learning algorithm is described for adaptive function approximation.

The EM method [299] is an efficient ML-based method for parameter estimation. It splits a complex problem into many separate small-scale subproblems, which can be solved using a simple method. The EM method has also been applied for RBFN learning [651, 649, 1108]. The shadow targets algorithm [1090], which employs a philosophy similar to that of the EM method, is an efficient RBFN training algorithm for topographic feature extraction.

The RBFN using regression weights can significantly reduce the number of hidden units, and is effectively used for approximating nonlinear dynamic systems [649, 1004, 957]. For a J_1-J_2-1 RBFN, the linear regression weights are defined by [649]

$$w_i = \mathbf{a}_i^{\mathrm{T}} \tilde{\mathbf{x}} + \xi_i \tag{6.77}$$

for $i = 1, \cdots, J_2$, where w_i is the weight from the ith hidden unit to the output unit, $\mathbf{a}_i = (a_{i,0}, a_{i,1}, \cdots, a_{i,J_1})^{\mathrm{T}}$ is the regression parameter vector, $\tilde{\mathbf{x}} = (1, x_1, \cdots, x_{J_1})^{\mathrm{T}}$ is the augmented input vector, and ξ_i is a zero-mean Gaussian white-noise process. For the Gaussian RBFN, the RBF centers $\mathbf{c}_i$ and their widths σ_i can be selected by the C-means and the nearest-neighbor heuristic [817], while the parameters of the regression weights are estimated by the EM method [649]. The RBFN with linear regression weights has also been studied [1004], where a simple but fast computational procedure is achieved by using a high-dimensional raised-cosine RBF. Storage space is also reduced by allowing the RBF centers to be situated at a nonuniform grid of points.

Some regularization techniques for improving the generalization capability of the MLP and the RBFN have been discussed in [790]. As in the MLP, the favored weight quadratic penalty term $\sum w_{ij}^2$ is also appropriate for the RBFN. The widths of the RBFs are widely known to be a major source of ill-conditioning in RBFN training, and large width parameters are desirable for better generalization. Some suitable penalty terms for widths are given in [790].

In the case of approximating a given function $f(x)$, a parsimonious design of the Gaussian RBFN can be based on the Gaussian spectrum of the known function, $\gamma_{\mathrm{G}}(f; \mathbf{c}, \sigma)$ [37]. According to the Gaussian spectrum, one can estimate the necessary number of RBF units and evaluate how appropriate the use of the Gaussian RBFN is. Gaussian RBFs are selected according to the peaks (negative as well as positive) of the Gaussian spectrum. Only the weights of the RBFN are needed to be tuned. Analogous to the PCA of the data sets, the principal Gaussian components of the target function are extracted. If there are a few sharp peaks on the spectrum surface, the Gaussian RBFN is suitable for approximation with a parsimonious architecture. On the other hand, if there are many peaks with similar importance or small peaks situated in large flat regions, it is advisable to choose an alternative way for the approximation of the function.

6.10 Normalized Radial Basis Function Networks

The normalized RBFN is defined by normalizing the vector composing the responses of all the RBF units [817]

$$y_i(\mathbf{x}) = \sum_{k=1}^{J_2} w_{ki} \widehat{\phi}_k(\mathbf{x}) \tag{6.78}$$

for $i = 1, \cdots, J_3$, where

$$\widehat{\phi}_k(\mathbf{x}) = \frac{\phi(\mathbf{x} - \mathbf{c}_k)}{\sum_{j=1}^{J_2} \phi(\mathbf{x} - \mathbf{c}_j)} \qquad (6.79)$$

The normalization operation is nonlocal, since each hidden node is required to know about the outputs of other hidden nodes. Hence, the convergence process is computationally costly.

A simple algorithm, called the *weighted averaging (WAV)* [700], is inspired by the functional equivalence of the normalized RBFN of the form (6.78) and FISs [543]. When interpreting a J_1-J_2-J_3 normalized RBFN as a fuzzy system, the WAV can restrict the weights to lie in the range of possible output values and preserve the linguistic meaning of fuzzy IF-THEN rules. Assuming that all the training examples are roughly equally reliable, the WAV algorithm is given by

$$w_{ij} = \frac{\sum_{p=1}^{N} \widehat{\phi}_i(\mathbf{x}_p) y_{p,j}}{\sum_{p=1}^{N} \widehat{\phi}_i(\mathbf{x}_p)} \qquad (6.80)$$

for $i = 1, \cdots, J_2$, $j = 1, \cdots, J_3$. Thus, w_{ij} always lies in the range of $y_{p,j}$, and the linguistic interpretation, "IF $\mathbf{x}$ is close to $\mathbf{c}_i$ THEN y_j is close to w_{ij}", is valid. In the RLS and LMS methods, the weights may be negative and the linguistic meaning is no longer preserved. If the initial training data is poor and the data quality improves as time goes on, a tracking algorithm is applied so as to keep track of the time variation of the system after the training stage. The WAV has a complexity close to the LMS and has an MSE close to the RLS.

The normalized RBFN given by (6.78) can be presented in another form [145, 649]. The network output is defined by

$$y_i(\mathbf{x}) = \frac{\sum_{j=1}^{J_2} w_{ji} \phi(\mathbf{x} - \mathbf{c}_j)}{\sum_{j=1}^{J_2} \phi(\mathbf{x} - \mathbf{c}_j)} \qquad (6.81)$$

Now, normalization is performed in the output layer. As it already receives information from all the hidden units, the locality of the computational processes is preserved. The two forms of the normalized RBFN, (6.78) and (6.81), are equivalent, and their similarity with FISs has been pointed out in [543].

In the normalized RBFN of the form (6.81), the traditional roles of the weights and activities in the hidden layer are exchanged. In the RBFN, the weights determine as to how much each hidden node contributes to the output, while in the normalized RBFN, the activities of the hidden nodes determine which weights contribute most to the output. The normalized RBFN provides better smoothness than the RBFN. Due to the localized property of the receptive fields, for most data points, there is usually only one hidden node that contributes significantly to (6.81). The normalized RBFN (6.81) can be trained using a procedure similar to that for the RBFN. The normalized Gaussian RBFN exhibits superiority in supervised classification due to its soft modification rule [855]. It is also a universal approximator in the space of continuous functions with compact support in the space $L^p(R^p, \mathrm{d}\mathbf{x})$ [96].

The normalized RBFN loses the localized characteristics of the localized RBFN and exhibits excellent generalization properties, to the extent that hidden nodes need to be recruited only for training data at the boundaries of the class domains. This obviates the need for a dense coverage of the class domains, in contrast to the RBFN. Thus, the normalized RBFN softens the curse of dimensionality associated with the localized RBFN [145]. The normalized Gaussian RBFN outperforms the Gaussian RBFN in terms of the training and generalization errors, and exhibits a more uniform error over the training domain. In addition, the normalized Gaussian RBFN is not sensitive to the RBF widths.

The normalized RBFN is an RBFN with a quasilinear activation function with a squashing coefficient decided by the actvations of all the hidden units. The output units of the RBFN can also employ the sigmoidal activation function. The RBFN with the sigmoidal function at the output nodes outperforms the case of linear or quasilinear functions at the output nodes in terms of sensitivity to learning parameters, convergence speed as well as accuracy [1114].

6.11 Optimizing Network Structure

Since the number of RBF centers is usually determined *a priori*, the structure of an RBFN may not be optimum and learning may be inefficient. Alternatives for RBFN learning are to determine the number and locations of the RBF centers automatically by using constructive and pruning methods.

6.11.1 Constructive Methods

The philosophy of the constructive approach is to gradually increase the number of RBF centers until a criterion is satisfied. In [584], a new prototype is created in a region of the input space by splitting an existing prototype c_j selected by a splitting criterion, and splitting is performed by adding the perturbation vectors $\pm \epsilon_j$ to c_j. The resulting vectors $c_j \pm \epsilon_j$ together with the existing centers form the initial set of centers for the next growing cycle. The perturbation vector ϵ_j can be obtained in terms of a deviation measure computed from c_j and the input vectors represented by c_j

$$\epsilon_{j,i} = \rho_0 \left(\frac{1}{N_j} \sum_{\mathbf{x}_k \in \mathcal{C}_j} (x_{k,i} - c_{j,i})^2 \right)^{\frac{1}{2}} \tag{6.82}$$

for $i = 1, \cdots, J_1$, where $\epsilon_j = (\epsilon_{j,1}, \cdots, \epsilon_{j,J_1})^{\mathrm{T}}$, ρ_0 is a scaling factor to guarantee $\|\epsilon_j\| \ll \|c_j\|$, $x_{k,i}$ and $c_{j,i}$ are, respectively, the ith components of $\mathbf{x}_k$ and c_j, and N_j is the size of $\mathcal{C}_j$, which is the set of input vectors represented by c_j according to the nearest-neighbor paradigm. Feedback splitting and purity splitting are two criteria for the selection of splitting centers. Existing algorithms for updating the centers $\mathbf{c}_j$, widths σ_j, and weights can be used. The process continues until a stopping criterion is satisfied.

In a heuristic incremental algorithm [341], the training phase is an iterative process that adds a hidden node $\mathbf{c}_t$ at each epoch t by an error-driven rule. Each epoch t consists of three phases. The data point with the worst approximation, denoted $\mathbf{x}_s$, is first recruited as a new hidden node $\mathbf{c}_t = \mathbf{x}_s$, and its weight to each output node j, w_{tj}, is fixed at the error at the jth output node performed by the network at the $(t-1)$th epoch on $\mathbf{x}_s$

$$w_{tj} = y_{s,j} - \sum_{k=1}^{t-1} w_{kj} \phi \left(\|\mathbf{x}_s - \mathbf{c}_k\| \right) \tag{6.83}$$

The next two phases are, respectively, local tuning and fine tuning of the variances of the RBFs. In local tuning, only the variance of the new node is learnt using the information from all the pattern pairs by minimizing the global error, whereas all the other parameters remain fixed. In fine tuning, all the variances of the RBFs are learnt in order to further minimize the global error. The ES approach is performed to optimize the variances in both the steps. The network allows a fast and accurate reconstruction of continuous or discontinuous function starting from randomly sampled data sets.

The incremental RBFN architecture using hierarchical gridding of the input space [125] allows for a uniform approximation without wasting resources. The centers and variances of the added nodes are fixed through heuristic considerations. Additional layers of Gaussians at lower scales are added where the residual error is higher. The number of Gaussians of each layer and their variances are computed from considerations based on linear filtering theory. The weight of each Gaussian is estimated through a maximum *a posteriori* estimate carried out locally on a subset of the data points. The method shows a high accuracy in the reconstruction, and it can deal with nonevenly spaced data points and is fully parallelizable.

Hierarchical RBFN [351] is a multiscale version of the RBFN. It is constituted by hierarchical layers, each containing a Gaussian grid at a decreasing scale. The grids are not completely filled, but units are inserted only where the local error is over a threshold. This guarantees a uniform residual error and the allocation of more units with smaller scales where the data contain higher frequencies. The constructive approach is based only on the local operations, which do not require any iteration on the data. It allows for an effective network to be built in a very short time. Like a traditional wavelet-based multiresolution analysis (MRA), the hierarchical RBFN employs Riesz bases and enjoys asymptotic approximation properties for a very large class of functions. While multiscale approximations in the MRA are created from fine scales to coarse scales, the hierarchical RBFN operates the other way round. The coarse-to-fine approach enables the hierarchical RBFN to grow until the reconstructed surface meets the required quality.

The forward OLS algorithm [206] described in Sect. 6.5 is a well-known constructive algorithm. Based on the OLS algorithm, a constructive algorithm for the generalized Gaussian RBFN is given in [1158]. RBFN learning based on a modification to the cascade-correlation algorithm [347] works in a way similar to the OLS method, but with a significantly faster convergence [679]. The OLS incorporated with the sensitivity analysis has also been employed to search for the optimal RBF centers [1029].

The dynamic decay adjustment (DDA) algorithm is a fast constructive training method for the RBFN when used for classification [102]. The DDA is motivated from the probabilistic nature of the PNN [1050], the constructive nature of the restricted Coulomb energy (RCE) networks [945] as well as the independent adjustment of the decay factor or width σ_i of each prototype. The RCE network is inspired by systems of charged particles in the three-dimensional space, and can be treated as a special kind of RBFN for classification tasks. The RCE network constructs its architecture dynamically during the training, and it has a very fast learning speed. The radial adjustment is class dependent and it distinguishes between the different neighbors. The DDA algorithm relies on two thresholds of σ_i, namely, the lower threshold σ^- and the upper threshold σ^+, in order to decide on the introduction of the RBF units in the networks. The DDA method is faster and also achieves a higher classification accuracy than the conventional RBFN [817], the MLP trained with the Rprop, and the RCE.

Incremental RBFN learning algorithms are derived based on the GCS model [378] and based on a Hebbian learning rule adapted from the NG model [379]. The insertion strategy is on accumulated error of a subset of the data set. The resource-allocating network (RAN) [919] is a well-known RBFN construction method, which is described in Subsect. 6.11.2. Another example of the constructive approach is the competitive RBF algorithm based on the ML classification approach [1259]. A review of constructive approaches to structural learning of FNNs is given in [648].

6.11.2 Resource-allocating Networks

The RAN [919], proposed by Platt, is a sequential learning method for the localized RBFN such as the Gaussian RBFN, which is suitable for online modeling of non-stationary processes. The network begins with no hidden units. As the pattern pairs are received during the training, a new hidden unit may be added according to the novelty in the data. The novelty in the data is decided by two conditions

$$\|\mathbf{x}_t - \mathbf{c}_i\| > \varepsilon(t) \tag{6.84}$$

$$\|\mathbf{e}(t)\| = \|\mathbf{y}_t - f(\mathbf{x}_t)\| > e_{\min} \tag{6.85}$$

where $\mathbf{c}_i$ is the center nearest to $\mathbf{x}_t$, the prediction error $\mathbf{e} = (e_1, \cdots, e_{J_3})^{\mathrm{T}}$, and $\varepsilon(t)$ and $e_{\min}$ are thresholds to be selected appropriately. The algorithm starts with $\varepsilon(t) = \varepsilon_{\max}$, where $\varepsilon_{\max}$ is chosen as the largest scale in the input space, typically the entire input space of nonzero probability. The distance $\varepsilon(t)$ shrinks exponentially as

$$\varepsilon(t) = \max\left\{\varepsilon_{\max}e^{-\frac{t}{\tau}}, \varepsilon_{\min}\right\} \qquad (6.86)$$

where τ is a decay constant. The value for $\varepsilon(t)$ is decayed until it reaches $\varepsilon_{\min}$. Assuming that there are k nodes at time $t-1$, for the Gaussian RBFN, the newly added hidden unit at time t can be initialized as

$$\mathbf{c}_{k+1} = \mathbf{x}_t \qquad (6.87)$$
$$w_{(k+1)j} = e_j(t) \qquad (6.88)$$
$$\sigma_{k+1} = \alpha\,\|\mathbf{x}_t - \mathbf{c}_i\| \qquad (6.89)$$

for $j = 1, \cdots, J_3$, where the value for σ_{k+1} is based on the nearest-neighbor heuristic and α is a parameter defining the size of neighborhood. If a pattern pair $(\mathbf{x}_t, \mathbf{y}_t)$ does not pass the novelty criteria, no hidden unit is added and the existing network parameters are adapted using the LMS method [1182]. The RAN method performs much better than the RBFN learning algorithm using random centers and that using the centers clustered by the C-means [817] in terms of network size and MSE. The RAN method achieves roughly the same performance as the MLP trained with the BP, but with much less computation.

Instead of starting from zero hidden nodes, an agglomerative clustering algorithm is used for the initialization of the RAN [1138]. In [566], the LMS method is replaced by the EKF method for the network parameter adaptation so as to generate a more parsimonious network, although the LMS method is considerably faster to implement than the EKF method. Two geometric criteria, namely the *prediction error criterion*, which is the same as (6.85), and the *angle criterion*, are also obtained from a geometric viewpoint. The angle criterion attempts to assign RBFs that are nearly orthogonal to all the other existing RBFs. These criteria are proved to be equivalent to Platt's criteria [919]. In [565], the statistical novelty criterion is defined by

$$\frac{\|\mathbf{e}(t)\|}{\sqrt{\operatorname{tr}(\boldsymbol{\Xi}(t))}} > z_\alpha \qquad (6.90)$$

where the estimation error $\mathbf{e}(t)$ and the matrix $\boldsymbol{\Xi}(t)$ are, respectively, computed as part of the EKF algorithm, as given in (3.161) through (3.164), and z_α is the value of the Z-statistic at $\alpha\%$ level of significance, that is, for $\alpha\%$ of the data, which does not cause any computation overhead. By using the EKF method and using the criterion (6.90) to replace the criteria (6.84) and (6.85), for a given task, more compact networks and smaller MSEs are achieved than the RAN [919] and the EKF-based RAN [566].

Resource Allocating Networks with Pruning

Numerous improvements on the RAN [919] have been made by integrating node pruning procedure [1234, 1235, 964, 990, 1091, 511, 512]. The minimal RAN [1234, 1235] is based on the EKF-based RAN [566], and achieves a more compact network with equivalent or better accuracy by incorporating a pruning strategy to remove inactive nodes and augmenting the basic growth criterion of the RAN. The output of each RBF unit is scaled as

$$\hat{o}_i(\mathbf{x}) = \frac{|o_i(\mathbf{x})|}{\max_{1 \leq j \leq J_2} \{|o_j(\mathbf{x})|\}} \tag{6.91}$$

for $i = 1, \cdots, J_2$. If $\hat{o}_i(\mathbf{x})$ is below a predefined threshold δ for a given number of iterations, this node is idle and can be removed. For a given accuracy, the minimal RAN achieves a smaller complexity than the MLP trained with RProp [949], and achieves a more compact network and requiring less training time than the MLP constructed by the DI algorithm [818].

In [964], the RAN is improved by using the Givens QR decomposition-based RLS for the adaptation of the weights and integrating a node-pruning strategy. The ERR criterion in [208] is used to select the most important regressors. In [990], the RAN [919] is improved by using in each iteration the combination of the SVD and QR-cp methods for determining the structure as well as for pruning the network. The SVD and the QR-cp determine a subset of RBFs that is relevant to the linear output combination. In the early phase of learning, the addition of RBFs is in small groups, and this leads to an increased rate of convergence. If a particular RBF is not considered for a given number of iterations, it is removed. The size of the network is more compact than the RAN.

The growing and pruning algorithm for RBF (GAP-RBF) [511] and the generalized GAP-RBF (GGAP-RBF) [512] are sequential learning algorithms for realizing parsimonious RBFNs. These algorithms make use of the notion of *significance* of a hidden neuron, which is defined as a neuron's statistical contribution over all the inputs seen so far to the overall performance of the network. The GAP-RBF is based on the RAN [919]. In addition to the two growing criteria of the RAN, a new neuron is added only when its significance is also above a chosen learning accuracy. If during the training the significance of a neuron becomes less than the learning accuracy, that neuron will be pruned. For each new pattern, only its nearest neuron is checked for growing, pruning, or updating using the EKF. The GGAP-RBF enhances the significance criterion such that it is applicable for training examples with arbitrary sampling density. The GAP-RBF and the GGAP-RBF outperform several other sequential learning algorithms such as the RAN [919], the EKF-based RAN [566], and the minimal RAN [1235] in terms of the learning speed, network size and generalization performance. The GAP-RBF and the GGAP-RBF have a low computational complexity, since only the nearest neuron of an input pattern is checked for significance or is updated.

In [1091], the EKF and statistical novelty criterion-based method [565] is extended by incorporating an online pruning procedure, which is derived using the parameters and innovation statistics estimated from the EKF. The online pruning method is analogous to the saliency-based OBS [460] and OBD [656]. The IncNet and IncNet Pro [547] are RAN-EKF networks with statistically controlled growth criterion. The pruning method is similar to the OBS, but based on the result of the EKF algorithm.

6.11.3 Constructive Methods with Pruning

The RAN algorithms with pruning strategy [1234, 1235, 964, 990, 1091, 511, 512] introduced in Subsect. 6.11.2 belong to the category of constructive methods with pruning. Some other algorithms of this category are introduced below.

The normalized RBFN [957] can be trained by using a constructive method with pruning strategy based on the novelty of the data and the overall behavior of the network. The network starts from one neuron, and adds a new neuron if an example passes two novelty criteria, until a specified maximum number of neurons is reached. The first criterion is the same as (6.85), and the second one deals with the activation of the nonlinear neurons

$$\max_i \phi_i\left(\mathbf{x}_t\right) < \zeta \tag{6.92}$$

where ζ is a threshold. The pseudo-Gaussian function given by (6.11) and (6.12) is used as the RBF. A sequential learning algorithm is derived from the gradient-descent method. As in [649], RBF weights are linear regression functions of the input variables. After the whole pattern set is presented at an epoch, the algorithm starts to remove those neurons that meet any of the three cases, namely, neurons with a very small mean activation for the whole pattern set, neurons with a very small activation region, or neurons having an activation very similar to that of other neurons. The combination of the regression weights, the normalized activation function and the pseudo-Gaussian function significantly reduces the number of hidden units within an RBFN.

Another sequential training technique is given in [22]. It starts with zero hidden node and progressively builds the model as new data become available. The algorithm is based on a fuzzy partition of the input space that defines a multidimensional grid, from which the RBF centers are selected. The centers of the fuzzy subspaces are considered as candidates for RBF centers, so that at least one selected RBF center is close enough to each input example. The method is capable of adapting online the structure of the RBFN, by adding new units when an input example does not belong to any of the subspaces that have been selected, or deleting old ones when no data have been assigned to the respective fuzzy subspaces for a long period of time. In this way, the network retains a reasonable size, and at the same time describes well the operating region of the system at any instant. The weights are updated using the RLS algorithm. The method avoids selecting the centers only among the available data.

Despite its advantage as an efficient and fast growing RBFN algorithm, the DDA [102] may result in too many neurons. The DDA with temporary neurons improves the DDA by introducing online pruning of neurons after each DDA training epoch [879]. After each training epoch, if the individual neurons cover a sufficient number of samples, they are marked as *permanent*; otherwise, they are deleted. This mechanism results in a significant reduction in the number of neurons, but having a similar classification performance. The DDA with selective pruning and model selection is another extension to the DDA [870]. In contrast to the DDA with temporary neurons [879], only a portion of the neurons that cover only one training example are pruned and pruning is carried out only after the last epoch of the DDA training. The method improves the generalization performance of the DDA [102] and the DDA with temporary neurons [879], but yields a larger network size than the DDA with temporary neurons.

6.11.4 Pruning Methods

The pruning methods discussed in Chapter 3 can be applied to the RBFN since the RBFN is a kind of FNN. Well-known pruning methods are the OBD [656] and OBS [460]. Pruning algorithms based on the regularization technique are also popular since additional terms that penalize the complexity of the network are incorporated into the MSE criterion.

The pruning method proposed in [684] starts from a big RBFN, and achieves a compact network through an iterative procedure of training and selection. The training procedure adaptively changes the centers and the width of the RBFs and trains the linear weights. The selection procedure performs the elimination of the redundant RBFs using an objective function based on the MDL principle [951]. The approach is quite general, can also be used for types of models other than neural networks, and can easily be adapted to other types of networks.

In [833], all the data vectors are initially selected as centers. Redundant centers in the RBFN are eliminated by merging two centers at each adaptation cycle by using an iterative clustering method. The technique is superior to the traditional RBFN algorithms, particularly in terms of the processing speed and solvability of nonlinear patterns. In [627], two methods are described for reducing the size of the PNN while preserving the classification performance as good as possible.

6.12 Radial Basis Function Networks for Modeling Dynamic Systems

The sequential RBFN learning algorithms, such as the RAN family described in Subsect. 6.11.2 and the studies in [957, 22], are capable of modifying both the network structure and the output weights online; thus, these algorithms are particularly suitable for modeling dynamical time-varying systems, where not only the dynamics but also the operating region changes with time.

In order to model complex nonlinear dynamical systems, the state-dependent autoregressive (AR) model with functional coefficients is often used. The RBFN can be used as a nonlinear AR time-series model for forecasting application [1027]. The RBFN can also be used to approximate the coefficients of a state-dependent AR model, thus yielding the RBF-AR model [1130]. The RBF-AR model has the advantages of both the state-dependent AR model for describing nonlinear dynamics and the RBFN for function approximation. The RBF-ARX model is an RBF-AR model with an exogenous variable [904]. The RBF-ARX model usually uses far fewer RBF centers when compared to the RBFN.

The TDNN model can be an MLP-based or an RBFN-based temporal neural network for nonlinear dynamics and time-series learning. The RBFN-based TDNN [101] uses the same spatial representation of time as the MLP-based TDNN [1137]. The TDNN is described in Chapter 3, and its architecture was shown in Fig. 3.8. The learning of the RBFN-based TDNN uses the RBFN learning algorithms.

For time-series applications, the input to the network is $\mathbf{x}(t) = (y(t-1), \cdots, y(t-n_y))^{\mathrm{T}}$, and the network output is $y(t)$. There are some problems with the RBFN when used as a time-series predictor [113]. First, the Euclidean distance measure is not always appropriate for measuring the similarity between the input vector $\mathbf{x}_t$ and the prototype $\mathbf{c}_i$. The Euclidean distance measure is strictly precise only when the components of the data are uncorrelated. However, $\mathbf{x}_t$ is itself highly autocorrelated. Second, the node response is radially symmetrical, whereas the data may be distributed differently in each dimension. Third, when the minimum lag n_y is a large number, if all, up to n_y, lagged versions of the output $y(t)$ are concatenated as the input vector, the network is too complex and the performance deteriorates due to irrelevant inputs and an oversized structure. The dual-orthogonal RBFN algorithm overcomes most of these limitations for nonlinear time-series prediction [113]. Motivated by the LDA technique, a distance metric is defined based on a classification function of the set of input vectors in order to achieve improved clustering. The forward OLS is used first to determine the significant lags and then to select the RBF centers. In both steps, the ERR is used for the selection of significant nodes [205]. The forward OLS procedure for the selection of RBF centers can be terminated using the AIC criterion [18] when the optimal number of centers is found.

For online adaptation of nonlinear systems, a constant exponential forgetting factor is commonly applied to all the past data uniformly. This is incorrect for nonlinear systems whose dynamics are different in different operating regions. In [1240], online adaptation of the Gaussian RBFN is implemented using a localized forgetting method, which sets different forgetting factors in different regions according to the response of the local prototypes to the current input vector. The method is applied in conjunction with the ROLS in [1241] and the computing is very efficient.

The spatial representation of time in the TDNN model is inconvenient and also the use of temporal window imposes a limit on the sequence length. Recurrent RBFNs, which combine features from the RNN and the RBFN, are suitable for the modeling of nonlinear dynamic systems [372, 1092, 985]. In [985], a recurrent RBFN is used for the recognition of a simple temporal sequence. Time is an internal mechanism, which is implicit via recurrent connection. Each input-layer node of the RBFN is changed to a neuron with the sigmoidal activation function and self-connection. The training process is flexible and is the same as in the algorithm for the RCE network [945]. In [1092], the EKF method is applied for online parameter and structure adaptation. The recurrent RBFN has the adaptive ability to deal with large noise in nonlinear system identification. The recurrent RBFN introduced in [372] has a four-layer architecture, with the input layer, the RBF layer, the state layer, and the one-neuron output layer. The state and output layers use the sigmoidal activation function. This network is well suited for dealing with automata. Some techniques for injecting finite state automata into the network have also been proposed in [372].

6.13 Hardware Implementations of Radial Basis Function Networks

Hardware implementations of neural networks are commonly based on building blocks, and thus allow for the inherent parallelism of neural networks. Most hardware implementations for the RBFN are developed for the Gaussian RBFN, since it is biologically plausible.

The properties of the MOS transistor are desirable for analog designs of the Gaussian RBFN. In the subthreshold or weak-inversion region, the drain current of the MOS transistor has an exponential dependence on the gate bias and dissipates very low power. This exponential characteristic of the MOS devices is usually exploited for designing the Gaussian function [1160, 747, 154, 713]. In [154], a compact analog Gaussian cell is designed whose core takes only eight transistors and can be supplied with a high number of input pairs. On the other hand, the MOS transistor has a square-law dependence on the bias voltages in its *strong-inversion* or *saturation* region. This is also desirable since it provides a high-current driving, a large dynamic range as well as high noise immunity. Based on the property in the strong-inversion region, a compact programmable analog Gaussain synapse cell has been designed in [233], where programmability is realized by changing the stored weight w_{ji}, the reference current and the sizes of the transistors in the differential pairs.

Similarity measure, typically the Euclidean distance measure, is essential in many neural-network models such as the clustering networks and the RBFN. The circuits for the Euclidean distance measure are usually based on the square-law property of the strong-inversion region [245, 713, 781, 1233, 780]. The generalized measure of similarity between two voltage inputs can be implemented by using the Gaussian-type or *bump* circuit and by using the *bump-antibump* circuit. These circuits are based on the concept of the *current correlator* developed for weak-inversion operation [293]. Based on the current correlator [293], an analog Gaussian/square function computation circuit is given in [713]. This circuit exhibits independent programmability for the center, width, and peak amplitude of the dc transfer curve. When operating in the strong-inversion region, it calculates the squared difference, whereas in the weak-inversion region it realizes a Gaussian-like function. In [245], circuits for calculating Euclidean distance and computing programmable Gaussian units with tunable center and variance are also designed based on the square-law of the strong-inversion region.

In [181], the building blocks of the Guassian RBFN hardware consist mainly of analog circuits, namely, operational amplifier, multiplier, multiplying digital-to-analog converter (DAC), floating resistor, summer and exponentiator. Parameters of the RBFN are represented digitally for convenient interfacing. Individual Guassian RBF units allow independent tuning of the centers, widths and amplitudes. A pulsed VLSI RBFN chip is fabricated in [781, 780] where a collection of pulse width modulation (PWM) analog circuits are combined on a single RBFN chip. The PWM is a hybrid pulse stream technique that combines the advantages of both the analog and digital VLSI. The distance metric is based on the square-law property of the strong-inversion region, and a Gaussian-like RBF is produced using two MOS transistors.

An analog VLSI circuit that computes the RBFN and MLP propagation rules on a single chip is designed in [1233] to form the synapse and neuron of a conic-section function network [311]. The circuit computes both the weighted sum for the MLP and the Euclidean distance for the RBF. The two propagation rules are then aggregated as the design of synapse and neuron of the conic-section function network. The circuit can also operate in either the MLP or the RBFN mode.

Analog implementation results in low-cost parallelism. However, inaccurate circuit parameters affect the computational accuracy. Digital implementation can redress the accuracy problem and can have more flexibilities. Nevertheless, digital designs are slow in computing. In digital designs, the nonlinear activation function is approximated by using an LUT that stores many input-output associations. This method needs an external memory to store the LUT, thus simultaneous evaluation of the activation function for multiple neurons is not possible. Some direct digital VLSI implementations of the

RBFN are also available in the literature such as the RBF generator [748] and the RBFN with DDA [6]. A hybrid VLSI/digital design of the RBFN, which integrates a custom analog VLSI circuit and a commercially available digital signal processor, is also developed in [1160].

6.14 Complex Radial Basis Function Networks

Many signal-processing applications are performed in a multidimensional complex space. In communication systems, signals are typically transmitted in the form of complex digital symbols. Complex RBFNs are more efficient than the RBFN, in the case of nonlinear signal processing involving complex signals, such as equalization and modeling of nonlinear channels. Digital channel equalization can be treated as a classification problem.

In the complex RBFN, the input and the output of the network are complex values, whereas the activation function of the hidden nodes is the same as that for the RBFN. The Euclidean distance in the complex domain is defined by [211]

$$d\left(\mathbf{x}_t, \mathbf{c}_i\right) = \left[\left(\mathbf{x}_t - \mathbf{c}_i\right)^{\mathrm{H}}\left(\mathbf{x}_t - \mathbf{c}_i\right)\right]^{\frac{1}{2}} \tag{6.93}$$

where $\mathbf{c}_i$ is a J_1-dimensional complex center vector. The output weights are complex valued. Most existing RBFN learning algorithms can be easily extended for training various versions of the complex RBFN [211, 182, 669, 552]. When using clustering techniques to determine the RBF centers, the similarity measure can be based on the distance defined by (6.93). As in the RBFN, the Gaussian RBF is usually used in the complex RBFN. In [552], the minimal RAN algorithm [1234] has been extended to its complex-valued version.

The Mahalanobis distance (6.68) defined for the Gaussian RBF can be extended to the complex domain [669]

$$d\left(\mathbf{x}_t, \mathbf{c}_i\right) = \left(\left[\mathbf{x}_t - \mathbf{c}_i(t-1)\right]^{\mathrm{H}} \mathbf{\Sigma}_i^{-1}(t-1)\left[\mathbf{x}_t - \mathbf{c}_i(t-1)\right]\right)^{\frac{1}{2}} \tag{6.94}$$

for $i = 1, \cdots, J_2$. Note that the transpose T in (6.68) is changed into the Hermitian transpose H.

Learning of the complex Gaussian RBFN can be performed in two phases, where the RBF centers are first selected by using the incremental C-means algorithm [746] and the weights are then solved by fixing the RBF parameters [669]. At each iteration t, the C-means first finds the winning node, indexed by w, by using the nearest-neighbor paradigm, and then updates both the center and the variance of the winning node by

$$\mathbf{c}_w(t) = \mathbf{c}_w(t-1) + \eta\left[\mathbf{x}_t - \mathbf{c}_w(t-1)\right] \tag{6.95}$$

$$\mathbf{\Sigma}_w(t) = \mathbf{\Sigma}_w(t-1) + \eta\left[\mathbf{x}_t - \mathbf{c}_w(t-1)\right]\left[\mathbf{x}_t - \mathbf{c}_w(t-1)\right]^{\mathrm{H}} \tag{6.96}$$

where η is the learning rate. The C-means is repeated until the changes in all $\mathbf{c}_i(t)$ and $\boldsymbol{\Sigma}_i(t)$ are within a specified accuracy, that is,

$$\|\mathbf{c}_i(t) - \mathbf{c}_i(t-1)\| \leq \varepsilon_0 \tag{6.97}$$

$$\|\boldsymbol{\Sigma}_i(t) - \boldsymbol{\Sigma}_i(t-1)\|_{\mathrm{F}} \leq \varepsilon_1 \tag{6.98}$$

where ε_0 and ε_1 are predefined small positive numbers. After complex RBF centers are determined, the weight matrix $\mathbf{W}$ is determined using the LS or the RLS algorithm.

6.15 Properties of Radial Basis Function Networks

6.15.1 Receptive-field Networks

The RBFN with a localized RBF is a receptive-field or localized network. The localized approximation method provides the strongest output when the input is near the prototype of a node. For a suitably trained localized RBFN, similar input vectors, namely, input vectors that are close to each other, always generate similar outputs, while distant input vectors produce nearly independent outputs. This is the intrinsic local generalization property. A receptive-field network is an associative neural network in that only a small subspace is determined by the input to the network. The domain of receptive-field functions is practically a finite real interval defined by the parameters of the function. This property is particularly attractive since the modification of the receptive-field function produces a local effect. Thus, receptive-field networks can be conveniently constructed by adjusting the parameters of the receptive-field functions and/or adding or removing neurons.

The Gaussian RBFN is a popular receptive-field network. Another well-known receptive-field network is the cerebellar model articulation controller (CMAC) [21, 806]. The CMAC is a distributed LUT system and is also suitable for VLSI realization. It can approximate slow-varying functions, and is orders of magnitude faster than the BP. However, the CMAC may fail in approximating highly nonlinear or rapidly oscillating functions [264, 135].

6.15.2 Generalization Error and Approximation Error

A SURE-based model selection criterion for the RBFN is given by [410]

$$Err\,(J_2) = err\,(J_2) - N\sigma_{\mathrm{n}}^2 + 2\sigma_{\mathrm{n}}^2\,(J_2 + 1) \tag{6.99}$$

where Err is the generalization error on the new data, err denotes the training error for each model, N is the size of the pattern set, and σ_{n}^2 is the noise variance, which can be estimated from the MSE of the model.

An empirical comparison among the SURE-based method [600], crossvalidation, and the BIC [1010], has been made in [410]. The generalization error of the models by the SURE-based method can be less than that of the models selected by crossvalidation, but with much less computation. In addition, the SURE-based method does not require an extra validation set to choose the number of RBFs. Thus, the utility of the SURE-based method is greatest when there is insufficient data for a validation set. The SURE-based method and the BIC behave in a similar manner. However, the BIC generally gives preference to simpler models because it penalizes complex models more harshly.

A bound on the generalization error for FNNs is given by (2.18) [853, 852]. This bound has been considerably improved to $O\left(\left(\frac{\ln N}{N}\right)^{\frac{1}{2}}\right)$ in [637] for RBFN regression with the MSE function.

An upper bound for the error in function approximation using the Gaussian RBFN is also given in [720]. For any C^2 function[3] with support on the unit hypercube $[0, 1]^{J_1}$, the Gaussian RBFN has an approximation error with an upper bound of $O\left(n^{-2}\right)$ [720], if there are n^{J_1} RBF units with centers defined on a uniform mesh of the hypercube and the width σ is selected as $\sigma \propto n^{-1}$.

6.16 Radial Basis Function Networks vs. Multilayer Perceptrons

Both the MLP and the RBFN are used for supervised learning. In the RBFN, the activation of an RBF unit is determined by the distance between the input and prototype vectors. For classification problems, RBF units map input patterns from a nonlinearly separable space to a linearly separable space, and the responses of the RBF units form new feature vectors. Each RBF prototype is a cluster serving mainly a certain class. When the MLP with a linear output layer is applied to classification problems, minimizing the error at the output of the network is equivalent to maximizing the so-called *network discriminant function* at the output of the hidden units [1162].

A comparison between the MLP and the localized RBFN is as follows.

Global method vs. local method

The use of the sigmoidal activation function makes the MLP a global method. For an input pattern, many hidden units will contribute to the network output. On the other hand, the localized RBFN is a local method with respect to the input space in that each localized RBF covers a very small local zone. The local method satisfies the minimal disturbance principle [1183], that is, the adaptation not only reduces the output error for the current example, but also minimizes disturbance to those already learned. The localized RBFN is biologically plausible.

---

[3] A C^k function is continuously differentiable up to order k.

Local minima

Due to the sigmoidal function, the crosscoupling between hidden units of the MLP results in high nonlinearity in the error surface, resulting in the problem of local minima or nearly flat regions. This problem gets worse as the network size increases. A similar situation holds for RNNs. In contrast, the RBFN has a simple architecture with linear weights and the LMS adaptation rule is equivalent to a gradient search of a quadratic surface, therefore the RBFN has a unique solution to the weights.

Approximation and generalization

Due to the global activation function, the MLP has greater generalization for each training example, and thus the MLP is a good candidate for extrapolation. On the contrary, the extension of a localized RBF to its neighborhood is determined by its variance. This localized property prevents the RBFN from extrapolation beyond the training data.

Network resources and curse of dimensionality

The localized RBFN, like most kernel-type approximation methods, suffers from the problem of curse of dimensionality. The localized RBFN typically requires much more data and more hidden units to achieve an accuracy similar to that of the MLP. In order to approximate a wide class of smooth functions, the number of hidden units required for the three-layer MLP is polynomial with respect to the input dimensions, while the counterpart for the localized RBFN is exponential [71]. The curse-of-dimensionality problem can be alleviated by using smaller networks with more adaptive parameters [923] or by progressive learning [322].

Hyperplanes vs. hyperellipsoids

For the MLP, the representation of a hidden unit is a weighted linear summation of the input components, transformed by a monotonic activation function. Thus, the response of a hidden unit is constant on a surface that consists of parallel $(J_1 - 1)$-dimensional hyperplanes in the J_1-dimensional input space. As a result, the MLP is preferable for linearly separable problems. On the other hand, in the RBFN the hidden units use distance for the prototype vectors followed by transformation of a localized function. Accordingly, the activation of the hidden units is constant on concentric $(J_1 - 1)$-dimensional hyperspheres or hyperellipsoids. Thus, the RBFN may be more efficient for linearly inseparable classification problems.

Training speed and performing speed

The error surface of the MLP has many local minima or large flat regions called *plateaus*, which lead to slow convergence of the training process. Due to the gradient-search technique, the MLP requires many iterations, each containing a large amount of computation, to converge, and it also frequently gets trapped at local minima. For the localized RBFN, only a few hidden units have significant activations for a given input, thus the network modifies the weights only in the vicinity of the sample point and retains constant weights in the other regions. The RBFN requires orders of magnitude less training time for convergence than the MLP trained with the BP rule to achieve comparable performance [133, 817, 623]. For equivalent generalization performance, the MLP typically requires far fewer hidden units than the localized RBFN, thus the trained MLP is much faster in performing.

Remarks

Generally speaking, the MLP is a better choice if the training data is expensive. However, when the training data is cheap and plentiful or online training is required, the RBFN is most desirable. In addition, the RBFN is insensitive to the order of the appearance of the adjusted signals, and hence more suitable for online or subsequent adaptive adjustment [344].

Hybridization of the Two Models

Aside from the above comparison, some properties of the MLP and the RBFN are combined for modeling purposes [677, 311]. In the centroid-based MLP [677], a centroid layer is inserted into the MLP as the second layer. The output of each unit in this new layer is the Euclidean distance between a centroid and the input, where the centroid is located somewhere in the input space. Due to the simplicity of these units only minor modifications are needed for the BP algorithm. The convergence can be significantly improved and a significantly more efficient structure, as compared to the conventional MLP, is formed. The conic-section function network [311] is a neural-network model motivated by the observation that both the hyperplane and the hypersphere are special cases of conic-section functions. This network generalizes the activation function to include both the bounded (hypersphere) and unbounded (hyperplane) decision regions in one network. It can make automatic decisions with respect to the two decision regions. Learning is faster by properly initializing the nodes as RBF or sigmoidal nodes. It combines the speed of the RBFN and the error minimization of the MLP, and is suitable for complex problems with a high-dimensional input space.

6.17 Computer Experiments

We conclude this chapter by illustrating as to how to apply the RBFN to beamforming and DoA estimation.

6.17.1 Example 6.1: Radial Basis Function Networks for Beamforming

This example is taken from [332]. The RBFN is first trained to learn the linear constrained minimum variance (LCMV) beamformer[4], and the trained network is then used as the beamformer for real-time tracking.

For a linear array of L elements, the beamforming weights of the LCMV beamformer $\mathbf{W}_{\mathrm{LCMV}}$ for multiple desired signals is used to generate the training pairs. The input to the RBFN is the correlation between antenna measurements, $\mathbf{R}$, while the output is $\mathbf{W}_{\mathrm{LCMV}}$. The input $\mathbf{R}$ is preprocessed according to case (a) in Example 3.2, and this leads to $2L(L-1)$ input nodes. The output layer consists of $2L$ nodes.

The training data were generated by assuming that the sources were located at elevation angles θ ranging from $-90°$ to $90°$ with increments of $\Delta\theta$ for the one-dimensional case. For a two-dimensional array, in addition to elevation angles θ_i, azimuth angles ϕ_i can be made to range from 0 to $360°$ in order to span the field of view of the antenna. Once the RBFN is trained with a representative set of training pairs, it is ready to function in the performance phase.

For a uniform linear array of ten elements, the results by the RBFN and LCMV methods are compared in Fig. 6.4, for angular signal separations of $\Delta\theta - 15°$ and $10°$, respectively. It can be seen from the figure that the RBFN produces a solution for the beamformer, which is very close to that for the LCMV solution. The RBFN successfully tracks the multiple desired signals and places nulls in the direction of the interfering signals.

6.17.2 Example 6.2: Radial Basis Function Networks Based DoA Estimation

We now revisit Example 3.2. As an illustration, we consider case (b), where the phase differences between the array elements and the reference element are used as the network input. The performance of the RBFN is compared to that of the MLP. The network size is specified as 6-30-1. The training of the MLP is based on the BFGS algorithm[5], and we set the maximum epoch as 1000 and the training accuracy as $\varepsilon = 10^{-4}$. For the training of the RBFN, the centers are clustered using the FCM, the widths are decided by the nearest-neighbor rule, and the weight learning is based on the RLS algorithm with $\mu = 1$. We note that since the preprocessing to original antenna measurements is nonlinear and nonmonotonical, unsupervised clustering generates an unreasonable center distribution and thus, the performance of the training is poor. We conduct an form of supervised clustering, where the output in degrees is

---

[4] The LCMV beamformer is also called the *Wiener beamformer*.

[5] The computation is conducted on a PC with a Pentium III 600 MHz CPU and 128M RAM based on the Matlab© Neural Networks Toolbox.

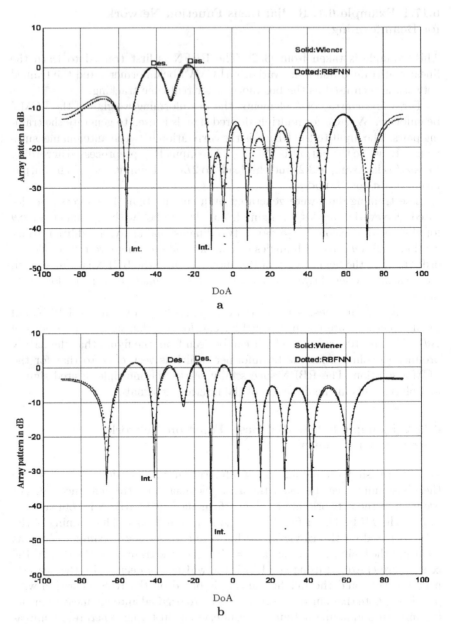

Fig. 6.4. Adaptive beampattern of a uniform linear array with ten elements. (**a**) Two desired and two interferring signals with $\Delta\theta = 15°$. (**b**) Two desired and two interferring signals with $\Delta\theta = 10°$. (From [332].)

first scaled into radians and then treated as an extended dimension of the input. The maximum epoch for the FCM is 100. The sizes of the training set and the testing set are 1000 and 100, respectively. The training and testing errors of the MLP and the RBFN are, respectively, shown in Figs. 6.5a and 6.5b.

The training time for the MLP is 37.41 s and the number of epochs is 108. The training and testing MSEs are, respectively, 0.9948 and 1.2733. The testing time for one pattern is 5×10^{-4} s. If we improve the accuracy of the MLP to $\varepsilon = 10^{-5}$, the training time is 291.33 s at the 1000th epoch with a training performance 1.2435×10^{-5}, and this is a ten-fold improvement. The training and testing MSEs are, respectively, reduced to 0.1243 and 0.08. The simulation result is shown in Fig. 6.5c.

The time for training the RBFN is 16.15 s, where 12.58 s is for the clustering of the centers, 0.06 s for calculating σ_i, and 3.51 s for RLS iteration. The training and testing MSEs are, respectively, 11.0617 and 10.8386. The testing time for each pattern is 0.0017 s.

When we increase the number of hidden units of the RBFN to 90, the simulation result is illustrated in Fig. 6.5d. The training time becomes 51.3 s, where 33.83 s is used for clustering, 0.44 s for calculating σ_i, and 17.03 s for RLS iterations. The training and testing MSEs are, respectively, 7.8578 and 7.7705. The testing time for each pattern is 0.0049s, which is three times that for the 6-30-1 RBFN.

From this example, we can see that the training of the RBFN is much faster than the training of the MLP if the centers are randomly selected from the training set. However, the MLP has better generalization, and is also faster in performing.

Training a single neural network to detect the DoAs of multiple sources requires exhaustive combinations of the source angle separations. It is a prohibitive task for more than three sources. Also, the resulting network size is huge, and the performance delay is intolerable. The coarse-fine strategy is appealing [321].

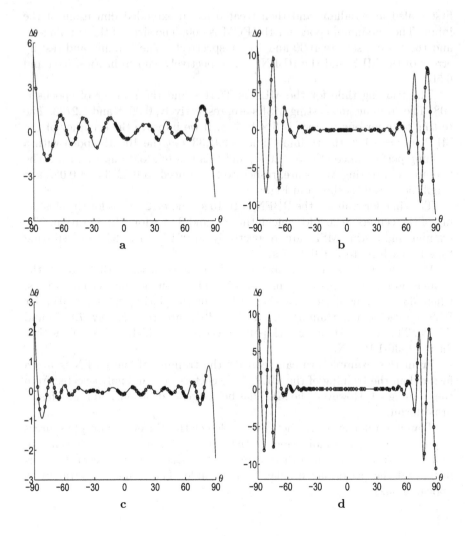

Fig. 6.5. Training and testing errors: A comparison between the MLP and the RBFN. (a) 6-30-1 MLP, $\varepsilon = 10^{-4}$. (b) 6-30-1 RBFN. (c) 6-30-1 MLP, $\varepsilon = 10^{-5}$. (d) 6-90-1 RBFN.

Principal Component Analysis Networks

Most signal-processing problems can be reduced to some form of eigenvalue or singular-value problems. EVD and SVD are usually used for solving these problems. The PCA is a statistical method, which is directly related to the EVD and the SVD. The minor component analysis (MCA) is a variant of the PCA, which is most useful for solving total least squares (TLS) problems. The independent component analysis (ICA) is a generalization of the PCA. The PCA and the ICA are usually used for feature extraction and blind signal separation (BSS).

Neural networks and algorithms for the PCA, MCA, ICA, and SVD are described in this chapter. Algorithms given in this chapter are typical unsupervised learning methods. Some other topics associated with the PCA are also described.

7.1 Stochastic Approximation Theory

The stochastic approximation theory [730], first introduced by Robbins and Monro in 1951, is now an important tool for analyzing stochastic discrete-time systems including the classical gradient-descent method.

Given a stochastic discrete-time system of the form

$$\Delta \mathbf{z}(t) = \mathbf{z}(t+1) - \mathbf{z}(t) = \eta(t) \left(\mathbf{f}(\mathbf{z}, t) + \mathbf{n}(t) \right) \qquad (7.1)$$

where $\mathbf{z}$ is the state vector, $\mathbf{f}(\mathbf{z}, t)$ is a finite nonzero vector with functions as entries, and $\mathbf{n}(t)$ is an unbiased noisy term at a particular instant. The continuous-time representation is very useful for analyzing the asymptotic behavior of the algorithm. According to the stochastic approximation theory, assuming that $\{\eta(t)\}$ is a sequence of positive numbers satisfying the Robbins–Monro conditions [730]

$$\sum_{t=1}^{\infty} \eta(t) = \infty, \qquad \sum_{t=1}^{\infty} \eta^2(t) < \infty \qquad (7.2)$$

the analysis of the stochastic system (7.1) can be transformed into the analysis of a deterministic differential equation

$$\frac{d\mathbf{z}}{dt} = \mathbf{f}(\mathbf{z}, t) \tag{7.3}$$

If all the trajectories of (7.3) converge to a fixed point $\mathbf{z}^*$, then the discrete-time system $\mathbf{z}(t) \to \mathbf{z}^*$ as $t \to \infty$ with probability one.

By (7.2), $\eta(t)$ is always required to approach zero, as $t \to \infty$. $\eta(t)$ is typically selected as $\eta(t) = \frac{1}{\alpha+t}$, where $\alpha \geq 0$ is a constant, or as $\eta(t) = \frac{1}{t^\beta}$, $\frac{1}{2} \leq \beta \leq 1$ [866, 994].

7.2 Hebbian Learning Rule

The classical Hebbian synaptic modification rule was introduced in [473]. Biological synaptic weights change in proportion to the correlation between the pre- and postsynaptic signals. For a single neuron, the Hebbian rule can be written as

$$\mathbf{w}(t+1) = \mathbf{w}(t) + \eta y(t)\mathbf{x}_t \tag{7.4}$$

where the learning rate $\eta > 0$, $\mathbf{w} \in R^n$ is the weight vector, $\mathbf{x}_t \in R^n$ is an input vector at time t, and $y(t)$ is the output of the neuron defined by

$$y(t) = \mathbf{w}^{\mathrm{T}}(t)\mathbf{x}_t \tag{7.5}$$

For a stochastic input vector $\mathbf{x}$, assuming that $\mathbf{x}$ and $\mathbf{w}$ are uncorrelated, the expected weight change is given by

$$E[\Delta \mathbf{w}] = \eta E[y\mathbf{x}] = \eta E\left[\mathbf{x}\mathbf{x}^{\mathrm{T}}\mathbf{w}\right] = \eta \mathbf{C}E[\mathbf{w}] \tag{7.6}$$

where $E[\cdot]$ is the expectation operator, and $\mathbf{C} = E\left[\mathbf{x}\mathbf{x}^{\mathrm{T}}\right]$ is the autocorrelation matrix of $\mathbf{x}$.

At equilibrium, $E[\Delta \mathbf{w}] = \mathbf{0}$, and hence, we have the deterministic equation $\mathbf{C}\mathbf{w} = \mathbf{0}$. Due to the effect of noise terms, $\mathbf{C}$ is a full-rank positive-definite Hermitian matrix with positive eigenvalues λ_i, $i = 1, 2, \ldots, n$, and the corresponding orthogonal eigenvectors $\mathbf{c}_i$, where $n = \mathrm{rank}(\mathbf{C})$. Thus, $\mathbf{w} = \mathbf{0}$ is the only equilibrium state.

Equation (7.4) can be further represented in the continuous-time form

$$\dot{\mathbf{w}} = y\mathbf{x} \tag{7.7}$$

Taking the statistical average, we have

$$E[\dot{\mathbf{w}}] = E[y\mathbf{x}] = \mathbf{C}E[\mathbf{w}] \tag{7.8}$$

This can be derived by minimizing the average instantaneous criterion function [461]

$$E\left[E_{\text{Hebb}}\right] = -\frac{1}{2}E\left[y^2\right] = -\frac{1}{2}E\left[\mathbf{w}^{\mathrm{T}}\right]C E[\mathbf{w}] \tag{7.9}$$

where E_{Hebb} is the instantaneous criterion function. At equilibrium, $E\left[\frac{\partial E_{\text{Hebb}}}{\partial \mathbf{w}}\right]$ $= -CE[\mathbf{w}] = 0$, thus $\mathbf{w} = \mathbf{0}$. Since the Hessian $E[\mathbf{H}(\mathbf{w})] = E\left[\frac{\partial E_{\text{Hebb}}^2}{\partial^2 \mathbf{w}}\right] = -\mathbf{C}$ is nonpositive for all $E[\mathbf{w}]$, the solution $\mathbf{w} = \mathbf{0}$ is unstable, which drives $\mathbf{w}$ to infinite magnitude, with a direction parallel to that of the eigenvector of $\mathbf{C}$ corresponding to the largest eigenvalue [461].

To prevent the divergence of the Hebbian rule, one can normalize $\|\mathbf{w}\|$ to unity after each iteration [1133, 974], and this leads to the normalized Hebbian rule. Other methods such as Oja's rule [864], Yuille's rule [1247], Linsker's rule [715, 716], and Hassoun's rule [461] add a weight-decay term to the Hebbian rule to stabilize the algorithm.

7.3 Oja's Learning Rule

Oja's learning rule introduces a weight decay term into the Hebbian rule [864] and is given by

$$\mathbf{w}(t+1) = \mathbf{w}(t) + \eta y(t)\mathbf{x}_t - \eta y^2(t)\mathbf{w}(t) \tag{7.10}$$

Oja's rule converges to a state that minimizes (7.9) subject to $\|\mathbf{w}\| = 1$. The solution is the principal eigenvector of $\mathbf{C}$ [864]. For small η, Oja's rule is proved to be equivalent to the normalized Hebbian rule [864].

The continuous-time version of Oja's rule is given by a nonlinear stochastic differential equation

$$\dot{\mathbf{w}} = \eta\left(y\mathbf{x} - y^2\mathbf{w}\right) \tag{7.11}$$

The corresponding deterministic equation based on statistical average is thus derived as

$$\dot{\mathbf{w}} = \eta\left[\mathbf{C}\mathbf{w} - \left(\mathbf{w}^{\mathrm{T}}\mathbf{C}\mathbf{w}\right)\mathbf{w}\right] \tag{7.12}$$

At equilibrium,

$$\mathbf{C}\mathbf{w} = \left(\mathbf{w}^{\mathrm{T}}\mathbf{C}\mathbf{w}\right)\mathbf{w} \tag{7.13}$$

It is easily seen that the solutions are $\mathbf{w} = \pm\mathbf{c}_i$, $i = 1, 2, \ldots, n$, whose corresponding eigenvalues λ_i are arranged in a descending order as $\lambda_1 \geq \lambda_2 \geq \cdots \lambda_n \geq 0$.

Note the average Hessian

$$\mathbf{H}(\mathbf{w}) = \frac{\partial}{\partial \mathbf{w}}\left[-\mathbf{C}\mathbf{w} + \left(\mathbf{w}^{\mathrm{T}}\mathbf{C}\mathbf{w}\right)\mathbf{w}\right]$$
$$= -\mathbf{C} + \mathbf{w}^{\mathrm{T}}\mathbf{C}\mathbf{w}\mathbf{I} + 2\mathbf{w}\mathbf{w}^{\mathrm{T}}\mathbf{C} \tag{7.14}$$

is positive-definite only at $\mathbf{w} = \pm\mathbf{c}_1$, where $\mathbf{I}$ is an $n \times n$ identity matrix, if $\lambda_1 \neq \lambda_2$ [461]. This can be seen from

$$\mathbf{H}\left(\mathbf{c}_i\right)\mathbf{c}_j = (\lambda_i - \lambda_j)\mathbf{c}_j + 2\lambda_j \mathbf{c}_i \mathbf{c}_i^{\mathrm{T}} \mathbf{c}_j$$

$$= \begin{cases} 2\lambda_i \mathbf{c}_i, & i = j \\ (\lambda_i - \lambda_j)\,\mathbf{c}_j, & i \neq j \end{cases} \tag{7.15}$$

Thus, Oja's rule always converges to the principal component of $\mathbf{C}$.

We would like to make a note here. The convergence analysis of the stochastic discrete-time algorithms such as the gradient-descent method is conventionally based on the stochastic approximation theory [730]. A stochastic discrete-time algorithm is first converted into deterministic continuous-time ODEs, and then analyzed by using Lyapunov's second theorem. This conversion is based on the Robbins–Monro conditions [730], which require the learning rate to gradually approach zero as $t \to \infty$. This limitation is not practical for implementation, especially for the learning of nonstationary data. Recently, Zufiria [1268] has proposed to convert the stochastic discrete-time algorithms into their deterministic discrete-time formulations that characterize their average evolution from a conditional expectation perspective. This method has been applied to Oja's rule. The dynamics have been analyzed, and chaotic behavior has been observed in some invariant subspaces. Analysis based on this method guarantees the convergence of the Oja's rule by selecting some constant learning rate. It has been further proved in [1232] that Oja's rule almost always converges exponentially to the unit eigenvector associated with the largest eigenvalue of $\mathbf{C}$, starting from points in an invariant set. The initial vectors have been suggested to be selected from the domain of a unit hypersphere to guarantee convergence. A constant learning rate for fast convergence has also been suggested as $\eta = \frac{0.618}{\lambda_1}$ [1232].

7.4 Principal Component Analysis

The PCA is a well-known statistical technique. It is widely used in engineering and scientific disciplines, such as pattern recognition, data compression and coding, image processing, high-resolution spectrum analysis, and adaptive beamforming. The PCA is based on the spectral analysis of the second-order moment matrix that statistically characterizes a random vector. In the zero-mean case, this matrix, called the *correlation matrix*, becomes the covariance matrix. In the area of image coding, the PCA is known by the name of the *Karhunen–Loeve transform (KLT)* [733], which is an optimal scheme for data compression based on the exploitation of correlation between neighboring pixels or groups of pixels. The PCA is directly related to the SVD, and the most common way to perform the PCA is via the SVD of the data matrix. However, the capability of the SVD is limited for very large data sets.

As discussed in Chapter 2, preprocessing usually maps a high-dimensional space to a low-dimensional space with minimum information loss. The process is known as *feature extraction*. The PCA is a well-known feature-extraction method. The PCA allows the removal of the second-order correlation among given random processes. By calculating the eigenvectors of the covariance matrix of the input vector, the PCA linearly transforms a high-dimensional input vector into a low-dimensional one whose components are uncorrelated.

The PCA is often based on optimization of some information criterion, such as the maximization of the variance of the projected data or the minimization of the reconstruction error. The aim of the PCA is to extract m orthonormal directions $\bar{\bar{\mathbf{w}}}_i \in R^n$, $i = 1, 2, \ldots, m$, in the input space that account for as much of the data's variance as possible. Subsequently, an input vector $\mathbf{x} \in R^n$ may be transformed into a lower m-dimensional space without losing essential intrinsic information. The vector $\mathbf{x}$ can be represented by being projected onto the m-dimensional subspace spanned by $\bar{\bar{\mathbf{w}}}_i$ using the inner products $\mathbf{x}^T \bar{\bar{\mathbf{w}}}_i$. This achieves dimensionality reduction.

The PCA finds those unit directions $\bar{\bar{\mathbf{w}}} \in R^n$ along which the projections of the input vectors, known as the *principal components (PCs)*, $y = \mathbf{x}^T \bar{\bar{\mathbf{w}}}$, have the largest variance

$$E_{\text{PCA}}(\mathbf{w}) = \mathrm{E}\left[y^2\right] = \bar{\bar{\mathbf{w}}}^T \mathbf{C} \bar{\bar{\mathbf{w}}} = \frac{\mathbf{w}^T \mathbf{C} \mathbf{w}}{\|\mathbf{w}\|^2} \qquad (7.16)$$

where $\bar{\bar{\mathbf{w}}} = \frac{\mathbf{w}}{\|\mathbf{w}\|}$. $E_{\text{PCA}}(\mathbf{w})$ is a positive-semidefinite function. Setting $\frac{\partial E_{\text{PCA}}}{\partial \mathbf{w}} = \mathbf{0}$, we get

$$\mathbf{C}\mathbf{w} = \frac{(\mathbf{w}^T \mathbf{C} \mathbf{w})}{\|\mathbf{w}\|^2} \mathbf{w} \qquad (7.17)$$

It can be verified that the solutions to (7.17) are $\mathbf{w} = \alpha \mathbf{c}_i$, $i = 1, 2, \ldots, n$, where $\alpha \in R$. When $\alpha = 1$, $\mathbf{w}$ becomes a unit vector.

We now examine the positive-definiteness of the Hessian of $E_{\text{PCA}}(\mathbf{w})$ at $\mathbf{w} = \mathbf{c}_i$. Multiplying the Hessian by $\mathbf{c}_j$ leads to [461]

$$\mathbf{H}(\mathbf{c}_i)\mathbf{c}_j = \begin{cases} \mathbf{0} & i = j \\ (\lambda_i - \lambda_j)\mathbf{c}_j & i \neq j \end{cases} \qquad (7.18)$$

Thus, $\mathbf{H}(\mathbf{w})$ has the same eigenvectors as $\mathbf{C}$ but with different eigenvalues. $\mathbf{H}(\mathbf{w})$ is positive-semidefinite only when $\mathbf{w} = \mathbf{c}_1$. As a result, $\mathbf{w}$ will eventually point in the direction of $\mathbf{c}_1$ and $E_{\text{PCA}}(\mathbf{w})$ takes its maximum value.

By repeating maximization of $E_{\text{PCA}}(\mathbf{w})$ but limiting $\mathbf{w}$ orthogonal to $\mathbf{c}_1$, the maximum of $E_{\text{PCA}}(\mathbf{w})$ is equal to λ_2 at $\mathbf{w} = \alpha \mathbf{c}_2$. Following this deflation procedure, all the m principal directions $\bar{\bar{\mathbf{w}}}_i$ can be derived [461]. The projection $y_i = \mathbf{x}^T \bar{\bar{\mathbf{w}}}_i$, $i = 1, 2, \ldots, m$, are the PCs of $\mathbf{x}$. This linear dimensionality reduction procedure is the KLT [733], and the result for two-dimensional input data is illustrated in Fig. 7.1.

A linear LS estimate $\hat{\mathbf{x}}$ can be constructed for the original input $\mathbf{x}$

$$\hat{\mathbf{x}} = \sum_{i=1}^{m} y_i \bar{\bar{\mathbf{w}}}_i \qquad (7.19)$$

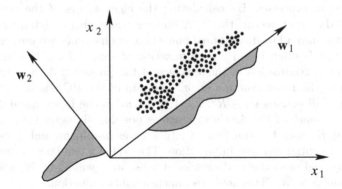

Fig. 7.1. Illustration of the PCA in two dimensions. Each data point is accurately characterized by its projections on the two principal directions $\overline{\overline{\mathbf{w}}}_1$ and $\overline{\overline{\mathbf{w}}}_2$, where $\overline{\overline{\mathbf{w}}}_1 = \mathbf{w}_1/\|\mathbf{w}_1\|$ and $\overline{\overline{\mathbf{w}}}_2 = \mathbf{w}_2/\|\mathbf{w}_2\|$. If the data is compressed to the one-dimensional space, each data point is then represented by its projection on the eigenvector $\overline{\overline{\mathbf{w}}}_1$.

The process can be treated as data reconstruction. The reconstruction error $\mathbf{e}$ is the difference between the original and reconstructed data

$$\mathbf{e} = \mathbf{x} - \hat{\mathbf{x}} = \sum_{i=m+1}^{n} y_i \overline{\overline{\mathbf{w}}}_i \tag{7.20}$$

Naturally, $\mathbf{e}$ is orthogonal to $\hat{\mathbf{x}}$. Each principal component y_i is a Gaussian with zero mean and variance $\sigma_i^2 = \lambda_i$. The variances of $\mathbf{x}$, $\hat{\mathbf{x}}$, and $\mathbf{e}$ can be, respectively, expressed as

$$\mathrm{E}\left[\|\mathbf{x}\|^2\right] = \sum_{i=1}^{n} \sigma_i^2 = \sum_{i=1}^{n} \lambda_i \tag{7.21}$$

$$\mathrm{E}\left[\|\hat{\mathbf{x}}\|^2\right] = \sum_{i=1}^{m} \sigma_i^2 = \sum_{i=1}^{m} \lambda_i \tag{7.22}$$

$$\mathrm{E}\left[\|\mathbf{e}\|^2\right] = \sum_{i=m+1}^{n} \sigma_i^2 = \sum_{i=m+1}^{n} \lambda_i \tag{7.23}$$

When we use only the first m_1 among the extracted m PCs to represent the raw data, we need to evaluate the error by replacing m by m_1.

7.5 Hebbian Rule-based Principal Component Analysis

Neural PCA originates from the seminal work by Oja [864]. Oja's single-neuron PCA model is illustrated in Fig. 7.2 [864]. The output of the neuron is updated by

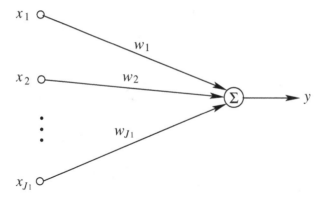

Fig. 7.2. The single-neuron PCA model extracts the first PC of **C**

$$y = \mathbf{w}^{\mathrm{T}}\mathbf{x} \tag{7.24}$$

where $\mathbf{w} = (w_1, \cdots, w_{J_1})^{\mathrm{T}}$. Note that the activation function is the linear function $\phi(x) = x$. The PCA turns out to be closely related to the Hebbian rule. We describe various linear PCA networks and algorithms in Sects. 7.5 to 7.8.

The PCA algorithms discussed in this section are based on the Hebbian rule [473]. The network model was first proposed by Oja [866], where a J_1-J_2 FNN is used to extract the first J_2 PCs. The architecture of the PCA network is shown in Fig. 7.3, which is a simple expansion of the single-neuron PCA model. The output of the network is given by

$$\mathbf{y} = \mathbf{W}^{\mathrm{T}}\mathbf{x} \tag{7.25}$$

where $\mathbf{y} = (y_1, y_2, \cdots, y_{J_2})^{\mathrm{T}}$, $\mathbf{x} = (x_1, x_2, \cdots, x_{J_1})^{\mathrm{T}}$, $\mathbf{W} = [\mathbf{w}_1 \ \mathbf{w}_2 \ \cdots \ \mathbf{w}_{J_2}]$, and $\mathbf{w}_i = (w_{1i}, w_{2i}, \cdots, w_{J_1i})^{\mathrm{T}}$.

7.5.1 Subspace Learning Algorithms

By using Oja's learning rule (7.10), $\mathbf{w}$ will converge to a unit eigenvector of the correlation matrix $\mathbf{C}$, and the variance of the output y is maximized. For zero-mean input data, this extracts the first PC [864, 247]. We rewrite (7.10) here for the convenience of presentation

$$\mathbf{w}(t+1) = \mathbf{w}(t) + \eta \left[y(t)\mathbf{x}_t - y^2(t)\mathbf{w}(t)\right] \tag{7.26}$$

where the term $y(t)\mathbf{x}_t$ is the Hebbian term, and $-y^2(t)\mathbf{w}(t)$ is a decaying term, which is used to prevent instability. In order to keep the algorithm convergent, $0 < \eta(t) < \frac{1}{1.2\lambda_1}$ is required [866], where λ_1 is the largest eigenvalue of $\mathbf{C}$. If $\eta(t) \geq \frac{1}{\lambda_1}$, $\mathbf{w}$ will not converge to $\pm\mathbf{c}_1$ even if it is initially close to the target [202]. One can select $\eta(t) = 0.5 \left[\mathbf{x}_t^{\mathrm{T}}\mathbf{x}_t\right]$ at the beginning and gradually decrease η [866].

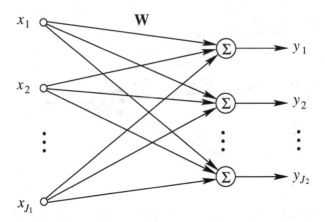

Fig. 7.3. Architecture of the PCA network

Symmetrical Subspace Learning Algorithm

Oja proposed a learning algorithm for the PCA network, referred to as the symmetrical *subspace learning algorithm (SLA)* [866]. The SLA can be derived by maximizing

$$E_{\mathrm{SLA}} = \frac{1}{2}\mathrm{tr}\left(\mathbf{W}^{\mathrm{T}}\mathbf{R}\mathbf{W}\right) \qquad (7.27)$$

subject to

$$\mathbf{W}^{\mathrm{T}}\mathbf{W} = \mathbf{I} \qquad (7.28)$$

where $\mathbf{I}$ is the $J_2 \times J_2$ identity matrix.

The SLA is given as follows [866]

$$\mathbf{w}_i(t+1) = \mathbf{w}_i(t) + \eta(t)y_i(t)\left[\mathbf{x}_t - \hat{\mathbf{x}}_t\right] \qquad (7.29)$$

$$\hat{\mathbf{x}} = \mathbf{W}\mathbf{y} \qquad (7.30)$$

After the algorithm converges, $\mathbf{W}$ is roughly orthonormal and the columns of $\mathbf{W}$, namely, $\mathbf{w}_i$, $i = 1, \cdots, J_2$, converge to some linear combination of the first J_2 principal eigenvectors of $\mathbf{C}$ [866, 635], which is a rotated basis of the dominant eigenvector subspace. This analysis is called the *principal subspace analysis (PSA)*. The value of $\mathbf{w}_i$ is dependent on the initial conditions and training samples.

The corresponding eigenvalues λ_i, $i = 1, \cdots, J_2$, approximate $E\left[y_i^2\right]$, which can be adaptively estimated by

$$\hat{\lambda}_i(t+1) = \left(1 - \frac{1}{t+1}\right)\hat{\lambda}_i(t) + \frac{1}{t+1}y_i^2(t+1) \qquad (7.31)$$

The PCA performs optimally when there is no noise process involved. The SLA has been extended in [1164] so as to extract a noise robust projection when PCA projection is disturbed by noise.

Weighted Subspace Learning Algorithm

The weighted SLA can be derived by maximizing the same criterion (7.27), but the constraint (7.28) can be modified as [868]

$$\mathbf{W}^{\mathrm{T}}\mathbf{W} = \boldsymbol{\alpha} \tag{7.32}$$

where $\boldsymbol{\alpha} = \mathrm{diag}\,(\alpha_1, \alpha_2, \cdots, \alpha_{J_2})$, is an arbitrary diagonal matrix with $\alpha_1 > \alpha_2 > \cdots > \alpha_{J_2} > 0$.

The weighted SLA is given by [865, 868]

$$\mathbf{w}_i(t+1) = \mathbf{w}_i(t) + \eta(t)y_i(t)\left[\mathbf{x}_t - \gamma_i \hat{\mathbf{x}}_t\right] \tag{7.33}$$

$$\hat{\mathbf{x}} = \mathbf{W}\mathbf{y} \tag{7.34}$$

for $i = 1, \cdots, J_2$, where γ_i, $i = 1, \cdots, J_2$, are any coefficients that satisfy $0 < \gamma_1 < \gamma_2 < \cdots < \gamma_{J_2}$. Due to the asymmetry introduced by γ_i, $\mathbf{w}_i$ almost surely converges to the eigenvectors of $\mathbf{C}$. The weighted subspace algorithm can perform the PCA, however, norms of the weight vectors are not equal to unity.

The subspace and weighted subspace algorithms are nonlocal algorithms that rely on the calculation of the errors and the backward propagation of the values between the layers.

Converting PSA into PCA

By adding one more term to the PSA algorithm, a PCA algorithm can be obtained [545]. This additional term rotates the basis vectors in the principal subspace toward the principal eigenvectors. The PCA derived from the SLA [866] is given as [545]

$$\mathbf{w}_i(t+1) = \mathbf{w}_i(t) + \eta(t)y_i(t)\left[\mathbf{x}_t - \hat{\mathbf{x}}_t\right] \tag{7.35}$$

$$+\eta(t)\rho_i\left(y_i(t)\mathbf{x}_t - \mathbf{w}_i(t)y_i^2(t)\right) \tag{7.36}$$

where $1 > |\rho_1| > |\rho_2| > \cdots > |\rho_{J_2}|$. This PCA algorithm generates weight vectors of unit length.

The adaptive learning algorithm (ALA) [202] is a PCA algorithm based on the SLA. In the ALA, each neuron adaptively updates its learning rate by

$$\eta_i(t) = \frac{\beta_i(t)}{\hat{\lambda}_i(t)} \tag{7.37}$$

where $\hat{\lambda}_i(t)$ is the estimated eigenvalue, which can be estimated using (7.31), $\beta_i(t)$ is set to be smaller than $2(\sqrt{2}-1)$ and decreases to zero as $t \to \infty$. If $\beta_i(t)$ is the same for all i, $\mathbf{w}_i(t)$ will quickly converge, at nearly the same rate, to $\mathbf{c}_i$ for all i in the order of descending eigenvalues. The ALA converges to the desired target both in the large eigenvalue case as well as in the small eigenvalue case. The performance is better than that of the generalized Hebbian algorithm (GHA) [994].

7.5.2 Generalized Hebbian Algorithm

By combining Oja's rule and the GSO procedure, Sanger proposed the GHA for extracting the first J_2 PCs [994]. The GHA can extract the first J_2 eigenvectors in the order of decreasing eigenvalues.

The GHA is given by [994]

$$\mathbf{w}_i(t+1) = \mathbf{w}_i(t) + \eta_i(t)y_i(t)\left[\mathbf{x}_t - \hat{\mathbf{x}}_i(t)\right] \tag{7.38}$$

$$\hat{\mathbf{x}}_i(t) = \sum_{j=1}^{i} \mathbf{w}_j(t)y_j(t) \tag{7.39}$$

for $i = 1, 2, \ldots, J_2$. The GHA becomes a local algorithm by solving the summation term in (7.39) in a recursive form

$$\hat{\mathbf{x}}_i(t) = \hat{\mathbf{x}}_{i-1}(t) + \mathbf{w}_i(t)y_i(t) \tag{7.40}$$

for $i = 1, \cdots, J_2$, where $\hat{\mathbf{x}}_0(t) = 0$. $\eta_i(t)$ is usually selected as the same for all neurons. When $\eta_i = \eta$ for all i, the algorithm can be written in a matrix form

$$\mathbf{W}(t+1) = \mathbf{W}(t) - \eta\mathbf{W}(t)\text{LT}\left[\mathbf{y}(t)\mathbf{y}^{\text{T}}(t)\right] + \eta\mathbf{x}_t\mathbf{y}^{\text{T}}(t) \tag{7.41}$$

where the operator LT[·] selects the lower triangle of the matrix contained within. In the GHA, the mth neuron converges to the mth PC, and all the neurons tend to converge together. $\mathbf{w}_i$ and $\text{E}\left[y_i^2\right]$ approach $\mathbf{c}_i$ and λ_i, respectively, as $t \to \infty$.

Both the SLA [866, 865] and the GHA [994] algorithms employ implicit or explicit GSO to decorrelate the connection weights from one another. The weighted subspace algorithm [865] performs well for extracting less-dominant components.

7.5.3 Other Hebbian Rule-based Algorithms

In addition to the popular SLA, weighted SLA and GHA, there are also some other Hebbian rule-based PCA algorithms such as the local LEAP[1] [198], the nonlocal dot-product-decorrelation (DPD) rule [905], and the local invariant-norm PCA [739].

The LEAP algorithm

The LEAP algorithm [198] is another local PCA algorithm for extracting all the J_2 PCs and their corresponding eigenvectors. The LEAP is given by

$$\mathbf{w}_i(t+1) = \mathbf{w}_i(t) + \eta\left\{\mathbf{B}_i(t)y_i(t)\left[\mathbf{x}_t - \hat{\mathbf{x}}_i(t)\right] - \mathbf{A}_i(t)\mathbf{w}_i(t)\right\} \tag{7.42}$$

---

[1] Learning machine for adaptive feature extraction via principal component analysis.

$$\hat{\mathbf{x}}_i(t) = \mathbf{w}_i(t)y_i(t) \tag{7.43}$$

for $i = 1, \cdots, J_2$, where η is the learning rate, $y_i(t)\mathbf{x}_t$ is a Hebbian term, and

$$\mathbf{A}_i(t) = \begin{cases} \mathbf{0}, & i = 1 \\ \mathbf{A}_{i-1}(t) + \mathbf{w}_{i-1}(t)\mathbf{w}_{i-1}^{\mathrm{T}}(t), & i = 2, \cdots, J_2 \end{cases} \tag{7.44}$$

$$\mathbf{B}_i(t) = \mathbf{I} - \mathbf{A}_i(t), \quad i = 1, \cdots, J_2 \tag{7.45}$$

The $J_1 \times J_1$ matrices $\mathbf{A}_i$ and $\mathbf{B}_i$ are important decorrelating terms for performing the GSO among all weights at each iteration.

Unlike the SLA [866] and GHA [994] algorithms, whose stability analyses are based on the stochastic approximation theory [730], the stability analysis of the LEAP algorithm is based on Lyapunov's first theorem, and η can be selected as a small positive constant. Due to the use of a constant learning rate, the LEAP is capable of tracking nonstationary processes. The LEAP can satisfactorily extract PCs even for ill-conditioned autocorrelation matrices [198].

The Dot-product-decorrelation Algorithm

The DPD algorithm is a nonlocal PCA algorithm [905]. The algorithm moves $\mathbf{w}_i$, $i = 1, \cdots, J_2$, towards the J_2 principal eigenvectors $\mathbf{c}_i$, ordered arbitrarily

$$\mathbf{w}_i(t+1) = \mathbf{w}_i(t) + \eta(t) \left[\mathbf{x}_t y_i(t) - \left(\sum_{j=1}^{J_2} \mathbf{w}_j(t)\mathbf{w}_j^{\mathrm{T}}(t) \right) \frac{\mathbf{w}_i(t)}{\|\mathbf{w}_i(t)\|} \right] \tag{7.46}$$

where $\eta(t)$ satisfies the Robbins–Monro conditions (7.2). The algorithm induces the norms of the weight vectors towards the corresponding eigenvalues, that is,

$$\|\mathbf{w}_i(t)\| \to \lambda_i(t) \tag{7.47}$$

as $t \to \infty$.

A recursive approximation of $r_i(t) = \frac{1}{\|\mathbf{w}_i(t)\|}$ is given for VLSI implementation, which results in only a slightly slower convergence

$$r_i(t+1) = r_i(t) + \eta(t) \left[1 - y_i^2(t)r_i^3(t) \right] \tag{7.48}$$

The network breaks the symmetry in its learning process by the difference in the norms of the weight vectors while keeping the symmetry in its structure. The algorithm is as fast as the GHA [994], weighted SLA [865], and least mean squared error reconstruction (LMSER) [1203] algorithms.

7.6 Least Mean Squared Error-based Principal Component Analysis

Existing PCA algorithms including the Hebbian rule-based algorithms can be derived by optimizing an objective function using the gradient-descent method. The least mean squared error (LMSE)-based methods are derived from the modified MSE function

$$E(\mathbf{W}) = \sum_{t_1=1}^{t} \mu^{t-t_1} \left\| \mathbf{x}_{t_1} - \mathbf{W}\mathbf{W}^{\mathrm{T}}\mathbf{x}_{t_1} \right\|^2 \tag{7.49}$$

where $0 < \mu \leq 1$ is a forgetting factor used for nonstationary observation sequences, and t is the current instant. Many adaptive PCA algorithms actually optimize (7.49) by using the gradient-descent method [1203, 1217] and the RLS method [64, 1217, 799, 876, 877].

The gradient descent or Hebbian rule-based algorithms are highly sensitive to parameters such as η. It is difficult to choose proper parameters guaranteeing both a small misadjustment and a fast convergence. To overcome these drawbacks, applying the RLS to the minimization of (7.49) yields RLS-based algorithms such as the adaptive principal components extraction (APEX) [642], the Kalman-type RLS [64], the projection approximation subspace tracking (PAST) [1217], the PAST with deflation (PASTd) [1217], and the robust RLS algorithm (RRLSA) [876].

All RLS-based PCA algorithms exhibit fast convergence and high tracking accuracy, and are suitable for slowly varying nonstationary vector stochastic processes. All these algorithms correspond to a three-layer J_1-J_2-J_1 linear autoassociative network model, and they can extract all the J_2 PCs in the descending order of the eigenvalues, where a GSO-like orthonormalization procedure is used.

In [688], a regularization term $\mu^t \overline{\mathbf{w}}^{\mathrm{T}} \mathbf{P}_0^{-1} \overline{\mathbf{w}}$ is added to (7.49), where $\overline{\mathbf{w}}$ is a stack vector of $\mathbf{W}$ and $\mathbf{P}_0$ is a diagonal matrix with dimension $J_1 J_2 \times J_1 J_2$. As t is sufficiently large, this term is negligible. This term ensures that the entries of $\mathbf{W}$ do not become too large. Without this term, some matrices in the recursive updating equations may become indefinite. Two PCA algorithms called the *Gauss–Seidel recursive PCA* and *Jacobi recursive PCA* are derived in [688].

7.6.1 The Least Mean Square Error Reconstruction Algorithm

The LMSER algorithm is derived on the MSE criterion using the gradient-descent method [1203]. The LMSER algorithm can be written as

$$\mathbf{w}_i(t+1) = \mathbf{w}_i(t) + \eta(t)\Big\{ 2\mathbf{A}(t) - \mathbf{C}_i(t)\mathbf{A}(t) - \mathbf{A}(t)\mathbf{C}_i(t)$$

$$- \gamma \left[\mathbf{B}_i(t)\mathbf{A}(t) + \mathbf{A}(t)\mathbf{B}_i(t) \right] \Big\} \mathbf{w}_i(t) \tag{7.50}$$

for $i = 1, \cdots, J_2$, where

$$\mathbf{A}(t) = \mathbf{x}_t \mathbf{x}_t^{\mathrm{T}} \tag{7.51}$$

$$\mathbf{C}_i(t) = \mathbf{w}_i(t) \mathbf{w}_i^{\mathrm{T}}(t), \ i = 1, \cdots, J_2 \tag{7.52}$$

$$\mathbf{B}_i(t) = \sum_{j=1}^{i-1} \mathbf{C}_j(t) = \mathbf{B}_{i-1}(t) + \mathbf{C}_{i-1}(t), \ i = 2, \cdots, J_2 \tag{7.53}$$

and $\mathbf{B}_1(t) = \mathbf{0}$. The selection of $\eta(t)$ is based on the Robbins–Monro conditions (7.2) [730] and $\gamma \geq 1$.

The LMSER reduces to Oja's algorithm when $\mathbf{W}(t)$ is orthonormal, namely, $\mathbf{W}^{\mathrm{T}}(t)\mathbf{W}(t) = \mathbf{I}$. In this sense, Oja's algorithm can be treated as an approximate stochastic gradient rule to minimize the MSE.

The LMSER [1203] has been compared with the weighted SLA [865] and the GHA [994] in [191]. The learning rates for all the algorithms are selected as $\eta(t) = \frac{\delta}{t^\alpha}$, where $\delta > 0$ and $\frac{1}{2} < \alpha \leq 1$. A tradeoff is obtained: Increasing the values of γ and δ results in a larger asymptotic MSE but faster convergence and *vice versa*, namely, the stability–speed problem. The LMSER [1203] uses nearly twice as much computation as the weighted SLA [865] and the GHA [994], for each update of the weight. However, it leads to a smaller asymptotic MSE and faster convergence for the minor eigenvectors [191].

7.6.2 The PASTd Algorithm

The PASTd [1217] is a well-known subspace tracking algorithm updating the signal eigenvectors and eigenvalues. The PASTd is based on the PAST. Both the PAST and the PASTd are derived for complex-valued signals, which are common in the signal-processing area. At iteration t, the PASTd algorithm is given as [1217]

$$y_i(t) = \mathbf{w}_i^{\mathrm{H}}(t-1)\mathbf{x}_i(t) \tag{7.54}$$

$$\delta_i(t) = \mu \delta_i(t-1) + |y_i(t)|^2 \tag{7.55}$$

$$\hat{\mathbf{x}}_i(t) = \mathbf{w}_i(t-1)y_i(t) \tag{7.56}$$

$$\mathbf{w}_i(t) = \mathbf{w}_i(t-1) + [\mathbf{x}_i(t) - \hat{\mathbf{x}}_i(t)]\frac{y_i^*(t)}{\delta_i(t)} \tag{7.57}$$

$$\mathbf{x}_{i+1}(t) = \mathbf{x}_i(t) - \mathbf{w}_i(t)y_i(t) \tag{7.58}$$

for $i = 1, \cdots, J_2$, where $\mathbf{x}_1(t) = \mathbf{x}_t$, and the superscript $*$ denotes the conjugate operator.

$\mathbf{w}_i(0)$ and $\delta_i(0)$ should be suitably selected. $\mathbf{W}(0)$ should contain J_2 orthonormal vectors, which can be calculated from an initial block of data or from arbitrary initial data. A simple way is to set $\mathbf{W}(0)$ to the J_2 leading unit vectors of the $J_1 \times J_1$ identity matrix. $\delta_i(0)$ can be set as unity. The choice of these initial values affects the transient behavior, but not the steady-state performance of the algorithm. $\mathbf{w}_i(t)$ provides an estimate of the ith eigenvector, and $\delta_i(t)$ is an exponentially weighted estimate of the corresponding eigenvalue.

Both the PAST and the PASTd have linear computational complexity, that is, $O\left(J_1 J_2\right)$ operations every update, as in the cases of the SLA [864], the GHA [994], the LMSER [1203], and the novel information criterion (NIC) algorithm [799]. The PAST computes an arbitrary basis of the signal subspace, while the PASTd is able to update the signal eigenvectors and eigenvalues. Both the algorithms produce nearly orthonormal, but not exactly orthonormal, subspace basis or eigenvector estimates. If perfectly orthonormal eigenvector estimates are required, an orthonormalization procedure is necessary.

The Kalman-type RLS [64] combines the basic RLS algorithm with the GSO procedure in a manner similar to that of the GHA. The Kalman-type RLS and the PASTd are exactly identical if the inverse of the covariance of the output of the ith neuron, $P_i(t)$, in the Kalman-type RLSA is set as $\frac{1}{\delta_i(t)}$ in the PASTd.

In the one-unit case, both the PAST and PASTd are identical to Oja's learning rule [864] except that the PAST and the PASTd have a self-tuning learning rate $\frac{1}{\delta_1(t)}$. Both the PAST and the PASTd provide much more robust estimates than the EVD, and converge much faster than the SLA [864]. The PASTd has been extended for the tracking of both the rank and the subspace by using information theoretic criteria such as the AIC and the MDL [1218].

7.6.3 The Robust RLS Algorithm

The RRLSA [876] is more robust than the PASTd [1217]. The RRLSA can be implemented in a sequential or parallel form. Given the ith neuron, the sequential algorithm is given for all the patterns as [876]

$$\overline{\overline{\mathbf{w}}}_i(t-1) = \frac{\mathbf{w}_i(t-1)}{\|\mathbf{w}_i(t-1)\|} \tag{7.59}$$

$$y_i(t) = \overline{\overline{\mathbf{w}}}_i^{\mathrm{T}}(t-1)\mathbf{x}_t \tag{7.60}$$

$$\hat{\mathbf{x}}_i(t) = \sum_{j=1}^{i-1} y_j(t)\overline{\overline{\mathbf{w}}}_j(t-1) \tag{7.61}$$

$$\mathbf{w}_i(t) = \mu\mathbf{w}_i(t-1) + [\mathbf{x}_t - \hat{\mathbf{x}}_i(t)]\, y_i(t) \tag{7.62}$$

$$\hat{\lambda}_i(t) = \frac{\|\mathbf{w}_i(t)\|}{t} \tag{7.63}$$

for $i = 1, \cdots, J_2$, where y_i is the output of the ith hidden unit, and $\mathbf{w}_i(0)$ is initialized as a small random value. By changing (7.61) into a recursive form, the RRLSA becomes a local algorithm.

The RRLSA has the same flexibility as the Kalman-type RLS [64], the PASTd [1217], and the APEX [642], in that increasing the number of neurons does not affect the previously extracted principal components. The RRLSA naturally selects the inverse of the output energy to be the adaptive learning rate for the Hebbian rule. The Hebbian and Oja rules are closely related to the RRLSA algorithm by suitable selection of the learning rates [876].

The RRLSA [876] is also robust to the error accumulation from the previous components, which exists in the sequential PCA algorithms like the Kalman-type RLS [64] and the PASTd [1217]. The RRLSA converges rapidly, even if the eigenvalues extend over several orders of magnitude. According to the empirical results [876], the RRLSA provides the best performance in terms of the convergence speed as well as the steady-state error, whereas the Kalman-type RLS and the PASTd have similar performance, which is inferior to that of the RRLSA, and the ALA [202] exhibits the poorest performance.

7.7 Other Optimization-based Principal Component Analysis

The PCA can be derived by many optimization methods based on a properly defined objective function. This leads to many other algorithms, including gradient descent-based algorithms [193, 715, 716, 898, 1247], the CG method [384], and the quasi-Newton method [574, 878]. The gradient-descent method usually converges to a local minimum. Some adaptive algorithms derived from the gradient-descent, conjugate direction, and Newton–Raphson methods, whose simulation results are better than that of the gradient-descent method [1203], have also been proposed in [192]. Second-order algorithms such as the CG [384] and quasi-Newton [574] methods typically converge much faster than first-order methods, but have a computational complexity of $O\left(J_1^2 J_2\right)$ per iteration.

The infomax principle [715, 716] was first proposed by Linsker to describe a neural-network algorithm. The principal subspace is derived by maximizing the mutual information criterion. Other examples of information-criterion-based algorithms are the NIC algorithm [799] and the coupled PCA [814].

7.7.1 Novel Information Criterion Algorithm

The NIC algorithm [799] is obtained by applying the gradient-descent method to maximize the NIC. The NIC is a cost function very similar to the mutual information criterion [921, 716], but integrates a soft constraint on the weight orthogonalization

$$E_{\text{NIC}} = \frac{1}{2} \left\{ \ln \left(\det \left(\mathbf{W}^{\text{T}} \mathbf{R} \mathbf{W} \right) \right) - \text{tr} \left(\mathbf{W}^{\text{T}} \mathbf{W} \right) \right\} \qquad (7.64)$$

Unlike the MSE, the NIC has a steep landscape along the trajectory from a small weight matrix to the optimum one. E_{NIC} has a single global maximum, and all the other stationary points are unstable saddle points. At the global maximum

$$E^*_{\text{NIC}} = \frac{1}{2} \left(\sum_{i=1}^{J_2} \ln \lambda_i - J_2 \right) \tag{7.65}$$

while $\mathbf{W}$ yields an arbitrary orthonormal basis of the principal subspace.

The NIC algorithm is derived from E_{NIC} by using the gradient-descent method

$$\mathbf{W}(t+1) = (1-\eta)\mathbf{W}(t) + \eta\widehat{\mathbf{C}}(t+1)\mathbf{W}(t)\left[\mathbf{W}^{\text{T}}(t)\widehat{\mathbf{C}}(t+1)\mathbf{W}(t)\right]^{-1} \tag{7.66}$$

where $\widehat{\mathbf{C}}(t)$ is the estimate of the covariance matrix $\mathbf{C}(t)$

$$\widehat{\mathbf{C}}(t) = \frac{1}{t}\sum_{i=1}^{t} \mu^{t-i}\mathbf{x}_i\mathbf{x}_i^{\text{T}} = \mu\frac{t-1}{t}\widehat{\mathbf{C}}(t-1) + \frac{1}{t}\mathbf{x}_t\mathbf{x}_t^{\text{T}} \tag{7.67}$$

and $\mu \in (0,1]$ is the forgetting factor. The NIC algorithm has a computational complexity of $O\left(J_1^2 J_2\right)$ for each iteration.

Like the PAST algorithm [1217], the NIC algorithm is a PSA method. It can extract the principal eigenvectors when the deflation technique is incorporated. The NIC algorithm converges much faster than the SLA [865] and the LMSER [1203], and is able to globally converge to the PSA solution from almost any weight initialization. Reorthormalization can be applied so as to perform true PCA [1217, 799].

By selecting a well-defined adaptive learning rate, the NIC algorithm also generalizes some well-known PSA/PCA algorithms. For online implementation, an RLS version of the NIC algorithm has also been given in [799]. The PAST algorithm [1217] is a special case of the NIC algorithm when η takes the value of unity, and the NIC algorithm essentially represents a robust improvement of the PAST.

In order to break the symmetry in the NIC, the weighted information criterion (WINC) [877] is obtained by adding a weight to the NIC. Two WINC algorithms are, respectively, derived by using the gradient-ascent and the RLS. The gradient-ascent-based WINC algorithm can be viewed as an extended weighted SLA [868] with an adaptive step size, leading to a much faster convergence speed. The RLS-based WINC algorithm not only provides a fast convergence and a high accuracy, but also has a low computational complexity.

7.7.2 Coupled Principal Component Analysis

The most popular PCA or MCA algorithms do not consider eigenvalue estimates in the update equations of the weights, and they suffer from the stability–speed problem, since the eigenmotion depends on the eigenvalues of the covariance matrix. The convergence speed of a system depends on the eigenvalues of its Jacobian. In PCA algorithms, the eigenmotion depends on the principal eigenvalue of the covariance matrix, while in MCA algorithms it depends on all the eigenvalues [814].

Coupled learning rules can be derived by applying the Newton method to a common information criterion. In coupled PCA/MCA algorithms, both the eigenvalues and the eigenvectors are simultaneously adapted. The Newton method yields averaged systems with identical speed of convergence in all eigendirections. The Newton descent-based PCA and MCA algorithms, respectively called *nPCA* and *nMCA*, are derived by using the information criterion [814]

$$E_{\text{coupled}}(\mathbf{w}, \lambda) = \frac{\mathbf{w}^{\mathrm{T}} \mathbf{C} \mathbf{w}}{\lambda} - \mathbf{w}^{\mathrm{T}} \mathbf{w} + \ln \lambda \qquad (7.68)$$

where λ is the eigenvalue estimate.

By approximation $\mathbf{w}^{\mathrm{T}} \mathbf{w} \simeq 1$, the nPCA is reduced to the ALA [202]. Further approximating the ALA by $\mathbf{w}^{\mathrm{T}} \mathbf{C} \mathbf{w} \simeq \lambda$ leads to an algorithm called *cPCA*. The cPCA is a stable PCA algorithm, but there may be fluctuation of the weight vector length in the iteration process. This problem can be avoided by explicitly renormalizing the weight vector at every iteration, and this leads to the robust PCA (rPCA) algorithm [814]

$$\mathbf{w}(t+1) = \mathbf{w}(t) + \eta(t) \left(\frac{\mathbf{x}_t y(t)}{\lambda(t)} - \mathbf{w}(t) \right) \qquad (7.69)$$

$$\mathbf{w}(t+1) = \frac{\mathbf{w}(t+1)}{\|\mathbf{w}(t+1)\|} \qquad (7.70)$$

$$\lambda(t+1) = \lambda(t) + \eta(t) \left(y^2(t) - \lambda(t) \right) \qquad (7.71)$$

where $\eta(t)$ is a small positive number, and can be selected according to the Robbins–Monro conditions (7.2) [730]. Note that (7.71) is similar to (7.31) with $\eta(t) = \frac{1}{1+t}$ and the two are the same as $t \to \infty$. The rPCA is shown to be closely related to the RRLSA algorithm [876] by applying the first-order Taylor approximation on the rPCA. The RRLSA can also be derived from the ALA algorithm by using the first-order Taylor approximation.

In order to extract multiple PCs, one has to apply an orthonormalization procedure, which is the GSO, or its first-order approximation as used in the SLA [866, 865], or deflation as in the GHA [994]. In the coupled learning rules, multiple PCs are simultaneously estimated by a coupled system of equations. It has been reported in [815] that in the coupled learning rules a first-order approximation of the GSO is superior to the standard deflation procedure in terms of the orthonormality error and the quality of the eigenvectors and eigenvalues generated. An additional normalization step that enforces unit length of the eigenvectors further improves the orthonormality of the weight vectors [815].

7.8 Anti-Hebbian Rule-based Principal Component Analysis

When the update of a synaptic weight is proportional to the correlation of the pre- and postsynaptic activities, but the direction of the change is opposite to that in the Hebbian rule [473], the new learning rule is called an *anti-Hebbian learning rule*. The anti-Hebbian rule can be used to remove correlations between units receiving correlated inputs [368, 973, 974]. The anti-Hebbian rule is inherently stable [973, 974].

Anti-Hebbian rule-based PCA algorithms can be derived by using a network architecture of the J_1-J_2 FNN with lateral connections among the output units [973, 974, 368]. The lateral connections can be in a symmetrical or hierarchical topology. A hierarchical lateral connection topology is illustrated in Fig. 7.4, based on which the Rubner–Tavan PCA algorithm [973, 974] and the APEX [641] are proposed. In [368], the local PCA algorithm is based on a full lateral connection topology. The feedforward weight matrix $\mathbf{W}$ is described in the preceding sections, and the lateral weight matrix $\mathbf{U} = [\mathbf{u}_1 \cdots \mathbf{u}_{J_2}]$ is a $J_2 \times J_2$ matrix, where $\mathbf{u}_i = (u_{1i}, u_{2i}, \cdots, u_{J_2i})^{\mathrm{T}}$ includes all the lateral weights connected to neuron i and u_{ji} denotes the lateral weight from neuron j to neuron i.

7.8.1 Rubner–Tavan Principal Component Analysis Algorithm

The Rubner–Tavan PCA algorithm is based on the PCA network with hierarchical lateral connection topology [973, 974]. The algorithm extracts the first J_2 PCs in decreasing order of the eigenvalues. The output of the network is given as [973, 974]

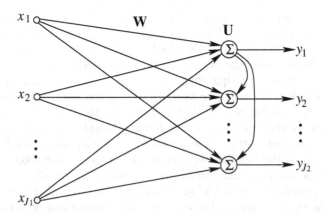

Fig. 7.4. Architecture of the PCA network with hierarchical lateral connections. The lateral weight matrix $\mathbf{U}$ is an upper triangular matrix with the diagonal elements being zero.

$$y_i = \mathbf{w}_i^{\mathrm{T}}\mathbf{x} + \mathbf{u}_i^{\mathrm{T}}\mathbf{y}, \quad i = 1, \cdots, J_2 \tag{7.72}$$

Note that $u_{ji} = 0$ for $j \geq i$ and $\mathbf{U}$ is a $J_2 \times J_2$ upper triangular matrix.

The weights $\mathbf{w}_i$ are trained by Oja's rule, and the lateral weights $\mathbf{u}_i$ are updated by the anti-Hebbian rule

$$\mathbf{w}_i(t+1) = \mathbf{w}_i(t) + \eta_1(t)y_i(t)\left[\mathbf{x}_t - \hat{\mathbf{x}}(t)\right] \tag{7.73}$$

$$\hat{\mathbf{x}} = \mathbf{W}^{\mathrm{T}}\mathbf{y} \tag{7.74}$$

$$\mathbf{u}_i(t+1) = \mathbf{u}_i(t) - \eta_2 y_i(t)\mathbf{y}(t) \tag{7.75}$$

This is a nonlocal algorithm. Typically, the learning rate $\eta_1 = \eta_2 > 0$ is selected as a small number between 0.001 and 0.1 or according to a heuristic derived from the Robbins–Monro conditions [866].

During the training process, the outputs of the neurons are gradually uncorrelated and the lateral weights approach zero. The network should be trained until the lateral weights $\mathbf{u}_i$ are below a specified level. The PCA algorithm proposed in [368] has the same form as the Rubner–Tavan PCA given by (7.72) through (7.74), but $\mathbf{U}$ is a full matrix.

7.8.2 APEX Algorithm

The APEX algorithm is used to adaptively extract the PCs [641]. The algorithm is recursive and adaptive, namely, given $i-1$ PCs, it can produce the ith PC iteratively. The hierarchical structure of lateral connections among the output units serves the purpose of weight orthogonalization. This structure also allows the network to grow or shrink without retraining the old units. The convergence analysis of the APEX algorithm is based on the stochastic approximation theory [730], and the APEX is proved to have the property of exponential convergence.

Assuming that the correlation matrix $\mathbf{C}$ has distinct eigenvalues arranged in the decreasing order as $\lambda_1 > \lambda_2 > \cdots > \lambda_{J_2}$ with the corresponding eigenvectors $\mathbf{w}_1, \cdots, \mathbf{w}_{J_2}$, the algorithm is given as [641, 642]

$$\mathbf{y} = \mathbf{W}^{\mathrm{T}}\mathbf{x} \tag{7.76}$$

$$y_i = \mathbf{w}_i^{\mathrm{T}}\mathbf{x} + \mathbf{u}^{\mathrm{T}}\mathbf{y} \tag{7.77}$$

where $\mathbf{y} = (y_1, \cdots, y_{i-1})^{\mathrm{T}}$ is the output vector, $\mathbf{u} = \left(u_{1i}, u_{2i}, \cdots, u_{(i-1)i}\right)^{\mathrm{T}}$, and $\mathbf{W} = [\mathbf{w}_1 \cdots \mathbf{w}_{i-1}]$ is the weight matrix of the first $i-1$ neurons. These definitions are for the first i neurons, which are different from their respective definitions given in preceding sections. The iteration is given as [641, 642]

$$\mathbf{w}_i(t+1) = \mathbf{w}_i(t) + \eta_i(t)\left[y_i(t)\mathbf{x}_t - y_i^2(t)\mathbf{w}_i(t)\right] \tag{7.78}$$

$$\mathbf{u}(t+1) = \mathbf{u}(t) - \eta_i(k)\left[y_i(t)\mathbf{y}(t) + y_i^2(t)\mathbf{u}(t)\right] \tag{7.79}$$

Equations (7.78) and (7.79) are respectively the Hebbian and anti-Hebbian parts of the algorithm. y_i tends to be orthogonal to all the previous components due to the anti-Hebbian rule, also called the *orthogonalization rule*.

The APEX can also be derived from the RLS method using the MSE criterion. Based on the RLS method, an optimum learning rate in terms of convergence speed is given by [642]

$$\eta_i(t) = \frac{1}{\sum_{l=0}^{t} \mu^{t-l} y_i^2(l)} = \frac{\eta_i(t-1)}{\mu - y_i^2(t)\eta_i(t-1)} \tag{7.80}$$

where $0 < \mu \le 1$ is a forgetting factor, which induces an effective time window of size $M = \frac{1}{1-\mu}$. In practice, we can use a hard-limiting window of size M, and set

$$\eta_i(t) = \frac{1}{\sum_{l=t-M+1}^{t} y^2(l)} \tag{7.81}$$

This is natural in the case where the number of samples is M. The optimal learning rate can also be written as [641]

$$\eta_i(t) = \frac{1}{M\sigma_i^2} \tag{7.82}$$

where $\sigma_i^2 = E\left[y_i^2(t)\right]$ is the average output power or variance of neuron i. According to [641]

$$\sigma_i^2(t) \to \lambda_i, \quad \text{as } n \to \infty \tag{7.83}$$

A practical value of η_i is selected by

$$\eta_i(t) = \frac{1}{M\lambda_{i-1}} \tag{7.84}$$

since $\lambda_{i-1} > \lambda_i$ and λ_i is not easy to get.

Both sequential and parallel APEX algorithms have been presented in [642]. In the parallel APEX, all the J_2 output neurons work simultaneously. In the sequential APEX, the output neurons are added one by one. The sequential APEX is more attractive in practical applications, since one can decide a desirable number of neurons during the learning process. The APEX algorithm is especially useful when the number of required PCs is not known *a priori*. When the environment is changing over time, a new PC can be added to compensate for the change without affecting the previously computed principal components. Thus, the network structure can be expanded if necessary. The sequential APEX is given by Algorithm 7.8.2 [642].

The stopping criterion can be that for each i the changes in $\mathbf{w}_i$ and $\mathbf{u}$ are below a threshold. At this time, $\mathbf{w}_i$ converges to the eigenvector of the correlation matrix $\mathbf{C}$ corresponding to the ith largest eigenvalue, and $\mathbf{u}$ converges to zero. The stopping criterion can also be the change of the average output variance $\sigma_i^2(t)$ being sufficiently small.

Algorithm 7.8.2 (APEX) For every neuron $i = 1, \cdots, J_2$:

1. Initialize $\mathbf{w}_i(0)$ and $\mathbf{u}(0)$ as small random numbers, and $\eta_i(0)$ as a small positive value.
2. Select μ or M.
3. Set $t = 1$.
4. Repeat:
   a) Present example $\mathbf{x}_t$.
   b) Choose η_i according to (7.80) or (7.81).
   c) Update $\mathbf{w}_i(t)$ and $\mathbf{u}(t)$ according to (7.78) and (7.79).
   d) Set $t = t + 1$.
   until a stopping criterion is met.

Most existing linear complexity methods including the GHA [994], the SLA [866], and the PCA with the lateral connections [973, 974, 368, 641, 642] require a computational complexity of $O(J_1 J_2)$ per iteration. For the recursive computation of each additional PC, the APEX requires $O(J_1)$ operations per iteration, while the GHA utilizes $O(J_1 J_2)$ per iteration.

In contrast to the heuristic derivation of the APEX [642], a class of learning algorithms, called the $\psi\text{-}APEX$, is presented based on criterion optimization [358, 355]. ψ can be selected as any function that guarantees the stability of the network. Some members in the class have better numerical performance and require less computational effort compared to that of both the GHA and the APEX.

7.9 Nonlinear Principal Component Analysis

The aforementioned PCA algorithms apply a linear transform to the input data. The PCA is based on the Gaussian assumption for data distribution, and the optimality of the PCA results from taking into account only the second-order statistics, namely, the covariances. For non-Gaussian data distributions, the PCA is not able to capture complex nonlinear correlations, and nonlinear processing of the data is usually more efficient. Nonlinearities introduce higher-order statistics into the computation in an implicit way. Higher-order statistics, defined by cumulants or higher-than-second moments, are needed for a good characterization of non-Gaussian data.

The Gaussian distribution is only one of the canonical exponential distributions, and it is suitable for describing real-value data. In the case of binary-valued, integer-valued, or non-negative data, the Gaussian assumption is inappropriate, and a family of exponential distributions can be used. For example, the Poisson distribution is better suited for integer data and the Bernoulli distribution to binary data, and an exponential distribution to non-negative data. All these distributions belong to the exponential family. The

PCA can be generalized to distributions of the exponential family [257]. This generalization is based on a generalized linear model and criterion functions using the Bregman distance. This approach permits hybrid dimensionality reduction in which different distributions are used for different attributes of the data.

When the feature space is nonlinearly related to the input space, we need to use nonlinear PCA. The outputs of nonlinear PCA networks are usually more independent than their respective linear cases. For non-Gaussian input data, the PCA may fail to provide an adequate representation, while a nonlinear PCA permits the extraction of higher-order components and provides a sufficient representation. Nonlinear PCA networks and learning algorithms can be classified into symmetric and hierarchical ones similar to those for the PCA networks. After training, the lateral connections between output units are not needed, and the network becomes purely feedforward.

7.9.1 Kernel Principal Component Analysis

Kernel PCA [1008, 832] is a special, linear algebra-based nonlinear PCA, which introduces kernel functions into the PCA. The kernel PCA first maps the original input data into a high-dimensional feature space using the kernel method and then calculates the PCA in the high-dimensional feature space. The linear PCA in the high-dimensional feature space corresponds to a nonlinear PCA in the original input space.

Given an input pattern set $\left\{ \mathbf{x}_i \in R^{J_1} \,\middle|\, i = 1, \cdots, N \right\}$, $\boldsymbol{\varphi} : R^{J_1} \to R^{J_2}$ is a nonlinear map from the J_1-dimensional input to the J_2-dimensional feature space. A J_2-by-J_2 correlation matrix in the feature space is defined by

$$\mathbf{C}_1 = \frac{1}{N} \sum_{i=1}^{N} \boldsymbol{\varphi}\left(\mathbf{x}_i\right) \boldsymbol{\varphi}^{\mathrm{T}}\left(\mathbf{x}_i\right) \tag{7.85}$$

Like the PCA, the set of feature vectors is limited to zero-mean

$$\frac{1}{N} \sum_{i=1}^{N} \boldsymbol{\varphi}\left(\mathbf{x}_i\right) = \mathbf{0} \tag{7.86}$$

A procedure to select $\boldsymbol{\varphi}$ satisfying (7.86) is given in [1007].

The PCs are then computed by solving the eigenvalue problem [1008, 832]

$$\lambda \mathbf{v} = \mathbf{C}_1 \mathbf{v} = \frac{1}{N} \sum_{j=1}^{N} \left(\boldsymbol{\varphi}\left(\mathbf{x}_j\right)^{\mathrm{T}} \mathbf{v} \right) \boldsymbol{\varphi}\left(\mathbf{x}_j\right) \tag{7.87}$$

Thus, $\mathbf{v}$ must be in the span of the mapped data

$$\mathbf{v} = \sum_{i=1}^{N} \alpha_i \boldsymbol{\varphi}\left(\mathbf{x}_i\right) \tag{7.88}$$

After premultiplying both sides of (7.88) by $\varphi(\mathbf{x}_j)$ and performing mathematical manipulations, the kernel PCA problem reduces to

$$\mathbf{K}\boldsymbol{\alpha} = \lambda\boldsymbol{\alpha} \tag{7.89}$$

where λ and $\boldsymbol{\alpha} = (\alpha_1, \cdots, \alpha_N)^\mathrm{T}$ are, respectively, the eigenvalues and the corresponding eigenvectors of $\mathbf{K}$, and $\mathbf{K}$ is an $N \times N$ kernel matrix with

$$K_{ij} = \kappa(\mathbf{x}_i, \mathbf{x}_j) = \varphi^\mathrm{T}(\mathbf{x}_i)\varphi(\mathbf{x}_j) \tag{7.90}$$

$\kappa(\cdot)$ being the kernel function.

Popular kernel functions used in the kernel method are the polynomial, Gaussian kernel, and sigmoidal kernels, which are respectively given by [1119, 832]

$$\kappa(\mathbf{x}_i, \mathbf{x}_j) = \left(\mathbf{x}_i^\mathrm{T}\mathbf{x}_j + \theta\right)^{a_0} \tag{7.91}$$

$$\kappa(\mathbf{x}_i, \mathbf{x}_j) = e^{-\frac{\|\mathbf{x}_i - \mathbf{x}_j\|^2}{2\sigma^2}} \tag{7.92}$$

$$\kappa(\mathbf{x}_i, \mathbf{x}_j) = \tanh\left(c_0\left(\mathbf{x}_i^\mathrm{T}\mathbf{x}_j\right) + \theta\right) \tag{7.93}$$

where a_0 is a positive integer, $\sigma > 0$, and $c_0, \theta \in R$. Even if the exact form of $\varphi(\cdot)$ does not exist, any symmetric function $\kappa(\mathbf{x}_i, \mathbf{x}_j)$ satisfying Mercer's theorem can be used as a kernel function [832].

Arrange the eigenvalues in the descending order $\lambda_1 \geq \lambda_2 \geq \cdots \geq \lambda_{J_2} > 0$, and denote their corresponding eigenvectors as $\boldsymbol{\alpha}_1, \cdots, \boldsymbol{\alpha}_{J_2}$. The eigenvectors are further normalized as

$$\boldsymbol{\alpha}_k^\mathrm{T}\boldsymbol{\alpha}_k = \frac{1}{\lambda_k} \tag{7.94}$$

The nonlinear PCs of $\mathbf{x}$ can be extracted by projecting the mapped pattern $\varphi(\mathbf{x})$ onto $\mathbf{v}_k$ [1008, 832]

$$\mathbf{v}_k^\mathrm{T}\varphi(\mathbf{x}) = \sum_{j=1}^{N} \alpha_{k,j}\kappa(\mathbf{x}_j, \mathbf{x}) \tag{7.95}$$

for $k = 1, 2, \ldots, J_2$, where $\alpha_{k,j}$ is the jth element of $\boldsymbol{\alpha}_k$.

The kernel PCA algorithm is much more complicated and may sometimes be caught more easily in local minima. The PCA needs to deal with an eigenvalue problem of a $J_1 \times J_1$ matrix, while the kernel PCA needs to solve an eigenvalue problem of an $N \times N$ matrix. Sparse approximation methods can be applied to reduce the computational cost [832].

7.9.2 Robust/Nonlinear Principal Component Analysis

In order to increase the robustness of the PCA against outliers, a simple way is to eliminate the outliers or replace them by more appropriate values. A better alternative is to use a robust version of the covariance matrix based on the

M-estimator. The data from which the covariance matrix is constructed may be weighted such that the samples far from the mean have less importance.

Several popular PCA algorithms have been generalized into robust versions by applying a statistical-physics approach [1206], where the defined objective function can be regarded as a soft generalization of the M-estimator [514].

In this subsection, robust PCA algorithms are defined so that the optimization criterion grows less than quadratically and the constraint conditions are the same as for the PCA algorithms, which are based on the quadratic criterion [587]. The robust PCA problem usually leads to mildly nonlinear algorithms, in which the nonlinearities appear at selected places only and at least one neuron produces the linear response $y_i = \mathbf{x}^T \mathbf{w}_i$. When all the neurons generate nonlinear responses $y_i = \phi \left(\mathbf{x}^T \mathbf{w}_i \right)$, the algorithm is referred to as the nonlinear PCA.

Variance Maximization-based Robust Principal Component Analysis

The PCA is to maximize the output variances $\mathrm{E}\left[y_i^2\right] = \mathrm{E}\left[\left(\mathbf{w}_i^T \mathbf{x}\right)^2\right] = \mathbf{w}_i^T \mathbf{C} \mathbf{w}_i$ of the linear network under orthonormality constraints. In the hierarchical case, the constraints take the form $\mathbf{w}_i^T \mathbf{w}_j = \delta_{ij}$, $j \leq i$, δ_{ij} being the Kronecker delta. In the symmetric case, symmetric orthonormality constraints $\mathbf{w}_i^T \mathbf{w}_j = \delta_{ij}$ are applied. The SLA [866, 865] and GHA [994] algorithms correspond to the symmetric and hierarchical network structures, respectively.

To derive robust PCA algorithms, the variance maximization criterion is generalized as $\mathrm{E}\left[\sigma(\mathbf{w}_i^T \mathbf{x})\right]$ for the ith neuron, subject to hierarchical or symmetric orthonormality constraints, where $\sigma(x)$ is the M-estimator assumed to be a valid differentiable cost function that grows less than quadratically, at least for large values of x. Examples of such functions are $\sigma(x) = \ln \cosh(x)$ and $\sigma(x) = |x|$. The robust PCA in general does not coincide with the corresponding PCA solution, although it can be close to it. The robust PCA is derived by applying the gradient-descent method [587]

$$\mathbf{w}_i(t+1) = \mathbf{w}_i(t) + \eta(t)\varphi\big(y_i(t)\big)\mathbf{e}_i(t) \tag{7.96}$$

$$\mathbf{e}_i(t) = \mathbf{x}_t - \hat{\mathbf{x}}_i(t) \tag{7.97}$$

$$\hat{\mathbf{x}}_i(t) = \sum_{j=1}^{I(i)} y_j(t)\mathbf{w}_j(t) \tag{7.98}$$

where $\mathbf{e}_i(t)$ is the instantaneous representation error vector, and the influence function $\varphi(x) = \frac{\mathrm{d}\sigma(x)}{\mathrm{d}x}$.

In the symmetric case, $I(i) = J_2$ and the errors $\mathbf{e}_i(t) = \mathbf{e}(t)$, $i = 1, \cdots, J_2$. When $\varphi(x) = x$, the algorithm is simplified to the SLA [866, 865]. Otherwise, it defines a robust generalization of Oja's rule, first proposed quite heuristically in [867].

In the hierarchical case, $I(i) = i$, $i = 1, \cdots, J_2$. If $\varphi(x) = x$, the algorithm coincides exactly with the GHA [994]; Otherwise, it defines a robust generalization of the GHA. In the hierarchical case, $\mathbf{e}_i(t)$ can be calculated in a recursive form

$$\mathbf{e}_i(t) = \mathbf{e}_{i-1}(t) - y_i(t)\mathbf{w}_i(t) \tag{7.99}$$

with $\mathbf{e}_0(t) = \mathbf{x}_t$.

Mean Squared Error Minimization-based Robust Principal Component Analysis

PCA algorithms can also be derived by minimizing the MSE $\mathrm{E}\left[\|\mathbf{e}_i\|^2\right]$, where $\mathbf{e}_i(t)$ is given as (7.97). Accordingly, robust PCA can be obtained by minimizing $\mathbf{1}^{\mathrm{T}}\mathrm{E}\left[\sigma\left(\mathbf{e}_i\right)\right] = \mathrm{E}\left[\|h\left(\mathbf{e}_i\right)\|^2\right]$, where $\mathbf{1}$ is a J_2-dimensional vector, all of whose entries are unity, and $\sigma(\cdot)$ and $h(\cdot)$ are applied componentwise on the vector within. Here, $h(x) = \sqrt{\sigma(x)}$. When $\sigma(x) = x^2$, it corresponds to the MSE. A robust PCA is defined if $\sigma(x)$ grows less than quadratically. Using the gradient-descent method leads to [587]

$$\mathbf{w}_i(t+1) = \mathbf{w}_i(t) + \eta(t)\left[\mathbf{w}_i(t)^{\mathrm{T}}\varphi\left(\mathbf{e}_i(t)\right)\mathbf{x}_t + \mathbf{x}_t^{\mathrm{T}}\mathbf{w}_i(t)\varphi\left(\mathbf{e}_i(t)\right)\right] \tag{7.100}$$

$\mathbf{w}_i$ estimates the robust counterparts of the principal eigenvectors $\mathbf{c}_i$. The first term in the bracket is very small and can be neglected, and we get the simplified algorithm

$$\begin{aligned}\mathbf{w}_i(t+1) &= \mathbf{w}_i(t) + \eta(t)\mathbf{x}_t^{\mathrm{T}}\mathbf{w}_i(t)\varphi\left(\mathbf{e}_i(t)\right) \\ &= \mathbf{w}_i(t) + \eta(t)y_i(t)\varphi\left(\mathbf{e}_i(t)\right)\end{aligned} \tag{7.101}$$

Algorithms (7.101) and (7.96) resemble each other. However, algorithm (7.101) generates a linear final input-output mapping, while in algorithm (7.96) the input-output mapping is nonlinear.

When $\varphi(x) = x$, algorithms (7.101) and (7.96) are the same as the SLA [864, 865] in the symmetric case, and the same as the GHA [994] in the hierarchical case. The optimal PCA subspace estimation algorithm is given as (7.100) with $\varphi(x) = x$ [587, 1203].

Another Nonlinear Extension to Principal Component Analysis

A nonlinear PCA algorithm may be derived by the gradient-descent method for minimizing the MSE $\mathrm{E}\left[\|\boldsymbol{\epsilon}_i\|^2\right]$ [587, 1203], where the error vector $\boldsymbol{\epsilon}_i$ is a nonlinear extension to $\mathbf{e}_i$, which is defined by (7.97). The nonlinear PCA so obtained has a form similar to the robust PCA given by (7.96) through (7.98)

$$\mathbf{w}_i(t+1) = \mathbf{w}_i(t) + \eta(t)\varphi\left(y_i(t)\right)\boldsymbol{\epsilon}_i(t) \tag{7.102}$$

$$\epsilon_i(t) = \mathbf{x}_t - \sum_{j=1}^{I(i)} \varphi\big(y_j(t)\big)\mathbf{w}_j(t) \tag{7.103}$$

for $i = 1, \cdots, J_2$.

The symmetric case of the algorithm was first proposed in [867]. In this case, $I(i) = J_2$ and all $\epsilon_i(t)$ take the same value. The nonlinear PCA in the hierarchical case, initially introduced in [995], is a direct nonlinear generalization of the GHA. In the hierarchical case, $I(i) = i$ and (7.103) can be computed recursively

$$\epsilon_i(t) = \epsilon_{i-1}(t) - \varphi\big(y_i(t)\big)\mathbf{w}_i(t) \tag{7.104}$$

with $\epsilon_0(t) = \mathbf{x}(t)$.

It has been pointed out in [587] that robust and nonlinear PCA algorithms have better stability properties than the corresponding PCA algorithms if the (odd) nonlinearity $\varphi(x)$ grows less than linearly, namely, $|\varphi(x)| < |x|$. On the contrary, nonlinearities growing faster than linearly cause stability problems easily and are not recommended.

7.9.3 Autoassociative Network-based Nonlinear Principal Component Analysis

The MLP can be used to perform nonlinear dimensionality reduction and hence, nonlinear PCA. Both the input and output layers of the MLP have J_1 units, and one of its hidden layers, known as the *bottleneck* or *representation layer*, has J_2 units, $J_2 < J_1$. The network is trained to reproduce its input vectors. This kind of network is called the *autoassociative MLP*, which has been discussed in Chapter 4 for associative memory of bipolar vectors. After the network is trained, it performs a projection onto the J_2-dimensional subspace spanned by the first J_2 PCs of the data. The vectors of weights leading to the hidden units form a basis set that spans the principal subspace, and data compression therefore occurs in the bottleneck layer. Many applications of the MLP in the autoassociative mode for the PCA are available in the literature [129, 58, 628, 117].

In linear neural networks, the activation functions are all linear. The cost function E is characterized by a unique global minimum, and all the other critical points are saddle points [58]. Thus, the network has no processing advantage over the aforementioned unsupervised linear PCA network. If nonlinear activation functions are applied in the hidden layer, the network performs as a nonlinear PCA network.

The three-layer autoassociative J_1-J_2-J_1 FNN or MLP network can also be used to extract the first J_2 PCs of J_1-dimensional data. In the case of nonlinear units, local minima certainly appear. However, if linear units are used in the output layer, nonlinearity in the hidden layer is theoretically meaningless [129, 58]. This is due to the fact that the network tries to approximate a linear mapping.

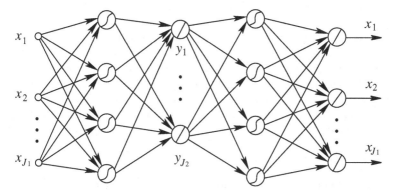

Fig. 7.5. Architecture of Kramer's nonlinear PCA network. The S shape in a circle denotes a nonlinear activation function such as the sigmoidal function and the slash (/) in a circle denotes a linear function. y_i, $i = 1, \cdots, J_2$, is the ith output of the bottleneck layer.

Kramer's Nonlinear Principal Component Analysis Network

Kramer's nonlinear PCA network [628] is a five-layer autoassociative MLP, whose architecture is illustrated in Fig. 7.5. It has J_1 input and J_1 output nodes. The third layer has J_2 nodes. Nonlinear activation functions such as the sigmoidal functions are used in the second and fourth layers, while the nodes in the bottleneck and output layers usually have linear activation functions, although they can be nonlinear. The network is trained by the BP algorithm. Kramer's nonlinear PCA fits a lower-dimensional surface through the training data.

The three-layer MLP can approximate arbitrarily well any continuous function [273, 508]. The input, second, and bottleneck layers constitute a three-layer MLP, which projects the training data onto the surface giving PCs. Likewise, the combination of the bottleneck, fourth, and output layers models defines the surface that inversely maps the PCs into the training data. The outputs of the network are trained to approximate the inputs. After the network is trained, the nodes in the bottleneck layer give a lower-dimensional representation of the inputs.

Usually, the data compression achieved in the bottleneck layer in such networks is somewhat better than that provided by the respective PCA solution [571]. This is actually a nonlinear PCA network. However, the BP algorithm is prone to local minima and often requires excessive time for convergence.

Hierarchical Nonlinear Principal Component Analysis Network

A hierarchical nonlinear PCA network composed of a number of independent subnetworks can extract ordered nonlinear PCs [987]. Each subnetwork extracts one PC, and has at least five layers. The subnetworks can be selected as the Kramer's nonlinear PCA network. The subnetworks are hierarchically arranged and trained. In each subnetwork, the number of input and output units are set to J_1, and the number of units in the bottleneck layer corresponds to the index of the PC it extracts.

Online BP learning is used to independently adjust the weights of each subnetwork. The adjustment of the weights in the ith subnetwork is dependent on the bottleneck layer output of the $i - 1$ trained subnetworks. This network constructs the extraction functions in the order of the reconstruction efficiency as to the objective data. The number of PCs to be extracted is not required to be known in advance.

7.9.4 Other Networks for Dimensionality Reduction

In contrast to autoassociative networks, the output pattern in heteroassociative networks is not the same as the input pattern for each training pair. Heteroassociative networks develop arbitrary internal representations in the hidden layers to associate inputs to class identifiers, usually in the context of pattern classification. Training can be easily trapped into local minima, especially when there are overlaps among classes and training starts from patterns in the overlapping area. The rate of dimensionality reduction is not as high as that of autoassociative networks.

A hybrid hetero-/autoassociative network [176] is constructed with a set of autoassociative outputs and a set of heteroassociative outputs. Both sets of output nodes are fully connected to the same bottleneck layer. In this way, more information is embedded in the bottleneck layer, which is adjusted to jointly satisfy both the requirements of dimensionality reduction and the similarity in the application domain. The improvements over an autoassociative network or the PCA by using heteroassociative training can be attributed to the reorganization done by the network in the representation layer space.

As discussed in Chapter 5, SOMs [610] are competitive learning-based neural networks. Lateral inhibitory connections for output neurons are usually used to induce the WTA competition among all the output neurons. It is capable of performing dimensionality reduction on the input. The SOM is inherently nonlinear, and is viewed as a nonlinear PCA [953]. The ASSOM [613, 616] can be treated as a hybrid of the VQ and the PCA.

7.10 Minor Component Analysis

In contrast to the PCA, the MCA, as a variant of the PCA, is to find the smallest eigenvalues and their corresponding eigenvectors of the autocorrelation matrix $\mathbf{C}$ of the signals. The MCA is closely associated with the curve and surface fitting under the TLS criterion [1204]. The MCA is useful in many fields including spectrum estimation, optimization, TLS parameter estimation in adaptive signal processing, and eigen-based bearing estimation.

The PCA can be used as a solution to the maximization problem when the weight vector $\mathbf{w}$ is the eigenvector corresponding to the largest eigenvalue of $\mathbf{C}$. On the other hand, the MCA provides an alternative solution to the TLS problem [400]. The TLS technique achieves a better global optimal objective than the LS technique [424]. Both the solutions of the TLS and LS problems can be obtained by the SVD. However, the TLS technique is computationally much more expensive than the LS technique [350].

Minor components (MCs) can be extracted in ways similar to that for PCs. A simple idea is to reverse the sign of the PCA algorithms. This is in view of the fact that in many algorithms PCs correspond to the maximum of a cost function, while MCs correspond to the minimum of the same cost function. However, this idea does not work in general and has been discussed in [865].

7.10.1 Extracting the First Minor Component

The anti-Hebbian learning rule and its normalized version can be used for the MCA [1205]. The anti-Hebbian algorithm tends rapidly to infinite magnitudes of the weights. The normalized anti-Hebbian algorithm leads to better convergence, but it may also lead to infinite magnitudes of weights before the algorithm converges [1076]. To avoid this, one can renormalize the weight vector at each iteration.

The constrained anti-Hebbian learning algorithm [399, 400] has a simple structure, and requires a low computational complexity per update. The constrained anti-Hebbian learning algorithm has been applied to adaptive FIR and IIR filtering. It can be used to solve the TLS parameter estimation [400], and has been extended for complex-valued TLS problem [399]. However, as in the anti-Hebbian algorithm [1205], the convergence of the magnitudes of the weights cannot be guaranteed unless the initial weights take special values.

The total least mean squares (TLMS) algorithm [350] is a random adaptive algorithm for extracting the MC, which has an equilibrium point under persistent excitation conditions. The TLMS algorithm requires about $4J_1$ multiplications per iteration, which is twice the complexity of the LMS [1182] algorithm. An adaptive step-size learning algorithm [875] has been derived for extracting the MC by introducing information criterion. The algorithm globally converges asymptotically to a stable equilibrium point, which corresponds to the MC and its corresponding eigenvector. The algorithm outperforms the TLMS [350] in terms of both the convergence speed and the estimation accuracy.

7.10.2 Oja's Minor Subspace Analysis

Oja's minor subspace analysis (MSA) algorithm can be formulated by reversing the sign of the learning rate of the SLA for the PSA [866]

$$\mathbf{W}(t+1) = \mathbf{W}(t) - \eta \left[\mathbf{x}_t - \hat{\mathbf{x}}(t)\right] \mathbf{y}^{\mathrm{T}}(t) \qquad (7.105)$$

$$\hat{\mathbf{x}}_t = \mathbf{W}(t)\mathbf{y}(t) \tag{7.106}$$

where $\mathbf{y}(t) = \mathbf{W}^{\mathrm{T}}(t)\mathbf{x}_t$, $\eta > 0$ is the learning rate. This algorithm requires the assumption that the smallest eigenvalue of the autocorrelation matrix $\mathbf{C}$ is less than unity. However, Oja's MSA algorithm is known to diverge [865, 316, 47]. The bigradient PSA algorithm [212] is a modification to the SLA [866] and is obtained by introducing an additional bigradient term embodying the orthonormal constraints of the weights, and it can be used for the MSA by reversing the sign of η.

7.10.3 Self-stabilizing Minor Component Analysis

A general algorithm that can extract, in parallel, principal and minor eigenvectors of arbitrary dimensions is derived based on the natural-gradient method [213]. The difference between the PCA and the MCA lies in the sign of the learning rate. The MCA algorithm proposed in [213] can be written as

$$\mathbf{W}(t+1) = \mathbf{W}(t) - \eta\left[\mathbf{x}_t\mathbf{y}^{\mathrm{T}}(t)\mathbf{W}^{\mathrm{T}}(t)\mathbf{W}(t) - \mathbf{W}(t)\mathbf{y}(t)\mathbf{y}^{\mathrm{T}}(t)\right] \tag{7.107}$$

At initialization, $\mathbf{W}^{\mathrm{T}}(0)\mathbf{W}(0)$ is required to be diagonal. Algorithm (7.107) suffers from a marginal instability, and thus it requires intermittent normalization such that $\|\mathbf{w}_i\| = 1$ [316].

A self-stabilizing MCA algorithm is given in [316] as

$$\mathbf{W}(t+1) = \mathbf{W}(t) - \eta\left[\mathbf{x}_t\mathbf{y}^{\mathrm{T}}(t)\mathbf{W}^{\mathrm{T}}(t)\mathbf{W}(t)\mathbf{W}^{\mathrm{T}}(t)\mathbf{W}(t) - \mathbf{W}(t)\mathbf{y}(t)\mathbf{y}^{\mathrm{T}}(t)\right] \tag{7.108}$$

Algorithm (7.108) is self-stabilizing, such that none of $\|\mathbf{w}_i(t)\|$ deviates significantly from unity. Algorithm (7.108) diverges for the PCA when $-\eta$ is changed to $+\eta$. Both the algorithms (7.107) and (7.108) have a complexity of $O\left(J_1 J_2\right)$.

7.10.4 Orthogonal Oja Algorithm

The orthogonal Oja (OOja) algorithm consists of Oja's MSA (7.105) [866] plus an orthogonalization of $\mathbf{W}(t)$ at each iteration [5]

$$\mathbf{W}^{\mathrm{T}}(t)\mathbf{W}(t) = \mathbf{I} \tag{7.109}$$

In this case, the above algorithms (7.105) [866], (7.107) [213], and (7.108) [316] are equivalent. A Householder transform-based implementation of the MCA algorithm is given as [5]

$$\hat{\mathbf{x}}(t) = \mathbf{W}(t)\mathbf{y}(t) \tag{7.110}$$

$$\mathbf{e}(t) = \mathbf{x}_t - \hat{\mathbf{x}}(t) \tag{7.111}$$

$$\vartheta(t) = \frac{1}{\sqrt{1 + \eta^2 \|\mathbf{e}(t)\|^2 \|\mathbf{y}(t)\|^2}} \tag{7.112}$$

$$\mathbf{u}(t) = \frac{1 - \vartheta(t)}{\eta \|\mathbf{y}(t)\|^2} \hat{\mathbf{x}}(t) + \vartheta(t)\mathbf{e}(t) \tag{7.113}$$

$$\overline{\overline{\mathbf{u}}}(t) = \frac{\mathbf{u}(t)}{\|\mathbf{u}(t)\|} \tag{7.114}$$

$$\mathbf{v}(t) = \mathbf{W}^{\mathrm{T}}(t)\overline{\overline{\mathbf{u}}}(t) \tag{7.115}$$

$$\mathbf{W}(t+1) = \mathbf{W}(t) - 2\overline{\overline{\mathbf{u}}}(t)\mathbf{v}^{\mathrm{T}}(t) \tag{7.116}$$

where $\mathbf{W}$ is initialized as any arbitrary orthogonal matrix and $\mathbf{y}$ is given by (7.25). The OOja is numerically very stable. By reversing the sign of η, we extract J_2 PCs.

The normalized Oja (NOja) is derived by optimizing the MSE criterion subject to an approximation to the orthonormal constraint (7.109) [47]. This leads to the optimal learning rate. The normalized orthogonal Oja (NOOja) is an orthogonal version of the NOja such that (7.109) is perfectly satisfied [47]. Both algorithms offer, as compared to Oja's SLA, a faster convergence, orthogonality, and a better numerical stability with a slight increase in the computational complexity. By switching the sign of η in given learning algorithms, both the NOja and the NOOja can be used for the estimation of minor and principal subspaces of a vector sequence.

All the algorithms (7.105), (7.107), (7.108), the OOja, the NOja, and the NOOja have a complexity of $O(J_1 J_2)$ [5, 316]. The OOja, the NOjia, and the NOOjia require less computation load than algorithms (7.107) and (7.108) [5, 47].

7.10.5 Other Developments

By using the Rayleigh quotient as an energy function, the invariant-norm MCA [740] is analytically proved to converge to the first MC of the input signals. The MCA algorithm has been extended to sequentially extract multiple MCs in the ascending eigenvalue order by using the idea of sequential elimination in [738]. However, the invariant-norm MCA [740] leads to divergence in finite time [1076], and this drawback can be eliminated by renormalizing the weight vector at each iteration. In addition, it has been proved in [214] that the MCA algorithm given in [738] does not work at all whenever the dimension of the subspace is larger than unity. In [214], an alternative MCA algorithm for extracting multiple MCs is described by using the idea of sequential addition, and a conversion method between the MCA and the PCA is also discussed.

Based on a generalized differential equation for the generalized eigenvalue problem, a class of algorithms can be obtained for extracting the first PC or MC by selecting different parameters and functions [1260]. Many existing PCA algorithms such as the ones described in [864, 1247, 1146] and the MCA algorithms such as the one in [1146] are special cases of this class. Some new and simpler MCA algorithms are also given in [1260]. All the algorithms of this class have the same order of convergence speed and are robust to implementation error.

A rapidly convergent quasi-Newton method has been applied to extract multiple MCs in [775]. The proposed algorithm has a complexity of $O\left(J_2 J_1^2\right)$ but with a quadratic convergence. The algorithm makes use of the implicit orthogonalization procedure that is built into it through an inflation technique.

7.11 Independent Component Analysis

ICA [258], as an extension of the PCA, is a recently developed statistical model. The ICA was originally introduced for BSS, and was later generalized for feature extraction. The ICA has now been widely used for BSS, feature extraction, and signal detection. For BSS applications, the ICA model is required to have model identifiability and separability [258]. Instead of obtaining uncorrelated components as in the PCA, the ICA attempts to linearly transform the original inputs into features that are statistically mutually independent. The first neural-network model with a heuristic learning algorithm, which is related to the ICA, was developed for online BSS of linearly mixed signals in [558].

7.11.1 Formulation of Independent Component Analysis

Let a J_1-vector $\mathbf{x}$ denote a linear mixture and a J_2-vector $\mathbf{s}$, whose components have zero mean and are statistically mutually independent, denote the original source signals. The ICA data model can be written as

$$\mathbf{x} = \mathbf{A}\mathbf{s} + \mathbf{n} \tag{7.117}$$

where $\mathbf{A}$ is a constant full-rank $J_1 \times J_2$ mixing matrix whose elements are the unknown coefficients of the mixtures, and $\mathbf{n}$ denotes the additive noise term, which is often omitted since it is usually impossible to separate noise from the sources. The ICA takes one of three forms, namely, the square ICA for $J_1 = J_2$, the overcomplete ICA for $J_1 < J_2$, and the undercomplete ICA for $J_1 > J_2$. While the undercomplete ICA is useful for feature extraction, the overcomplete ICA may be applied to signal and image-processing methods based on multiscale and redundant basis sets.

The goal of the ICA is to estimate $\mathbf{s}$ by

$$\mathbf{y} = \mathbf{W}^{\mathrm{T}}\mathbf{x} \tag{7.118}$$

such that the components of $\mathbf{y}$, which is the estimate of $\mathbf{s}$, are statistically as independent as possible. $\mathbf{W}$ is a $J_1 \times J_2$ demixing matrix.

The statistical independence property implies that the joint probability density of the components of $\mathbf{s}$ equals the product of the marginal densities of the individual components. Each component of **s** is a stationary stochastic process and only one of the components is allowed to be Gaussian distributed. The higher-order statistics of the original inputs is required for estimating **s**, rather than the second-order moment or covariance of the samples as used in the PCA. The Cramer–Rao bound[2] on estimating the source signals in the ICA is derived in [618], based on the assumption that all independent components have finite variance.

Two distinct characteristics exist between the PCA and the ICA. The components of the signal extracted by the ICA are statistically independent, not merely uncorrelated as in the PCA. The demixing matrix **W** of the ICA is not orthogonal, while in the PCA the components of the weights are represented on an orthonormal basis. The ICA provides in many cases a more meaningful representation of the data than the PCA. The ICA can be realized by adding nonlinearity to linear PCA networks such that they are able to improve the independence of their outputs. In [591], an efficient ICA algorithm is derived by minimizing a nonlinear PCA criterion using the RLS approach. A conceptual comparison of the PCA and the ICA is illustrated in Fig. 7.6.

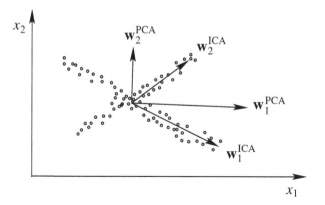

Fig. 7.6. An illustration of the PCA and the ICA for a two-dimensional non-Gaussian data set. The PCA extracts components with maximal variance, while the ICA extracts components with maximal independence. $\mathbf{w}_i^{\mathrm{ICA}}$ and $\mathbf{w}_i^{\mathrm{PCA}}$, $i = 1, 2$, are the ith principal and ith independent directions, respectively.

---

[2] The MSE for any estimate of a nonrandom parameter has a lower bound, called the *Cramer–Rao bound*. This lower bound defines the ultimate accuracy of any estimator, and is closely related to the ML estimator. To estimate a vector of parameters $\boldsymbol{\theta}$ from a data vector **x** that has a probability density, by using some unbiased estimator $\hat{\boldsymbol{\theta}}$, the Cramer–Rao bound is the lower bound for the variance of $\hat{\boldsymbol{\theta}}$.

7.11.2 Independent Component Analysis and Regression

The ICA can be used for regression, *i.e.* to predict the missing observations [525]. Regression by the ICA is closely related to regression by the MLP [525]. If linear dependencies are removed from the data at the preprocessing phase, regression by the ICA is, as a first-order approximation, equivalent to regression by a three-layer MLP. The output of each hidden unit in the MLP corresponds to the estimate of one independent component. Analogous to the ICA, the number of hidden units in the MLP can be treated as the model order in the ICA, the nonlinearity corresponds to the probability densities of the independent components, and overlearning in the MLP corresponds to modeling the data with too many independent components [525]. Learning is to find the optimal weights that enable the hidden neurons to extract the prominent-feature components from the input data set. If the initial weights enable the hidden neurons to extract as many prominent components as possible, network training may substantially speed up. In [1215], the fastICA algorithm [526] has been applied to determine the optimal initial weights of the three-layer MLP.

7.11.3 Approaches to Independent Component Analysis

A well-known two-phase approach to the ICA is to preprocess the data by the PCA, and then to estimate the necessary rotation matrix. A generic approach to the ICA consists of preprocessing the data, defining measures of non-Gaussianity, and optimizing an objective function, known as a *contrast function* [993]. The two most common data preprocessing methods are data centering and data whitening. Some measures of non-Gaussianity are kurtosis, differential entropy, negentropy, and mutual information, which can be derived from one another [993]. For example, one approach is to minimize the mutual information between the components of the output vector

$$I(\mathbf{y}) = \sum_{i=1}^{J_2} H(y_i) - H(\mathbf{y}) \qquad (7.119)$$

where $H(\mathbf{y}) = -\int p(\mathbf{y}) \ln p(\mathbf{y}) d\mathbf{y}$ is the joint entropy, and $H(y_i) = -\int p_i(y_i) \ln p_i(y_i) dy_i$ is the marginal entropy of component i, $p(\mathbf{y})$ being the joint PDF of all the elements of $\mathbf{y}$ and $p_i(y_i)$ the marginal PDF of y_i. The mutual information I is non-negative and is zero only when the components are mutually independent.

Kurtosis is the degree of peakedness of a distribution, based on the fourth central moment of the distribution. One commonly used definition of kurtosis is given by

$$\kappa(y_i) = E[y_i^4] - 3(E[y_i^2])^2 \qquad (7.120)$$

If $\kappa(y_i) < 0$, $y_i(t)$ is a sub-Gaussian source, while for super-Gaussian sources $\kappa(y_i) > 0$. The kurtosis of Gaussian sources is zero. In contrast to the PCA, nonlinearity is introduced in ICA algorithms, and is necessary to represent sub-Gaussian and super-Gaussian sources.

Some of the popular ICA algorithms are the infomax [94], the natural-gradient [31], the equivariant adaptive separation via independence (EASI) [156], and the FastICA algorithms [526]. These methods can be easily extended to the complex domain by using Hermitian transpose and complex nonlinear functions.

It is worth mentioning in passing that, in the context of BSS, the higher-order statistics are necessary only for temporally uncorrelated stationary sources. Second-order statistics-based source separation exploits temporally correlated stationary sources [235] and the nonstationarity of the sources [778, 235]. Many natural signals are inherently nonstationary with time-varying variances, since the source signals incorporate time delays into the basic BSS model. A recent review of the ICA is given in [993].

7.11.4 FastICA Algorithm

The FastICA algorithm is a well-known fixed-point ICA algorithm [526, 523, 411]. It is derived from the optimization of the kurtosis or the negentropy measure by using Newton's method. The FastICA algorithm achieves a reliable and at least a quadratic convergence.

The FastICA algorithm [523] estimates multiple independent components one by one using a GSO-like deflation scheme. The FastICA algorithm includes two phases, namely, prewhitening the observed data to remove any second-order correlations, and performing an orthogonal rotation of the whitened data to find the directions of the sources.

The mixtures $\mathbf{x}$ are first prewhitened according to Subsect. 2.8.2

$$\mathbf{v}(t) = \mathbf{V}^{\mathrm{T}} \mathbf{x}_t \tag{7.121}$$

where $\mathbf{v}(t)$ is the whitened mixture and $\mathbf{V}$ denotes a $J_1 \times J_2$ whitening matrix. The components of $\mathbf{v}(t)$ are mutually uncorrelated with unit variances, namely, $\mathrm{E}\left[\mathbf{v}(t)\mathbf{v}(t)^{\mathrm{T}}\right] = \mathbf{I}_{J_2}$.

The demixing matrix $\mathbf{W}$ is factorized by

$$\mathbf{W}^{\mathrm{T}} = \mathbf{U}^{\mathrm{T}} \mathbf{V}^{\mathrm{T}} \tag{7.122}$$

where $\mathbf{U} = [\mathbf{u}_1 \cdots \mathbf{u}_{J_2}]$ is the $J_2 \times J_2$ orthogonal separating matrix, that is, $\mathbf{U}^{\mathrm{T}}\mathbf{U} = \mathbf{I}_{J_2}$. The vectors $\mathbf{u}_i$ can be obtained by iterating the whitened mixtures $\mathbf{v}(t)$

$$\widetilde{\mathbf{u}}_i = \mathrm{E}\left[\mathbf{v}\phi\left(\mathbf{u}_i^{\mathrm{T}}\mathbf{v}\right)\right] - \mathrm{E}\left[\dot{\phi}\left(\mathbf{u}_i^{\mathrm{T}}\mathbf{v}\right)\right]\mathbf{u}_i \tag{7.123}$$

$$\mathbf{u}_i = \frac{\widetilde{\mathbf{u}}_i}{\|\widetilde{\mathbf{u}}_i\|} \tag{7.124}$$

for $i = 1, \cdots, J_2$, where nonlinearity $\phi(\cdot)$ can be selected as $\phi_1(x) = \tanh(ax)$ and $\phi_2(x) = xe^{-\frac{x^2}{2}}$. The independent components can be estimated in a hierarchical fashion, that is, estimated one by one. After the ith independent component is estimated, the vector $\mathbf{u}_i$ is orthogonalized by an orthogonalization procedure. The FastICA can also be implemented in a symmetric mode, where all the independent components are extracted and orthogonalized at the same time [523]. A similar fixed-point algorithm based on the nonstationary property of signals has been proposed in [524].

In [411], a comprehensive experimental comparison has been conducted on different classes of ICA algorithms including the fixed-point FastICA [526], the infomax [94], the natural-gradient [31], the EASI [156], and an RLS-based nonlinear PCA [590]. The fixed-point FastICA with symmetric orthogonalization and tanh nonlinearity is concluded as the best tradeoff for estimating the ICA, since it provides results similar to that of the infomax and the natural-gradient, which are optimal with respect to minimizing the mutual information, but with a clearly smaller computational load. When $\phi(x) = \phi_3(x) = x^3$, the algorithm achieves cubic convergence; however, the algorithm is less accurate than the case when tanh nonlinearity is used [411].

The FastICA is easy to use and there are no step-size parameters to choose[3], while gradient descent-based algorithms seem to be preferable only if fast adaptivity in a changing environment is required. The FastICA algorithm directly finds the independent components of practically any non-Gaussian distribution using any nonlinearity $\phi(\cdot)$ [523].

The FastICA can approach the Cramer–Rao bound in two situations [618], namely, when the distribution of the sources is nearly Gaussian and the algorithm is in the symmetric mode using the nonlinear function $\phi_1(x)$, $\phi_2(x)$ or $\phi_3(x)$, and when the distribution of the sources is very different from Gaussian and the nonlinear function equals the score function of each independent component.

7.11.5 Independent Component Analysis Networks

The ICA does not require a nonlinear network for linear mixtures, but its basis vectors are usually nonorthogonal and the learning algorithm must contain some nonlinearities for higher-order statistics.

As we discussed previously, a three-layer J_1-J_2-J_1 linear autoassociative network can be used as a PCA network. It can also be used as an ICA network, as long as the outputs of the hidden layer are independent. For the ICA network, the weight matrix between the input and hidden layers corresponds to the $J_1 \times J_2$ demixing matrix $\mathbf{W}$, and the weight matrix from the hidden layer to the output layer corresponds to the $J_2 \times J_1$ mixing matrix $\mathbf{A}$. In [589], $\mathbf{W}$ is further factorized into two parts according to (7.122) and the network

---

[3] The fixed-point FastICA algorithm can be downloaded at the web site *http://www.cis.hut.fi/projects/ica/fastica/*.

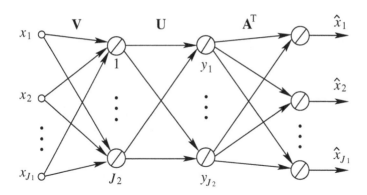

Fig. 7.7. Architecture of the ICA network. The slash (/) in a circle denotes a linear activation function, and the weight matrices between the layers are, respectively, $\mathbf{V}$, $\mathbf{U}$, and $\mathbf{A}^{\mathrm{T}}$. y_i, $i = 1, \cdots, J_2$, is the i output of the bottleneck layer, and $\hat{x}_j$, $j = 1, \cdots, J_1$, is the estimate of x_j.

becomes a four-layer J_1-J_2-J_2-J_1 network, as shown in Fig. 7.7. The weight matrices between the layers are, respectively, $\mathbf{V}$, $\mathbf{U}$, and $\mathbf{A}^{\mathrm{T}}$.

Each of the three weight matrices performs one of the processing tasks required for the ICA, namely, whitening, separation, and estimation of the basis vectors of the ICA. It can be used for both BSS and estimation of the basis vectors of the ICA, which is useful, for example, in projection pursuit. If the task is merely BSS, the last ICA basis vector estimation layer is not needed in the network.

The weights between the input and second layers perform prewhitening

$$\mathbf{v}(t) = \mathbf{V}^{\mathrm{T}}\mathbf{x}_t \tag{7.125}$$

The weights between the second and third layers perform separation

$$\mathbf{y}(t) = \mathbf{U}^{\mathrm{T}}\mathbf{v}(t) \tag{7.126}$$

and the weights between the last two layers estimate the basis vectors of the ICA

$$\hat{\mathbf{x}}(t) = \mathbf{A}\mathbf{y}(t) \tag{7.127}$$

When the three-layer linear autoassociative network is used for the PCA, we have the relations $\hat{\mathbf{x}} = \mathbf{A}\mathbf{W}^{\mathrm{T}}\mathbf{x}$ and $\mathbf{A} = \mathbf{W}(\mathbf{W}^{\mathrm{T}}\mathbf{W})^{-1}$ [58]. For the PCA, $\mathbf{W}^{\mathrm{T}}\mathbf{W} = \mathbf{I}_{J_2}$, thus $\hat{\mathbf{x}} = \mathbf{W}\mathbf{W}^{\mathrm{T}}\mathbf{x}$. The ICA solution can be obtained by imposing the additional constraint that the components of the bottleneck layer output vector $\mathbf{y} = \mathbf{W}^{\mathrm{T}}\mathbf{x}$ must be mutually independent or as independent as possible. For the ICA, $\hat{\mathbf{x}} = \mathbf{A}\mathbf{W}^{\mathrm{T}}\mathbf{x} = \mathbf{A}\left(\mathbf{A}^{\mathrm{T}}\mathbf{A}\right)^{-1}\mathbf{A}^{\mathrm{T}}\mathbf{x}$ is the LS approximation [589]. Note that $\mathbf{A}$ and $\mathbf{W}$ are the pseudoinverses of each other.

Prewhitening

Prewhitening is introduced in Subsect. 2.8.2. Accordingly

$$\mathbf{V}^{\mathrm{T}} = \mathbf{\Lambda}^{\frac{1}{2}}\mathbf{E}^{\mathrm{T}} \tag{7.128}$$

where $\mathbf{\Lambda} = \mathrm{diag}\,(\lambda_1, \cdots, \lambda_{J_2})$, $\mathbf{E} = [\mathbf{c}_1, \cdots, \mathbf{c}_{J_2}]$, with λ_i and $\mathbf{c}_i$ as the ith largest eigenvalue and the corresponding eigenvector of the covariance matrix $\mathbf{C}$. The PCA can be applied to solve for the eigenvalues and eigenvectors.

A simple local algorithm for learning the whitening matrix is given by [156]

$$\mathbf{V}(t+1) = \mathbf{V}(t) - \eta(t)\mathbf{V}(t)\left[\mathbf{v}(t)\mathbf{v}^{\mathrm{T}}(t) - \mathbf{I}\right] \tag{7.129}$$

It is used as a part of the EASI separation algorithm [156]. This algorithm does not have any optimality properties in data compression, and it sometimes suffers from stability problems. The validity of the algorithm can be justified by observing the whiteness condition $\mathrm{E}\left[\mathbf{v}(t)\mathbf{v}^{\mathrm{T}}(t)\right] = \mathbf{I}_{J_2}$ after convergence.

Prewhitening usually makes separation algorithms converge faster and often have better stability properties. However, if the mixing matrix $\mathbf{A}$ is ill-conditioned, whitening can make the separation of sources more difficult or even impossible [156, 589].

Separating Algorithms

The separating algorithms can be based on the robust PCA algorithm [867], the nonlinear PCA algorithm [867, 587], or the bigradient nonlinear PCA algorithm [589]. $\mathbf{W}$ can also be calculated iteratively without prewhitening as in the EASI algorithm [156], or the generalized EASI algorithm [589]. The nonlinear PCA algorithm is given as [589, 867, 587]

$$\mathbf{U}(t+1) = \mathbf{U}(t) + \eta(t)\left[\mathbf{v}(t) - \mathbf{U}(t)\phi(\mathbf{y}(t))\right]\phi\left(\mathbf{y}^{\mathrm{T}}(t)\right) \tag{7.130}$$

where $\phi(\cdot)$, such as $\phi(t) = t^3$, is usually selected to be odd for stability and separation reasons, and η is positive and slowly reducing to zero, or is a small constant.

Estimation of Basis Vectors

The estimation of the basis vectors can be based on the LS solution

$$\hat{\mathbf{x}}(t) = \widehat{\mathbf{A}}\mathbf{y}(t) = \mathbf{W}\left(\mathbf{W}^{\mathrm{T}}\mathbf{W}\right)^{-1}\mathbf{y}(t) \tag{7.131}$$

where $\widehat{\mathbf{A}} = [\hat{\mathbf{a}}_1 \cdots \hat{\mathbf{a}}_{J_2}] = \mathbf{W}\left(\mathbf{W}^{\mathrm{T}}\mathbf{W}\right)^{-1}$ is a $J_1 \times J_2$ matrix. If prewhitening, (7.128), is applied, $\widehat{\mathbf{A}}$ can be simplified as

$$\widehat{\mathbf{A}} = \mathbf{E}\mathbf{\Lambda}^{\frac{1}{2}}\mathbf{U} \tag{7.132}$$

Thus, the unnormalized ith basis vector of the ICA is $\hat{\mathbf{a}}_i = \mathbf{E}\boldsymbol{\Lambda}^{\frac{1}{2}}\mathbf{u}_i$ where $\mathbf{u}_i$ is the ith column of $\mathbf{U}$, and its squared norm becomes $\|\hat{\mathbf{a}}_i\|^2 = \mathbf{u}_i^{\mathrm{T}}\boldsymbol{\Lambda}\mathbf{u}_i$. Local algorithms for estimating the basis vectors can be derived by minimizing the MSE $\mathrm{E}\left[\|\mathbf{x} - \mathbf{A}\mathbf{y}\|^2\right]$ using the gradient-descent method [589] or faster convergent methods. The gradient descent-based algorithm is given by

$$\mathbf{A}(t+1) = \mathbf{A}(t) + \eta\mathbf{y}(t)\left[\mathbf{x}_t^{\mathrm{T}} - \mathbf{y}^{\mathrm{T}}(t)\mathbf{A}(t)\right] \tag{7.133}$$

Remarks

For any of the last three layers of the ICA network, it is possible to use either a local or a nonlocal learning method. Usually, local algorithms are advised to learn only the critical part, source separation, since efficient standard numerical methods are available for whitening and estimation of the basis vectors of the ICA. One can also use simple local learning algorithms in each layer.

7.11.6 Nonlinear Independent Component Analysis

ICA algorithms discussed so far are linear ICA methods for separating original sources from linear mixtures. Blind separation of the original signals in nonlinear mixtures has many difficulties such as the intrinsic indeterminacy, the unknown distribution of the sources as well as the mixing conditions, and the presence of noise. It is impossible to separate the original sources using only the source independence assumption of some unknown nonlinear transformations of the sources [527]. In spite of these difficulties, some methods are effective for nonlinear BSS. Nonlinear ICA can be modeled by a parameterized neural network whose parameters can be determined under the criterion of independence of its outputs.

The inverse of the nonlinear mixing model can be modeled by using the three-layer MLP [147, 1219, 1079] or the RBFN [1080]. In [1219], the natural-gradient method is applied to entropy maximization and mutual information minimization, and a sigmoidal function is adopted as the nonlinear function based on the work proposed in [147]. In [1079], the GA is used to replace the gradient-descent method. Two cost functions based on higher-order statistics are established to measure the statistical dependence of the outputs. The RBFN is applied for nonlinear BSS in [1080]. A contrast function comprised of mutual information and cumulants is defined, and this results in diminishing the indeterminacies caused by nonlinearity. Two learning algorithms are developed by using the gradient-descent method and an unsupervised clustering method.

The SOM [611] provides a parameter-free, nonlinear method to the nonlinear ICA problem [479, 709, 456]. It is a model-free method, but suffers from the exponential growth of the network complexity with the dimensions of the output lattice. In [456], an extended SOM-based technique is used to perform nonlinear ICA for the denoising of images.

The kernel-based nonlinear BSS method [770] exploits second-order statistics in a kernel-induced feature space. This method extends a linear algorithm to the nonlinear domain using the kernel-based method of extracting nonlinear features applied in the support vector machine (SVM) [1120] and the kernel PCA [1008]. The algorithm exploits second-order statistics rather than statistical independence and high-order statistics. The technique could likewise be applied to other linear covariance-based BSS algorithms.

Non-negative ICA

Non-negativity is a natural condition for many real-world applications, for example in the analysis of images, text, or air quality. Neural networks can be suggested by imposing a non-negativity constraint on the outputs [1203] or weights. Non-negative PCA and non-negative ICA algorithms are given in [922], where the sources s_i must be non-negative. The proposed non-negative ICA algorithms are all based on a two-stage process common to many other ICA algorithms, prewhitening and rotation. However, instead of using the usual non-Gaussianity measures such as the kurtosis in the rotation stage, the non-negativity constraint is used.

Constrained ICA

Constrained ICA (CICA) is a framework that incorporates additional requirements and prior information in the form of constraints into the ICA contrast function [735]. It is an optimization problem that can be solved by traditional optimization methods. Adaptive solutions using Newton-like learning are given in [735].

7.12 Constrained Principal Component Analysis

When certain subspaces are less preferred than others, this yields the constrainted PCA (CPCA) [640]. The optimality criterion for the CPCA is variance maximization, as in the PCA, but with an external subspace othogonality constraint that extracted PCs are orthogonal to some undesired subspace.

Given a J_1-dimensional stationary stochastic input vector process $\{\mathbf{x}_t\}$, and an l-dimensional ($l < J_1$) constraint process, $\{\mathbf{q}(t)\}$, such that

$$\mathbf{q}(t) = \mathbf{Q}\mathbf{x}_t \qquad (7.134)$$

where $\mathbf{Q}$ is an orthonormal constraint matrix, spanning an undesirable subspace $\mathcal{L}$. The task is to find, in the PC sense, the most representative J_2-dimensional subspace $\mathcal{L}^{J_2}$ that is constrained to be orthogonal to $\mathcal{L}$, where $l + J_2 \leq J_1$. That is, we are required to find the optimal linear transform

$$\mathbf{y}(t) = \mathbf{W}^{\mathrm{T}}\mathbf{x}_t \qquad (7.135)$$

where $\mathbf{W}$ is orthonormal, by minimizing

$$E_{\text{CPCA}} = \text{E}\left[\|\mathbf{x} - \hat{\mathbf{x}}\|^2\right] = \text{E}\left[\|\mathbf{x} - \mathbf{W}\mathbf{y}\|^2\right] \tag{7.136}$$

under the constraint

$$\mathbf{Q}\mathbf{W} = \mathbf{0} \tag{7.137}$$

The optimal solution to the CPCA problem is given by [640, 642]

$$\mathbf{W}^* = [\tilde{\mathbf{c}}_1 \cdots \tilde{\mathbf{c}}_{J_2}] \tag{7.138}$$

where $\tilde{\mathbf{c}}_i$, $i = 1, \cdots, J_2$, are the principal eigenvectors of the skewed autocorrelation matrix

$$\mathbf{C}_{\text{s}} = \left(\mathbf{I} - \mathbf{Q}\mathbf{Q}^{\text{T}}\right)\mathbf{C} \tag{7.139}$$

At the optimum, E_{CPCA} takes its minimum

$$E_{\text{CPCA}}^* = \sum_{i=J_2+1}^{J_1} \tilde{\lambda}_i \tag{7.140}$$

where $\tilde{\lambda}_i$, $i = 1, \cdots, J_1$, are the eigenvalues of $\mathbf{C}_{\text{s}}$ in descending order. Like the PCA, the components now maximize the output variance, but under the additional constraint (7.137).

The PCA usually obtains the best fixed-rank approximation to the data in the LS sense. On the other hand, the CPCA allows specifying metric matrices that modulate the effects of rows and columns of a data matrix. This actually is the weighted LS estimation. The CPCA first decomposes the data matrix by projecting the data matrix onto the spaces spanned by matrices of external information, and then applies the PCA to decomposed matrices, which involves the generalized SVD. The APEX algorithm has been applied to recursively solve the CPCA problem [642].

7.13 Localized Principal Component Analysis

The nonlinear PCA problem can be attacked in an alternative way. The data space is partitioned into a number of disjunctive regions, followed by the estimation of the principal subspace within each partition by linear PCA. This method can be called *localized PCA*[4], and the distribution is collectively modeled by a collection or a mixture of linear PCA models, each characterizing a partition. Most natural data sets have large eigenvalues in only a few eigendirections, while the variances in other eigendirections are so small as to be

---

[4] The localized PCA is different from the local PCA. A local PCA algorithm is a PCA algorithm in which the update at each node makes use of only the local information.

considered as noise. The localized PCA method provides an efficient means to decompose high-dimensional data-compression problems into low-dimensional data-compression problems.

The VQ-PCA [571] is a locally linear model that uses VQ to define the Voronoi regions for the localized PCA. The algorithm builds a piecewise linear model of the data. It performs better than the global models implemented by the linear PCA model and Kramer's nonlinear PCA, and is significantly faster than Kramer's nonlinear PCA [571].

The localized PCA method is commonly used in image compression. An image is often first transformation-coded by the PCA and then the coefficients are quantized. Adaptive combination of the PCA and VQ networks has been proposed in [1163], where an autoassociative network is used to perform the PCA and the SCL is used to perform the VQ. The error between the input and output of the autoassociative network is fed to the VQ network, and a simple learning algorithm is given. The network produces better results than by using the two algorithms successively.

An online localized PCA algorithm [813] has been developed by extending the NG method [768]. Instead of the Euclidean distance measure, a combination of a normalized Mahalanobis distance and the squared reconstruction error guides the competition between the units. The weighting between the two measures is determined from the residual variance in the minor subspace of each submodel. The unit centers are updated as in the NG, while the subspace learning is based on the RRLSA algorithm [876].

Under the same philosophy as for the localized PCA, the localized ICA is used to characterize nonlinear ICA. Clustering is first used for an overall coarse nonlinear representation of the underlying data and the linear ICA is then applied in each cluster so as to describe local features of the data [588]. This leads to a better representation of the data than in the linear ICA in a computationally feasible manner.

The ASSOM [614] is another localized PCA for unsupervised extraction of invariant local features from the input data. The ASSOM associates a subspace instead of a single weight vector to each node of the SOM. The subspaces in the ASSOM can be formed by applying the ICA [869].

7.14 Extending Algorithms to Complex Domain

The complex PCA is a generalization of the PCA for nonlinear feature extraction and dimensionality reduction in complex-valued data sets [506]. The complex PCA has been widely applied to complex-valued data and two-dimensional vector fields.

The complex PCA employs the same neural-network architecture as that of the PCA, but with complex weights. The objective functions for the PCA can also be adapted to the complex PCA by changing the transpose into the

Hermitian transpose. For example, for the complex PCA, one can minimize the MSE function

$$E = \frac{1}{N} \sum_{i=1}^{N} \left\| \mathbf{z}_i - \mathbf{W}\mathbf{W}^{\mathrm{H}}\mathbf{z}_i \right\|^2 \tag{7.141}$$

where $\mathbf{z}_i$, $i = 1, \cdots, N$, are the input complex vectors. By minimizing (7.141), the first complex PC is extracted.

In [939], a complex-valued neural-network model is developed for the nonlinear complex PCA (NLCPCA)[5]. The NLCPCA has the ability to extract nonlinear features missed by the PCA. The NLCPCA uses the architecture of Kramer's nonlinear PCA network [628], but with complex weights and biases. For a similar number of model parameters, the NLCPCA captures more variance of a data set than the alternative real approach, where each complex variable is replaced by two real variables and is applied to Kramer's nonlinear PCA. The complex hyperbolic tangent $\tanh(z)$ with $|z| < \frac{\pi}{2}$ [605] is selected as the nonlinear transfer function. Since the modulus of the net input of a neuron $\left|\mathbf{w}^{\mathrm{H}}\mathbf{z}\right|$ may be $\geq \frac{\pi}{2}$, a restriction on the magnitudes of the input and weights has to be considered. This is satisfied by initializing with weights and biases of small magnitudes, and using a weight penalty in the objective function. A complex-valued BP or quasi-Newton algorithm can be used for training.

Other examples of complex PCA algorithms may be mentioned here. Both the PAST and the PASTd are, respectively, the PSA and the PCA algorithms derived for complex-valued signals [1217]. A heuristic complex extension of the GHA [994] and the APEX [642] are, respectively, given in [288] and [215]. The robust complex PCA algorithms have also been derived in [248] for hierarchically extracting PCs of complex-valued signals based on a robust statistics-based loss function.

As far as the complex MCA is concerned, the constrained anti-Hebbian learning algorithm [399, 400] has been extended for the complex-valued TLS problem [399] and has been applied to adaptive FIR and IIR filtering. The adaptive invariant-norm MCA algorithm [740] has been generalized to the case for complex-valued input signal vector $\mathbf{x}(t)$.

For ICA algorithms, the FastICA algorithm has been applied to complex signals [115]. The ψ-APEX algorithms and the GHA are, respectively, extended to the complex-valued case [356, 357]. Based on a suitably selected nonlinear function, these algorithms can be used for BSS of complex-valued circular source signals.

---

[5] The MATLAB code for the NLCPCA can be downloaded from *http://www.ocgy.ubc.ca/projects/clim.pred/download.html.*

7.15 Other Generalizations of the PCA

In [1231, 727], simple neural-network models, described by differential equations, have been proposed to calculate the largest and smallest eigenvalues as well as their corresponding eigenvectors of any real symmetric matrix. Some other generalizations of the PCA are listed below.

Generalized Eigenvalue Decomposition

The generalized EVD (GEVD) is a statistical tool that is extremely useful in feature extraction, pattern recognition as well as signal estimation and detection.

The GEVD problem involves the matrix equation

$$\mathbf{R}_1 \mathbf{w}_i = \lambda_i \mathbf{R}_2 \mathbf{w}_i \tag{7.142}$$

where $\mathbf{R}_1, \mathbf{R}_2 \in R^{J_1 \times J_1}$, and λ_i, $\mathbf{w}_i$, $i = 1, \cdots, J_2$, are, respectively, the ith generalized eigenvalue and its corresponding generalized eigenvector. For real symmetric and positive-definite matrices, all the generalized eigenvectors are real and the corresponding generalized eigenvalues are positive.

The GEVD achieves simultaneous diagonalization of $\mathbf{R}_1$ and $\mathbf{R}_2$

$$\mathbf{W}^T \mathbf{R}_1 \mathbf{W} = \mathbf{\Lambda}, \quad \mathbf{W}^T \mathbf{R}_2 \mathbf{W} = \mathbf{I} \tag{7.143}$$

where $\mathbf{W} = [\mathbf{w}_1, \cdots, \mathbf{w}_{J_2}]$ and $\mathbf{\Lambda} = \text{diag}(\lambda_1, \cdots, \lambda_{J_2})$. Typically, $\mathbf{R}_1$ and $\mathbf{R}_2$ are, respectively, the full covariance matrices of zero-mean stationary random signals $\mathbf{x}_1, \mathbf{x}_2 \in R^{J_2}$. In this case, iterative GEVD algorithms can be obtained by using two PCA steps. When $\mathbf{R}_2$ becomes an identity matrix, the GEVD reduces to the PCA.

Any generalized eigenvector $\mathbf{w}_i$ is a stationary point of the criterion function

$$E_{\text{GEVD}}(\mathbf{w}) = \frac{\mathbf{w}^T \mathbf{R}_1 \mathbf{w}}{\mathbf{w}^T \mathbf{R}_2 \mathbf{w}} \tag{7.144}$$

The LDA problem is a typical GEVD problem. The three-layer LDA network [760] is obtained by the concatenation of two Rubner–Tavan PCA subnetworks. Each subnetwork is trained by the Rubner–Tavan PCA algorithm [974, 973]. Based on the Rubner–Tavan PCA network architecture, online local learning algorithms for LDA and GEVD are given in [297].

A number of methods for GEVD have been proposed in the literature. These methods are typically adaptive ones for online implementation. These include the LDA-based gradient-descent algorithm [190, 1202], a quasi-Newton type GEVD algorithm [774], an RLS-like fixed-point GEVD algorithm [936], the error correction learning [297], and the Hebbian learning [297]. All these algorithms first extract the principal generalized eigenvector and then estimate the minor generalized eigenvectors using a deflation procedure.

Two-dimensional PCA

Two-dimensional PCA [1220] is especially designed for image representation. An image covariance matrix is constructed directly using the original image matrices instead of the transformed vectors, and its eigenvectors are derived for image-feature extraction.

For $m \times n$ images, the size of the image covariance (scatter) matrix using the two-dimensional PCA is $n \times n$, whereas for the PCA the size is $mn \times mn$. This results in considerable computational advantage in the two-dimensional PCA. The two-dimensional PCA evaluates the covariance matrix more accurately over the PCA. When used for face recognition, the two-dimensional PCA results in a better recognition accuracy.

The two-dimensional PCA is a row-based PCA, and it only reflects the information between rows. The diagonal PCA [1254] improves the two-dimensional PCA by defining the image scatter matrix as the covariances between the variations of the rows and those of the columns of the images, and is shown to be more accurate than the PCA and the two-dimensional PCA.

Supervised PCA

Like the supervised clustering discussed in Sect. 5.10, the supervised PCA [216] is achieved by augmenting the input of the PCA with the class label of the data set.

7.16 Crosscorrelation Asymmetric Networks

Given two sets of random vectors with zero mean, $\{\mathbf{x}_t \in R^{n_1}\}$ and $\{\mathbf{y}_t \in R^{n_2}\}$, the crosscorrelation matrix is defined by

$$\mathbf{C}_{xy} = \mathrm{E}\left[\mathbf{x}_t \mathbf{y}_t^{\mathrm{T}}\right] = \sum_{i=1}^{n} \sigma_i \mathbf{v}_i^x \left(\mathbf{v}_i^y\right)^{\mathrm{T}} \tag{7.145}$$

where $\sigma_i > 0$ is the ith singular value, $\mathbf{v}_i^x$ and $\mathbf{v}_i^y$ are its corresponding left and right singular vectors, and $n = \min\{n_1, n_2\}$.

The crosscorrelation asymmetric PCA/MCA networks can be used to extract the singular values of the crosscorrelation matrix of two stochastic signal vectors, or to implement the SVD of a general matrix.

7.16.1 Extracting Multiple Principal Singular Components

The crosscorrelation asymmetric PCA (APCA) network consists of two sets of neurons that are laterally hierarchically connected [305]. The topology of the network is shown in Fig. 7.8. The vectors $\mathbf{x}$ and $\mathbf{y}$ are, respectively, the

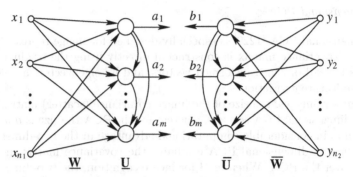

Fig. 7.8. Architecture of the crosscorrelation APCA network. The APCA network is composed of two hierarchical PCA networks. $\mathbf{x} \in R^{n_1}$ and $\mathbf{y} \in R^{n_2}$ are input vectors, $\mathbf{a}, \mathbf{b} \in R^m$ are the output vectors of the hidden layers, $\underline{\mathbf{W}}$ and $\overline{\mathbf{W}}$ are feedforward weight matrices, and $\underline{\mathbf{U}}$ and $\overline{\mathbf{U}}$ are lateral weight matrices.

n_1-dimensional and n_2-dimensional input signals, the $n_1 \times m$ matrix $\underline{\mathbf{W}} = [\underline{\mathbf{w}}_1 \ldots \underline{\mathbf{w}}_m]$ and the $n_2 \times m$ matrix $\overline{\mathbf{W}} = [\overline{\mathbf{w}}_1 \ldots \overline{\mathbf{w}}_m]$ are the feedforward weights, while the $n_2 \times m$ matrices $\underline{\mathbf{U}} = [\underline{\mathbf{u}}_1 \ldots \underline{\mathbf{u}}_m]$ and $\overline{\mathbf{U}} = [\overline{\mathbf{u}}_1 \ldots \overline{\mathbf{u}}_m]$ are the lateral connection weights, where $\underline{\mathbf{u}}_i = (\underline{u}_{1i}, \cdots, \underline{u}_{mi})^{\mathrm{T}}$, $\overline{\mathbf{u}}_i = (\overline{u}_{1i}, \cdots, \overline{u}_{mi})^{\mathrm{T}}$, and $m \leq \min\{n_1, n_2\}$. This model is to perform the SVD of $\mathbf{C}_{xy}$ [305].

The network has the following relations:

$$\mathbf{a} = \underline{\mathbf{W}}^{\mathrm{T}}\mathbf{x} \tag{7.146}$$

$$\mathbf{b} = \overline{\mathbf{W}}^{\mathrm{T}}\mathbf{y} \tag{7.147}$$

where $\mathbf{a} = (a_1, \cdots, a_m)^{\mathrm{T}}$ and $\mathbf{b} = (b_1, \cdots, b_m)^{\mathrm{T}}$.

The objective function for extracting the first principal singular value of the covariance matrix is given by

$$E_{\mathrm{APCA}}(\underline{\mathbf{w}}, \overline{\mathbf{w}}) = \frac{\mathrm{E}\left[a_1(t)b_1(t)\right]}{\|\underline{\mathbf{w}}\|\,\|\overline{\mathbf{w}}\|} = \frac{\underline{\mathbf{w}}^{\mathrm{T}}\mathbf{C}_{xy}\overline{\mathbf{w}}}{\|\underline{\mathbf{w}}\|\,\|\overline{\mathbf{w}}\|} \tag{7.148}$$

It is an indefinite function. When $\mathbf{y} = \mathbf{x}$, it reduces to the PCA [864]. After the principal singular component is extracted, a deflation transformation is introduced to nullify the principal singular value so as to make the next singular value principal. Thus, $\mathbf{C}_{xy}$ in the criterion (7.148) can be replaced by one of the following three transformed forms so as to extract the $(i+1)$th principal singular component

$$\mathbf{C}_{xy}^{(i+1)} = \mathbf{C}_{xy}^{(i)}\left(\mathbf{I} - \mathbf{v}_i^y\left(\mathbf{v}_i^y\right)^{\mathrm{T}}\right) \tag{7.149}$$

$$\mathbf{C}_{xy}^{(i+1)} = \left(\mathbf{I} - \mathbf{v}_i^x\left(\mathbf{v}_i^x\right)^{\mathrm{T}}\right)\mathbf{C}_{xy}^{(i)} \tag{7.150}$$

$$\mathbf{C}_{xy}^{(i+1)} = \left(\mathbf{I} - \mathbf{v}_i^x\left(\mathbf{v}_i^x\right)^{\mathrm{T}}\right)\mathbf{C}_{xy}^{(i)}\left(\mathbf{I} - \mathbf{v}_i^y\left(\mathbf{v}_i^y\right)^{\mathrm{T}}\right) \tag{7.151}$$

for $i = 1, \cdots, m - 1$, where $\mathbf{C}_{xy}^{(1)} = \mathbf{C}_{xy}$. These are, respectively, obtained by the transforms on the data:

$$\mathbf{x} \leftarrow \mathbf{x}, \qquad \mathbf{y} \leftarrow \mathbf{y} - \mathbf{v}_i^x \left(\mathbf{v}_i^x\right)^{\mathrm{T}} \mathbf{y} \qquad (7.152)$$

$$\mathbf{x} \leftarrow \mathbf{x} - \mathbf{v}_i^y \left(\mathbf{v}_i^y\right)^{\mathrm{T}} \mathbf{x}, \qquad \mathbf{y} \leftarrow \mathbf{y} \qquad (7.153)$$

$$\mathbf{x} \leftarrow \mathbf{x} - \mathbf{v}_i^y \left(\mathbf{v}_i^y\right)^{\mathrm{T}} \mathbf{x}, \qquad \mathbf{y} \leftarrow \mathbf{y} - \mathbf{v}_i^x \left(\mathbf{v}_i^x\right)^{\mathrm{T}} \mathbf{y} \qquad (7.154)$$

Using a deflation transformation, the two sets of neurons are trained with the crosscoupled Hebbian learning rules, which are given by [305]

$$\underline{\mathbf{w}}_j(t+1) = \underline{\mathbf{w}}_j(t) + \eta \left[\mathbf{x}(t) - \underline{\mathbf{w}}_j(t) a_j(t)\right] b_j'(t) \qquad (7.155)$$

$$\overline{\mathbf{w}}_j(t+1) = \overline{\mathbf{w}}_j(t) + \eta \left[\mathbf{y}(t) - \overline{\mathbf{w}}_j(t) b_j(t)\right] a_j'(t) \qquad (7.156)$$

for $j = 1, \cdots, m$, where η is the learning rate selected as a small constant or according to the Robbins–Monro conditions,

$$a_j' = a_j - \sum_{i=1}^{j-1} u_{ij} a_i \qquad (7.157)$$

$$b_j' = b_j - \sum_{i=1}^{j-1} \overline{u}_{ij} b_i \qquad (7.158)$$

$$a_i = \underline{\mathbf{w}}_i^{\mathrm{T}} \mathbf{x}, \quad b_i = \overline{\mathbf{w}}_i^{\mathrm{T}} \mathbf{y}, \quad i = 1, \cdots, j \qquad (7.159)$$

and the lateral weights should be equal to

$$\underline{u}_{ij} = \underline{\mathbf{w}}_i^{\mathrm{T}} \underline{\mathbf{w}}_j, \quad \overline{u}_{ij} = \overline{\mathbf{w}}_i^{\mathrm{T}} \overline{\mathbf{w}}_j, \quad i = 1, \cdots, j - 1 \qquad (7.160)$$

A set of lateral connections among the units is called a *lateral othogonaliztion network*. Hence, $\overline{\mathbf{U}}$ and $\underline{\mathbf{U}}$ are upper triangular matrices. The stability of the algorithm is proved based on the Lyapunov's second theorem.

A local algorithm for calculating $\underline{u}_{ij}$ and $\overline{u}_{ij}$ is derived by premultiplying (7.155) with $\underline{\mathbf{w}}_i^{\mathrm{T}}$, premultiplying (7.156) with $\overline{\mathbf{w}}_i^{\mathrm{T}}$ (7.157) and then employing (7.160)

$$\underline{u}_{ij}(t+1) = \underline{u}_{ij}(t) + \eta \left[a_i(t) - \overline{u}_{ij}(t) a_j(t)\right] b_j'(t) \qquad (7.161)$$

$$\overline{u}_{ij}(t+1) = \overline{u}_{ij}(t) + \eta \left[b_i(t) - \overline{u}_{ij}(t) b_j(t)\right] a_j'(t) \qquad (7.162)$$

The initial values can be selected as $\underline{u}_{ij}(0) = \underline{\mathbf{w}}_i^{\mathrm{T}}(0) \underline{\mathbf{w}}_j(0)$ and $\overline{u}_{ij}(0) = \overline{\mathbf{w}}_i^{\mathrm{T}}(0) \overline{\mathbf{w}}_j(0)$. However, this initial condition is not critical to the convergence of the algorithm.

$\underline{\mathbf{w}}_i$ and $\overline{\mathbf{w}}_i$ approximate the ith left and right principal singular vectors of $\mathbf{C}_{xy}$, respectively, and σ_i approximates its corresponding criterion E_{APCA}, as $t \to \infty$; that is, the algorithm extracts the first m principal singular values in the descending order and their corresponding left and right singular vectors. Like the APEX, the APCA algorithm incrementally adds nodes without retraining the learned nodes. Exponential convergence has been demonstrated by simulation [305].

7.16.2 Extracting the Largest Singular Component

When m in the APCA network is selected as unity, the principal singular component of $\mathbf{C}_{xy}$ can be extracted by using a modification to the crosscoupled Hebbian rule [349]

$$\underline{\mathbf{w}}_1(t+1) = \underline{\mathbf{w}}_1(t) + \eta \left[b_1(t)\mathbf{x}(t) - \|\underline{\mathbf{w}}_1(t)\|^2 \underline{\mathbf{w}}_1(t) \right] \qquad (7.163)$$

$$\overline{\mathbf{w}}_1(t+1) = \overline{\mathbf{w}}_1(t) + \eta \left[a_1(t)\mathbf{y}(t) - \|\overline{\mathbf{w}}_1(t)\|^2 \overline{\mathbf{w}}_1(t) \right] \qquad (7.164)$$

This algorithm can efficiently extract the principal singular component, and has been proved to have global asymptotic convergence. When $\mathbf{y}(t) = \mathbf{x}(t)$, the algorithm is reduced to the PCA algorithm given in [864].

When the crosscorrelation matrix is replaced by a general nonsquare matrix, (7.163) and (7.164) can be directly transformed into the algorithm for extracting the principal singular component of a general matrix $\mathbf{A} \in R^{n_1 \times n_2}$ [349]

$$\underline{\mathbf{w}}_1(t+1) = \underline{\mathbf{w}}_1(t) + \eta \left[\mathbf{A}\overline{\mathbf{w}}_1(t) - \|\underline{\mathbf{w}}_1(t)\|^2 \underline{\mathbf{w}}_1(t) \right] \qquad (7.165)$$

$$\overline{\mathbf{w}}_1(t+1) = \overline{\mathbf{w}}_1(t) + \eta \left[\mathbf{A}^{\mathrm{T}}\underline{\mathbf{w}}_1(t) - \|\overline{\mathbf{w}}_1(t)\|^2 \overline{\mathbf{w}}_1(t) \right] \qquad (7.166)$$

7.16.3 Extracting Multiple Principal Singular Components for Nonsquare Matrices

Using the algorithm given by (7.165) and (7.166) and a deflation transformation, one can extract multiple principal singular components for the nonsquare matrix $\mathbf{A}$ [349]

$$\underline{\mathbf{w}}_i(t+1) = \underline{\mathbf{w}}_i(t) + \eta_i \left[\mathbf{A}_i\overline{\mathbf{w}}_i(t) - \|\underline{\mathbf{w}}_i(t)\|^2 \underline{\mathbf{w}}_i(t) \right] \qquad (7.167)$$

$$\overline{\mathbf{w}}_i(t+1) = \overline{\mathbf{w}}_i(t) + \eta_i \left[\mathbf{A}_i^{\mathrm{T}}\underline{\mathbf{w}}_i(t) - \|\overline{\mathbf{w}}_i(t)\|^2 \overline{\mathbf{w}}_i(t) \right] \qquad (7.168)$$

for $i = 1, \cdots, m$, and

$$\mathbf{A}_{i+1} = \mathbf{A}_i - \underline{\mathbf{w}}_i\overline{\mathbf{w}}_i^{\mathrm{T}}, \quad i = 1, \cdots, m-1 \qquad (7.169)$$

where $\mathbf{A}_1 = \mathbf{A}$, and the learning rates η_i are suggested to be $\eta_i < \frac{1}{4\sigma_i}$. As $t \to \infty$, $\underline{\mathbf{w}}_i(t)$ and $\overline{\mathbf{w}}_i(t)$ represent the left and right singular vectors corresponding to the ith singular value, arranged in the descending order $\sigma_1 \geq \sigma_2 \geq \cdots \geq \sigma_m > 0$.

The measures of stopping iteration can be given as

$$e_i(t) = \left\| \mathbf{A}_i\overline{\mathbf{w}}_i(t) - \|\underline{\mathbf{w}}_i(t)\|^2 \underline{\mathbf{w}}_i(k) \right\|^2$$
$$+ \left\| \mathbf{A}_i^{\mathrm{T}}\underline{\mathbf{w}}_i(t) - \|\overline{\mathbf{w}}_i(t)\|^2 \overline{\mathbf{w}}_i(t) \right\|^2 < \varepsilon \qquad (7.170)$$

for $i = 1, \cdots, m$, where ε is a small number such as 10^{-20}. The above algorithm has been proved to be convergent.

The algorithm can efficiently perform the SVD of an ill-posed matrix. It can be used to solve the smallest singular component of the general matrix $\mathbf{A}$. Although the method is indirect for computing the smallest singular component of a nonsquare matrix, it is efficient and robust. The algorithm is especially useful for TLS problems. Some adaptive SVD algorithms for subspace tracking of a recursively updated data matrix have been surveyed and proposed in [55].

7.17 Computer Experiments

The concept of subspace is involved in many signal-processing problems. This requires the EVD of the autocorrelation matrix of a data set or the SVD of the crosscorrelation matrix of two data sets. In the area of ASP, typical applications are cyclostationary beamforming algorithms using the APCA algorithm [323], and DoA estimation using an MCA or PCA algorithm. Interested readers are referred to [321]. In the following, we first give a simulation comparison of three popular PCA algorithms, and then sample an application of the PCA for image compression.

7.17.1 Example 7.1: A Comparison of the Weighted SLA, the GHA, and the APEX

This example illustrates the use of three PCA algorithms, namely, the weighted SLA, the GHA, and the APEX, for extracting the multiple PCs. The computation is conducted on a PC with a Pentium III 600 MHz CPU and 128M RAM using the Matlab©.

Given a data set of 500 vectors $\{\mathbf{x}_p \in R^3\}$, where $\mathbf{x}_p = (x_{p,1}, x_{p,2}, x_{p,3})^{\mathrm{T}}$. We take $x_{p,1} = 1 + N(0, 1)$, $x_{p,2} = 2N(0, 1)$, and $x_{p,3} = 4N(0, 1)$, where $N(0, 1)$ denotes a normal distribution. The autocorrelation matrix is calculated as $\mathbf{C} = \frac{1}{500}\sum_{p=1}^{500} \mathbf{x}_p \mathbf{x}_p^{\mathrm{T}}$. Theoretically, for infinite number of samples, $\mathbf{C}$ has 3 eigenvalues, in the descending order: $\lambda_1 = 16$, $\lambda_2 = 4$, and $\lambda_3 = 2$, and the corresponding eigenvectors are $\mathbf{c}_1 = (0, 0, 1)^{\mathrm{T}}$, $\mathbf{c}_2 = (0, 1, 0)^{\mathrm{T}}$, and $\mathbf{c}_3 = (1, 0, 0)^{\mathrm{T}}$. In this example, the simulation results slightly deviate from these values since we use only 500 samples.

We conduct simulation for three popular PCA algorithms: the weighted SLA, the GHA, and the APEX. For the weighted SLA, we select $\boldsymbol{\gamma} = (1, 2, 3)^{\mathrm{T}}$. Our numerical experiments show that for the APEX the optimal learning rate given by (7.84) is undesirable in implementation. We select for the three algorithms the same learning rate $\eta_i = \frac{1}{100+t}$, where each t corresponds to the presentation of a new sample. The data set is repeated 10 times, and the training samples are provided in a fixed deterministic sequence. For each

Table 7.1. Comparison of PCA algorithms: the weighted SLA, the GHA, the APEX, and the EVD. The results are based on the average of 50 random runs. The EVD method is used as the theoretical solution to the PCA problem.

	Weighted SLA	GHA	APEX	EVD
Training time (s)	1.8806	2.4990	2.6746	0.0292
λ_1	15.7263	15.7707	15.7685	15.8097
λ_2	1.8831	3.9223	3.8878	3.9411
λ_3	0.7017	1.8842	1.9278	1.9942
$\cos\theta_1$	1.0000	0.9999	0.9999	—
$\cos\theta_2$	0.9736	1.0000	0.9995	—
$\cos\theta_3$	0.9750	0.9999	0.9949	—
$\|\mathbf{w}_1\|$	1.0000	1.0003	0.9999	1.0000
$\|\mathbf{w}_2\|$	0.7062	1.0007	1.0003	1.0000
$\|\mathbf{w}_3\|$	0.5756	0.9999	1.0046	1.0000

algorithm, we calculate the adaptations for $\|\mathbf{w}_i\|$, λ_i, and the cosine of the angle θ_i between $\mathbf{w}_i$ and $\mathbf{c}_i$.

The performances of the algorithms are evaluated by averaging 50 random runs. The results at the end of 10 epochs are listed in Table 7.1. The adaptations for a random run are shown in Figs. 7.9, 7.10, and 7.11. Analysis based on Table 7.1 shows that the GHA and the APEX have similar performance. Our empirical results also show that for the GHA a larger starting η can be used.

From Figs. 7.9, 7.10, and 7.11, we see that for all the algorithms, the convergence to smaller eigenvalues is slow. The strategy used in the ALA [202] can be applied to make the algorithms converge to all the eigenvalues at the same speed. For all three algorithms, $\mathbf{w}_i$ converges to the directions of $\mathbf{c}_i$. However, for the weighted SLA, only $\|\mathbf{w}_1\|$ converges to unity, while the remaining $\|\mathbf{w}_i\|$ do not converge to unity, and accordingly λ_i, $i = 2, 3$, do not converge to their theoretical values.

For the weighted SLA, we also test for different γ_i. When all γ_i are selected as unity, the weighted SLA reduces to the SLA. $\cos\theta_i$ could be any value between $[-1, +1]$. For the SLA, all the $\|\mathbf{w}_i\|$ converge to unity very rapidly, but the λ_is converge to some values different from their theoretical eigenvalues and also not in the descending order. When γ_is are selected as sufficiently distinct values, the weight vectors converge to the directions or directions opposite to those of the principal eigenvectors, that is, $\cos\theta_i \to \pm 1$ as $t \to \infty$. The converging λ_i and $\|\mathbf{w}_i\|$ do not converge to their respective theoretical values λ_i^{EVD} and $\|\mathbf{w}_i^{\mathrm{EVD}}\|$, which are calculated from the EVD. Empirical results show that $\lambda_i^{\mathrm{EVD}} = \gamma_i \lambda_i$ and $\|\mathbf{w}_i^{\mathrm{EVD}}\| = \sqrt{\gamma_i}\|\mathbf{w}_i\|$.

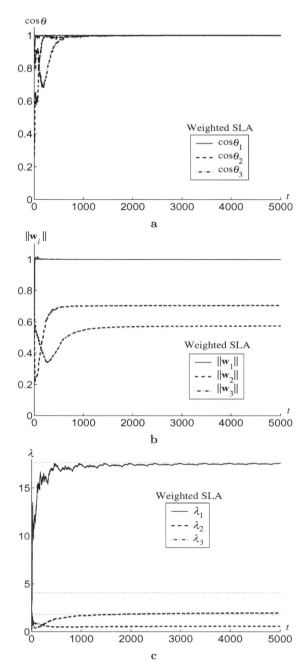

Fig. 7.9. Adaptation of the weighted SLA. (**a**) The cosine of the angle θ_i between $\mathbf{w}_i$ and $\mathbf{c}_i$. (**b**) The normal of weight vectors $\|\mathbf{w}_i\|$. (**c**) Eigenvalues λ_i. The dotted lines in (**c**) correspond to the theoretical eigenvalues.

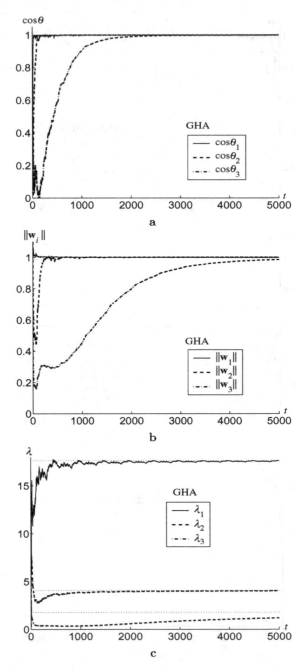

Fig. 7.10. Adaptation of the GHA. (**a**) The cosine of the angle θ_i between $\mathbf{w}_i$ and $\mathbf{c}_i$. (**b**) The normals of weight vectors $\|\mathbf{w}_i\|$. (**c**) Eigenvalues λ_i. The dotted lines in (**c**) correspond to the theoretical eigenvalues.

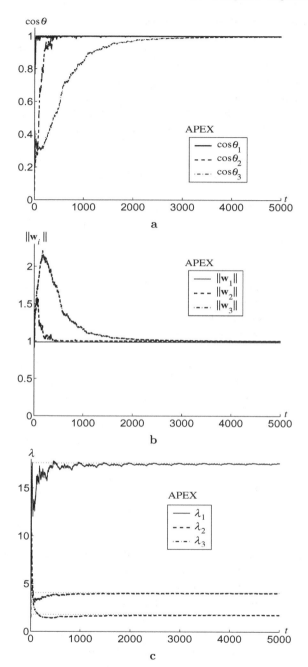

Fig. 7.11. Adaptation of the APEX. (**a**) The cosine of the angle θ_i between $\mathbf{w}_i$ and $\mathbf{c}_i$. (**b**) The normal of weight vectors $\|\mathbf{w}_i\|$. (**c**) Eigenvalues λ_i. The dotted lines in (**c**) correspond to the theoretical eigenvalues.

7.17.2 Example 7.2: Image Compression

Image compression is to remove the redundancy in an image for storage and/or transmission purposes. It is an optimization problem so that the decoded image has the best quality. For a large picture, one usually partitions the picture into many nonoverlapping 8×8-pixel regions, and then compresses each region one by one. For example, if we compress each of the 64-pixel patch into 8 data, we achieve a compression ratio of 1:8. This work can be performed by using a PCA network.

Here we reproduce the example given by Sanger [994]. An image of three children, as shown in Fig. 7.12a, is used for training. It is digitized as a 256×256 image with 256 gray levels.

A linear 64-8 PCA network is used to learn the image. By the 8×8 partitioning, we get $32 \times 32 = 1024$ samples. The samples are fed twice. Each of the output nodes is connected by 64 weights, denoted by an 8×8 mask. The codebook for the coding is comprised of all the 64-weight vectors. The training results are illustrated in Fig. 7.13, where the positive weights are shown as white, the negative weights as black, and zero weights as gray.

After the learning algorithm converges, the PCA network can be used to code the image. An 8×8 block is multiplied by each of the eight weight masks, and this yields 8 coefficients. Sanger further uniformly quantizes each of the 8 coefficients with a number of bits, which are proportional to the logarithm of the variance of that coefficient over the whole image. The first two coefficients require 5 bits each, the third coefficient requires 3 bits, and the remaining five coefficients require 2 bits each, and a total of 23 bits are used by Sanger to code each 8×8 block, that is, a bit rate of 0.36 bits per pixel. This achieves a compression ratio of $\frac{64 \times 8}{23} = 22.26$ to 1.

The reconstruction of the image from the quantized coefficients can be conducted by multiplying the weights by those quantized coefficients, and combining the reconstructed blocks into an image. The reconstructed image is illustrated in Fig. 7.12b. The reconstructed image is sufficiently good to the eye, and is as good as the network without any quantization.

The PCA network trained using the image of three children can be used to code images with similar statistical properties. Sanger uses the image of a dog, shown in Fig. 7.14a, to test the generalization capability of the trained network. The coding of the dog requires a total of 35 bits for each 8×8 block of pixels, resulting in a bit rate of 0.55 bits per pixel. The reconstructed image of the dog is shown in Fig. 7.14b, which is surprisingly of good quality to the human eye.

Similar results for image compression using a three-layer autoassociative network with BP learning has been reported in [265, 461].

The PCA as well as the LDA achieves the same results for an original data set and its orthonormally transformed version [209]. Thus, the PCA and the LDA can be directly implemented in the DCT domain and the results are exactly the same as that obtained from the spatial domain. For images

a

b

Fig. 7.12. Image compression of the image of three children. (**a**) The original image. (**b**) The reconstructed image. (From [994].)

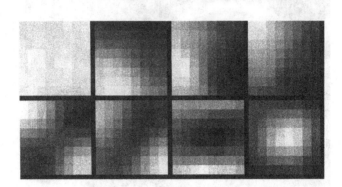

Fig. 7.13. Network weights after training with the image of children. Positive weights are shown white, negative weights are shown as black, and zero weights are shown as gray. (From [994].)

a

Fig. 7.14. Image compression of the image of a dog. (a) The original image. (b) The reconstructed image. (From [994].) (To be continued.)

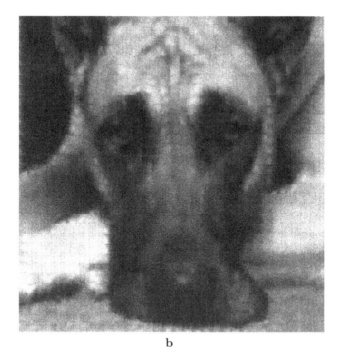

b

Fig. 7.14. continued (from [994])

compressed using the DCT, such as in the JPEG or MPEG standard, the PCA and and the LDA can be directly implemented in the DCT domain such that the inverse DCT transform can be avoided and computation conducted on a reduced data dimensionality.

8

Fuzzy Logic and Neurofuzzy Systems

The field of fuzzy sets and logic is a rigorous mathematical one, and provides an effective tool for modeling the uncertainty in human reasoning. In fuzzy logic, the knowledge of experts is modeled by linguistic rules represented in the form of IF-THEN logic. Like the MLP and the RBFN, some fuzzy inference systems (FISs) have universal function approximation capability. FISs can be used in many areas where neural networks are applicable, such as in data analysis, control and signal processing.

In this chapter, we first introduce fundamentals of fuzzy sets and logic. A comparison between FISs and neural networks is then made. This is followed by the conversions between neural networks and FISs as well as rule extraction from numerical data. Finally, a discussion on the synergy of neural and fuzzy systems is included and some neurofuzzy models described.

8.1 Fundamentals of Fuzzy Logic

The concept of fuzzy sets was first proposed by Zadeh [1248] as a method for modeling the uncertainty in human reasoning. Rather than the binary logic, fuzzy logic uses the notion of membership. Fuzzy logic is most suitable for the representation of vague data and concepts on an intuitive basis, such as human linguistic description, *e.g.* the expressions *approximately, good, tall*. The conventional or *crisp* set can be treated as a special case of the concept of a fuzzy set. A fuzzy set is uniquely determined by its membership function (MF), and it is also associated with a linguistically meaningful term.

Fuzzy logic provides a systematic framework to incorporate human experience. Fuzzy logic is based on three core concepts, namely, fuzzy sets, linguistic variables, and possibility distributions. A fuzzy set is an effective means to represent linguistic variables. A linguistic variable is a variable whose value can be described qualitatively using a linguistic expression and quantitatively using an MF [1249]. Linguistic expressions are useful for communicating concepts and knowledge with human beings, whereas MFs are useful for processing numeric input data. When a fuzzy set is assigned to a linguistic variable, it imposes an elastic constraint, called a *possibility distribution*, on the possible values of the variable.

Fuzzy logic is a rigorous mathematical discipline. Fuzzy reasoning is a straightforward formalism for encoding human knowledge or common sense in a numerical framework, and FISs can approximate arbitrarily well any continuous function on a compact domain [622, 1149]. FISs and FNNs can approximate each other to any degree of accuracy [143].

Fuzzy logic first found popular applications in control systems. In fuzzy control, human knowledge is codified by means of linguistic IF-THEN rules, which build up an FIS. An exact model is not needed for controller design. Since its first reported industrial application in 1982 [498], Japanese industry has produced numerous consumer appliances using fuzzy controllers. This has aroused global interest in the industrial and scientific community, and fuzzy logic has also been widely applied in data analysis, regression, and signal and image processing. To meet the strict demand on real-time processing, many application-specific integrated circuit (ASIC) designs have been advanced [324].

8.1.1 Definitions and Terminologies

In this section, we list some definitions and terminologies used in the fuzzy logic literature.

Universe of Discourse

The universal set $\mathcal{X} : \mathcal{X} \to [0,1]$ is called the *universe of discourse*, or simply the *universe*. The implication $\mathcal{X} \to [0,1]$ is the abbreviation for the IF-THEN rule:

$$\text{IF } x \text{ is in } \mathcal{X}, \text{ THEN its MF } \mu_{\mathcal{X}}(x) \text{ is in } [0,1].$$

where $\mu_{\mathcal{X}}(x)$ is the MF of x. The universe $\mathcal{X}$ may contain either discrete or continuous values.

Linguistic Variable

A linguistic variable is a variable whose values are linguistic terms in a natural or artificial language. For example, the size of an object is a linguistic variable, whose value can be *small*, *medium*, and *big*.

Fuzzy Set

A *fuzzy set* $\mathcal{A}$ in $\mathcal{X}$ is defined by

$$\mathcal{A} = \{ (x, \mu_A(x)) \mid x \in \mathcal{X} \} \tag{8.1}$$

where $\mu_A(x) \in [0, 1]$ is the MF of x in $\mathcal{A}$. For $\mu_A(x)$, the value 1 stands for complete membership of the set $\mathcal{A}$, while 0 represents that x does not belong to the set at all.

A *fuzzy set* can also be represented by

$$A = \begin{cases} \sum_{x_i \in \mathcal{X}} \frac{\mu_A(x_i)}{x_i}, & \text{if } \mathcal{X} \text{ is discrete} \\ \int_{\mathcal{X}} \frac{\mu_A(x)}{x}, & \text{if } \mathcal{X} \text{ is continuous} \end{cases} \tag{8.2}$$

The summation, integral, and division signs syntactically denote the union of $(x, \mu_A(x))$ pairs.

Support

The elements on the fuzzy set $\mathcal{A}$ whose membership is larger than zero are called the *support* of the fuzzy set

$$\text{sp}(\mathcal{A}) = \{x \in \mathcal{A} | \mu_A(x) > 0\} \tag{8.3}$$

Height

The *height* of a fuzzy set $\mathcal{A}$ is defined by

$$\text{hgt}(\mathcal{A}) = \sup\{\mu_A(x) | x \in \mathcal{X}\} \tag{8.4}$$

Normal Fuzzy Set

If $\text{hgt}(\mathcal{A}) = 1$, then the fuzzy set $\mathcal{A}$ is said to be *normal*.

Non-normal Fuzzy Set

If $0 < \text{hgt}(\mathcal{A}) < 1$, the fuzzy set $\mathcal{A}$ is said to be *non-normal*. It can be normalized by dividing it by the height of $\mathcal{A}$

$$\bar{\mu}_A(x) = \frac{\mu_A(x)}{\text{hgt}(\mathcal{A})} \tag{8.5}$$

Fuzzy Subset

A fuzzy set $\mathcal{A} = \{(x, \mu_A(x)) | x \in \mathcal{X}\}$ is said to be a *fuzzy subset* of $\mathcal{B} = \{(x, \mu_B(x)) | x \in \mathcal{X}\}$ if $\mu_A(x) \leq \mu_B(x)$, denoted by $\mathcal{A} \subseteq \mathcal{B}$, where $\subseteq$ is the inclusion operator.

Fuzzy Partition

For a linguistic variable, a number of fuzzy subsets are enumerated as the value of the variable. This collection of fuzzy subsets is called a *fuzzy partition*. Each fuzzy subset has a MF. For a finite fuzzy partition $\{A_1, A_2, \cdots, A_n\}$ of a set A, the MF for each $x \in A$ satisfies

$$\sum_{i=1}^{n} \mu_{A_i}(x) = 1 \qquad (8.6)$$

and A_i is normal, that is, the height of A_i is unity. A fuzzy partition is illustrated in Fig. 8.1.

Empty Set

The subset of $\mathcal{X}$ having no element is called the *empty set*, denoted by $\emptyset$.

Complement

The *complement* of A, written $\overline{A}$, $\neg A$ or NOT A, is defined as $\mu_{\overline{A}}(x) = 1 - \mu_A(x)$. Thus, $\overline{\mathcal{X}} = \emptyset$ and $\overline{\emptyset} = \mathcal{X}$.

α-cut

The *α-cut* or *α-level set* of a fuzzy set A, written $\mu_A[\alpha]$, is defined as the set of all elements in A whose degree of membership is not less than α

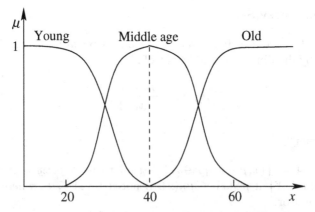

Fig. 8.1. A fuzzy partition of human age. The fuzzy set for representing the linguistic variable *human age* is partitioned into three fuzzy subsets, namely, *young*, *middle age*, and *old*. Each fuzzy subset is characterized by an MF.

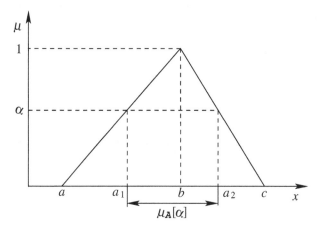

Fig. 8.2. Schematic of the triangular MF and its α-cut. a, b, and c are shape parameters. $\mu_{\mathcal{A}}[\alpha] = [a_1, a_2]$.

$$\mu_{\mathcal{A}}[\alpha] = \{x \in \mathcal{A} \,|\, \mu_{\mathcal{A}}(x) \geq \alpha\} \tag{8.7}$$

for $\alpha \in [0, 1]$. This is illustrated in Fig. 8.2, where $\mu_{\mathcal{A}}[\alpha]$ can be written as an interval $[a_1, a_2]$.

A fuzzy set $\mathcal{A}$ is usually represented as $\mu = \mu(x)$, where μ is the fuzzy logic, $\mu(\cdot)$ is the MF, and $x \in \mathcal{A}$. The inverse of $\mu = \mu(x)$ can be represented by $x = \mu^{-1}(\alpha)$, $\alpha \in [0, 1]$, where each value of α may correspond to one or more values of x. A fuzzy set is usually represented by a finite number of its membership values.

Kernel or Core

All the elements in a fuzzy set $\mathcal{A}$ with membership degree 1 constitute a subset called the *kernel* or *core* of the fuzzy set, written as $\mathrm{co}(\mathcal{A}) = \mu_{\mathcal{A}}[1]$.

Convex Fuzzy Set

A fuzzy set $\mathcal{A}$ is said to be *convex* if and only if

$$\mu_{\mathcal{A}}\left(\lambda x_1 + (1 - \lambda)x_2\right) \geq \mu_{\mathcal{A}}\left(x_1\right) \wedge \mu_{\mathcal{A}}\left(x_2\right) \tag{8.8}$$

for $\lambda \in [0, 1]$, and $x_1, x_2 \in \mathcal{X}$, where $\wedge$ denotes the minimum operation. Any α-cut set of a convex fuzzy set is a closed interval.

Concave Fuzzy Set

A fuzzy set $\mathcal{A}$ is said to be *concave* if and only if

$$\mu_{\mathcal{A}}\left(\lambda x_1 + (1 - \lambda)x_2\right) \leq \mu_{\mathcal{A}}\left(x_1\right) \vee \mu_{\mathcal{A}}\left(x_2\right) \tag{8.9}$$

for $\lambda \in [0, 1]$, and $x_1, x_2 \in \mathcal{X}$, where $\vee$ denotes the maximum operation.

Fuzzy Number

A fuzzy number $\mathcal{A}$ is a fuzzy set of the real line with a normal, convex and continuous MF of bounded support. A fuzzy number is usually represented by a family of α-level sets or by a discretized MF, as illustrated in Fig. 8.3.

Fuzzy Singleton

A fuzzy set $\mathcal{A} = \{(x, \mu_{\mathcal{A}}(x)) \,|\, x \in \mathcal{X}\}$ is said to be a *fuzzy singleton* if $\mu_{\mathcal{A}}(x) = 1$ for $x \in \mathcal{X}$ and $\mu_{\mathcal{A}}(x') = 0$ for all $x' \in \mathcal{X}$ with $x' \neq x$.

Hedge

A hedge transforms a fuzzy set into a new fuzzy set. A hedge operator is comparable to an adverb in English. Hedges are used to intensify or dilute the characteristic of a fuzzy set such as *very* and *quite*, or to approximate a fuzzy set or convert a scalar to a fuzzy set such as *roughly*. The use of hedge enables the dynamical creation of fuzzy sets and this also helps to reduce the complexity of rules. For example, for a fuzzy set *tall* with membership degree $\mu_{\mathcal{A}}(x)$, *very tall* can be described using the membership degree $\mu_{\mathcal{A}}^{2}(x)$, while *quite tall* can be described using the membership degree $\mu_{\mathcal{A}}^{\frac{1}{2}}(x)$. An illustration of hedge operations is given in Fig. 8.4.

Extension Principle

Given mapping $f : \mathcal{X} \to \mathcal{Y}$, if we have a fuzzy set $\mathcal{A} = \{(x, \mu_{\mathcal{A}}(x)) | x \in \mathcal{X}\}$, $\mu_{\mathcal{A}}(x) \in [0, 1]$, the *extension principle* is defined by the operation

$$f(\mathcal{A}) = f\left(\{(x, \mu_{\mathcal{A}}(x)) | x \in \mathcal{X}\}\right) = \{(f(x), \mu_{\mathcal{A}}(x)) | x \in \mathcal{X}\} \tag{8.10}$$

The application of the extension principle transforms x into $f(x)$, but does not affect the membership function $\mu_{\mathcal{A}}(x)$.

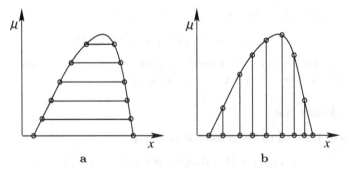

Fig. 8.3. Representations of a fuzzy number. **(a)** α-level sets. **(b)** Discretized MF.

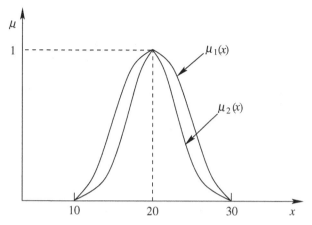

Fig. 8.4. An illustration of hedge operations. The MF $\mu_1(x)$ is a hedge operation that transforms a scalar 20 into a fuzzy number *close to* 20. The MF $\mu_2(x) = \mu_1^2(x)$ is a hedge operator that realizes *very close to* 20.

Cartesian Product

If $\mathcal{X}$ and $\mathcal{Y}$ are two universal sets, then $\mathcal{X} \times \mathcal{Y}$ is the set of all ordered pairs (x, y) for $x \in \mathcal{X}$ and $y \in \mathcal{Y}$. Let $\mathcal{A}$ be a fuzzy set of $\mathcal{X}$ and $\mathcal{B}$ a fuzzy set of $\mathcal{Y}$. The Cartesian product is defined as

$$\mathcal{A} \times \mathcal{B} = \{ (z, \mu_{\mathcal{A} \times \mathcal{B}}(z)) | \, z = (x, y) \in \mathcal{Z}, \mathcal{Z} = \mathcal{X} \times \mathcal{Y} \} \qquad (8.11)$$

where $\mu_{\mathcal{A} \times \mathcal{B}}(z) = \mu_{\mathcal{A}}(x) \wedge \mu_{\mathcal{B}}(y)$, where $\wedge$ denotes the t-norm operation.

Fuzzy Relation

Fuzzy relation is used to describe the association between two things. If $\mathcal{R}$ is a subset of $\mathcal{X} \times \mathcal{Y}$, then $\mathcal{R}$ is said to be a relation between $\mathcal{X}$ and $\mathcal{Y}$, or a relation on $\mathcal{X} \times \mathcal{Y}$. Mathematically,

$$\mathcal{R}(x, y) = \{ ((x, y), \mu_{\mathcal{R}}(x, y)) | \, (x, y) \in \mathcal{X} \times \mathcal{Y}, \mu_{\mathcal{R}}(x, y) \in [0, 1] \} \qquad (8.12)$$

where $\mu_{\mathcal{R}}(x, y)$ is the degree of membership for association between x and y. A fuzzy relation is also a fuzzy set.

Fuzzy Matrix and Fuzzy Graph

Given finite, discrete fuzzy sets $\mathcal{X} = \{x_1, x_2, \cdots, x_m\}$ and $\mathcal{Y} = \{y_1, \cdots, \cdots, y_n\}$, a fuzzy relation on $\mathcal{X} \times \mathcal{Y}$ can be represented by an $m \times n$ matrix $\mathbf{R} = [R_{ij}] = [\mu_{\mathcal{R}}(x_i, y_j)]$. This matrix is called a *fuzzy matrix*.

The fuzzy relation $\mathcal{R}$ can be represented by a fuzzy graph. In a fuzzy graph, all x_i and y_j are vertices, and the grade $\mu_{\mathcal{R}}(x_i, y_j)$ is added to the connection from x_i and y_j.

8.1.2 Membership Function

A fuzzy set $\mathcal{A}$ over the universe of discourse $\mathcal{X}$, $\mathcal{A} \subseteq \mathcal{X} \to [0, 1]$, is described by the degree of membership $\mu_{\mathcal{A}}(x) \in [0, 1]$ for each $x \in \mathcal{X}$. Unimodality and normality are two important aspects of the MFs [236]. Piecewise-linear functions such as triangles and trapezoids are often used as MFs in applications. The triangular MF can be defined by

$$\mu(x; a, b, c) = \begin{cases} \frac{x-a}{b-a}, & a \le x \le b \\ \frac{c-x}{c-b}, & b < x \le c \\ 0, & \text{otherwise} \end{cases} \tag{8.13}$$

where the shape parameters satisfies $a \le b \le c$, and $b \in \mathcal{X}$. It is shown in Fig. 8.2. The triangular MF is useful for modeling linguistic terms such as "The value is close to 10". The trapezoid MF can be defined by

$$\mu(x; a, b, c, d) = \begin{cases} 0, & x \le a \text{ or } x \ge d \\ \frac{x-a}{b-a}, & a < x < b \\ 1, & b \le x \le c \\ \frac{d-x}{d-c}, & c < x < d \end{cases} \tag{8.14}$$

where a, b, c, and d are shape parameters. The trapezoid MF is suitable for modeling such linguistic terms as "He looks like a teenager".

The Gaussian and bell-shaped functions have continuous derivatives, and are usually used to replace the triangular MF when shape parameters are adapted using a gradient-descent procedure. The Gaussian function is given by

$$\mu(x; c, \sigma) = e^{-\frac{(x-c)^2}{2\sigma^2}} \tag{8.15}$$

and the bell-shaped function is defined by

$$\mu(x; c, a, b) = \frac{1}{1 + \left[\left(\frac{x-c}{a}\right)^2\right]^b} \tag{8.16}$$

In (8.15) and (8.16), c is the center of the curves, and a, b, and σ are their shape parameters.

Another popular MF is the sigmoidal function of the form

$$\mu(x; c, \beta) = \frac{1}{1 + e^{-\beta(x-c)}} \tag{8.17}$$

where c shifts the function to the left or to the right, and β controls the shape of the function. When $\beta > 1$ it is an S-shaped function, and when $\beta < -1$ it is a Z-shaped function.

When an S-shaped function is multiplied by a Z-shaped function, a π-shaped function is obtained:

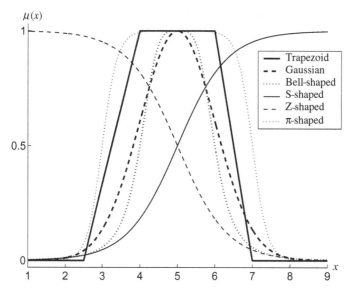

Fig. 8.5. Shapes of some popular MFs. The parameters for each MF are selected as: (a) Trapezoid, $a = 2.5$, $b = 4$, $c = 6$, $d = 7$. (b) Gaussian, $\sigma = 1$, $c = 5$. (c) Bell-shaped, $a = 1$, $b = 2$, $c = 5$. (d) S-shaped, $\beta = 1.5$, $c = 5$. (e) Z-shaped, $-\beta = 1.5$, $c = 5$. (f) π-shaped, $\beta_1 = 5$, $c_1 = 3$, $\beta_2 = 5$, $c_2 = 7$.

$$\mu\left(x; c_1, \beta_1, c_2, \beta_2\right) = \frac{1}{1 + e^{-\beta_1(x-c_1)}} \cdot \frac{1}{1 + e^{-\beta_2(x-c_2)}} \qquad (8.18)$$

where $\beta_1 > 1$, $\beta_2 < -1$, and $c_1 < c_2$. π-shaped MFs can be used in situations similar to that where trapezoid MFs are used. These commonly used MFs are illustrated in Fig. 8.5.

8.1.3 Intersection and Union

The set operations *intersection* and *union* correspond to the logic operations *conjunction* (*AND*) and *disjunction* (*OR*), respectively. *Intersection* is described by the so-called *triangular norm* (*t-norm*), denoted by $T(x, y)$, whereas *union* is described by the so-called *triangular conorm* (*t-conorm*), denoted by $C(x, y)$.

If $\mathcal{A}$ and $\mathcal{B}$ are fuzzy subsets of $\mathcal{X}$, then intersection $\mathcal{I} = \mathcal{A} \cap \mathcal{B}$ is defined by

$$\mu_{\mathcal{I}}(x) = T\left(\mu_{\mathcal{A}}(x), \mu_{\mathcal{B}}(x)\right) \qquad (8.19)$$

Definition 8.1 (t-norm). *A mapping* $T : [0, 1] \times [0, 1] \rightarrow [0, 1]$ *with the following four properties is called t-norm. For all* $x, y, z \in [0, 1]$,

- *Commutativity:* $T(x, y) = T(y, x)$.
- *Monotonicity:* $T(x, y) \leq T(x, z)$, *if* $y \leq z$.

- *Associativity:* $T\big(x, T(y, z)\big) = T\big(T(x, y), z\big)$.
- *Linearity:* $T(x, 1) = x$.

Basic t-norms are listed below [144]

$$
\begin{aligned}
T_{\mathrm{m}}(x, y) &= \min(x, y), && \text{standard intersection} && (8.20)\\
T_{\mathrm{b}}(x, y) &= \max(0, x + y - 1), && \text{bounded sum} && (8.21)\\
T_{\mathrm{p}}(x, y) &= xy, && \text{algebraic product} && (8.22)
\end{aligned}
$$

$$
T^*(x, y) = \begin{cases} x, & \text{if } y = 1 \\ y, & \text{if } x = 1 \\ 0, & \text{otherwise} \end{cases}, \qquad \text{drastic intersection} \qquad (8.23)
$$

For all $x, y \in [0, 1]$, there is relation

$$
T^*(x, y) \le T_b(x, y) \le T_p(x, y) \le T_m(x, y) \tag{8.24}
$$

$T^*(x, y)$ and $T_{\mathrm{m}}(x, y)$ are, respectively, the lower and upper bounds of any t-norm.

Similarly, union $\mathcal{U} = \mathcal{A} \cup \mathcal{B}$ is defined by

$$
\mu_{\mathcal{U}}(x) = C\left(\mu_{\mathcal{A}}(x), \mu_{\mathcal{B}}(x)\right) \tag{8.25}
$$

Definition 8.2 (t-conorm). *A mapping* $C : [0, 1] \times [0, 1] \to [0, 1]$ *having the following four properties is called* t-conorm. *For all* $x, y, z \in [0, 1]$,

- *Commutativity:* $C(x, y) = C(y, x)$.
- *Monotonicity:* $C(x, y) \le C(x, z)$, *if* $y \le z$.
- *Associativity:* $C\left(x, C(y, z)\right) = C\left(C(x, y), z\right)$.
- *Linearity:* $C(x, 0) = x$.

The corresponding basic t-conorms are defined by [144]

$$
\begin{aligned}
C_{\mathrm{m}}(x, y) &= \max(x, y), && \text{standard union} && (8.26)\\
C_{\mathrm{b}}(x, y) &= \min(1, x + y), && \text{bounded sum} && (8.27)\\
C_{\mathrm{p}}(x, y) &= x + y - xy, && \text{algebraic sum} && (8.28)
\end{aligned}
$$

$$
C^*(x, y) = \begin{cases} x, & \text{if } y = 0 \\ y, & \text{if } x = 0 \\ 1, & \text{otherwise} \end{cases}, \qquad \text{drastic union} \qquad (8.29)
$$

For all $x, y \in [0, 1]$,

$$
C_{\mathrm{m}}(x, y) \le C_{\mathrm{p}}(x, y) \le C_{\mathrm{b}}(x, y) \le C^*(x, y) \tag{8.30}
$$

$C_{\mathrm{m}}(x, y)$ and $C^*(x, y)$ are, respectively, the lower and upper bounds of any t-contorm, respectively.

When the t-norm and the t-conorm satisfy

$$1 - T(x, y) = C(1 - x, 1 - y) \tag{8.31}$$

T and C are said to be *dual*. This makes De Morgan's laws $\overline{\mathcal{A} \cap \mathcal{B}} = \overline{\mathcal{A}} \cup \overline{\mathcal{B}}$ and $\overline{\mathcal{A} \cup \mathcal{B}} = \overline{\mathcal{A}} \cap \overline{\mathcal{B}}$ still hold in the fuzzy set theory. The above basic t-norms and t-conorms with the same subscripts are dual. To satisfy the principle of duality, they are usually used in pairs.

8.1.4 Aggregation, Fuzzy Implication, and Fuzzy Reasoning

Aggregation

Aggregation or *composition* operations on fuzzy sets provide a means for combining several sets in order to produce a single fuzzy set. T-conorms are usually used as aggregation operators. Consider the relations

$$\mathcal{R}_1(x, y) = \{ ((x, y), \mu_{\mathcal{R}_1}(x, y)) | \, (x, y) \in \mathcal{X} \times \mathcal{Y}, \mu_{\mathcal{R}_1}(x, y) \in [0, 1] \}$$
$$\mathcal{R}_2(y, z) = \{ ((y, z), \mu_{\mathcal{R}_2}(y, z)) | \, (y, z) \in \mathcal{Y} \times \mathcal{Z}, \mu_{\mathcal{R}_2}(y, z) \in [0, 1] \} \tag{8.32}$$

The max-min composition, denoted by $\mathcal{R}_1 \circ \mathcal{R}_2$ with MF $\mu_{\mathcal{R}_1 \circ \mathcal{R}_2}$, is defined by

$$\mathcal{R}_1 \circ \mathcal{R}_2 = \left\{ ((x, z), \max_y \{ \min (\mu_{\mathcal{R}_1}(x, y), \mu_{\mathcal{R}_2}(y, z)) \}) \big| (x, z) \in \mathcal{X} \times Z, y \in \mathcal{Y} \right\} \tag{8.33}$$

There are some other composition operations, such as the min-max composition, denoted by $\mathcal{R}_1 \diamond \mathcal{R}_2$ with the difference that the role of max and min are interchanged. The two compositions are related by $\overline{\mathcal{R}_1 \diamond \mathcal{R}_2} = \overline{\mathcal{R}_1} \circ \overline{\mathcal{R}_2}$.

Fuzzy Implication

Fuzzy implication is used to represent fuzzy rules. It is a mapping f of an input fuzzy region $\mathcal{A}$ onto an output fuzzy region $\mathcal{B}$ according to the defined fuzzy relation $\mathcal{R}$ on $\mathcal{A} \times \mathcal{B}$

$$(y, \mu_{\mathcal{B}}(y)) = f((x, \mu_{\mathcal{A}}(x))) \tag{8.34}$$

Denote p as "x is $\mathcal{A}$" and q as "y is $\mathcal{B}$", then (8.34) can be stated as $p \to q$ (if p then q). For a fuzzy rule expressed as a fuzzy implication using the defined fuzzy relation $\mathcal{R}$, the output linguistic variable $\mathcal{B}$ is denoted by

$$\mathcal{B} = \mathcal{A} \circ \mathcal{R} \tag{8.35}$$

which is characterized by $\mu_{\mathcal{B}}(y) = \vee_x (\mu_{\mathcal{A}}(x) \wedge \mu_{\mathcal{R}}(x, y))$.

Fuzzy Reasoning

Fuzzy reasoning, also called *approximate reasoning*, is an inference procedure for deriving conclusions from a set of fuzzy rules and one or more conditions [544]. The compositional rule of inference is the essential rational behind fuzzy reasoning.

A simple example of fuzzy reasoning is described here. Consider the fuzzy set $\mathcal{A}$ and the fuzzy relation $\mathcal{R}$ on $\mathcal{A} \times \mathcal{B}$ given by

$$\mathcal{A} = \left\{ (x, \mu_{\mathcal{A}}(x)) \, \big| \, x \in \mathcal{X} \right\}$$
$$\mathcal{R}(x, y) = \left\{ ((x, y), \mu_{\mathcal{R}}(x, y)) \, \big| \, (x, y) \in \mathcal{X} \times \mathcal{Y} \right\} \tag{8.36}$$

Fuzzy set $\mathcal{B}$ can be inferred from fuzzy set $\mathcal{A}$ and their fuzzy relation $\mathcal{R}(x, y)$ according to the max-min composition

$$\mathcal{B} = \mathcal{A} \circ \mathcal{R} = \left\{ \left(y, \max_{x} \left\{ \min \left(\mu_{\mathcal{A}}(x), \mu_{\mathcal{R}}(x, y) \right) \right\} \right) \, \Big| \, x \in \mathcal{X}, y \in \mathcal{Y} \right\} \tag{8.37}$$

8.1.5 Fuzzy Inference Systems and Fuzzy Controllers

In control systems, the inputs to the systems are the error and the change in the error of the feedback loop, while the output is the control action. Fuzzy logic-based controllers are popular control systems. The general architecture of a fuzzy controller is depicted in Fig. 8.6. Fuzzy controllers are knowledge-based, where knowledge is defined by fuzzy IF-THEN rules. The core of a fuzzy controller is an FIS, in which the data flow involves fuzzification, knowledge-base evaluation, and defuzzification.

In an FIS[1], the knowledge base is comprised of the fuzzy rule base and the database. The database contains the linguistic term sets considered in the linguistic rules and the MFs defining the semantics of the linguistic variables, and information about domains. The rule base contains a collection of linguistic rules that are joined by the ALSO operator. An expert provides his knowledge in the form of linguistic rules. The fuzzification process collects the inputs and then converts them into linguistic values or fuzzy sets. The decision logic, called *fuzzy inference engine*, generates output from the input, and finally the defuzzification process produces a crisp output for control action.

FISs are universal approximators capable of performing nonlinear mappings between inputs and outputs. The interpretations of a certain rule and the rule base depend on the FIS model. The Mamdani [753] and the TSK [1075] models are two popular FISs. The Mamdani model is a nonadditive fuzzy model that aggregates the output of fuzzy rules using the maximum operator, while the TSK model is an additive fuzzy model that aggregates the output of rules using the addition operator. Kosko's standard

---

[1] An FIS is sometimes termed a *fuzzy system* or a *fuzzy model* in the literature.

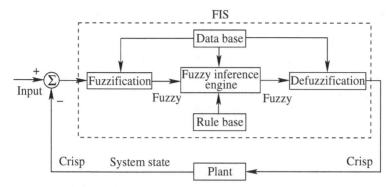

Fig. 8.6. The architecture of a fuzzy controller. The core of the fuzzy controller is an FIS.

additive model (SAM) [624] is another additive fuzzy model. All these models can be derived from fuzzy graph [1229], and are universal approximators [622, 1149, 142, 143, 173, 840].

Both neural networks and fuzzy logic can be used to approximate an unknown control function. Neural networks achieve a solution using the learning process, while FISs apply a vague interpolation technique. FISs are appropriate for modeling nonlinear systems whose mathematical models are not available. Unlike neural networks and other numerical models, fuzzy models operate at a level of information granules—fuzzy sets.

8.1.6 Fuzzy Rules and Fuzzy Interference

There are two types of fuzzy rules, namely, *fuzzy mapping rules* and *fuzzy implication rules* [1229]. A fuzzy mapping rule describes a functional mapping relationship between inputs and an output using linguistic terms, while a fuzzy implication rule describes a generalized logic implication relationship between two logic formulas involving linguistic variables. Fuzzy implication rules generalize set-to-set implications, whereas fuzzy mapping rules generalize set-to-set associations. The former was motivated to allow intelligent systems to draw plausible conclusions in a way similar to human reasoning, while the latter was motivated to approximate complex relationships such as nonlinear functions in a cost-effective and easily comprehensible way. The foundation of fuzzy mapping rule is fuzzy graph, while the foundation of fuzzy implication rule is a generalization to two-valued logic.

A rule base consists of a number of rules in the IF-THEN logic

$$\text{IF } condition, \text{ THEN } action.$$

The condition, also called *premise*, is made up of a number of *antecedents* that are negated or combined by different operators such as AND or OR computed with t-norms or t-conorms. In a fuzzy-rule system, some variables are linguistic variables and the determination of the MF for each fuzzy subset is critical. MFs can be selected according to human intuition, or by learning from training data.

Fuzzy logic can be used as the basis for inference systems. A fuzzy inference is made up of several rules with the same output variables. Given a set of fuzzy rules, the inference result is a combination of the fuzzy values of the conditions and the corresponding actions. For example, we have a set of N_r rules

$$R_i: \text{IF } (condition = \mathcal{C}_i) \text{ THEN } (action = \mathcal{A}_i)$$

for $i = 1, \ldots, N_r$, where $\mathcal{C}_i$ is a fuzzy set. Assuming that a condition has a membership degree of μ_i associated with the set $\mathcal{C}_i$. The condition is first converted into a fuzzy category using a syntactical representation

$$condition = \frac{\mathcal{C}_1}{\mu_1} + \frac{\mathcal{C}_2}{\mu_2} + \cdots + \frac{\mathcal{C}_{N_r}}{\mu_{N_r}} \tag{8.38}$$

Note the difference from the definition of a finite fuzzy set in (8.2). We can see each rule is valid to a certain extent. A fuzzy inference is the combination of all the possible consequences. The action coming from a fuzzy inference is also a fuzzy category

$$action = \frac{\mathcal{A}_1}{\mu_1} + \frac{\mathcal{A}_2}{\mu_2} + \cdots + \frac{\mathcal{A}_{N_r}}{\mu_{N_r}} \tag{8.39}$$

This is also a syntactical representation. The inference procedure depends on fuzzy reasoning. This result can be further processed or transformed into a crisp value.

8.1.7 Fuzzification and Defuzzification

Fuzzification is to transform crisp inputs into fuzzy subsets. Given crisp inputs $x_i, \ i = 1, \cdots, n$, fuzzification is to construct the same number of fuzzy sets $\mathcal{A}^i$,

$$\mathcal{A}^i = \text{fuzz} \, (x_i) \tag{8.40}$$

where fuzz$(\cdot)$ is a fuzzification operator. Fuzzification is determined according to the defined MFs.

Defuzzification is to map fuzzy subsets of real numbers into real numbers. In an FIS, defuzzification is applied after aggregation. Defuzzification is necessary in fuzzy controllers, since the machines cannot understand control signals in the form of a complete fuzzy set.

Popular defuzzification methods include the centroid defuzzifier [753], and the mean-of-maxima defuzzifier [753]. The centroid defuzzifier is the best-known method, which is to find the centroid of the area surrounded by the MF and the horizontal axis. A discrete centroid defuzzifier is given by [555]

$$\text{defuzz}(\mathcal{B}) = \frac{\sum_{i=1}^{K} \mu_{\mathcal{B}}(y_i) y_i}{\sum_{i=1}^{K} \mu_{\mathcal{B}}(y_i)} \tag{8.41}$$

where K is the number of quantization steps by which the universe of discourse $\mathcal{Y}$ of the MF $\mu_{\mathcal{B}}(y)$ is discretized.

Aggregation and defuzzification can be combined into a single phase, such as the weighted-mean method [353]

$$\text{defuzz}(\mathcal{B}) = \frac{\sum_{i=1}^{N_r} \mu_i b_i}{\sum_{i=1}^{N_r} \mu_i} \tag{8.42}$$

where N_r is the number of rules, μ_i is the degree of activation of the ith rule, and b_i is a numerical value associated with the consequent of the ith rule, $\mathcal{B}_i$. The parameter b_i can be selected as the mean value of the α-level set when α is equal to μ_i [353].

8.1.8 Mamdani Model and Takagi–Sugeno–Kang Model

Given a set of N examples $\left\{ (\mathbf{x}_p, \mathbf{y}_p) \, \middle| \, \mathbf{x}_p \in R^n, \mathbf{y}_p \in R^m \right\}$, the underlying system can be identified by using some fuzzy models. Two popular FIS models are the Mamdani and TSK models.

Mamdani Model

For the Mamdani model with N_r rules, the ith rule is given by

$$R_i: \text{IF } \mathbf{x} \text{ is } \mathcal{A}_i, \text{ THEN } \mathbf{y} \text{ is } \mathcal{B}_i$$

for $i = 1, \ldots, N_r$, where $\mathcal{A}_i = \{\mathcal{A}_i^1, \mathcal{A}_i^2, \cdots, \mathcal{A}_i^n\}$, $\mathcal{B}_i = \{\mathcal{B}_i^1, \mathcal{B}_i^2, \cdots, \mathcal{B}_i^m\}$, and $\mathcal{A}_i^j$ and $\mathcal{B}_i^k$ are, respectively, fuzzy sets that define an input and output space partitioning.

For an n-tuple input in the form of "$\mathbf{x}$ is $\mathcal{A}'$", the system output "$\mathbf{y}$ is $\mathcal{B}'$" is characterized by combining the rules according to

$$\mu_{\mathcal{B}'}(\mathbf{y}) = \bigvee_{i=1}^{N_r} \left(\mu_{\mathcal{A}_i'}(\mathbf{x}) \wedge \mu_{\mathcal{B}_i}(\mathbf{y}) \right) \tag{8.43}$$

where $\mathcal{A}' = \{\mathcal{A}'^1, \mathcal{A}'^2, \cdots, \mathcal{A}'^n\}$, $\mathcal{B}' = \{\mathcal{B}'^1, \mathcal{B}'^2, \cdots, \mathcal{B}'^m\}$, $\mathcal{A}'^j$ and $\mathcal{B}'^k$ are, respectively, fuzzy sets that define an input and output space partitioning,

$$\mu_{\mathcal{A}_i'}(\mathbf{x}) = \mu_{\mathcal{A}'}(\mathbf{x}) \wedge \mu_{\mathcal{A}_i}(\mathbf{x}) = \bigwedge_{j=1}^{n} \left(\mu_{\mathcal{A}'^j} \wedge \mu_{\mathcal{A}_i^j} \right) \tag{8.44}$$

$\mu_{\mathcal{A}'}(\mathbf{x}) = \bigwedge_{j=1}^{n} \mu_{\mathcal{A}'^j}$ and $\mu_{\mathcal{A}_i}(\mathbf{x}) = \bigwedge_{j=1}^{n} \mu_{\mathcal{A}_i^j}$ being, respectively, the membership degrees of $\mathbf{x}$ to the fuzzy sets $\mathcal{A}'$ and $\mathcal{A}_i$, $\mu_{\mathcal{B}_i}(\mathbf{y}) = \bigwedge_{k=1}^{m} \mu_{\mathcal{B}_i^k}$ is the membership degree of $\mathbf{y}$ to the fuzzy set $\mathcal{B}_i$, $\mu_{\mathcal{A}_i^{\prime j}}$ is the association between the jth input of $\mathcal{A}'$ and the ith rule, $\mu_{\mathcal{B}_i^k}$ is the association between the kth input of $\mathcal{B}$ and the ith rule, $\wedge$ is the intersection operator, and $\vee$ is the union operator.

Minimum and product are the most common intersection operators. When minimum and maximum are, respectively, used as the intersection and union operators, the Mamdani model is called a *max-min model*. Kosko's SAM [624] has the same rule form, but the SAM uses the product operator for the fuzzy intersection operation, and uses sup-product as well as addition as the composition operators.

We now illustrate the inference procedure for the Mamdani model. Assume that we have a two-rule Mamdani FIS with the rules of the form

$$R_i:\ \text{IF } x_1 \text{ is } \mathcal{A}_i \text{ and } x_2 \text{ is } \mathcal{B}_i,\ \text{THEN } y \text{ is } \mathcal{C}_i.$$

for $i = 1, 2$. When the max-min composition is employed, for the inputs "x_1 is $\mathcal{A}'$" and "x_2 is $\mathcal{B}'$", the fuzzy reasoning procedure for the output y is illustrated in Fig. 8.7. When two crisp inputs x_1' and x_2' are fed, the derivation of the output y' is illustrated in Fig. 8.8. As a comparison with Fig. 8.7, Fig. 8.9 illustrates the result when the max-product composition is used to replace the max-min composition. A defuzzification strategy is needed to get a crisp output value.

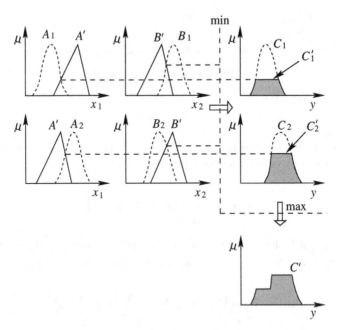

Fig. 8.7. The inference procedure of the Mamdani model with the max-min composition and fuzzy inputs

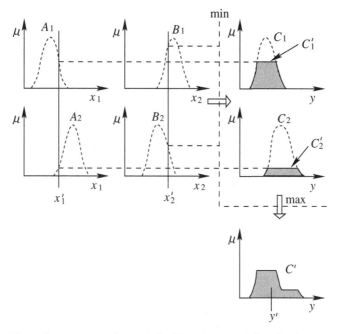

Fig. 8.8. The inference procedure of the Mamdani model with the max-min composition and crisp inputs

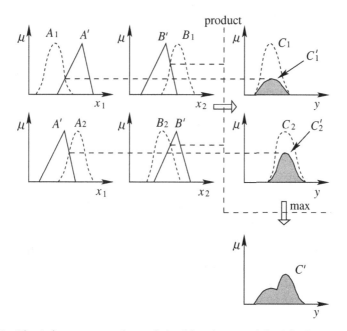

Fig. 8.9. The inference procedure of the Mamdani model with the max-product composition and fuzzy inputs

The Mamdani model offers a high semantic level and a good generalization capability. It contains fuzzy rules built from expert knowledge. However, FISs based only on expert knowledge may result in insufficient accuracy. For accurate numerical approximation, the TSK model can usually generate a better performance.

Takagi–Sugeno–Kang Model

In the TSK model [1075], for the same set of examples $\{(\mathbf{x}_p, \mathbf{y}_p)\}$, fuzzy rules are given in the form

$$R_i: \text{IF } \mathbf{x} \text{ is } \mathcal{A}_i, \text{ THEN } \mathbf{y} = \mathbf{f}_i(\mathbf{x})$$

for $i = 1, 2, \ldots, N_r$, where $\mathbf{f}_i(\mathbf{x}) = \left(f_i^1(\mathbf{x}), f_i^2(\mathbf{x}), \cdots, f_i^m(\mathbf{x})\right)^{\mathrm{T}}$ is a crisp vector function of $\mathbf{x}$; usually $f_i^j(\mathbf{x})$ is selected as a linear relation of $\mathbf{x}$

$$f_i^j(\mathbf{x}) = a_{i,0}^j + a_{i,1}^j x_1 + \cdots + a_{i,n}^j x_n \tag{8.45}$$

where $a_{i,k}^j$, $k = 0, 1, \cdots, n$, are adjustable parameters.

For an n-tuple input in the form of "$\mathbf{x}$ is $\mathcal{A}'$", the output $\mathbf{y}'$ is obtained by combining the rules according to

$$\mathbf{y}' = \frac{\sum_{i=1}^{N_r} \mu_{\mathcal{A}_i'}(\mathbf{x}) \mathbf{f}_i(\mathbf{x})}{\sum_{i=1}^{N_r} \mu_{\mathcal{A}_i'}(\mathbf{x})} \tag{8.46}$$

where $\mu_{\mathcal{A}_i'}(\mathbf{x})$ is defined by (8.44). This model produces a real-valued function, and it is essentially a model-based fuzzy control method. The stability analysis of the TSK model is given in [1081]. When $f_i^j(\cdot)$ are first-order polynomials, the model is termed the *first-order TSK model*, which is the typical form of the TSK model. When $f_i^j(\cdot)$ are constants, it is called the *zero-order TSK model*, which can be viewed as a special case of the Mamdani model.

Similarly, we illustrate the inference procedure of the TSK model. Given a two-rule TSK FIS with the rules of the form

$$R_i: \text{IF } x_1 \text{ is } \mathcal{A}_i \text{ and } x_2 \text{ is } \mathcal{B}_i, \text{ THEN } y = f(x_1, x_2).$$

for $i = 1, 2$. When two crisp inputs x_1' and x_2' are fed, the inference for the output y' is as illustrated in Fig. 8.10.

In comparison with the Mamdani model, the TSK model, which is based on automatic learning from the data, can accurately approximate a function using fewer rules. It has a stronger and more flexible representation capability than the Mamdani mode. In the TSK model, rules are extracted from the data, but the generated rules may have no meaning for experts. The TSK model has found more successful applications in building fuzzy systems.

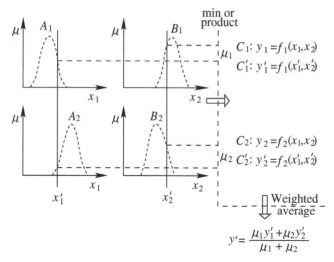

Fig. 8.10. The inference procedure for the TSK model with the min or product operator

8.1.9 Complex Fuzzy Logic

Complex fuzzy sets and logic are mathematical extensions of fuzzy sets and logic from the real domain to the complex domain [933, 932]. A complex fuzzy set S is characterized by a complex-valued MF, and the membership degree of any element x in S is given by a complex value of the form

$$\mu_S(x) = r_S(x)e^{j\varphi_S(x)} \tag{8.47}$$

where the amplitude $r_S(x) \in [0,1]$, and φ_S is the phase, that is, $\mu_S(x)$ is within a unit circle in the complex plane.

As in the development of the fuzzy logic theory, the design of the set-theoretic operations is of vital importance in the complex fuzzy logic. In [933, 932], basic set operators for fuzzy logic have been extended for the complex fuzzy logic, and some additional operators such as the vector aggregation, set rotation and set reflection, are also defined. The operations of intersection, union and complement for complex fuzzy sets are defined on the modulus of the complex membership degree without a consideration of its phase information. In [306], the complex fuzzy logic is extended to a logic of vectors in the plane, rather than scalar quantities. In [825], a complex fuzzy set is defined as an MF mapping the complex plane into $[0,1] \times [0,1]$.

Complex fuzzy sets are superior to the Cartesian products of two fuzzy sets. Complex fuzzy logic maintains both the advantages of the fuzzy logic and the properties of complex fuzzy sets. In complex fuzzy logic, rules constructed are strongly related and a relation manifested in the phase term is associated with complex fuzzy implications. In a complex FIS, the output of each rule is a complex fuzzy set, and phase terms are necessary when combining multiple rules so as to generate the final output. Complex FISs are useful for solving problems in which rules are related to one another with the nature of the relation varying as a function of the input to the system [932]. These problems may be very difficult or impossible to solve using traditional fuzzy methods.

The fuzzy complex number [141] is a different concept from the complex fuzzy set [933]. The fuzzy complex number was introduced by incorporating the complex number into the support of the fuzzy set. A fuzzy complex number is a fuzzy set of complex numbers, which have real-valued membership degree in the range $[0, 1]$. An α-cut of a fuzzy complex number is based on the modulus of the complex numbers in the fuzzy set. The operations of addition, subtraction, multiplication and division for fuzzy complex numbers are derived using the extension principle, and closure of the set of fuzzy complex numbers is proved under each of these operators. In a nutshell, a fuzzy complex number is a fuzzy set in one dimension, while a complex fuzzy set or number is a fuzzy set in two dimensions.

8.2 Fuzzy Logic vs. Neural Networks

Like FNNs, many different fuzzy systems have been proved to be universal approximators in various articles, for example, see [693, 543, 143, 352, 629, 1155], and the references in [629]. In [693], the Mamdani model and FNNs are shown to be able to approximate each other to an arbitrary accuracy. The equivalence between the TSK model and the RBFN under certain conditions has been established in [543, 518]. The equivalence between fuzzy expert systems and neural networks has been proved in [143]. Gaussian-based Mamdani systems have the ability of approximating any sufficiently smooth function and reproducing its derivatives up to any order [352]. In [629], fuzzy systems with Gaussian MFs are proved to be universal approximators for a smooth function and its derivatives. In [1155], the fuzzy system with nth-order B-spline MFs and the CMAC network [21] with nth-order B-spline basis functions are proved to be universal approximators for a smooth function and its derivatives up to the $(n-2)$th order.

From the point of view of an expert system, fuzzy systems and neural networks are quite similar as inference systems. An inference system involves knowledge representation, reasoning, and knowledge acquisition:

- **Knowledge representation.** A trained neural network represents knowledge using connection weights and neurons in a distributed manner. In a fuzzy system, knowledge is represented using IF-THEN rules.
- **Reasoning.** When an input is presented to a neural network, an output is generated. This pure numerical process can be treated as a reasoning process. In a fuzzy system, reasoning is logic based.

- **Knowledge acquisition.** Knowledge acquisition is via learning in a neural network, while for a fuzzy system knowledge is encoded by a human expert.

Both neural networks and fuzzy systems are dynamic, parallel distributed processing systems that estimate functions. They estimate a function without any mathematical model and learn from experience with sample data.

Fuzzy systems can be applied to problems with knowledge represented in the form of IF-THEN rules. Problem-specific *a priori* knowledge can be integrated into the systems. Training pattern set and system modeling are not needed, and only heuristics are used. During the tuning process, one needs to add, remove, or change a rule, or even change the weight of a rule. This process, however, requires the knowledge of experts.

Neural networks are useful when we have training pattern set. We do not need any knowledge of the modeling of the problem. A trained neural network is a black box that represents knowledge in its distributed structure. However, any prior knowledge of the problem cannot be incorporated into the learning process. It is difficult for human beings to understand the internal logic of the system. Nevertheless, by extracting rules from neural networks, users can understand what neural networks have learned and how neural networks predict.

8.3 Fuzzy Rules and Multilayer Perceptrons

There are many techniques for extracting rules from trained neural networks (see [534, 177, 544, 175], and the survey paper [1088]). In this section, we discuss the relation between fuzzy rules and MLPs, namely, the extraction of rules from trained MLPs and the representation of fuzzy rules using MLPs. In Sect. 8.4, similar problems are dealt with for RBFNs.

8.3.1 Equality Between Multilayer Perceptrons and Fuzzy Inference Systems

For a three-layer $(J_1\text{-}J_2\text{-}J_3)$ MLP, let the input to the network be $\mathbf{x} = (x_1, \cdots, x_{J_1})^{\mathrm{T}}$, the output of the network be $\mathbf{y} = (y_1, \cdots, y_{J_3})^{\mathrm{T}}$, the output of the hidden layer be $\mathbf{o}^{(2)} = \left(o_1^{(2)}, \cdots, o_{J_2}^{(2)}\right)^{\mathrm{T}}$, the bias for the jth hidden neuron be $\theta_j^{(2)}$, the weight from the ith input node to the jth hidden neuron be $w_{ij}^{(1)}$, and the weight from the jth hidden neuron to the kth output neuron be $w_{jk}^{(2)}$. Then,

$$y_k = \phi^{(2)} \left(\sum_{j=1}^{J_2} x_j^{(2)} w_{jk}^{(2)} \right) \tag{8.48}$$

$$o_j^{(2)} = \phi^{(1)} \left(\sum_{i=1}^{J_1} x_i w_{ij}^{(1)} + \theta_j^{(2)} \right) \tag{8.49}$$

where $\phi^{(1)}(\cdot)$ and $\phi^{(2)}(\cdot)$ are the activation functions. $\phi^{(1)}(\cdot)$ is usually selected as the logistic function $\phi^{(1)}(net) = \frac{1}{1+e^{-net}}$, and $\phi^{(2)}(\cdot)$ can be a sigmoidal or a linear function. For a network with $\phi^{(1)}(\cdot)$ as the logistic function and $\phi^{(2)}(\cdot)$ as the linear function $\phi^{(2)}(net) = net$, there always exists a fuzzy additive system that calculates the same function as the network does [99].

In [99], a fuzzy logic operator, called *interactive-or (i-or)*, is defined by applying the concept of f-duality to the logistic function. The use of the i-or operator explains clearly the acquired knowledge of a trained MLP. The i-or operator is defined by [99]

$$a \otimes b = \frac{a \cdot b}{(1-a) \cdot (1-b) + a \cdot b} \tag{8.50}$$

The i-or operator works on $(0, 1)$. It is a hybrid between both a t-norm and a t-conorm. Based on the i-or operator, the equality between MLPs and FISs is thus established [99]. The equality proof also yields an automated procedure for knowledge acquisition. An extension of the method has been presented in [175], where the fuzzy rules obtained are in agreement with the domain of the input variables and a new logical operator, similar to, but with a higher representational power than the i-or, is defined.

In [325], relations between input uncertainties and fuzzy rules have been established. Sets of crisp logic rules applied to uncertain inputs have been shown to be equivalent to fuzzy rules with sigmoidal MFs applied to crisp inputs. Integration of a reasonable uncertainty distribution for a fixed rule threshold or interval gives a sigmoidal MF, and several new sigmoidal MFs have been introduced using this method. Crisp logic and fuzzy rule systems have been shown to be, respectively, equivalent to the logical network and the three-layer MLP. Keeping fuzziness on the input side enables easier understanding of the networks or the rule systems.

8.3.2 Extracting Rules According to Activation Functions

In [177, 1049], MLPs are interpreted by fuzzy rules in such a way that the sigmoidal activation function is decomposed into three TSK fuzzy rules with one TSK fuzzy rule for each partition. This is illustrated in Fig. 8.11. The three fuzzy rules are listed as follows.

R_1: IF x is in section S_1, THEN $y = 0$
R_2: IF x is in section S_2, THEN $y = \beta x + \frac{1}{2}$
R_3: IF x is in section S_3, THEN $y = 1$

Accordingly, the value of the activation function at point x_0 is given by

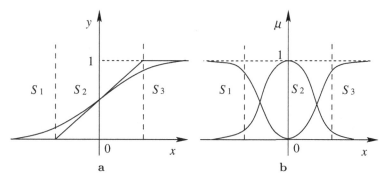

Fig. 8.11. The sigmoidal function is represented by three TSK rules. (**a**) The sigmoidal function and its partitioning. (**b**) The MFs for the partitions.

$$\phi(x_0) = 0 \cdot \mu_1(x_0) + \left(\beta x + \frac{1}{2}\right)\mu_2(x_0) + 1 \cdot \mu_3(x_0) \qquad (8.51)$$

where $\mu_i(x)$, $i = 1, 2, 3$, are MFs for the three partitions. $\mu_i(x)$, when approximated by simple linear functions such as the triangle function, is easily realizable in hardware by using threshold comparators. An algorithm for rule extraction is given in [177], in which product is used as the intersection operator for the combination of the MFs of sigmoidal neurons and the rules are extracted in the linear zone S_2. The algorithm extracts $O(N)$ rules, where N is the number of examples. Rule generation from a trained neural network can be done by analyzing the saturated zones S_1 and S_3 of the fuzzy activation functions [1049]. Two saturated zones can be obtained for every output node. Therefore, the number of rules is $2J_2$, J_2 being the number of hidden neurons. Rule extraction in the nonsaturated zones has not been solved.

8.3.3 Representing Fuzzy Rules Using Multilayer Perceptrons

A fuzzy set is usually represented by a finite number of its supports. In comparison with conventional MF-based FISs, α-cut based FISs [1107] have a number of advantages. They can considerably reduce the required memory and time complexity, since they depend on the number of membership-grade levels, and not on the number of elements in the universes of discourse. Secondly, the inference operations can be performed for each α-cut set independently, and this enables parallel implementation. An α-cut based FIS can also easily interface with two-valued logic since the α-level sets themselves are crisp sets. In addition, fuzzy set operations based on the extension principle can be performed efficiently using α-level sets [1107, 707].

For α-cut based FISs, fuzzy rules can be implemented by an MLP with the BP rule. For fuzzy rules of the form

$$R_i\colon \text{IF } x \text{ is } \mathcal{A}_i, \text{ THEN } y \text{ is } \mathcal{B}_i.$$

for $i = 1, \ldots, N_{\mathrm{r}}$, where $\mathcal{A}_i$ and $\mathcal{B}_i$ are fuzzy sets. Assume the support for $\mathcal{A}_i$ and $\mathcal{B}_i$ are, respectively, $[a_1, a_2]$, and $[b_1, b_2]$. We can divide the intervals linearly or by α-cut, and get $\{x_1, x_2, \ldots, x_n \in [a_1, a_2]\}$, and $\{y_1, y_2, \ldots, y_m \in [b_1, b_2]\}$, where $n, m \geq 2$. A training set of discretized samples is obtained as

$$(\mu_{\mathcal{A}_i}(x), \mu_{\mathcal{B}_i}(y)) = (\mu_{\mathcal{A}_i}(x_1), \ldots, \mu_{\mathcal{A}_i}(x_n); \mu_{\mathcal{B}_i}(y_1), \ldots, \mu_{\mathcal{B}_i}(y_m)) \quad (8.52)$$

for $i = 1, \ldots, N_{\mathrm{r}}$. This is a learning problem of N_{r} samples with n inputs and m outputs.

8.4 Fuzzy Rules and Radial Basis Function Networks

8.4.1 Equivalence Between Takagi–Sugeno–Kang Model and Radial Basis Function Networks

The normalized RBFN is found functionally equivalent to a class of TSK systems [543]. For the convenience of presentation, we reproduce the output of the J_1-J_2-J_3 normalized RBFN given by (6.81)

$$y_j = \frac{\sum_{i=1}^{J_2} w_{ij} \phi(\|\mathbf{x} - \mathbf{c}_i\|)}{\sum_{i=1}^{J_2} \phi(\|\mathbf{x} - \mathbf{c}_i\|)} \quad (8.53)$$

for $j = 1, \cdots, J_3$. When the t-norm in the TSK model is selected as algebraic product and the MFs are selected the same as RBFs of the RBFN, the two models are mathematically equivalent [543, 541]. Note that each hidden unit corresponds to a fuzzy rule. The normalized RBFN provides a localized solution that is amenable to rule extraction. The receptive fields of some RBFs should overlap to prevent incompleteness of fuzzy partitions.

To have a perfect match between $\phi(\|\mathbf{x} - \mathbf{c}_i\|)$ in (8.53) and $\mu_{\mathcal{A}_i'}(\mathbf{x})$ in (8.46), one is required to select a factorizable $\phi(\|\mathbf{x} - \mathbf{c}_i\|)$ in (8.53) such that

$$\mu_{\mathcal{A}_i'}(\mathbf{x}) = \prod_{j=1}^{J_1} \mu_{\mathcal{A}_i'^j}(x_j) \Longleftrightarrow \phi(\|\mathbf{x} - \mathbf{c}_i\|) = \prod_{j=1}^{J_1} \phi(|x_j - c_{i,j}|) \quad (8.54)$$

Each component $\phi(|x_j - c_{i,j}|)$ corresponds to an MF $\mu_{\mathcal{A}_i'^j}$. Note that the Gaussian RBF is the only strictly factorizable function.

In the normalized RBFN, w_{ij}s typically take constant values and the normalized RBFN corresponds to the zero-order TSK model. When the RBF weights are linear regression functions of the input variables [649, 957], the model is functionally equivalent to the first-order TSK model.

8.4.2 Fuzzy Rules and Radial Basis Function Networks: Representation and Extraction

In a practical implementation of the TSK model, one can select some $\mu_{\mathcal{A}_i'^j} = 1$ or some $\mu_{\mathcal{A}_i'^j} = \mu_{\mathcal{A}_k'^j}$ in order to increase the distinguishability of the fuzzy partitions. Correspondingly, one should share some component RBFs or set some component RBFs to unity [555]. This considerably reduces the effective number of free parameters in the RBFN.

When implementing a component RBF or MF sharing, a distance measure like the Euclidean distance is used to describe the similarity between two component RBFs. The Euclidean distance between two Gaussian RBFs $\phi(c_i, \sigma_i)$ and $\phi(c_j, \sigma_j)$, where c_i and c_j are centers and σ_i and σ_j are widths, is defined by [555]

$$d\left(\phi\left(c_i, \sigma_i\right), \phi\left(c_j, \sigma_j\right)\right) = \sqrt{\left(c_i - c_j\right)^2 + \left(\sigma_i - \sigma_j\right)^2} \qquad (8.55)$$

After applying a clustering technique to locate prototypes and adding a regularization term describing the total similarity between all the RBFs and the shared RBF to the MSE function (6.3), a gradient-descent procedure is conducted so as to extract interpretable fuzzy rules from a trained RBFN [555]. The method can be applied to RBFNs with constant or linear regression weights.

A fuzzy system can be first constructed according to heuristic knowledge and existing data, and then converted into an RBFN. This is followed by a refinement of the RBFN using a learning algorithm. Due to this learning procedure, the interpretability of the original fuzzy system may be lost. The RBFN is then again converted into interpretable fuzzy system, and knowledge is extracted from the network. This process, as illustrated in Fig. 8.12, refines the original fuzzy system design. The algorithm for rule extraction from the RBFN is given in [555].

In [1099], it is shown that the Gaussian RBFN can be generated from simple probabilistic rules and probabilistic rules can also be extracted from trained normalized Gaussian RBFNs. Methods for reducing network complexity have been presented in order to obtain concise and meaningful rules. Two algorithms for rule extraction from RBFNs, which, respectively, generate a single rule describing each class and a single rule from each hidden unit, have been presented in [785]. Under the framework, existing domain knowledge in rule format can be inserted into an RBFN as an initialization of optimal network training.

8.5 Rule Generation from Trained Neural Networks

In Sects. 8.3 and 8.4, we have described rule generation from trained MLPs and RBFNs. In this section, we describe rule generation from other trained neural networks.

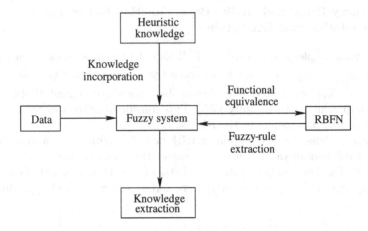

Fig. 8.12. Using the RBFN learning procedure to optimize a fuzzy TSK system

Rule generation encompasses both rule extraction and rule refinement. Rule extraction is to extract knowledge from trained neural networks using the network parameters, while rule refinement is to refine the rules that are extracted from neural networks and initialized with crude domain knowledge.

The extraction of M-of-N rules from a trained FNN with binary $\{-1, +1\}$ weights and inputs has been reported in [1016]. The generated rules are accurate and simple. The M-of-N rules are classification rules of the form "IF (M of the N antecedents are true) THEN $\cdots$". The values of discrete and continuous input attributes can be converted to binary before applying the method. The simplicity of the extracted rules is reflected in the small number of rules and the relatively small number of M-of-N conditions per rule.

FNNs generally do not have the capability to represent recursive rules when the depth of the recursion is not known *a priori*. RNNs have the ability to store information over indefinite periods of time, develop hidden states through learning, and thus conveniently represent recursive linguistic rules [810]. They are particularly well suited for problem domains, where incomplete or contradictory prior knowledge is available. In such cases, knowledge revision or refinement is also possible.

Discrete-time RNNs have been used to correctly classify strings of a regular language [871]. Rules defining the learned grammar can be extracted in the form of deterministic finite-state automata (DFAs) by applying clustering algorithms in the output space of neurons. Starting from an initial network state, the algorithm searches the equally partitioned output space of N state neurons in a breadth-first manner. A heuristic is used to choose among the consistent DFAs the model that best approximates the learned regular grammar. The extracted rules demonstrate high accuracy and fidelity and the algorithm is portable. Based on [871], an augmented RNN that encodes fuzzy finite-state automata (FFAs) and recognizes a given fuzzy regular language with an arbitrary accuracy has been constructed in [872]. FFAs are transformed into equivalent DFAs by using an algorithm that computes fuzzy-string membership. FFAs can model dynamical processes whose current state depends on the current input and previous states. The granularity within both extraction techniques is at the level of ensemble of neurons, and thus, the approaches are not strictly decompositional.

RNNs are suitable for crisp/fuzzy grammatical inference. A method that uses a SOM for extracting knowledge from an RNN [119] is able to infer a crisp/fuzzy regular language. Rule extraction has also been carried out upon Kohonen networks [1109]. A comprehensive survey on rule generation from trained neural networks has been provided from a softcomputing perspective in [810], where the optimization capability of EAs are emphasized for rule refinement. Rule extraction from RNNs aims to find models of an RNN, typically in the form of finite state machines. A recent overview of rule extraction from RNNs is given in [537].

8.6 Extracting Rules from Numerical Data

FISs can also be designed directly from expert knowledge and data. The design process is usually decomposed into two phases, namely, rule generation and system optimization [440]. Rule generation leads to a basic system with a given space partitioning and the corresponding set of rules, while system optimization can be the optimization of membership parameters and rule base. Design of fuzzy rules can be conducted in one of three ways, namely, all the possible combinations of fuzzy partitions, one rule for each data pair, or dynamically choosing the number of fuzzy sets.

For good interpretability, a suitable selection of variables and the reduction of the rule base are necessary. During the system optimization phase, merging techniques such as cluster merging and fuzzy-set merging are usually used for interpretability purposes. Fuzzy-set merging leads to a higher interpretability than cluster merging. The reduction of a set of rules results in a loss of numerical performance on the training data set, but a more compact rule base has a better generalization capability and is also easier for human understanding.

The optimization capability of EAs [982] or the learning capability of neural networks [543] are also used for extracting fuzzy rules and optimizing MFs and rule base. Methods for designing FISs from data are analyzed and surveyed in [440], with emphasis on clustering methods for rule generation and EAs on system optimization. They are grouped into several families and compared based on rule interpretability.

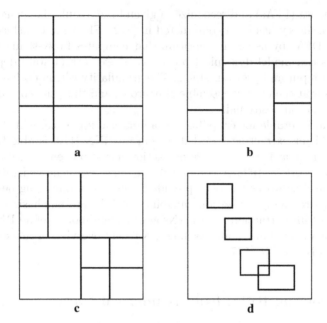

Fig. 8.13. Partitioning of the two-dimensional input space. (**a**) Grid partitioning. (**b**) k-d tree partitioning. (**c**) Multilevel grid partitioning. (**d**) Scatter partitioning.

8.6.1 Rule Generation Based on Fuzzy Partitioning

Rule generation can be based on a partitioning of the multidimensional space. Fuzzy partitioning corresponds to structure identification for FISs, followed by parameter identification using a learning algorithm. There are usually three methods for partitioning the input space, namely, grid partitioning, tree partitioning, and scatter partitioning. An illustration of these partitioning methods in the two-dimensional input space is given in Fig. 8.13.

Grid Partitioning

The grid structure has easy interpretability and is most widely used for generating fuzzy rules. Fuzzy sets of each variable are shared by all the rules. However, the number of fuzzy rules grows exponentially with input dimension, that is, the method is plagued by the problem of *curse of dimensionality*. For n input variables, each being partitioned into m_i fuzzy sets, a total of $\prod_{i=1}^{n} m_i$ rules are needed to cover the whole input space. Since each rule has a few parameters to adjust, there are too many parameters to adapt during the learning process. Too many fuzzy rules also harm the interpretability of the fuzzy system.

The grid structure is a static structure, and is appropriate for a small dimensional data set with a good coverage. The performance of the resultant model depends entirely on the initial definition of these grids. Thus, a training procedure can be applied to optimize the grid structure and the rule consequences. For a medium-dimensional data set, in order to obtain a compact rule base, one can use a learning procedure, such as the gradient-descent method to realize an adaptive grid partitioning [543]. The grid structure is illustrated in Fig. 8.13a.

Tree Partitioning

k-d tree and multilevel grid structures are two hierarchical partitioning techniques [1067]. The input space is first partitioned roughly, and a subspace is recursively divided until a desired approximation performance is achieved. The k-d tree results from a series of *guillotine cuts*. A guillotine cut is a cut that is entirely across the subspace to be partitioned. After the ith guillotine cut, the entire space is partitioned into $i + 1$ regions. Heuristics based on the distribution of training examples or parameter-identification methods can usually be employed to find a proper k-d tree structure [1067]. A k-d tree partitioning is illustrated in Fig. 8.13b.

For the multilevel grid structure [1067], the top-level grid coarsely partitions the whole space into equal-sized and evenly spaced fuzzy boxes, which are recursively partitioned into finer grids until a criterion is met. Hence, a multilevel grid structure is also called a *box tree*. The criterion can be that the resulting boxes have a similar number of training examples or that an application-specific evaluation in each grid is below a threshold. A multilevel grid partitioning is illustrated in Fig. 8.13c. A multilevel grid in the two-dimensional space is called a *quad tree*.

Tree partitioning significantly relieves the problem of rule explosion, but it needs some heuristics to extract rules and its application to high-dimensional problems faces practical difficulties.

Scatter Partitioning

Scatter partitioning uses multidimensional antecedent fuzzy sets. Scatter partitioning usually generates fewer fuzzy regions than the grid and tree partitioning techniques owing to the natural clustering property of training patterns. Fuzzy clustering algorithms form a family of rule-generation techniques. The training examples are gathered into homogeneous groups and a rule is associated to each group. The fuzzy sets are not shared by the rules, but each of them is tailored for one particular rule. Thus, the resulting fuzzy sets are usually difficult to interpret [440].

Clustering is well adapted for large work spaces with a small number of training examples. However, scatter partitioning of high-dimensional feature spaces is difficult, and some learning or evolutionary procedures may be necessary. Clustering algorithms, which are discussed in Chapter 5, can be applied for scatter partitioning. A scatter partitioning is illustrated in Fig. 8.13d.

The curse of dimensionality can also be alleviated by reducing the input dimensions. This is because some of the inputs are irrelevant to the output or some of the inputs are correlated with one another. Accordingly, we can discard some irrelevant inputs or compress the input space using feature-selection or feature-extraction techniques.

Some clustering-based methods for extracting fuzzy rule for function approximation have been proposed in [1208, 226, 227, 39]. These methods are based on the TSK model. Clustering can be used for identification of the antecedent part of the model such as determination of the number of rules and initial rule parameters. Clustering generates scatter partitioning of input space since a rule is generated only where there is a cluster of data. Each cluster center corresponds to a fuzzy rule. The consequent part of the model can be estimated by the linear LS method. In [227], the combination of the subtractive clustering with the linear LS method provides an extremely fast and accurate method for fuzzy-system identification, which is better than the adaptive-network-based FIS (ANFIS) [541]. Based on the Mamdani model, a clustering-based method for function approximation is also given in [1153].

8.6.2 Hierarchical Rule Generation

Hierarchical structure for fuzzy-rule systems can also effectively solve the rule-explosion problem [931, 1150, 725]. A hierarchical fuzzy system is comprised of a number of low-dimensional fuzzy systems connected in a hierarchical fashion. The low-dimensional fuzzy systems can be TSK systems, each constituting a level in the hierarchical fuzzy system. The total number of rules increases only linearly with the number of input variables. For example, for a hierarchical fuzzy system shown in Fig. 8.14, if there are n variables each of which is partitioned into m_i fuzzy subsets, the total number of rules is only $\sum_{i=1}^{n-1} m_i m_{i+1}$. Hierarchical TSK systems [1150] and generalized hierarchical TSK systems [725] have been shown to be universal approximators of any continuous function defined on a compact set.

In Fig. 8.14, the n input variables are x_i, $i = 1, \cdots, n$, and the output is denoted by y. There exist relations

$$y_i = f_i \left(y_{i-1}, x_{i+1} \right) \tag{8.56}$$

for $i = 1, \cdots, n - 1$, where f_i is the nonlinear relation described by the ith TSK system, y_i is the output of the ith TSK system, and $y_0 = x_1$. The final output is $y = y_{n-1}$. The output y is easily obtained by a recursive procedure. Thus, the inference in the hierarchical fuzzy system is in a recursive manner, which is different from the conventional fuzzy inference procedure.

The hierarchical fuzzy system reduces the number of rules, however, the curse of dimensionality is inherent in the system. In the standard fuzzy system, the degree of freedom is unevenly distributed over the IF and THEN parts of the rules, with a comprehensive IF part to cover the whole domain

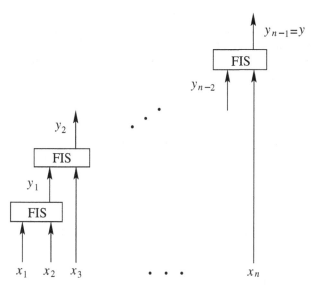

Fig. 8.14. Example of a hierarchical fuzzy system with n inputs and one output. The system is comprised of $n-1$ two-input TSK systems. The n input variables are x_i, $i = 1, \cdots, n$, the output is denoted by y, and y_i is the output of the ith TSK system.

and a simple THEN part. The hierarchical fuzzy system, on the other hand, provides with an incomplete IF part but a more complex THEN part. The gradient-descent method can be applied to parameter learning of these systems. Generally, conventional fuzzy systems achieve universal approximation using piecewise-linear functions, while the hierarchical fuzzy system achieves it through piecewise-polynomial functions [1150, 725].

8.6.3 Rule Generation Based on Look-up Table

Designing fuzzy systems from pattern pairs is a nonlinear regression problem. In the simple LUT technique [1151], each pattern pair generates one fuzzy rule and then a selection process determines the important rules, which are used to construct the final fuzzy system. In the LUT technique, the input MFs do not change with the sampling data, thus the designed fuzzy system uniformly covers the domain of interest.

In the LUT technique, the input and output spaces are first divided into fuzzy regions, then a fuzzy rule is generated from a given pattern pair, and finally a degree is assigned to each rule to resolve rule conflicts and reduce the number of rules. When a new pattern pair becomes available, a rule is created for this pattern pair and the fuzzy rule base is updated.

Given a set of N examples, $\{(\mathbf{x}_p, y_p)\}$, $\mathbf{x}_p = (x_{p,1}, \cdots, x_{p,n})^{\mathrm{T}}$. The domain intervals of these components are $\left[x_i^-, x_i^+\right]$, $i = 1, \cdots, n$, and $[y^-, y^+]$, respectively. A fuzzy-rule system is designed to determine the underlying mapping between the examples. The LUT technique is implemented in five steps given by Algorithm 8.6.3 [1151, 1153].

The denominator in (8.61) is zero when $\mathbf{x}$ is not covered by any rule in the fuzzy rule base. This problem can be solved by a modification. Let

$$\overline{\mathbf{k}}^* = \arg \min_{\overline{\mathbf{k}} \in \mathcal{K}} \left\| \mathbf{x} - \boldsymbol{\gamma}^k \right\| \tag{8.62}$$

be the index of the rule in the fuzzy rule base whose center is the closest one to $\mathbf{x}$, where $\boldsymbol{\gamma}^k = \left(\gamma_1^{k_1}, \cdots, \gamma_n^{k_n} \right)^{\mathrm{T}}$. If $\overline{\mathbf{k}}^*$ is not unique, choose an arbitrary one. Then if $\prod_{i=1}^n \mu_{A_i^{k_i^*}}(x_i) \neq 0$, the fuzzy system can be constructed by (8.61); otherwise, $\hat{f}(x) = \overline{y}(k^*)$ [1153]. If the triangular MFs are changed into Gaussian MFs, the member degrees will not be zero at any point, and thus (8.61) is well defined and no modification is necessary [1153].

The fuzzy system thus constructed is proved to be a universal approximator by using the Stone–Weierstrass theorem [1151]. The approach has the advantage that modification of the rule base is very easy as new examples are available. It is a simple and fast one-pass procedure, since no iterative training is required. Naturally, this algorithm produces an enormous number of rules, when the total input data is considerable. There also arises the problem of contradictory rules, and noisy data in the training examples will affect the consequence of a rule. A similar grid partitioning-based method in which each datum generates one rule has also been derived in [3].

8.6.4 Other Methods

Many other general methods can be used to automatically extract fuzzy rules from a set of numerical examples and to build a fuzzy system for function approximation; some of these are heuristics-based approaches [500, 958, 307, 1085], and hybrid neural-fuzzy approaches such as the ANFIS [541].

In [500], a framework for quickly prototyping an expert system from a set of numerical examples has been established. The algorithm is divided into six steps: cluster and fuzzify the output data, construct initial MFs for input attributes, construct the initial decision table, simplify the initial decision table, rebuild MFs in the simplification process, and derive decision rules from the decision table.

The fuzzy system can be built in a constructive way [958]. Starting from an initially simple system, the number of MFs in the input domain and the number of rules are adapted in order to reduce the approximation error. The algorithm is able to recognize as to where it is necessary to assign a larger number or density of rules and to increase the density of MFs in a specific input variable to improve the approximation accuracy.

A function approximation problem can be first converted into a pattern-classification problem, and then solved by using a fuzzy system [307, 1085]. In

Algorithm 8.6.3 (LUT)

1. *Define fuzzy sets to cover the input and output space.*
   Define m_i fuzzy sets $\mathcal{A}_i^{k_i}$, $k_i = 1, \cdots, m_i$, on each $\left[x_i^-, x_i^+\right]$ with the triangular MFs

$$\mu_{\mathcal{A}_i^1}(x_i) = \mu\left(x_i; \gamma_i^1, \gamma_i^1, \gamma_i^2\right)$$
$$\mu_{\mathcal{A}_i^{k_i}}(x_i) = \mu\left(x_i; \gamma_i^{k_i-1}, \gamma_i^{k_i}, \gamma_i^{k_i+1}\right), \quad k_i = 2, \cdots, m_i - 1$$
$$\mu_{\mathcal{A}_i^{m_i}}(x_i) = \mu\left(x_i; \gamma_i^{m_i-1}, \gamma_i^{m_i}, \gamma_i^{m_i}\right) \tag{8.57}$$

   for $i = 1, \cdots, n$, where $x_i^- = \gamma_i^1 < \gamma_i^2 < \cdots < \gamma_i^{m_i} = x_i^+$ and $\mu(\cdot)$ is defined by (8.13).
   Similarly, define m_y fuzzy sets $\mathcal{B}^l$, $l = 1, \cdots, m_y$, on $\left[y^-, y^+\right]$ with centers at $y^- = \varsigma^1 < \varsigma^2 < \cdots < \varsigma^{m_y} = y^+$.
2. *Generate one fuzzy rule from each example.*
   Given $(\mathbf{x}_p, y_p)$. Determine the fuzzy set $\mathcal{A}_i^{k_i^p}$ such that

$$k_i^p = \arg_{k_i} \max_{1 \leq k_i \leq m_i} \mu_{\mathcal{A}_i^{k_i}}(x_{p,i}) \tag{8.58}$$

   for $i = 1, \cdots, n$. Similarly, determine $\mathcal{B}^{l^p}$ such that

$$l^p = \arg_l \max_{1 \leq l \leq m_y} \mu_{\mathcal{B}^l}(y_p) \tag{8.59}$$

   A fuzzy rule is then generated as

   R_p: IF x_1 is $\mathcal{A}_1^{k_1^p}$ and $\cdots$ and x_n is $\mathcal{A}_n^{k_n^p}$, THEN y is $\mathcal{B}^{l^p}$.

3. *Assign a degree to each rule and resolve conflicting rules.*
   When the number of examples is large, there is a high probability of conflicting rules, *i.e.* rules with the same IF parts but different THEN parts. Each rule is assigned a degree of fulfillment, defined by

$$D(R_p) = \left(\prod_{i=1}^n \mu_{\mathcal{A}_i^{k_i^p}}(x_{p,i})\right) \mu_{\mathcal{B}^{l^p}}(y_p) \tag{8.60}$$

   For a group of conflicting rules, only the rule with the maximum degree is retained.
4. *Create the combined fuzzy rule base.*
   The generated rules as well as the human expert's knowledge in the form of linguistic rules can be combined so as to produce a fuzzy rule base

   $R^{(k)}$: IF x_1 is $\mathcal{A}_1^{k_1}$ and $\cdots$ and x_n is $\mathcal{A}_n^{k_n}$, THEN y is $\mathcal{B}^{(k)}$.

   where $\overline{\mathbf{k}} = (k, k_1, \cdots, k_n)^{\mathrm{T}} \in \mathcal{K}$, $\mathcal{K}$ being the index set of the fuzzy rule base.
5. *Construct the fuzzy system.*
   Finally, the fuzzy system can be built as

$$\hat{f}(x) = \frac{\sum_{\overline{\mathbf{k}} \in \mathcal{K}} \overline{y}^{(k)} \prod_{i=1}^n \mu_{\mathcal{A}_i^{k_i}}(x_i)}{\sum_{k \in \mathcal{K}} \prod_{i=1}^n \mu_{\mathcal{A}_i^{k_i}}(x_i)} \tag{8.61}$$

   where $\overline{y}^{(k)}$ is the center of $\mathcal{B}^{(k)}$.

[1085], the universe of discourse of the output variable is divided into multiple intervals, each regarded as a class, and then a class is assigned to each of the training data according to the desired value of the output variable. The data of each class are then partitioned in the input space to achieve a higher accuracy in the approximation of the class regions until a termination criterion is satisfied. Class regions can be represented using either hyperboxes or ellipsoidal regions. Fuzzy rules are then extracted from the approximated class regions. For a given input datum, the resulting vector of the classmembership degrees is defuzzified into a single real value, which represents the approximated final result. A similar idea is used in [307], where a clustering method is applied to estimate the local centroids and covariances of the pattern classes.

8.7 Interpretability

One of the motivations for using fuzzy systems is due to their interpretability. Interpretability helps to check the plausibility of a system, leading to easy maintenance of the system. It can also be used to acquire knowledge from a problem characterized by numerical examples. An improvement in interpretability can enhance the performance of generalization when the data set is small.

The interpretability of a rule base is usually related to continuity, consistency and completeness [440]. Continuity guarantees that small variations of the input do not induce large variations in the output. Consistency means that if two or more rules are simultaneously fired, their conclusions are coherent. Completeness means that for any possible input vector, at least one rule is fired and there is no inference breaking.

When neurofuzzy systems are used to model nonlinear functions described by training sets, the approximation accuracy can be optimized by the learning procedure. However, since learning is accuracy oriented, it usually causes a reduction in the interpretability of the generated fuzzy system. The loss of interpretability can be in the following forms [555]:

1. Incompleteness of fuzzy partitions—Two neighboring fuzzy subsets in a fuzzy partition have no overlap.
2. Indistinguishability of fuzzy partitions—The MFs of two fuzzy subsets are so similar that the fuzzy partition is indistinguishable.
3. Inconsistancy of fuzzy rules—The MFs lose their prescribed physical meaning.
4. Too fuzzy or too crisp fuzzy subsets.
5. Incompactness of the fuzzy system.

To improve the interpretability of neurofuzzy systems, one can add to the cost function, regularization terms that apply constraints on the parameters of fuzzy MFs. For example, the order of the centers of all the fuzzy subsets $\mathcal{A}^i$, which are partitions of the fuzzy set $\mathcal{A}$, should be specified and remain unchanged during learning. Similar MFs should be merged to improve the distinguishability of fuzzy partitions and to reduce the number of fuzzy subsets [1018]. One can also reduce the number of free parameters in defining fuzzy subsets. To increase the interpretability of the designed fuzzy system, the same linguistic term should be represented by the same MF. This results in weight sharing [840, 555]. For the TSK model, one practice for good interpretability is to keep the number of fuzzy subsets much smaller than N_{r}, the number of fuzzy rules, especially when N_{r} is large.

8.8 Fuzzy and Neural: A Synergy

While neural networks have strong learning capabilities at the numerical level, it is difficult for the users to understand them at the logic level. Fuzzy logic, on the other hand, has a good capability of interpretability and can also integrate expert's knowledge. The hybridization of both paradigms yields the capabilities of learning, good interpretation and incorporating prior knowledge.

The combination can be in different forms. The simplest form may be the concurrent neurofuzzy model, where a fuzzy system and a neural network work separately. The output of one system can be fed as the input of the other system. The cooperative neurofuzzy model corresponds to the case that one system is used to adapt the parameters of the other system. Neural networks can be used to learn the membership values for fuzzy systems, to construct IF-THEN rules [394, 988], or to construct a decision logic.

The true synergy of the two paradigms is a hybrid neural/fuzzy system, which captures the merits of both the systems. It can be in the form of either a fuzzy neural network or a neurofuzzy system. A hybrid neural-fuzzy system does not use multiplication, addition, or the sigmoidal function. Alternatively, fuzzy logic operations such as t-norm and t-conorm are used.

A fuzzy neural network is a neural network equipped with the capability of handling fuzzy information, and it is more like a neural network, where the input signals, activation functions, weights, and/or the operators are based on the fuzzy set theory. Thus, symbolic structure is incorporated [902]. The network can be represented in an equivalent rule-based format, where the premise is the concatenation of fuzzy AND and OR logic, and the consequence is the network output. Two types of fuzzy neurons, namely AND neuron and OR neuron, are defined. The NOT logic is integrated into the weights. Weights always have values in the interval $[0, 1]$, and negative weight is achieved by using the NOT operator. As shown in Fig. 8.15, the outputs of the AND and OR neurons are, respectively, defined by

$$y_{\mathrm{AND}} = \wedge \left(\vee \left(w_1, x_1 \right), \vee \left(w_2, x_2 \right) \right) = T \left(C \left(w_1, x_1 \right), C \left(w_2, x_2 \right) \right) \qquad (8.63)$$

$$y_{\mathrm{OR}} = \vee \left(\wedge \left(w_1, x_1 \right), \wedge \left(w_2, x_2 \right) \right) = C \left(T \left(w_1, x_1 \right), T \left(w_2, x_2 \right) \right) \qquad (8.64)$$

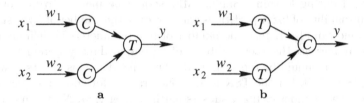

Fig. 8.15. Two types of fuzzy neurons. **(a)** AND neuron. **(b)** OR neuron.

The weights of the fuzzy neural network can be interpreted as calibration factors of the conditions and rules.

A neurofuzzy system is a fuzzy system whose parameters are learned by a learning algorithm obtained from neural networks. It can always be interpreted as a system of fuzzy rules. Learning is used to adaptively adjust the rules in the rule base, and to produce or optimize the MFs of a fuzzy system. A neurofuzzy system has a neural-network architecture constructed from fuzzy reasoning. Structured knowledge is codified as fuzzy rules, while the adapting and learning capabilities of neural networks are retained. Expert knowledge can increase learning speed and estimation accuracy.

Both fuzzy neural networks and neurofuzzy systems can be treated as neural networks, where the units employ the t-norm or t-conorm operator instead of an activation function. The hidden layers represent fuzzy rules. Since it may be difficult to judge whether a hybrid fuzzy-neural system is more like a neurofuzzy system or a fuzzy neural network, we call both types of synergisms neurofuzzy systems in this chapter.

Neurofuzzy systems can be obtained by representing some of the parameters of a neural network, such as the inputs, weights, outputs, and shift terms as continuous fuzzy numbers. When only the input is fuzzy, it is a Type-I neurofuzzy system. When everything except the input is fuzzy, we get a Type-II model. A Type-III model is defined as one where the inputs, weights, and shift terms are all fuzzy.

The functions realizing the inference process are usually nondifferentiable, and thus, the popular gradient-descent or BP algorithm cannot always be applied for training neurofuzzy systems. To make use of gradient-based algorithms, one has to select differential functions for the inference functions. For nondifferentiable inference functions, training can be performed by using EAs. The shape of the MFs, the number of fuzzy partitions, and the rule base can all be evolved by using EAs. Roughly speaking, the neurofuzzy method is superior to the neural-network method in terms of the convergence speed and compactness of the structure. Fundamentals in neurofuzzy synergism for modeling and control have been reviewed in [544].

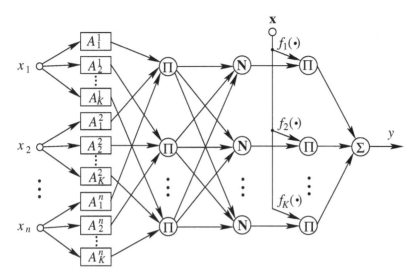

Fig. 8.16. ANFIS: graphical representation of the TSK model. The symbol N in the circles denotes the normalization operator, and $\mathbf{x} = (x_1, x_2, \cdots, x_n)^T$.

8.9 Neurofuzzy Models

In this section, we describe some popular neurofuzzy models and also give a brief survey of neurofuzzy models.

8.9.1 The ANFIS Model

Architecture of the ANFIS Model

The ANFIS is a well-known neurofuzzy model [543, 541, 544]. The ANFIS model, shown in Fig. 8.16, has a six-layer (n-nK-K-K-K-1) architecture, and is a graphical representation of the TSK model. The functions of the various layers are given below.

1. Layer 1 is the input layer with n nodes.
2. Layer 2 has nK nodes, each outputting the membership value of the ith antecedent of the jth rule

$$o_{ij}^{(2)} = \mu_{\mathcal{A}_j^i}(x_i) \qquad (8.65)$$

for $i = 1, \cdots, n$, $j = 1, \ldots, K$, where $\mathcal{A}_j^i$ defines a partition of the space of x_i, and $\mu_{\mathcal{A}_j^i}(x_i)$ is typically selected as a generalized bell MF defined by (8.16)

$$\mu_{\mathcal{A}_j^i}(x_i) = \mu\left(x_i; c_j^i, a_j^i, b_j^i\right) \qquad (8.66)$$

The parameters c_j^i, a_j^i, and b_j^i are referred to as *premise parameters*.

3. Layer 3 has K fuzzy neurons with the product t-norm as the aggregation operator. Each node corresponds to a rule, and the output of the jth neuron determines the degree of fulfillment of the jth rule

$$o_j^{(3)} = \prod_{i=1}^{n} \mu_{\mathcal{A}_j^i}(x_i) \tag{8.67}$$

for $j = 1, \cdots, K$.

4. Each neuron in layer 4 performs normalization, and the outputs are called *normalized firing strengths*

$$o_j^{(4)} = \frac{o_j^{(3)}}{\sum_{k=1}^{K} o_k^{(3)}} \tag{8.68}$$

for $j = 1, \cdots, K$.

5. The output of each node in layer 5 is defined by

$$o_j^{(5)} = o_j^{(4)} f_j(\mathbf{x}) \tag{8.69}$$

for $j = 1, \cdots, K$, where $f_j(\cdot)$ is given for the jth node in layer 5. Parameters in $f_j(\mathbf{x})$ are referred to as *consequent parameters*.

6. The outputs of layer 5 are summed and the output of the network gives the TSK model (8.46)

$$o^{(6)} = \sum_{j=1}^{K} o_j^{(5)} \tag{8.70}$$

Learning of the ANFIS Model

In the ANFIS model, functions used at all the nodes are differentiable, thus the BP algorithm can be used to train the network. Each MF $\mu_{\mathcal{A}_i^j}$ is specified by a predefined shape and its corresponding shape parameters. The shape parameters are adjusted by a learning algorithm using a sample set of size N, $\{(\mathbf{x}_t, y_t)\}$. For nonlinear modeling, the effectiveness of the model is dependent on the MFs used.

The TSK fuzzy rules are employed in the ANFIS model

$$R_i: \text{ IF } \mathbf{x} \text{ is } \mathcal{A}_i, \text{ THEN } y = f_i(\mathbf{x}) = \sum_{j=1}^{n} a_{i,j} x_j + a_{i,0}$$

for $i = 1, \cdots, K$, where $\mathcal{A}_i = \{\mathcal{A}_i^1, \mathcal{A}_i^2, \cdots, \mathcal{A}_i^n\}$ are fuzzy sets and $a_{i,j}$, $j = 0, 1, \cdots, n$, are consequent parameters. The output of the network at time t is thus given by

$$\hat{y}_t = \frac{\sum_{i=1}^{K} \mu_{\mathcal{A}_i}(\mathbf{x}_t) f_i(\mathbf{x})}{\sum_{i=1}^{K} \mu_{\mathcal{A}_i}(\mathbf{x}_t)} \tag{8.71}$$

where $\mu_{\mathcal{A}_i}(\mathbf{x}_t) = \bigwedge_{j=1}^{n} \mu_{\mathcal{A}_i^j}(x_{t,j}) = \prod_{j=1}^{n} \mu_{\mathcal{A}_i^j}(x_{t,j})$. Accordingly, the error measure at time t is defined by

$$E_t = \frac{1}{2} \left(\hat{y}_t - y_t \right)^2 \tag{8.72}$$

After the rule base is specified, the ANFIS adjusts only the MFs of the antecedents and the consequent parameters. The BP algorithm can be used to train both the premise and consequent parameters. A more efficient procedure is to learn the premise parameters by the BP, but to learn the linear consequent parameters $a_{i,j}$ by the RLS method [541]. The learning rate η can be adaptively adjusted by using a heuristic, for example, increase η by 10% if the error decreases in four consecutive steps, and decrease η by 10% if the error is subject to consecutive combinations of increase and decrease. It is reported in [541] that this hybrid learning method provides better results than the MLP trained by the BP method and the cascade-correlation network [347].

Second-order methods are also applied for the training of the ANFIS. In [542], the LM method is used for ANFIS training. Compared to the hybrid method, the LM method achieves a better precision, but the interpretability of the final MFs is quite weak. In [203], the RProp [949] and the RLS methods are used to learn the premise parameters and the consequent parameters, respectively. The ANFIS model has been generalized for classification by employing parameterized t-norms [1067]. The generalized model employs tree partitioning for structure identification and the Kalman filtering method for parameter learning.

The ANFIS is attractive for applications in view of its network structure and the standard learning algorithm[2]. Training of the ANFIS follows the spirit of the minimal disturbance principle and is thus more efficient than the MLP [544].

However, the ANFIS is computationally expensive due to the curse-of-dimensionality problem arising from grid partitioning. Constraints on MFs and initialization using prior knowledge cannot be provided to the ANFIS model due to the learning procedure. The learning results may be difficult to interpret. Thus, the ANFIS model is suitable for applications, where performance is more important than interpretation.

In order to preserve the plausibility of the ANFIS, one can add some regularization terms to the cost function so that some constraints on the interpretability are considered [544]. Tree or scattering partitioning can resolve the problem of the curse of dimensionality, but leads to a reduction in the interpretability of the generated rules.

Generalization of the ANFIS Model

The ANFIS has been extended to the coactive ANFIS [811] and to the generalized ANFIS [49]. The coactive ANFIS [811] is a generalization of the ANFIS

---

[2] The ANFIS model is available in the Matlab© Fuzzy Logic Toolbox. C codes of the ANFIS training algorithm and several examples can be retrieved via anonymous ftp from *user/ai/areas/fuzzy/systems/anfis/* at *ftp.cs.cmu.edu*.

by introducing nonlinearity into the TSK rules. The generalized ANFIS [49] is based on a generalization of the TSK model and a generalized Gaussian RBFN. The generalized fuzzy model is trained by using the generalized RBFN model, based on the functional equivalence between the two models.

The sigmoid-ANFIS [1253] is a special form of the ANFIS, where only sigmoidal MFs are employed in this model. The sigmoid-ANFIS is a combination of the additive TSK-type MLP and the additive TSK-type FIS. The additive TSK-type MLP, as an extended model of the MLP, is proved to be functionally equivalent to the TSK-type FIS and to be a universal approximator [1253]. The sigmoid-ANFIS model adopts the *interactive-or* operator [99] as its fuzzy connectives. The gradient-descent algorithm can also be directly applied to the TSK model without representing it in a network structure [854].

The unfolding-in-time is a method to transform an RNN into an FNN so that the BP algorithm can be used. The ANFIS-unfolded-in-time [1041] is a method that duplicates the ANFIS T times to integrate temporal information, where T is the number of time intervals needed in the specific problem. The ANFIS-unfolded-in-time is designed for prediction of time series data. Simulation results show that the recognition error is much smaller in the ANFIS-unfolded-in-time compared to that in the ANFIS.

8.9.2 Generic Fuzzy Perceptron

The generic fuzzy perceptron (GFP) [840] has a structure similar to that of the three-layer MLP. In contrast to the MLP, the network inputs and the weights are modeled as fuzzy sets, and t-norm or t-conorm is used as the activation at each unit. The hidden layer acts as the rule layer. The output units usually use a defuzzufication function. The GFP can interpret its structure in the form of linguistic rules and the structure of the GFP can be treated as a linguistic rule base, where the weights between the input and hidden (rule) layers are called *fuzzy antecedent weights* and the weights between the hidden (rule) and output layers *fuzzy consequent weights*. The GFP model is based on the Mamdani model.

Based on the GFP model, there are three fuzzy models, namely, neurofuzzy controller (NEFCON) [841, 840, 857], neurofuzzy classification (NEFCLASS) [840], and neuronfuzzy function approximation (NEFPROX) [840]. Due to the use of nondifferentiable t-norm and t-conorm, the gradient-descent method cannot be applied. A set of linguistic rules are used for describing the performance of the models. This knowledge-based fuzzy error is independent of the range of the output value. Learning algorithms for all these models are derived from the fuzzy error using simple heuristics.

Initial fuzzy partitions are needed to be specified for each input variable. Some connections that have identical linguistic values are forced to have the same weights so as to keep the interpretability. Prior knowledge can be integrated in the form of fuzzy rules to initialize the neurofuzzy systems, and the remaining rules are obtained by learning. Both the fuzzy sets and the fuzzy rules can be learned, and learning results in shifting the MFs and changing their supports.

The NEFCON has a single output node, and is used for control. A reinforcement learning algorithm is used for online learning. The NEFCON tries to integrate as many rules known *a priori* as possible. The NEFCLASS and the NEFPROX can learn rules by using supervised learning instead of reinforcement learning. The rule base of a NEFCLASS system approximates an unknown function that represents the classification problem and maps an input to its class. Compared to neural networks, the NEFCLASS uses a much simpler learning strategy, where no clustering is involved in finding the rules. The NEFCLASS does not use MFs in the rules' consequents.

The NETPROX is similar to the NEFCON and the NEFCLASS, but it is more general. As an example of the GFP model, the architecture of the NETPROX is shown in Fig. 8.17. If there is no prior knowledge, a NEFPROX system can be started with no hidden unit and rules can be incrementally learned. If the learning algorithm creates too many rules, only the best rules are kept by evaluating individual rule errors. Each rule represents a number of samples of the unknown function in the form of a fuzzy sample. Parameter learning is used to compensate for the error caused by rule removing.

The NETPROX is more important for function approximation. An empirical performance comparison between the ANFIS and the NETPROX has been made in [840]. For the problems considered therein, the NEFPROX is an

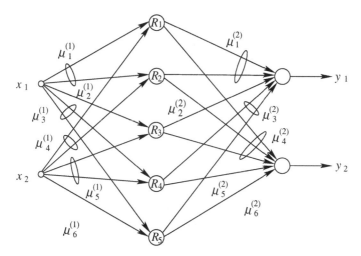

Fig. 8.17. The NEFPROX as an example of the GFP model. When there is only a single output, the NEFPROX has the same architecture as the NEFCON, and when no MFs are used in the consequent parts, the NEFPROX has the same architecture as the NEFCLASS. The hidden layer is the rule layer with each node corresponding to a rule, and the output layer is the defuzzification layer. All the $\mu_i^{(1)}$ and $\mu_i^{(2)}$ are, respectively, the MFs used in the premise and consequent parts.

order of magnitude faster than the ANFIS model of [541], but with a higher approximation error. This is due to the fact that the ANFIS uses a complex learning algorithm, which results in a longer computation time and small approximation error. Interpretation of the learning result is difficult for both the ANFIS and the NEFPROX: the ANFIS represents a TSK system, while the NEFPROX represents a Mamdani system with too many rules. To increase the interpretability of the NEFPROX, pruning strategies can be employed to reduce the number of rules.

8.9.3 Other Neurofuzzy Models

In Chapter 5, we have introduced fuzzy clustering algorithms, which are primarily based on competitive learning networks such as the Kohonen network and the ART models. Neurofuzzy systems can employ network topologies similar to those of the layered FNN architecture [469, 557, 236, 302], the RBFN model [19, 1195, 863], the SOM model [1136], and the RNN architecture [708, 723].

Neurofuzzy models discussed in this chapter are mainly used for function approximation. These models typically have a layered FNN architecture and are based on TSK-type FISs. Neurofuzzy systems are usually trained by using the gradient-descent method [884, 535, 469, 707, 604, 724, 1102]. Gradient descent in this case is sometimes termed as the *fuzzy BP algorithm*. CG algorithms are also used for training a neurofuzzy systems [724]. Based on the fuzzification of the linear autoassociative neural networks, the fuzzy PCA [302] can extract a number of relevant features from high-dimensional fuzzy data.

A typical architecture of a neurofuzzy system includes an input layer, an output layer, and several hidden layers. The weights are fuzzy sets, and the neurons apply t-norm or t-conorm operations. The hidden layers are usually used as rule layers. The layers before the rule layers perform as premise layers, while those after perform as consequent layers.

HyFIS

Hybrid neural FIS (HyFIS) [604] is a five-layer neurofuzzy model based on the Mamdani FIS. Layer 1 is the input layer with crisp values. Nodes in layers 2 and 4 act as MFs to express the input and output fuzzy linguistic variables using Gaussian MFs. Expert knowledge can be used for the initialization of these MFs. Each node in layer 3 performs the fuzzy AND operation, and is a rule node. Each rule node represents one fuzzy rule and a possible IF-part of the rule, and all the rule nodes form a fuzzy rule base. Each node in layer 4 represents a possible THEN-part of a rule, and performs the fuzzy OR operation to integrate the field rules leading to the same output linguistic variables. The nodes of layers 3 and 4 are fully connected with weights in the interval $[-1, +1]$, while the weights between all other adjacent layers are unity. The node in layer 5 acts as a defuzzifier and computes a crisp output signal.

The HyFIS employs a hybrid learning scheme comprised of two phases, namely, rule generation from data (structure learning) and rule tuning using BP learning (parameter learning). The HyFIS first extracts fuzzy rules from data by using the LUT technique [1151]. This is used as the initial structure so that the learning process can be fast, reliable and highly intuitive. The gradient-descent method is then applied to tune the MFs of input/output linguistic variables and the network weights by minimizing the error function. Only a few training iterations are needed for the model to converge, since the initial structure and weights of the model are set properly. The HyFIS model is comparable in performance with the ANFIS [541].

Fuzzy Min-max Neural-network Models

Fuzzy min-max neural networks are a class of neurofuzzy models using min-max hyperboxes for clustering, classification, and regression [1038, 1039, 391, 1073]. The max-min fuzzy Hopfield network [723] is a fuzzy RNN for fuzzy associative memory (FAM). The manipulations of the hyperboxes involve mainly comparison, addition and subtraction operations, thus learning is extremely efficient.

The fuzzy min-max classification network [1038] utilizes fuzzy sets as pattern classes. Each fuzzy set is an aggregate of min-max hyperboxes. The learning algorithm has the ability to learn online and in a single pass through the data. The fuzzy min-max clustering network [1039] provides the ability to incorporate new data and add new clusters without retraining. Both of the methods implement three steps for each example: Hyperboxes are first expanded for accommodating the new example, then an overlap test is conducted to determine overlaps among hyperboxes associated with different classes or clusters, and finally hyperboxes are contracted so as to eliminate overlaps. The fuzzy min-max clustering network stabilizes into pattern clusters in only a few passes through a data set. However, the training result is highly dependent on the presentation order of the examples and on the learning parameter, which also imposes the same constraint on coverage resolution in the whole input space [1073].

The general fuzzy min-max network [391] is a generalization of the fuzzy min-max clustering and classification algorithms [1038, 1039]. The method combines the supervised and unsupervised learning within a single training algorithm. The fusion of clustering and classification results in an algorithm that can be used as pure clustering, pure classification, or hybrid clustering/classification. Learning is usually completed in a few passes through the data and consists of placing and adjusting the hyperboxes in the pattern space.

In [1073], two batch learning algorithms for the fuzzy min-max neural network, namely, the top-down fuzzy min-max algorithm for classification and the top-down fuzzy min-max regressor algorithm for regression, are discussed.

The former algorithm improves the original learning algorithm [1038] by using the batch mode, while the latter extends the network for solving regression problems by using a hybrid fuzzy classifier and a gradient-descent algorithm.

A constructive training method called the *adaptive resolution classifier (ARC)* and its pruning version called the *pruning ARC (PARC)* [956] have been proposed in [956] for the fuzzy min-max neural classification network [1038]. The ARC and the PARC generate a regularized min-max network by a succession of hyperbox cuts. By using a recursive cutting procedure, it is possible to obtain a better generalization capability. These algorithms allow a fully automatic training process, since no critical parameters must be fixed in advance. In addition, the adaptive resolution mechanism significantly enhances the generalization capability of the trained models.

Fuzzy Radial Basis Function Networks

The fuzzy basis function network (FBFN) [1152] has a structure similar to that of the RBFN. It uses the singleton fuzzifier, product inference, the centroid defuzzifier, and Gaussian input MFs. Similar to the ANFIS model, it is also based on the TSK model. In addition to the simple structure, the FBFN possesses the advantage that it can readily adopt various learning algorithms already developed for the RBFN. The FBFN is capable of uniformly approximating any continuous nonlinear function to a specified accuracy with a finite number of basis functions [1152]. The OLS method [205, 208] is used for selecting significant fuzzy basis functions. In [662], two constructive algorithms, namely the LS and the GA-based adaptive LS algorithms, have been proposed for training the FBFN. Both of the algorithms generate results superior to those of the OLS and BP algorithms. The fuzzy RBFN [809] integrates the merits of the RBFN and the FCM algorithm. The architecture of the RBFN is suitably modified to incorporate FCM computation. It incorporates fuzzy-set concepts at the input, output and hidden layers. The model can handle both linguistic and numeric inputs, and provides a soft decision in the case of overlapping pattern classes at the output. The input is in terms of linguistic values *low, medium* and *high*. The output is provided as classmembership values. Thus, an n-dimensional input pattern may be represented as a $3n$-dimensional vector. The RBF-based adaptive fuzzy system [231] can automatically generate fuzzy rules by recruiting the basis function units gradually and adjusting the system parameters. Another fuzzy RBFN has been presented in [220], which is an RBFN with the connection weights between the hidden and output layers as well as the network output being fuzzy numbers.

Adaptive Neurofuzzy Systems

Self-organization has been introduced in hybrid systems to create adaptive models for adaptively representing time-varying systems and model identification. Adaptation can be through one of the two strategies: a fixed space partitioning with adaptive fuzzy rule parameters or a simultaneous adaptation of space partitioning and fuzzy rule parameters. While the ANFIS model belongs to the former category, adaptive parsimonious neurofuzzy systems can be achieved by using a constructive approach and the latter strategy [231, 1195].

The dynamic fuzzy neural network (DFNN) [1195, 336] is an online implementation of the TSK fuzzy system based on an extended RBFN and its learning algorithm. The extended RBFN has five layers and no bias, and the weights may be a linear regression of the input. The system starts with no hidden unit, and neurons can be added or deleted dynamically according to their significance to system performance. Both the parameters and the structure can be adjusted simultaneously. A parsimonious structure with high performance can be achieved by ERR-based [208] pruning technique. The method has a performance better than that of the RBFN trained by the OLS [208], the minimal RAN [1234], or the EKF-based RAN [566].

Similar to the ANFIS architecture, the self-organizing fuzzy neural network (SOFNN) [683] has a five-layer fuzzy neural network architecture. It is an online implementation of a TSK-type fuzzy model. The SOFNN is based on neurons with an ellipsoidal basis function, and the neurons are added or pruned dynamically in the learning process. The parameter learning makes the network converge quickly through the RLS algorithm. The pruning strategy is then applied using the OBS method [460] based on the error covariance matrix obtained during the RLS learning [689]. Similar MFs can be combined into one new MF. The SOFNN algorithm is superior to the DFNN in time complexity [1195].

8.10 Fuzzy Neural Circuits

To date, there are lots of fuzzy hardware patents and products in a wide variety of systems, e.g., consumer products, industrial controllers, and automotive electronics. Generally, fuzzy systems can be easily implemented in the digital form, which can be either general-purpose microcontrollers running fuzzy inference and defuzzification programs, or dedicated fuzzy coprocessors, or RISC processors with specialized fuzzy support, or fuzzy ASICs. The pros and cons of various digital fuzzy hardware implementation strategies are reviewed in [262].

A common approach to general-purpose fuzzy hardware is to use a software design tool to generate the program code for a target microcontroller. Examples include the Motorola-Aptronix fuzzy inference development language and Togai InfraLogic's MicroFPL system [520]. This approach leads to rapid design and testing, but has a low performance. On the other hand, dedicated fuzzy processors and ASICs have physical and performance characteristics that are closely matched to an application, and their performances would be optimized to suit a given problem. However, the design and test costs are high.

Fuzzy coprocessors work in conjunction with a host processor. They are general-purpose hardware, and thus have a lower performance compared to a custom fuzzy hardware. A number of commercially available fuzzy coprocessors are listed in [989]. Some issues arising from the implementation of such coprocessors are discussed in [891].

RISC processors with specialized fuzzy support are also available [262, 989]. A fuzzy-specific extension to the instruction set is defined and implemented using hardware/software codesign techniques. The fuzzy-specific instructions significantly speed up fuzzy computation with no increase in the processor cycle time and with only a minor increase in the chip area.

In [520], the tool TROUT was created to automate fuzzy neural ASIC design. The TROUT produces a specification for small, customized, application-specific circuits called *smart parts*. A smart part is a dedicated compact-size circuit customized to a single function. A designer can package a smart part in a variety of ways. The model library of the TROUT includes fuzzy or neural-network models for implementation as circuits. To synthesize a circuit, the TROUT takes as its input an application data set, optionally augmented with user-supplied hints. It delivers, as output, technology-independent very high level hardware description language (VHDL) code, which describes a circuit implementing a specific fuzzy or neural-network model optimized for the input data set. As an example, the TROUT has been used for the synthesis of the fuzzy min-max classification network [1038].

There are also many analog [681, 254, 643], and mixed-signal [84, 128] fuzzy circuits. Analog circuits usually operate in the current mode and are fabricated using the CMOS technology, and this leads to the advantages of high speed, small-circuit area, high performance, and low power dissipation. A design methodology for fuzzy ASICs and general-purpose fuzzy processors is given in [643], based on the LR^3 fuzzy implication cells and the LR fuzzy arithmetic cells. In [84, 128], the fabrication of mixed-signal CMOS chips for fuzzy controllers is considered; in these circuits, the computing power is provided by the analog part while the digital part is used for programmability.

An overview of the existing hardware implementations of neural and fuzzy systems has been made in [946], where limitations, advantages, and bottlenecks of analog, digital, pulse stream (spiking), and other techniques are discussed. The use of hardware/software codesign is concluded as a means of exploiting the best from both the hardware and software techniques, as it allows a fast design of complex systems with the highest performance–cost ratio. A survey of digital fuzzy logic controllers has also been presented in [891].

---

[3] LR denotes left-right. LR fuzzy sets are a special representation for fuzzy numbers. The MF of an LR fuzzy set is represented by a left part and a right part.

8.11 Computer Experiments

In this section, we give two examples based on the ANFIS model. The computation is conducted on a PC with a Pentium III 600 MHz CPU and 128M RAM using the Matlab© Fuzzy Logic Toolbox.

8.11.1 Example 8.1: Solve the DoA Estimation Using the ANFIS with Grid Partitioning

We now revisit Examples 3.2 and 6.2. This time we use the ANFIS model to solve the same DoA estimation problem.

The input and output preprocessing is the same as case (b) of Example 3.2, that is, there are six inputs and one output. For uniformly generated 500 training patterns, the ranges of the input and output variables are

$$x_1 \in [\ 0.99999847692826, 1.00000000000000],$$
$$x_2 \in [\ 0.99999390771768, 1.00000000000000],$$
$$x_3 \in [\ 0.99998629238218, 1.00000000000000],$$
$$x_4 \in [-0.00174525116003, 0.00174531978741],$$
$$x_5 \in [-0.00349049700419, 0.00349063425833],$$
$$x_6 \in [-0.00523573221663, 0.00523593809627],$$
$$y \ \in [-1.56766100478533, 1.56139036076618].$$

A set of 100 random samples is used for testing. Note that y is in radians. The MSE is calculated by converting y into degrees.

An initial TSK FIS is first generated by using grid partitioning. Since the ranges for x_1, x_2, and x_3 are very small, they each are partitioned into 2 subsets. The ranges of x_4, x_5, and x_6 are each partitioned into 3 subsets. The fuzzy partitioning for the input space is illustrated in Fig. 8.18. The Gaussian MF is selected.

The training of the ANFIS model generates best result with the MSE 0.3395 after one epoch. The ANFIS model generates 471 nodes, 1512 linear parameters, 30 nonlinear parameters, and 216 fuzzy rules. The training time is 478.51 s. The MSE for testing is 0.2992. The training error and the testing error are illustrated in Fig. 8.19.

By providing a pattern set of 2000 uniform patterns, the ANFIS model gets the best result after one epoch with an accuracy of 0.3187. These results are very close to that for 500 examples. The training time is 1820.73 s. The MSE for testing is 0.2792. Thus, we can see that the ANFIS model has a very good generalization capability, and a large pattern set is unnecessary for training.

The ANFIS training reaches the optimal performance after one epoch, and system performance degrades for more epochs. To improve the system performance, one has to resort to a finer partitioning of the system. This,

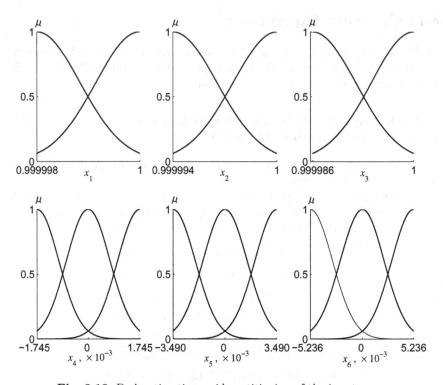

Fig. 8.18. DoA estimation: grid partitioning of the input space

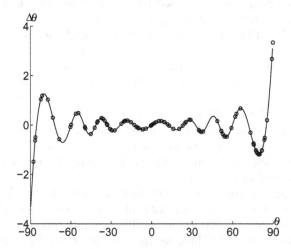

Fig. 8.19. DoA estimation using the ANFIS with grid partitioning: training and testing errors, for 500 training patterns and 100 testing patterns. All units are in degrees.

however, results in the curse-of-dimensionality problem, that is, a huge rule base and intolerable requirements on the training time and memory capacity.

8.11.2 Example 8.2: Solve the DoA Estimation Using the ANFIS with Scatter Partitioning

Clustering the input space is a desired method for generating fuzzy rules. This can significantly reduce the total number of fuzzy rules, hence offer a better generalization capability. The subtractive clustering [226] is now used for rule extraction so as to find an initial FIS for ANFIS training.

Radius r specifies the range of influence of the cluster center for each input and output dimension. The training error can be controlled by adjusting r, $r \in [0, 1]$. Specifying a smaller cluster radius usually yields more, smaller clusters in the data, and hence more rules.

Since the range of the input space is very small when compared with that of the output space, we select $r = 0.9$ for all the input dimensions and $r = 0.1$ for the output space. At the first epoch, training achieves the best performance with a training MSE of 0.3663, and the training time is only 2.69 s. The testing error is 0.4188. The ANFIS model has 219 nodes, 105 linear parameters, 180 nonlinear parameters, and 15 fuzzy rules. The scatter partitioning is shown in Fig. 8.20, and the training and testing errors are illustrated in Fig. 8.21.

For the 15 rules generated, each rule has its own MF for each input variable. For example, the ith rule is given by

$$R_i: \text{IF } x_1 \text{ is } \mu_{i,1} \text{ AND } x_2 \text{ is } \mu_{i,2} \text{ AND } x_3 \text{ is } \mu_{i,3} \text{ AND } x_4 \text{ is } \mu_{i,4} \text{ AND } x_5 \text{ is } \mu_{i,5} \text{ AND } x_6 \text{ is } \mu_{i,6} \text{ THEN } y \text{ is } \mu_{i,y}$$

where $\mu_{i,k}$, $k = 1, \cdots, 6$, and $\mu_{i,y}$ are MFs. These rules and the fuzzy inference process are illustrated in Fig. 8.22.

Similarly, if the training set is 2000 uniformly distributed patterns, the MSEs for training and testing are, respectively, 0.3738 and 0.3111. The computation time is 20.43 s. Thus, there is no significant improvement in training and generalization performance.

To improve the training accuracy, we need a finer clustering. According to the respective ranges of all the input components and the output, we select $\mathbf{r} = [0.8, 0.7, 0.6, 0.5, 0.4, 0.3, 0.05]$. A total of 27 rules are obtained. The training and testing MSEs are, respectively, 0.1048 and 0.1697. The training time is 8.21 s. This is shown in Fig. 8.23.

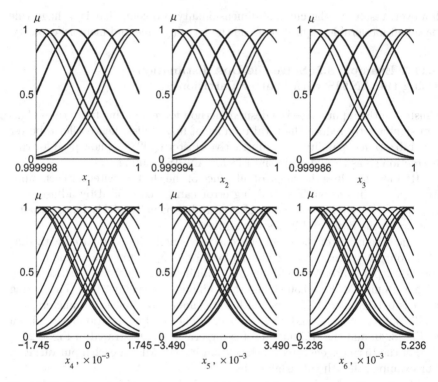

Fig. 8.20. DoA estimation: scatter partitioning of the input space. Note that some MFs coincide in the figure.

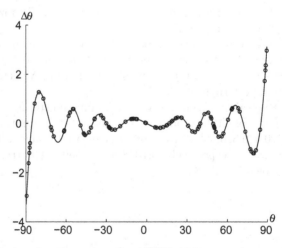

Fig. 8.21. DoA estimation using the ANFIS with scattering partitioning: training and testing errors, for 500 training patterns and 100 testing patterns. $\mathbf{r} = [0.9, 0.9, 0.9, 0.9, 0.9, 0.9, 0.1]$. All units are in degrees.

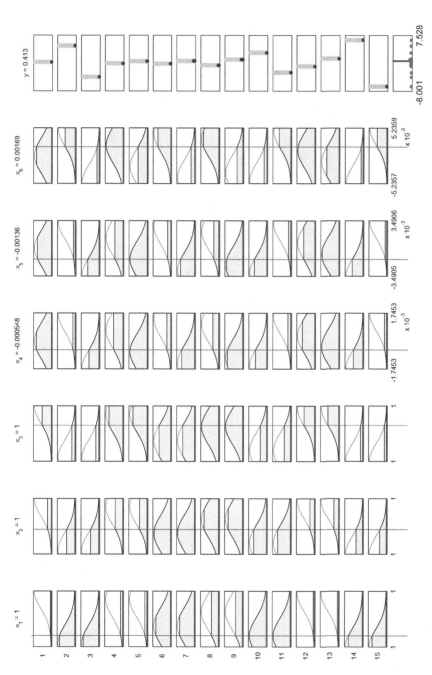

Fig. 8.22. The fuzzy rules for the DoA estimation using the ANFIS with scattering partitioning and the fuzzy-inference process from inputs to outputs. Each row of plots corresponds to one rule, and each column corresponds to either an input variable x_i or the output variable y.

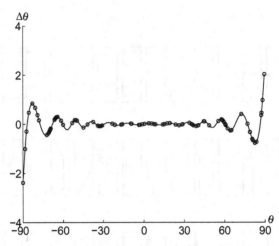

Fig. 8.23. DoA estimation using the ANFIS with scattering partitioning: training and testing errors, for 500 training patterns and 100 testing patterns. $\mathbf{r} = [0.8, 0.7, 0.6, 0.5, 0.4, 0.3, 0.05]$. All units are in degrees.

Evolutionary Algorithms and Evolving Neural Networks

Evolutionary algorithms are a class of general-purpose stochastic optimization algorithms under the universally accepted neo-Darwinian paradigm. The neo-Darwinian paradigm is a combination of the classical Darwinian evolutionary theory, the selectionism of Weismann, and the genetics of Mendel [366]. EAs are currently a major approach to adaptation and optimization.

This chapter is dedicated to EAs and their applications to neural and fuzzy systems. The fundamentals of EAs are first introduced, with emphasis on the GA and the ES. We also discuss some other methods that fall into this category, namely, the genetic programming (GP), the evolutionary programming (EP), the PSO, the immune algorithm, and the ant-colony optimization (ACO). This is followed by some topics pertaining to EAs and a comparison between EAs and the SA. Finally, we describe the application of EAs to the learning of neural networks as well as to the structural and parametric adaptations of fuzzy systems.

9.1 Evolution vs. Learning

The adaptation of creatures to their environments results from the interaction of two processes, namely evolution and learning. Evolution is a slow stochastic process at the population level that determines the basic structures of a species. Evolution operates on biological entities, rather than on the individuals themselves. At the other end, learning is a process of gradually improving an individual's adaptation capability to its environment by tuning the structure of the individual.

Evolution is based on the Darwinian model, also called the *principle of natural selection* or *survival-of-the-fittest*, while learning is based on the connectionist model of the human brain. In the Darwinian evolution, knowledge acquired by an individual during its lifetime cannot be transferred into its genome and subsequently passed on to the next generation. EAs are stochastic search methods that employ a search technique based on the Darwinian model, whereas neural networks are learning methods based on the connectionist model.

Combinations of learning and evolution, embodied by evolving neural networks, have better adaptability to a dynamic environment [1225, 638]. Evolution and learning can interact in the form of the Lamarckian evolution or be based on the Baldwin effect [59]. Both processes use learning to accelerate evolution.

The Lamarckian strategy allows the inheritance of the acquired traits during an individual's life into the genetic code so that the offspring can inherit its characteristics. Everything an individual learns during its life is encoded back into the chromosome and remains in the population. Although the Lamarckian evolution cannot be found in biological systems, EAs as artificial biological systems can benefit from the Lamarckian theory. Ideas and knowledge are passed from generation to generation, and the Lamackian theory can be used to characterize the evolution of human cultures.

The Baldwin effect is more biologically plausible. In the Baldwin effect, learning has an indirect influence, that is, learning makes individuals adapt better to their environments, thus increasing their reproduction probability. In effect, learning smoothes the fitness landscape and thus facilitates evolution. A parent cannot pass its learned traits to its offspring, instead only the fitness after learning is retained. In other words, the learned behaviors become instinctive behaviors in subsequent generations, and there is no direct alteration of the genotype. The acquired traits finally come under direct genetic control after many generations, namely, *genetic assimilation*. The Baldwin effect is purely Darwinian, not Lamarckian in its mechanism, although it has consequences that are similar to those of the Lamarckian evolution [1106].

9.2 Introduction to Evolutionary Algorithms

For an optimization problem in a domain, if the calculus is difficult to implement or is inapplicable, search methods such as EAs can be used. EAs are a class of stochastic search and optimization techniques guided obtained by natural selection and genetics. They are population-based algorithms by simulating the natural evolution of biological systems. Individuals in a population compete and exchange information with one another. There are three basic genetic operations, namely, *crossover*[1], *mutation*, and *selection*. The procedure of a typical EA is given by Algorithm 9.2.

EAs are stochastic processes performing searches over a complex and multimode space. They have the following advantages

1. EAs can solve hard problems reliably and fast. They are suitable for evaluation functions that are large, complex, noncontinuous, nondifferentiable, and multimodal.

---

[1] Also called *recombination*.

Algorithm 9.2 (EA)

1. Set $t = 0$.
2. Randomize initial population $\mathcal{P}(0)$.
3. Repeat:
   a) Evaluate fitness of each individual of $\mathcal{P}(t)$.
   b) Select individuals as parents from $\mathcal{P}(t)$ based on fitness.
   c) Apply search operators (crossover and mutation) to parents, and generate $\mathcal{P}(t+1)$.
   d) Set $t = t + 1$.
   until the termination criterion is satisfied.

---

2. The EA approach is a general-purpose one, that can be directly interfaced to existing simulations and models.
3. EAs are extendable and easy to hybridize.
4. EAs are directed stochastic global search. They can always reach the near-optimum or the global maximum.
5. EAs possess inherent parallelism by evaluating multipoints simultaneously.

EAs employ a structured, yet randomized, parallel multipoint search strategy that is biased toward reinforcing search points of high fitness. The evaluation function must be calculated for all the individuals of the population, thus resulting in a computation load that is much higher than that of a simple random search or a gradient search.

9.2.1 Terminologies

Some terminologies that are used in the EA literature are listed below. These terminologies are an analogy to their biological counterparts. The biological definitions given in this subsection are based on Wikipedia© [1185].

Population

A set of individuals in a generation is called a *population*, $\mathcal{P}(t) = \{\mathbf{x}_1, \mathbf{x}_2, \ldots, \mathbf{x}_{N_P}\}$, where $\mathbf{x}_i$ is the ith individual, and N_P is the size of the population. The initial populations is usually generated randomly, while the population of other generations are generated from some selection/reproduction procedure.

Chromosome

Each individual $\mathbf{x}_i$ in a population is a single *chromosome*. A chromosome, sometimes called a *genome*, is a set of parameters that define a solution to the problem under consideration. The chromosome is often represented as a string in EAs.

Biologically, a chromosome is a long, continuous piece of DNA, that contains many genes, regulatory elements and other intervening nucleotide sequences. Normal members of a particular species all have the same number of chromosomes. For example, human body cells contain 46 diploid chromosomes, that is, they have two set of chromosomes, one set of 46 chromosomes from the mother and the other set of 46 chromosomes from the father. Chromsomes are used to encode a biological organism.

Gene

In EAs, each chromosome $\mathbf{x}$ comprises of a string of elements x_i, called *genes*, i.e. $\mathbf{x} = [x_1 \ x_2 \ \cdots x_n]$, where n is the number of genes in the chromosome. Each gene encodes a parameter of the problem into the chromosome. A gene is usually encoded as a binary string or a real number.

In biology, genes are entities that parents pass to offspring during reproduction. These entities encode information essential for the construction and regulation of proteins and other molecules that determine the growth and functioning of the organism.

Allele

The biological definition for an *allele* is any one of a number of alternative forms of the same gene occupying a given position called a *locus* on a chromosome. In the EA terminology, the value of a gene is indicated as an *allele*.

Genotype

A *genotype* is biologically referred to the underlying genetic coding of a living organism, usually in the form of DNA. The genotype of each organism corresponds to an observable, known as a *phenotype*. In EAs, a genotype represents a coded solution, that is, an individual's chromosome.

Phenotype

Biologically, the *phenotype* of an organism is either its total physical appearance and constitution or a specific manifestation of a trait. A phenotype is determined by genotype or multiple genes and influenced by environmental factors. The concept of phenotypic plasticity describes the degree to which an organism's phenotype is determined by its genotype. A high level of plasticity means that environmental factors have a strong influence on the particular phenotype that develops. The ability to learn is the most obvious example of phenotypic plasticity. As another example of phenotypic plasticity, sports can strengthen muscles. However, some organs have very low phenotypic plasticity, for example, the color of human eyes cannot be changed by environment.

The mapping of a set of genotypes to a set of phenotypes is referred to as a *genotype–phenotype map*. In EAs, a phenotype represents a decoded solution.

Fitness

Fitness in biology refers to the ability of an individual of certain genotype to reproduce. The set of all possible genotypes and their respective fitness values is called a *fitness landscape*.

Fitness function is a particular type of objective function that quantifies the optimality of a solution, *i.e.* a chromosome, in an EA. It is used to map an individual's chromosome into a positive number. Fitness is the value of the objective function for a chromosome $\mathbf{x}_i$, namely $f(\mathbf{x}_i)$. After the genotype is decoded, the fitness function is used to convert the phenotype's parameter values into the fitness. Fitness is used to rate the solutions.

Natural Selection

Natural selection is believed to be the most important mechanism in the evolution of biological species. It alters biological populations over time by propagating heritable traits affecting individual organisms to survive and reproduce. It adapts a species to its environment. Natural selection does not distinguish between its two forms, namely, ecological selection and sexual selection, but it is concerned with those traits that help individuals to survive the environment and to reproduce. Natural selection causes traits to become more prevalent when they contribute to fitness.

Natural selection is different from artificial selection. Genetic drift and gene flow are two other mechanisms in biological evolution. Genetic flow, also known as *genetic migration*, is the migration of genes from one population to another.

Genetic Drift

Genetic drift is a contributing mechanism in biological evolution. As opposed to natural selection, genetic drift is a stochastic process that arises from random sampling in the reproduction. It changes allele frequencies (gene variations) in a population over many generations and affects traits that are more neutral. The genes of a new generation are a sampling from the genes of the successful individuals of the previous one, but with some statistical error. Drift is the cumulative effect over time of this sampling error on the allele frequencies in the population, and traits that do not affect reproductive fitness change in a population over time. Like selection, genetic drift acts on populations, altering allele frequencies and the predominance of traits. It occurs most rapidly in small populations and can lead some alleles to become extinct or become the only alleles in the population, thus reducing the genetic diversity in the population.

Termination Criterion

The search process of an EA will terminate when a certain termination criterion is met. Otherwise a new generation will be produced and the search process continues. The termination criterion can be selected as a maximum number of generations, or the convergence of the genotypes of the individuals. Convergence of the genotypes occurs when all the bits or values in the same positions of all the strings are identical, and crossover has no effect for further processes. Phenotypic convergence without genotypic convergence is also possible. For a given system, the objective values are required to be mapped into fitness values so that the domain of the fitness function is always greater than zero.

9.3 Genetic Algorithms

The GA [496] is the most popular form of EAs. A simple GA may consist of a population generator and selector, a fitness estimator, and three genetic operators, namely selection, mutation, and crossover. The mutation operator inverts randomly chosen bits with a certain probability. The crossover operator combines parts of the chromosomes of two individuals, and generates two new offspring, which are used to replace low fitness individuals in the population. After a certain number of generations, the search process will be terminated.

9.3.1 Encoding/Decoding

The GA uses binary coding. A chromosome $\mathbf{x}$ is a potential solution, denoted by a concatenation of the parameters $\mathbf{x} = [x_1, x_2, \cdots, x_n]$, where each x_i is a gene, and the value of x_i is an allele. $\mathbf{x}$ is encoded in the form

$$\underbrace{11...110}_{x_1}\underbrace{10...100}_{x_2}\cdots\underbrace{10...011}_{x_n} \tag{9.1}$$

If the chromosome is l-bit long, it has 2^l possible values. If the variable x_i is in the range $\left[x_i^-, x_i^+\right]$ with a coding $s_{l_i} \cdots s_2 s_1$, where l_i is its bit-length in the chromosome and $s_i \in \{0, 1\}$, then the decoding function is given by

$$x_i = x_i^- + \left(x_i^+ - x_i^-\right) \frac{1}{2^{l_i} - 1}\left(\sum_{j=0}^{l_i-1} s_j 2^j\right) \tag{9.2}$$

In binary coding, there is the so-called *Hamming cliffs* phenomenon, where large Hamming distances between the binary codes of adjacent integers occur. Gray coding is another approach to encoding the parameters into bits.

The decimal value of a Gray-encoded integer variable increases or decreases by 1 if only one bit is changed. However, the Hamming distance does not monotonously increase with the difference in integer values.

For a long period, Gray encoding was considered to outperform binary encoding in the GA. However, based on a Markov-chain analysis of the GA, there is little difference between the performance of binary and Gray codings for all possible functions [184]. Also, Gray coding does not necessarily improve the performance for functions that have fewer local minima in the Gray representation than in the binary representation. This reiterates the NFL theorem [1192], namely, no representation is superior for all classes of problems.

The conversion from binary coding to Gray coding is formulated as [1193, 184]

$$g_i = \begin{cases} b_1, & i = 1 \\ b_i \oplus b_{i-1}, & i > 1 \end{cases} \tag{9.3}$$

where g_i and b_i are, respectively, the ith Gray code bit and the ith binary code bit, which are numbered from 1 to n starting on the left, and $\oplus$ denotes addition mod 2, $i.e.$ exclusive-or. Gray coding can be converted into binary coding by [1193, 184]

$$b_i = \sum_{j=1}^{i} g_j \tag{9.4}$$

where the summation denotes summation mod 2. As an example, we can check the equivalence between the binary code 1011011011 and the gray code 1110110110. According to (9.3) and (9.4), the most significant i bits of the binary code determine the most significant i bits of the Gray code and $vice$ $versa$.

The performance of the GA depends on the choice of the encoding techniques. Binary coding is a nonlinear coding, which is undesirable when approaching the optimum. GAs usually use fixed-length binary coding. There are also some variable-length encoding methods [423, 776]. In the messy GA [423], both the value and the position of each bit are encoded in the chromosome. The delta coding was suggested in [776].

9.3.2 Selection/Reproduction

Selection embodies the principle of $survival$ of the $fittest$, which provides a driving force in the GA. Selection is based on the fitness of the individuals. From a population $\mathcal{P}(t)$, those individuals with strong fitness will be selected for reproduction so as to generate a population of the next generation, $\mathcal{P}(t + 1)$. Chromosomes with larger fitness are selected and are assigned a higher probability of reproduction.

Replacement Strategies

During the selection procedure, we need to decide as to how many individuals in one population will be replaced by the newly generated individuals so as to produce the population for the new generation. Thus, the selection mechanism is split into two phases, namely, parental selection and replacement strategy. There are many replacement strategies such as the complete generational replacement [496], replace-random [496], replace-worst, replace-oldest, and deletion by kill tournament [1043]. In the crowding strategy [291], an offspring replaces one of the parents whom it most resembles using the similarity measure of the Hamming distance. These replacement strategies may result in a situation where the best individuals in a generation may fail to reproduce. In [625], this problem is solved by introducing into the system a new variable that stores the best individuals obtained so far. Elitism strategy cures the same problem without changing the system state [285].

Sampling Mechanism

Sampling chromosomes from the sample space can be in a stochastic manner, a deterministic manner, or their mixed mode. Among conventional selection methods, the roulette-wheel selection [496] is a stochastic selection method, while the ranking selection [421] and the tournament selection [420] are mixed mode-selection methods.

Roulette-wheel Selection

The roulette-wheel selection [496, 420] is a simple and popular selection scheme. Segments of the roulette wheel are allocated to individuals of the population in proportion to the individuals' relative fitness scores. Selection of parents is carried out by successive spins of the roulette wheel, and an individual's possibility of being selected is based on its fitness. Thus, a chromosome with a higher fitness has a higher chance to be selected. The selection can be done by assigning a chromosome $\mathbf{x}_i$ a probability

$$P_i = \frac{f(\mathbf{x}_i)}{\sum_{i=1}^{N_P} f(\mathbf{x}_i)} \qquad (9.5)$$

for $i = 1, 2, \ldots, N_P$. Consequently, a chromosome with a larger fitness has a possibility of getting more offspring.

Only two chromosomes will be selected to undergo genetic operations. Typically, the population size N_P is relatively small, and this fitness-proportional selection may select a disportionately large number of unfit chromosomes. This easily induces premature convergence when all the individuals in the population become very similar after a few generations. The GA thus degenerates into a Monte Carlo-type search method. Scaling a raw objective function to some positive function is a method to mitigate these problems, and this can prevent the best chromosomes from producing too many expected offspring [420].

Ranking Selection

The ranking selection [421] can eliminate some of the problems inherent in the fitness-proportional selection. It can maintain a more constant selective pressure. Individuals are sorted according to their fitness values. The best individual is assigned the maximum rank N_P and the worst individual the lowest rank 1. The selection probability is linearly assigned according to their ranks

$$P_i = \frac{1}{N_P} \left(\beta - 2(\beta - 1) \frac{i-1}{N_P - 1} \right) \tag{9.6}$$

for $i = 1, 2, \ldots, N_P$, where β is selected in the interval $[0, 2]$.

Tournament Selection

The tournament selection [420] involves h individuals at a time. The h chromosomes are compared and a copy of the best performing individual becomes part of the mating pool. The tournament will be performed repeatedly N_P times until the mating pool is filled. Typically, the tournament size h is selected as 2.

Elitism Strategy

The elitism strategy for selecting the individual with best fitness can improve the convergence of the GA [977]. The elitism strategy always copies the best individual of a generation to the next generation. Although elitism may increase the possibility of premature convergence, it improves the performance of the GA in most cases and thus, is integrated in most GA implementations [285].

9.3.3 Crossover/Mutation

Both crossover and mutation are considered the driving forces of evolution. Crossover occurs when two parent chromosomes, normally two homologous instances of the same chromosome, break and then reconnect but to the different end pieces. Mutations can be caused by copying errors in the genetic material during cell division and by external environment factors. Although the overwhelming majority of mutations have no real effect, some can cause disease in organisms due to partially or fully nonfunctional proteins arising from the errors in the protein sequence.

Crossover

The primary exploration operator in the GA is crossover, which searches the range of possible solutions based on existing solutions. Crossover, as a binary operator, is to exchange information between two selected parent chromosomes at randomly selected positions and to produce two new offspring (individuals). Both the children will be different from either of their parents, yet retain some features of both.

The method of crossover is highly dependent on the method of the genetic coding. Some of the commonly used crossover techniques are the one-point crossover [496], the two-point crossover [291], the multipoint crossover [371], and the uniform crossover [1070]. The crossover points are typically at the same, random positions for both parent chromosomes. These crossover operators are illustrated in Fig. 9.1.

One-point Crossover

The one-point crossover requires one crossover point on the parent chromosomes, and all the data beyond that point are swapped between the two parent chromosomes. The one-point crossover is easy to model analytically, and it generates bias toward bits at the ends of the strings.

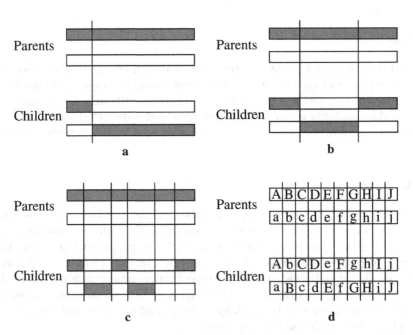

Fig. 9.1. Illustration of crossover operators. (**a**) One-point crossover. (**b**) Two-point crossover. (**c**) Multipoint crossover. (**d**) Uniform crossover. For the multipoint crossover and the uniform crossover, the exchange between crossover points takes place at a fixed probability. The alphabets denote the values of genes or bits.

Two-point Crossover

The two-point crossover selects two points on the parent chromosomes, and everything between the two points is swapped. The two-point crossover causes a smaller schema disruption than the one-point crossover. The two-point crossover eliminates this disadvantage of the one-point crossover, but generates bias at a different level.

The two-point crossover does not sample all regions of the string equally, and the ends of the string are rarely sampled. This problem can be solved by wrapping around the string. The substring from the first cut point to the second is crossed.

Multipoint Crossover

The multipoint crossover treats each string as a ring of bits divided by m crossover points into m segments, and each segment is exchanged at a fixed probability.

Uniform Crossover

The uniform crossover exchanges bits of a string rather than segments. Individual bits in the parent chromosomes are compared, and each of the nonmatching bits is probabilistically swapped with a fixed probability, typically 0.5. The uniform crossover is unbiased with respect to defining length. In the half-uniform crossover (HUX) [339], exactly half of the nonmatching bits are swapped.

Remarks

The one-point and two-point crossover operations preserve schemata due to low disruption rates, but are less exploratory. In contrast, the uniform crossover swaps are more exploratory, but have a high disruptive nature. The uniform crossover is more suitable for small populations, while the two-point crossover is better for large populations. The two-point crossover performs consistently better than the one-point crossover [1070].

When all the chromosomes are very similar or even the same in the population, it is difficult to generate a new structure by crossover only and premature convergence takes place. The mutation operation can introduce genetic diversity into the population. This prevents premature convergence from happening when all the individuals in the population become very similar.

Mutation

Mutation is a unary operator that requires only one parent to generate an offspring. A mutation operator typically selects a random position of a random chromosome and replaces the corresponding gene or bit by other information. Mutation helps to regain the lost alleles into the population.

Mutations can be classified into point mutations and large-scale mutations. Point mutations are changes to a single position, which can be substitutions, deletions, or insertions of a gene or a bit. Large-scale mutations can be similar to the point mutations, but operate at multiple positions simultaneously, or at one point with multiple genes or bits, or even on the chromosome scale. Functionally, mutations introduce the necessary amount of noise to do hill-climbing. Two additional large-scale mutation operators are the inversion and rearrangement operators.

The inversion operator [496, 420] picks up a portion between two randomly selected positions within a chromosome and then reverses it. Inversion reshuffles the order of the genes in order to achieve a better evolutionary potential. However, its computational advantage over other conventional genetic operators is not clear. The swap operator is the most primitive reordering operator, based on which many new unary operators including inversion can be derived. The rearrangement operator reshuffles a portion of a chromosome such that the juxtaposition of the genes or bits is changed. Some mutation operations are illustrated in Fig. 9.2.

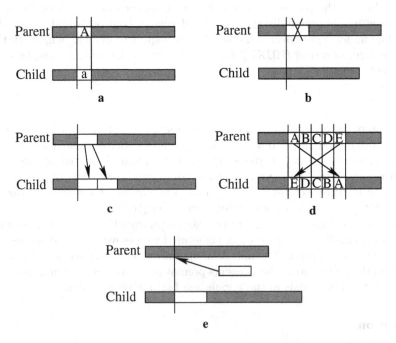

Fig. 9.2. Illustration of some mutation operators. (**a**) Substitution. (**b**) Deletion. (**c**) Duplication. (**d**) Inversion. (**e**) Insertion. The alphabets denote the values of genes or bits.

A high mutation rate can lead genetic search to random search. A high mutation rate may change the value of an important bit, and thus slow down the fast convergence of a good solution or slow down the process of convergence of the final stage of the iterations. Thus, mutation is made occasionally in the GA. In the simple GA, the mutation is typically selected as a substitution operation that changes one random bit in the chromosome at a time. An empirically derived formula that can be used as the probability of mutation P_m at a starting point is [1002]

$$P_m = \frac{1}{T\sqrt{l}} \qquad (9.7)$$

where T is the total number of generations and l is the string length.

The random nature of mutation and its low probability of occurrence make the convergence of the GA slow. The search process can be expedited by using the directed mutation technique [108] that deterministically introduces new points into the population by using gradient or extrapolation of the information acquired so far.

In passing, it may be mentioned that the relative importance of crossover and mutation has been discussed in the GA community and no compelling conclusion has been drawn. It is commonly agreed that crossover plays a more important role if the population size is large, and mutation is more important if the population size is small [829].

Other Genetic Operators

A lot of research effort has been directed towards improving the GA in terms of both the convergence speed and accuracy. A large number of noncanonical genetic operators are described in the literature. The best-known are hill-climbing operators [828], typically gradient-descent ones. In the parallel GA (PGA) [828], some or all individuals of the population improve their fitness by hill-climbing. Given an individual, this operator finds an alternative similar individual that represents a local minimum close to the original individual in the solution space. The combination of genetic operators and local search can be based on either the Lamarckian strategy or the Baldwin effect.

The bit climber [284] is a simple stochastic bit-flipping operator. The fitness is computed for an initial string. A bit of the string is randomly selected and flipped, and the fitness is computed at the new point. If the fitness is lower than its earlier value, the new string is updated as the current string. The operation repeats until no bit flip improves the fitness. The bit-based descent algorithm is several times faster than an efficient GA [284]. It can also be used as a local-search operator in the GA.

Most selection schemes are based on individuals' fitness. The entropy-Boltzmann selection method [663], stemming from the entropy, and the important sampling methods in the Monte Carlo simulation, tend to escape from local optima. It avoids the problem of premature convergence systematically. The adaptive fitness consists of the usual fitness together with the entropy change due to the environment, which may vary from generation to generation.

9.3.4 Real-coded Genetic Algorithms for Continuous Numerical Optimization

Although the GA is conventionally based on binary coding, for optimization problems, parameters are usually real numbers. The floating-point and fixed-point coding techniques are two methods for representing real numbers. The fixed-point coding allows more gradual mutation than the floating-point coding for the change of a single bit, and the fixed-point coding is sufficient for most cases.

The floating-point coding is widely used in continuous numerical optimization. It is capable of representing large or unknown domains, while the binary and fixed-point techniques may have to sacrifice accuracy for a large domain, since the decoding process does not permit an infinite increase in bit-lengths. The real-coded GA using the floating-point or the fixed-point coding has an advantage over the binary-coded GA in exploiting local continuities in function optimization [283]. The real-coded GA is faster, more consistent from run to run, and provides a higher precision than the binrary-coded GA [802]. Accordingly, genetic operators for the real-coded GA need to be defined.

Crossover

In analogy to crossover operators for the binary-coded GA, crossover operators for the real-coded GA such as the one-point, two-point, multipoint, and uniform crossover operators are also defined [1070]. Each gene (x_i) in a real-coded chromosome corresponds to a bit in a binary-coded chromosome. A blend of these crossover operators uniformly selects values that lie between the two points representing the two parents [340].

Crossover can also be defined as a linear combination of two parent vectors $\mathbf{x}_1$ and $\mathbf{x}_2$ and generates two offspring

$$\mathbf{x}_1' = \lambda \mathbf{x}_1 + (1 - \lambda)\mathbf{x}_2 \tag{9.8}$$

$$\mathbf{x}_2' = \lambda \mathbf{x}_2 + (1 - \lambda)\mathbf{x}_1 \tag{9.9}$$

where $0 < \lambda < 1$, and if λ is selected as 0.5 only one offspring is obtained [1011].

Assume that $\mathbf{x}_2$ is a better individual than $\mathbf{x}_1$. In order to generate offspring with a better fitness than their parents, crossover can be defined by extrapolation of the two points representing the two parents [1193, 802]

$$\mathbf{x}' = \lambda (\mathbf{x}_2 - \mathbf{x}_1) + \mathbf{x}_2 \tag{9.10}$$

where $0 < \lambda < 1$ is a random number. If this operator generates an infeasible offspring, another offspring is generated and tested by selecting another random value of λ. This procedure continues until a feasible solution is found or the specified number of attempts is reached. In the latter case, no offspring is produced. This crossover operator is suitable for locally fine tuning and for searching in a most promising direction [802].

In [690], the crossover operator is defined as that which generates four chromosomes from two parents according to a strategy of combining the maximum, minimum, or average of all the parameters encoded in the chromosome

$$\mathbf{x}_1'' = \frac{1}{2}\left(\mathbf{x}_1 + \mathbf{x}_2\right) \tag{9.11}$$

$$\mathbf{x}_2'' = \mathbf{x}^+ (1 - \lambda) + \max\left(\mathbf{x}_1, \mathbf{x}_2\right) \lambda \tag{9.12}$$

$$\mathbf{x}_3'' = \mathbf{x}^- (1 - \lambda) + \min\left(\mathbf{x}_1, \mathbf{x}_2\right) \lambda \tag{9.13}$$

$$\mathbf{x}_4'' = \frac{1}{2}\left[\left(\mathbf{x}^+ + \mathbf{x}^-\right)(1 - \lambda) + \left(\mathbf{x}_1 + \mathbf{x}_2\right)\lambda\right] \tag{9.14}$$

where $\mathbf{x}^+$ and $\mathbf{x}^-$ are vectors having elements as the lower and upper bounds for the corresponding parameters in $\mathbf{x}$, respectively, $\max\left(\mathbf{x}_1, \mathbf{x}_2\right)$ denotes a vector with each element obtained by taking the maximum among the corresponding elements of $\mathbf{x}_1$ and $\mathbf{x}_2$, and similarly $\min\left(\mathbf{x}_1, \mathbf{x}_2\right)$ gives a vector by taking the minimum value. These potential offspring spread over the domain. Only the one with the largest fitness, denoted $\mathbf{x}'$, is used as the offspring of the crossover operation. The three crossover operators defined by (9.8) and (9.9), (9.10), and (9.11) through (9.14) are illustrated in Fig. 9.3.

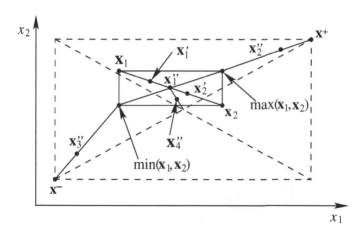

Fig. 9.3. A two-dimensional schematic of three crossover operators for the real-coded GA. The three crossover operators are, respectively, defined by (9.8) and (9.9), (9.10), and (9.11) through (9.14), $\lambda = \frac{1}{3}$. The symbols in the figure are the same as those in the equations.

Mutation

Mutation can be conducted by replacing one or more genes x_i, $i = 1, \cdots, n$, with a randomly selected real number x_i' from the domain of the corresponding parameter. The popular uniform mutation substitutes the value of a randomly selected gene with a random value between its upper and lower bounds [283].

A nonuniform mutation capable of fine tuning the system is defined by [802]

$$x_i' = \begin{cases} x_i + \Delta\left(t, x_i^+ - x_i\right), & r_1 \le 0.5 \\ x_i - \Delta\left(t, x_i - x_i^-\right), & r_1 > 0.5 \end{cases} \tag{9.15}$$

where r_1 is a uniform random number in $[0,1]$, x_i^- and x_i^+ are, respectively, the lower and upper bounds of x_i, $\Delta(t, y)$ is a function with domain $[0, y]$ whose probability of being close to 0 increases as t increases. This operator searches the space uniformly when t is small, and gradually searches locally as t increases. $\Delta(t, y)$ can be selected as

$$\Delta(t, y) = y \left[1 - r_2^{\left(1 - \frac{t}{T}\right)^b} \right] \tag{9.16}$$

where r_2 is a random number in $[0, 1]$, b is a parameter specifying the degree of nonuniformity, and T is the maximum number of generations. The function $\Delta(t, y)$ is illustrated in Fig. 9.4.

The Gausssian mutation [1011] is usually applied in the real-coded GA. The Gaussian mutation adds a Gaussian random number to one or multiple

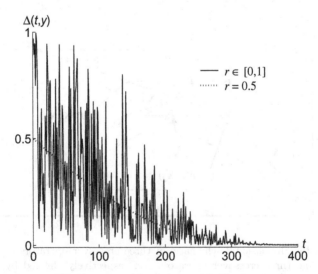

Fig. 9.4. Diagram of the function $\Delta(t, y)$ for $y = 1$. The parameters are selected as $T = 400$, and $b = 3$.

genes of the chromosome $\mathbf{x}$ and produces a new offspring $\mathbf{x}'$ with genes defined by

$$x_i' = x_i + N\left(0, \sigma_i\right) \tag{9.17}$$

for $i = 1, \cdots, n$, where $N\left(0, \sigma_i\right)$ is a random number drawn from a normal distribution with zero mean and standard deviation σ_i. The parameter σ_i is traditionally selected as a linearly or exponentially decreasing function such as $\sigma_i(t) = \frac{1}{\sqrt{1+t}}$.

The Cauchy mutation replaces the Gaussian distribution by the Cauchy distribution, and it is more likely to generate an offspring further away from its parent than the Gaussian mutation due to the long flat tails of the Cauchy distribution [1227]. The Cauchy mutation, however, has a weaker fine-tuning capability than the Gaussian mutation in small to mid-range regions. Thus, Cauchy mutation performs better when the current search point is far from the global minimum, while the Gaussian mutation is better at finding a local minimum in a good region. The two mutation operators are combined in [1227]. For each parent, two offspring are generated, each by one of the mutation methods, and only the better one is selected.

In [690], offspring obtained by crossover further undergo the mutation operation. Three new offspring are generated by allowing one parameter, some of the parameters, and all the parameters in the chromosome to change by a randomly generated number, subject to constraints on each parameter. Only one of the offspring will be used to replace the chromosome with the smallest fitness value, according to a predefined probability criterion that, as in the SA, allows uphill move in a controlled fashion. The probability of accepting a bad offspring is aimed at reducing the chance of converging to a local optimum. Hence, the search domain is significantly enlarged.

9.3.5 Genetic Algorithms for Sequence Optimization

For sequence optimization problems such as scheduling, the TSP and the MST, permutation encoding is a natural representation for a set of symbols, and each symbol can be identified by a distinct integer. This representation avoids missing or duplicate alleles [431].

Genetic operators should be defined so that infeasible solutions do not occur or a way is viable for repairing or rejecting infeasible solutions. Genetic operators for reordering a sequence of symbols can be unary operators such as inversion and swap [496], or binary operators that combine features of inversion and crossover, such as the partial matched crossover (PMX), the order crossover (OX), and the cycle crossover (CX) [420], edge recombination (ER) [1178], as well as intersection and union [370].

The random keys representation [89] encodes each symbol with a random number in $(0, 1)$. By sorting the random keys in a descending or ascending order, we can get a decoded solution. For example, assume that we are solving a TSP of 5 cities, with the chromosome for a route encoded as

$(0.52, 0.40, 0.81, 0.90, 0.23)$. If the genes are sorted in a descending order, the largest random key is 0.90, so the fourth city is the beginning of the route, and the whole route can be $4 \rightarrow 3 \rightarrow 1 \rightarrow 2 \rightarrow 5$. This representation avoids infeasible offspring by representing solutions in a soft manner, such that the real-coded GA and the ES can be applied directly for sequence-optimization problems. The random keys representation is simple and robust, and always allows simple crossover operations to generate feasible solutions. The ordering messy GA (OmeGA) [560] is specialized for solving sequence optimization problems. The OmeGA uses the mechanics of the fast messy GA [422] and represents the solutions using random keys.

The TSP is a benchmark for GAs. The performance of genetic operators for sequences has been empirically compared for the TSP in [370]. A general survey of various genetic operators for the TSP is given in [802].

9.3.6 Exploitation vs. Exploration

The convergence analysis of the simple GA is based on the concept of *schema* [496]. A schema is a bit pattern that functions as a set of binary strings. The schema theorem [496, 420] states that in the long run the best bit patterns will dominate the whole population. The schema theorem asserts that the proportions of the better schemata to the overall population increases as the generation progresses and eventually the search converges to the best solution with respect to the optimization function [420]. However, the GA often converges rather prematurely before the optimal solution is found.

Exploitation means taking advantage of the information already obtained, while *exploration* means searching new areas. Exploitation is achieved by the selection procedure, while exploration is achieved by genetic operators, which preserve genetic diversity in the population. The balance between exploitation and exploration controls the performance of the GA and is determined by the choice of the control parameters, namely, the probability of crossover P_c, the probability of mutation P_m, and the population size N_P. Some tradeoffs are made for selecting the optimal control parameters [890].

- Increasing P_c results in fast exploration, while it increases the disruption of good strings.
- Increasing P_m tends to transform the genetic search into a random search, while it helps reintroduce lost alleles into the population.
- Increasing N_P increases the genetic diversity in the population and reduces the probability of premature convergence, while it increases the time for convergence.

These control parameters depend on one another, and their choices are complex and depend on the nature of the problem. In the GA practice, for a small N_P one can select relatively large P_m and P_c, while for a large N_P smaller P_c and P_m are desirable. Typically, P_c is selected as 0.6 to 0.9, and P_m as 0.001 to 0.01 [420].

Empirical evidence shows that the optimal P_m in the GA differs according to whether or not crossover is used. When crossover is used, P_m should be selected as a small constant [860]. On the other hand, without crossover, the GA can start with a high P_m, decreasing towards the end of the run [480, 860].

In [419], the population size is suggested as $N_P = 1.65 \times 2^{0.21l}$, which is only dependent on the string length l. Empirical results show that the GA with N_P within 20 to 30, P_c in the range 0.75 to 0.95, and P_m in the range 0.005 to 0.01 performs well [1002].

9.3.7 Adaptation

Adaptation of control parameters is necessary for the best search process. At the beginning of a search process, the GA should have more emphasis on exploration, and at a later stage more emphasis should be on exploitation.

Increasing P_m and P_c promotes exploration at the expense of exploitation. A simple method to adapt P_m is given by [364]

$$P_m(t) = P_{m,0} - (P_{m,0} - P_{m,f}) \frac{t}{T} \tag{9.18}$$

where $P_{m,0}$ and $P_{m,f}$ are, respectively, the initial and final values for P_m, t is the generation index, and T is a predefined total number of generations. P_m can also be modified depending on the tradeoff between exploration and exploitation [890, 480]

$$P_m(t) = \frac{\alpha_0 e^{-\frac{\gamma_0 t}{2}}}{N_P \sqrt{l}} \tag{9.19}$$

where α_0 is a positive constant, γ_0 a non-negative constant, and l the length of the chromosome. In [1002], α_0 is selected as 1.76 and γ_0 as 0.

In an adaptive GA [890], P_c and P_m are not predefined but determined adaptively for each solution of the population. P_c and P_m range from 0 to 1.0 and 0 to 0.5, respectively. Low values of P_c and P_m are assigned to high-fitness solutions, while low-fitness solutions are assigned very high values of P_c and P_m. The best solution of every population is protected, not subjected to crossover but subjected to a minimal amount of mutation. All solutions with a fitness less than the average fitness have $P_m = 0.5$, that is, all subaverage solutions are completely disrupted and totally new solutions are created. This adaptive GA outperforms the simple GA significantly, and effectively avoids getting stuck at a local minimum.

The genetic diversity of the population can be easily improved so as to prevent premature convergence by adapting the size of the population [423, 42, 802] and using partial restart [339]. Partial restart is a simple approach to maintain genetic diversity [339]. Partial restart can be implemented by a fixed restart schedule at a fixed number of generations, or implemented when premature convergence occurs [650].

The binary coding of the GA results in limited accuracy and slow convergence. This drawback can also be eliminated by introducing adaptation into the GA. Examples of adaptive coding include the delta coding [776], the dynamic parameter encoding (DPE) [1009], and the fuzzy coding [1060, 1023].

9.3.8 Variants of the Genetic Algorithm

The messy GA [423] starts with a large initial population and halves it at regular intervals during the primordial stage of the messy GA. In the primordial stage only a selection operation is applied. This helps the population to get enriched with good building blocks. A dynamical population size also helps to reduce premature convergence. The fast messy GA [422] is an improved version of the messy GA.

The GENITOR [1176] employs an elitist selection that is a deterministic, rank-based selection method so that the best N_P individuals found so far are preserved by using a crossgenerational competition. Crossover produces only one offspring that immediately enters the population. Offspring do not replace their parents, but those least-fit individuals in the population. This selection strategy is similar to the $(\lambda + \mu)$ strategy of the ES.

The CHC^2 algorithm [339], like the GENITOR, also borrows from the $(\lambda + \mu)$ strategy of the ES. Incest prevention is introduced so that similar individuals are prevented from mating. The highly disruptive form of crossover, namely HUX, is applied, and mutation is not performed. Diversity is reintroduced by restarting partial population whenever convergence is detected. This is implemented by randomly flipping a fixed proportion of the best individual found so far as template, and introduce the better offspring into the population.

The GA with varying population size (GAVaPS) [42, 802] does not use any variation of selection mechanism discussed earlier, but introduces the concept of age of a chromosome in the number of generations. A chromosome will die off when its age reaches its lifetime decided by its fitness, and this serves as a selection mechanism. Each chromosome from the population has equal probability to reproduce, independent of its fitness. Some strategies for assigning lifetime values have been proposed to tune the size of the population to the current stage of search. Both the CHC and the GAVaPS significantly outperform the simple GA.

9.3.9 Parallel Genetic Algorithms

Parallel GAs can be in the form of global parallelized GAs, coarse-grained parallel GAs and fine-grained parallel GAs. Global parallelized GAs implement the GA by evaluating individuals and the genetic operations in explicitly parallel mode. The speedup is proportional to the number of processors.

---

^{2} CHC stands for *crossgenerational elitist selection, heterogeneous recombination,* and *cataclysmic mutation.*

In coarse-grained parallel GAs, the population is divided into a few isolated subpopulations, called *demes*. Individuals can migrate from one subpopulation to another. Fine-grained parallel GAs partition the population into many very small subpopulations, typically one individual per deme.

In the coarse-grained parallel GA, an existing GA such as the simple GA, GENITOR [1176] and CHC can be executed within each deme. The demes swap a few strings once in a few generations. The GENITOR II [1177] is a coarse-grained parallel version of the GENITOR algorithm. Individuals migrate at fixed intervals to neighboring nodes. Immigrants replace the worst individuals in the target deme. The PGA [828] is an asynchronous parallel GA, wherein each individual of the population improves its fitness by hill-climbing.

The fine-grained parallel GA [755, 256, 281], also called the *cellular GA* or *massively parallel GA*, organizes its population of chromosomes as a two-dimensional square grid with each grid point representing a chromosome. The processes of selection and mating are confined in a local area. This local reproduction has the effect of reducing selection pressure to achieve more exploration of the search space. Unlike the conventional panmictic mating, local mating can find very fast multiple optimal solutions in the same run, and is much more robust [256]. The performance of the algorithm degrades as the size of the neighborhood increases.

9.3.10 Two-dimensional Genetic Algorithms

Under the scenario of two-dimensional problems such as image processing, the conventional GA as well as EAs cannot be applied in a natural way, since linear encoding causes a loss of two-dimensional correlations. Encoding an image into a one-dimensional string increases the conceptual distance between the search space and its representation, and thus introduces extra problem-specific operators. If an image is encoded by concatenating horizontal lines, crossover operations result in a large vertical disruption. In two-dimensional GAs [169, 221, 998], each individual is a two-dimensional binary string.

In the two-dimensional GA, conventional mutation and reproduction operators can be applied in the normal way, but the conventional two-point crossover operator samples the matrix elements in a two-dimensional string very unevenly. Some genetic operators for two-dimensional strings are also defined, such as the crossover operator that exchanges rectangular blocks between pairs of matrices [221], and an unbiased crossover operator called *UN-BLOX (uniform block crossover)* [169]. The UNBLOX is a two-dimensional wraparound crossover and can sample all the matrix positions equally. The convergence rates of two-dimensional GAs are higher than that of the simple GA for bitmaps [169]. In [998], a two-dimensional crossover operator is defined for learning the architecture and weights of a neural network. A neural network is interpreted as an oriented graph, and the crossover operation is performed by swapping the subgraphs connected to a common selected neuron.

9.4 Evolutionary Strategies

The ES[3] [941, 1011] is another most popular EA. The ES was originally developed for numerical optimization problems [1011]. It was later extended to discrete optimization problems [476]. The objective parameters $\mathbf{x}$ and strategy parameters $\boldsymbol{\sigma}$ are directly encoded into the chromosome by using regular numerical representation, and thus no coding or decoding is necessary. Unlike the GA, the primary search operator in the ES is mutation.

9.4.1 Crossover, Mutation, and Selection Strategies

The canonical ES uses only mutation operations. Biologically, this corresponds to the asexual reproduction. However, as a heuristic, crossover operators used for the real-coded GA can be introduced into the ES. For example, the crossover operator can be defined by recombining two parents $\mathbf{x}_1$ and $\mathbf{x}_2$ such that the ith gene of the generated offspring $\mathbf{x}'$ takes the value

$$x_i' = \frac{1}{2} \left(x_{1,i} + x_{2,i} \right) \tag{9.20}$$

or selected as either $x_{1,i}$ or $x_{2,i}$. An offspring obtained from recombination is required to be mutated before it is evaluated and entered into the population.

Mutation can be applied to a parent or an offspring generated by crossover. For a chromosome $\mathbf{x} = [x_1, x_2, \cdots, x_n]$, a simple mutation is the Gaussian mutation that produces a new offspring $\mathbf{x}'$ with one or more genes defined by

$$x_i' = x_i + N\left(0, \sigma_i\right) \tag{9.21}$$

for $i = 1, \cdots, n$, where $N\left(0, \sigma_i\right)$ is a Gaussian distribution with zero mean and standard deviation σ_i, and $\boldsymbol{\sigma} = \left(\sigma_1, \cdots, \sigma_n\right)^{\mathrm{T}}$. The optimal σ_i is problem dependent, and is evolved automatically by encoding it into the chromosome. In practical implementations, σ_i is usually mutated first and then x_i is mutated by using σ_i'

$$\sigma_i' = \sigma_i e^{N\left(0, \delta\sigma_i\right)} \tag{9.22}$$

where $\delta\sigma_i$ is a parameter of the method.

For the ES, two major selection schemes are usually applied, namely, the $(\lambda + \mu)$ and (λ, μ) strategies, where μ is the population size and λ is the number of offspring generated from the population. As opposed to the GA, both selection schemes are deterministic sampling methods. These ranking-based selection schemes make the ES more robust than the GA. In the $(\lambda + \mu)$ strategy, μ fittest individuals are selected from the $(\lambda + \mu)$ candidates to form the next generation, while in the (λ, μ) scheme, μ fittest individuals are selected from λ $(\lambda \geq \mu)$ offspring to form the next generation. The $(\lambda + \mu)$ strategy is elitist and therefore guarantees a monotonically improving performance. This

---
[3] Also called *evolution strategy* in the literature.

selection strategy, however, is unable to deal with changing environments and jeopardizes the self-adaptation mechanism with respect to the strategy parameters, especially within small populations. Therefore, the (λ, μ) strategy is recommended, with a ratio of $\lambda/\mu = 7$ being optimal [53].

9.4.2 Evolutionary Strategies vs. Genetic Algorithms

The ES and the GA are the two most popular EAs for optimization. There are some major differences between them.

- **Selection procedure.** The selection procedure in the ES is deterministic: it always selects the specified number of best individuals as a population, and each individual in the population has the same mating probability. In contrast, the selection procedure in the GA is random and the chances of selection and mating are proportional to an individual's fitness.
- **Relative order of selection and genetic operations.** In the ES, the selection procedure is implemented after crossover and mutation, while in the GA, the selection procedure is carried out before crossover and mutation are applied.
- **Adaptation of control parameters.** In the ES, the strategy parameters σ are evolved automatically by encoding them into chromosomes. In contrast, the control parameters in the GA are problem specific and need to be prespecified.
- **Function of mutation.** In the GA, mutation is used to regain the lost genetic diversity, while in the ES, mutation functions as a hill-climbing search operator with adaptive step size σ. Due to the normal distribution nature in the Gaussian mutation, the tail part of the distribution may generate a chance for escaping from a local optimum.

Other differences are embodied in the encoding methods and genetic operators. The line between the two techniques is now being blurred, since both techniques are improved by borrowing the ideas from each other. For example, the CHC [339] has the properties of both the GA and the ES.

9.4.3 New Mutation Operators

New developments in the ES are mainly focused on designing new mutation operators. By suitably defining mutation operators, the ES can evolve significantly faster. The covariance matrix adaptation (CMA)-based mutation operator makes the ES two orders of magnitude faster than the conventional ES [453, 454].

The set of all mutation steps that yield improvements is called an *evolution path* of the ES [453]. The covariance matrix adaptation ES (CMA-ES) [453, 454] is such a technique that uses information embedded in the evolution path to accelerate the convergence. The CMA is a completely derandomized self-adaptation scheme. Subsequent mutation steps are uncorrelated with the previous ones. The mutation operator is defined by

$$\mathbf{x}' = \mathbf{x} + \delta \mathbf{B} \mathbf{z} \qquad (9.23)$$

where δ is a global step size, $\mathbf{z}$ is a random vector whose elements are drawn from a normal distribution $N(0, 1)$, and the columns of the rotation matrix $\mathbf{B}$ are the eigenvectors of the covariance matrix $\mathbf{C}$ of the distribution of mutation points. The step size δ is also adaptive. The CMA implements the PCA of the previously selected mutation steps to determine the new mutation distribution [454].

Some concepts of neural networks have been interestingly applied to the ES. Two self-organizing ESs have been defined in [805], which are, respectively, derived from the SOM [611] and the NG [768] based mutation operators. These ESs are not population based: The self-organizing networks are used to generate trial points, but no function values are computed for the network nodes. Instead, by using the network nodes one trial point is generated per iteration, and the function is computed for the generated trial points only. The network is used only to track the probability distribution of the trial points improving the current best-known function value. The SOM-type ES (SOM-ES) has the problem that the number of nodes increases exponentially with the number of function parameters due to its grid topology. The NG-type ES (NG-ES) does not have such a problem, but may suffer from premature convergence, which can be corrected by introducing an additional adaptation term. Both the ESs are empirically more reliable than the CMA-ES, without the necessity for parameter tuning.

9.5 Other Evolutionary Algorithms

Conventional EAs have their biological metaphor of the neo-Darwinian paradigm. In addition to the GA [496] and the ES [1011], the GP [625] and the EP [367] are also popular approaches to evolutionary computation[4]. The memetic algorithm [823] is also a popular method that uses the ideas in the neo-Darwinian paradigm and the Lamarckian strategy.

In the community of EAs, many researchers prefer to modify the canonical EAs by incorporating problem-specific knowledge and defining problem-specific genetic operators. Some even hybridize heuristics from different EAs. This can maximize the capability of EAs to practical applications. The boundaries between different branches of EAs overlap.

---

[4] C++ code libraries for GAs and EAs are available such as Wall's GALib at *http://www.mit.edu/people/moriken/doc/galib/*, EASEA-GALib and EASEA-EO at *http://www-rocq.inria.fr/EASEA/* and EOlib at *http://eodev.sourceforge.net*. Matlab© code for GAs and EAs can be located at the Matlab© Genetic Algorithm Toolbox and at *http://www.shef.ac.uk/~gaipp/ga-toolbox/*.

9.5.1 Genetic Programming

The canonical GA represents a solution by a binary string. This representation is discrete. The GP [625] is a variant of the GA for symbolic regression such as evolving computer programs, rather than for simple strings. It can also be used for automatic discovery of empirical laws. The major difference between the GA and the GP lies in coding. The GP has chromosomes of both variable length and data structure, and hierarchical trees are used to replace linear bit strings: Each tree represents the parse tree for a given program, hence providing a dynamic and variable representation.

The GP is particularly suitable for problems in which the optimal underlying structure must be discovered. The GP suffers from the so-called *bloat phenomenon*, resulting from the growth of noncoding branches in the individuals. The bloat phenomenon may cause an excessive consumption of computer resources and increase the cost of fitness computation.

9.5.2 Evolutionary Programming

The original EP technique [367] was presented for evolving artificial intelligence for predicting changes in an environment. The environment was coded as a sequence of symbols from a finite alphabet. Each chromosome is encoded as a finite state machine. The EP technique was later generalized for solving numerical optimization problems based on the Gaussian mutation [365, 366]. The EP and the ES are very similar to each other, and the EP corresponds to the $(\lambda + \lambda)$ strategy of the ES. The major differences between the two techniques are crossover and selection. The EP does not use crossover, but uses a probabilistic competition for selection.

More information on comparison between the ES and the EP is available in [54]. For continuous functional optimization, it is generally known that the EP or the ES works better than the GA [53].

9.5.3 Memetic Algorithms

In addition to the EAs based on the neo-Darwinian paradigm, some other optimization techniques using the concept of biological evolution are also interesting. The memetic algorithm [823, 947, 824], also called the *cultural algorithm* and *genetic local search*, was inspired by Dawkins' notion of a *meme* [286] defined as a unit of information that reproduces itself when people exchange ideas. Unlike genes, memes are typically adapted by the people who transmit them before they are passed on to the next generation. The memetic algorithm is a dual inheritance system that consists of a social population and a belief space, and models the evolution of culture or ideas. Their owner can improve upon the idea by incorporating local search. Although it was motivated by the evolution of ideas, the memetic algorithm can be considered as the GA or an EA making use of local search. Evolution and learning are combined using the Lamarckian strategy. The memetic algorithm is considerably faster than the simple GA.

Basically, the memetic algorithm combines local search heuristics with crossover operators. Crossover operators that introduce a high number of new alleles are often used in the memetic algorithm that have no biological analogy but mimic obsequious and rebellious behavior found in cultural systems. The problem-solving experience of individuals selected from the population space by an acceptance function is used to generate problem-solving knowledge that resides in the belief space. This knowledge can be viewed as a set of beacons that can control the evolution of individuals by means of an influence function. The influence function can use the knowledge in the belief space to modify any aspect of the individuals.

9.6 Theoretical Aspects

Although EAs are successfully used in various applications, there are few breakthroughs in theoretical aspects. The most important theoretical foundations are Holland's schama theorem [496] and Goldberg's building-block hypothesis [420], which capture the essence of the GA mechanism.

9.6.1 Schema Theorem and Building-block Hypothesis

A schema is a similarity template describing a subset of strings with the same bits (0 or 1) at certain positions. The combined effect of selection, crossover, and mutation gives the reproductive schema growth inequality [496]

$$m(H, t+1) \geq m(H, t) \cdot \frac{f(H)}{\bar{f}(t)} \left[1 - P_\mathrm{c} \cdot \frac{\delta(H)}{l-1} - o(H) P_\mathrm{m} \right] \qquad (9.24)$$

where H is a schema defined over the three-letter alphabet $\{0, 1, *\}$ of length l, $*$ is a don't care symbol, $m(H, t)$ is the number of examples of a particular schema H within a population at time t, $o(H)$ is the order of a schema H, namely the number of fixed positions (the number of 0s or 1s) present in the template, $\delta(H)$, the defining length of a schema H, is the distance between the outermost fixed positions (the first and last specific string positions), $f(H)$ is the average fitness of all strings in the population matched by the schema H, and $\bar{f}(t)$ is the average fitness of the whole population at time t. The schema theorem, given in Theorem 9.1, can be readily derived from (9.24) [496].

Theorem 9.1 (Schema Theorem). *Above-average schemata with short defining length and low order will receive exponentially increasing trials in subsequent generations of a GA.*

The building-block hypothesis is the assumption that strings with high fitness can be located by sampling building blocks with high fitness and combining the building blocks effectively. This is given in Theorem 9.2.

Theorem 9.2 (Building-Block Hypothesis). *A GA seeks near-optimal performance by the juxtaposition of short, low-order, and highly fit schemata, called* building blocks.

According to the schema theorem, schemata with high fitness and small defining lengths grow exponentially with time. Thus, the GA simultaneously processes a large number of schemata. For a population of N_P individuals, the GA implicitly evaluates approximately N_P^3 schemata in one generation [420].

The schema theorem and the building-block hypothesis constitute the theoretical foundation of the GA. However, there are a lot of criticisms on the schema theorem. The schema growth inequality provides a lower bound for one-generation transition of the GA. For multiple generations, the prediction of the schema may be useless or misleading due to the inexactness of the inequality [433].

9.6.2 Dynamics of Evolutionary Algorithms

Recently, more attempts have been made on characterizing the dynamics of EAs, which helps us to understand the conditions for EAs to converge to the global optimum [1135, 851, 1179, 1043].

The search process of EAs can be analyzed within the framework of Markov chains [366]. In [291], a Markov-chain analysis was conducted for a population of one-locus binary genes to reach different levels of convergence in an expected number of generations under random selection.

Despite the success of EAs, some problems, such as the selection of control parameters, the roles of crossover and mutation, replacement strategies, and convergence properties based on dynamics, are still unsolved. An exact model was introduced in [1135] to provide a complete model as to how all strings in the search space are processed by a simple GA using infinite population assumptions. In [851], another exact model for the simple GA has been obtained in the form of a Markov chain, and the trajectory followed by finite populations related to the evolutionary path predicted by the infinite population model.

An exact infinite population model of a simple GA for permutation-based representations has been developed in [1179]. Various permutation crossover operators can be represented by the mixing matrices. In [1043], a Markov-chain analysis has been made to model the expected time for a single member of the optimal class to take over finite populations in the case of different replacement strategies.

9.6.3 Deceptive Problems

GA-deceptive functions are a class of functions where low-order building blocks are misleading and their combinations cannot generate higher-order building blocks [420, 1180]. A fitness landscape with the global optimum surrounded by a part of the landscape of low average payoff is highly unlikely to be found by the GA, and thus the GA may converge to a suboptimal solution.

Deceptive problems remain as hard problems for EAs. The messy GA [423] was specifically designed to handle bounded deceptive problems. Due to deceptive problems, the building-block hypothesis is facing strong criticism [433]. In [433], the static building-block hypothesis was proposed as the underlying assumption for defining deception, and augmented GAs for deceptive problems were also proposed.

9.7 Other Population-based Optimization Methods

Recently, more and more computational techniques inspired by the metaphor of nature are emerging. Aside from neural networks and EAs, many other computational techniques are inspired by biological adaptive systems such as the collective behavior of animals and insects as well as the immune systems of mammals. In this section, we describe three well-known population-based optimization methods, namely the PSO, the immune algorithm, and the ACO. All these algorithms belong to a branch of swarm intelligence, an emergent collective intelligence of groups of simple agents [124]. They are general optimization methods and can be used for discrete and continuous function optimization.

9.7.1 Particle Swarm Optimization

The PSO originates from studies of synchronous bird flocking and fish schooling [598, 1030, 599, 250]. It is similar in some way to EAs, but requires only primitive mathematical operators, less computational bookkeeping and generally fewer lines of code, thus it is computationally inexpensive in terms of both memory requirements and speed. It evolves populations or swarms of individuals called *particles*.

For an optimization problem of n variables, a swarm of N_P particles is defined, where each particle is assigned a random position in the n-dimensional space as a candidate solution. Each particle has its own trajectory, namely, position $\mathbf{x}_i$ and velocity $\mathbf{v}_i$, and moves in the search space by successively updating its trajectory. Populations of particles modify their trajectories based on the best positions visited earlier by themselves and other particles. All particles have fitness values that are evaluated by the fitness function to be optimized. The particles are flown through the solution space by following the current optimum particles. The algorithm initializes a group of particles with random positions (solutions) and then searches for optima by updating generations (iterations). In every iteration, each particle is updated by following the two best values, namely the particle best *pbest*, denoted $\mathbf{x}_i^*$, $i = 1, \cdots, N_\mathrm{P}$, which is the best solution it has achieved so far, and the global best *gbest*, denoted $\mathbf{x}^\mathrm{g}$, which is the best value obtained so far by any particle in the population. The best value for the population in a generation is a local best, *lbest*.

At iteration $t + 1$, the swarm can be updated by the basic PSO algorithm [598]

$$\mathbf{v}_i(t + 1) = \mathbf{v}_i(t) + cr_1 \left[\mathbf{x}_i^*(t) - \mathbf{x}_i(t) \right] + cr_2 \left[\mathbf{x}^\mathrm{g}(t) - \mathbf{x}_i(t) \right] \qquad (9.25)$$
$$\mathbf{x}_i(t + 1) = \mathbf{x}_i(t) + \mathbf{v}_i(t + 1) \qquad (9.26)$$

for $i = 1, \cdots, N_\mathrm{P}$, where the acceleration constant $c > 0$, and r_1 and r_2 are uniform random numbers within $[0, 1]$. This basic PSO may lead to swarm explosion and divergence due to lack of control of the magnitude of the velocities. This can be solved by setting a threshold $v_{\max}$ on the absolute value of the velocity $\mathbf{v}_i$.

The PSO can locate the region of the optimum faster than EAs. However, once in this region the algorithm progresses slowly due to the fixed velocity stepsize. This can be addressed by introducing a weight parameter on the previous velocity of the particle into (9.25) [1030]

$$\mathbf{v}_i(t + 1) = \alpha \mathbf{v}_i(t) + c_1 r_1 \left[\mathbf{x}_i^*(t) - \mathbf{x}_i(t) \right] + c_2 r_2 \left[\mathbf{x}^\mathrm{g}(t) - \mathbf{x}_i(t) \right] \qquad (9.27)$$

where the parameter α is called the *inertia weight*, and the positive constants c_1 and c_2 are, respectively, *cognitive* and *social* parameters. Typically, both c_1 and c_2 are set as 2.0, and α gradually decreases from 1 to 0.

A number of heuristics on improving the basic PSO are available in the literature. The PSO has been extended to handle multiobjective optimization problems [255].

9.7.2 Immune Algorithms

In 1974, Jerne [549] proposed a network theory for the immune system based on the clonal selection theory [150, 10]. The biological immune system has the features of immunological memory and immunological tolerance. It has self-protection and self-regulation mechanisms. The basic components of the immune system are two types of lymphocytes, namely B-lymphocytes and T-lymphocytes, which are cells produced by bone marrow and by the thymus, respectively. B-lymphocytes generate antibodies on their surfaces to resist their specific antigens, while T-lymphocytes regulate the production of antibodies from B-lymphocytes. Roughly 10^7 distinct types of B-lymphocytes exist in a human body. An antibody recognizes and eliminates a specific type of antigen. The clonal selection theory describes the basic features of an immune response to an antigenic stimulus. The clonal operation is an antibody random map induced by the affinity including four steps, namely clone, clonal crossover, clonal mutation and clonal selection.

Artificial immune networks [46, 1122, 287] employ two types of dynamics. The short-term dynamics govern the increase or decrease of the concentration of a fixed set of lymphocyte clones and the corresponding immunoglobins. The metadynamics govern the recruitment of new species from an enormous pool of lymphocytes freshly produced by the bone marrow. The short-term dynamics correspond to a set of cooperating or competing agents, while the metadynamics refine the results of the short-term dynamics. As a result, the short-term dynamics are closely related to neural networks and the meta-dynamics are similar to the GA.

The immune algorithm, also called the *clonal selection algorithm*, intro-duces suppress cells to change search scope and memory cells to keep the candidate solutions. It is an EA inspired by the immune system, and is very similar to the GA. In the immune algorithm, *antigen* is defined as the problem to be optimized, and *antibody* is the solution to the objective function. Only those lymphocytes that recognize the antigens are selected to proliferate. The selected lymphocytes are subject to an affinity maturation process, which im-proves their affinity to the selective antigens. Learning in the immune system involves raising the relative population size and affinity of those lymphocytes. The immune algorithm first recognizes the antigen, and produces antibodies from memory cells. Then it calculates the affinity between antibodies, which can be treated as fitness. Antibodies are dispersed to the memory cell, and the concentration of antibodies is controlled by stimulating or suppressing anti-bodies. A diversity of antibodies for capturing unknown antigen is generated using genetic reproduction operators.

The immune mechanism can also be defined as a genetic operator and integrated into the GA [554]. The immune operator overcomes the blindness in action of the crossover and mutation and to make the fitness of population increase steadily. The immune operator is composed of two operations, namely a vaccination and an immune selection, and it utilizes reasonably selected vaccines to intervene in the variation of genes in an individual chromosome.

9.7.3 Ant-colony Optimization

The ACO [312, 111, 313, 314, 315] is a metaheuristic approach for solving dis-crete or continuous optimization problems such as COPs. The ACO heuristic was inspired by the foraging behavior of ants. Ants are capable of finding the shortest path between the food and the colony (nest) due to a simple pheromone-laying mechanism. The optimization is the result of the collective work of all the ants in the colony. Ants use their pheromone trails as a medium for communicating information. All the ants contribute to the pheromone reinforcement. Old trails will vanish due to evaporation. Different ACO algorithms arise from different pheromone value update rules[5].

The ant system [312] is an evolutionary approach, where several generations of artificial ants search for good solutions. Every ant of a generation builds up a complete solution, step by step, going through several decisions by choosing the nodes on a graph according to a probabilistic *state transition rule*, called the *random-proportional rule*. The probability for ant k at node i moving to node j at generation t is defined by

$$P_{i,j}^k(t) = \frac{\tau_{i,j}(t)d_{i,j}^{-\beta}}{\sum_{u \in \mathcal{J}_i^k} \tau_{i,u}d_{i,u}^{-\beta}} \tag{9.28}$$

for $j \in \mathcal{J}_i^k$, where $\tau_{i,j}$ is the intensity of the pheromone on edge $i \to j$, $d_{i,j}$ is the distance between nodes i and j, $\mathcal{J}_i^k$ is the set of nodes that remain to be visited by ant k positioned at node i to make the solution feasible, and $\beta > 0$ is a parameter that determines the relative importance of pheromone vs. distance. A tabu list is used to save the nodes already visited during each generation. When a tour is completed, the tabu list is used to compute the ant's current solution.

Once all the ants have built their tours, the pheromone is updated on all edges $i \to j$ according to a *global-pheromone updating rule*

$$\tau_{i,j}(t+1) = (1-\alpha)\tau_{i,j}(t) + \sum_{k=1}^{N_P} \tau_{i,j}^k(t) \tag{9.29}$$

where $\tau_{i,j}^k$ is the intensity of the pheromone on edge $i \to j$ contributed by ant k, taking $\frac{1}{L_k}$ if $i \to j$ is an edge used by ant k, and 0 otherwise, $\alpha \in (0,1)$ is a pheromone decay parameter, L_k is the length of the tour performed by ant k, and N_P is the number of ants. Thus, a shorter tour gets a higher reinforcement. Each edge has an LTM to store the pheromone.

The ant-colony system (ACS) [313] improves the ant system [312] by applying a *local pheromone updating rule* during the construction of a solution. The global updating rule is applied only to edges that belong to the best ant tour. Some important subsets of ACO algorithms, such as the most successful ACS and min-max ant system (MMAS) algorithms, have been proved to converge to the global optimum [315].

---

[5] An ACO website that provides information on publications, tutorials, software, and conferences is *http://iridia.ulb.ac.be/~mdorigo/ACO/ACO.html*.

9.8 Multiobjective, Multimodal, and Constraint-satisfaction Optimizations

9.8.1 Multiobjective Optimization

The multiobjective optimization problem is to optimize a system with m conflicting objectives, $f_i(\mathbf{x})$, $i = 1, \cdots, m$

$$\min \mathbf{f}(\mathbf{x}) = (f_1(\mathbf{x}), f_2(\mathbf{x}), \cdots, f_m(\mathbf{x}))^{\mathrm{T}} \tag{9.30}$$

where $\mathbf{x} = (x_1, x_2, \cdots, x_n)^{\mathrm{T}} \in \mathcal{X} \subset R^n$.

In order to optimize a system with conflicting objectives, it is necessary to establish a tradeoff between them. The weighted sum of these objectives is usually used as the compromise of the system. The weighted objective is defined by

$$F(\mathbf{x}) = \sum_{i=1}^{m} w_i \bar{\bar{f}}_i(\mathbf{x}) \tag{9.31}$$

where $\bar{\bar{f}}_i(\mathbf{x}) = \frac{f_i(\mathbf{x})}{|\max(f_i(\mathbf{x}))|}$ are normalized objectives, and $\sum_{i=1}^{m} w_i = 1$.

For many problems, there are difficulties in normalizing the individual objectives, and also in selecting the weights. The lexicographic order optimization is based on the ranking of the objectives in terms of their importance. Fuzzy logic can be used to define a tradeoff of multiple objectives [1265].

The Pareto method is a popular method for multiobjective optimization, and is based on the principle of nondominance. The Pareto optimum gives a set of solutions for which there is no way of improving one criterion without deteriorating another criterion [420, 247]. A solution $\mathbf{x}_1$ is said to dominate $\mathbf{x}_2$, denoted $\mathbf{x}_1 \succ \mathbf{x}_2$, if $\mathbf{x}_1$ is better than or equal to $\mathbf{x}_2$ in all attributes, and strictly better in at least one attribute. The space in R^n formed by Pareto optimal solutions is called the *Pareto optimal frontier*, $\mathcal{P}^*$. An illustration of Pareto optimal solutions for a two-dimensional problem with two objectives is given in Fig. 9.5.

The vector-evaluated GA (VEGA) [1001] is deemed the first GA for multiobjective optimization. The population is divided into equal-sized subpopulations, and each subpopulation is independently used for searching the optimum of a single objective. Crossover is performed across subpopulation boundaries. Some other heuristics are used to prevent the system from converging toward solutions that are not with respect to any criterion. This algorithm, however, has bias toward some regions [1052].

Goldberg [420] introduced nondominated sorting to rank a search population according to the Pareto optimality. This procedure of identifying nondominated sets of individuals is repeated until the whole population has been ranked. Niching and speciation techniques can be used to promote genetic diversity so that the entire Pareto frontier is covered. Equal probability of reproduction is assigned to all nondominated individuals in the population.

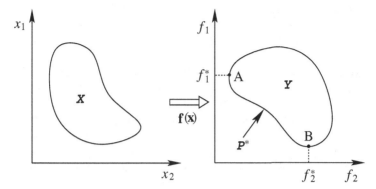

Fig. 9.5. An illustration of Pareto optimal solutions for a two-dimensional problem with two objectives. The frontier from point A to point B along the lower border of the domain $\mathcal{Y}$, denoted $\mathcal{P}^*$, contains all Pareto optimal solutions. $\mathcal{X} \subset R^n$ is the domain of $\mathbf{x}$, and $\mathcal{Y} \subset R^m$ is the domain of $\mathbf{f}(\mathbf{x})$.

The nondominated sorting GA (NSGA) [1052] implements a nondominated sorting in the GA along with a niching and speciation method [420] to find multiple Pareto-optimal points simultaneously. This ranking is supplemented by a GA technique known as *sharing*, which lowers the objective function values of designs that are too similar, keeping the algorithm from converging to a single optimum [420]. This returns a set of optimal tradeoffs between goals, instead of a single optimum. The NSGA-II [289, 290] improves the NSGA by introducing elitism and a crowed comparison operator.

Other popular EAs for multiobjective optimization may be based on the global nonelitist selection such as the niched Pareto GA (NPGA) [507], or the global elitist methods such as the strength Pareto EA (SPEA) [1266], Pareto envelope-based selection algorithm (PESA) [253], Pareto archived ES (PAES) [561] as well as SPEA-II [1267], or the local selection such as the evolutionary local selection algorithm (ELSA) [795]. All these algorithms have the ability to find multiple Pareto-optimal solutions in one single run.

9.8.2 Multimodal Optimization

Multimodal optimization is a generic and large class of optimization problems, where the search space is split into regions containing local optima. The objective of multimodal optimization is to identify a number of local optima and to maintain these solutions while continuing to search other local optima. Multimodality causes difficulty to any search method. Each peak in the solution landscape can be treated as a separate environment niche. A niche makes a particular subpopulation (species) unique. The niching mechanism embodies both cooperation and competition.

Crossover between individuals from different niches may lead to unviable offspring and is usually avoided [420]. It introduces a strong selection advantage to the niche with the largest population, and thus accelerates symmetry breaking of the search space and causes the population to become focused around one region of the search space. This, however, prevents a thorough exploration of the fitness landscape and makes it more likely for the population to find a suboptimal solution [928]. A finite population will quickly concentrate on one region of the search space. Increasing the population size, decreasing the crossover probability, biasing the crossover operator can also slow down symmetry breaking and thus, increase the likelihood of finding the global optimum, which, in turn, slows down the rate of exploitation.

For multimodal optimization, three techniques, namely niching, demes, and local search, are effective in identifying a number of local optima.

Niching

Niching can be implemented by preventing or suppressing the crossover of solutions that are dissimilar to each other. It is also necessary to prevent a large subpopulation from creating a disproportionate number of offspring. Niching techniques are usually based on the ideas of *crowding* [291] and/or *sharing* [420]. Crowding makes individuals within a single niche compete with each other over limited resources, and thus prevents a single genotype from dominating a population and allows other less-fit niches to form within the population. Sharing treats fitness as a shared resource among similar individuals, and thus prevents a single genotype from dominating the population and encourages the development of new niches. Crowding is simple, while sharing is far more complex, yet far more effective in multimodal optimization [420].

The sequential niching technique [90] modifies the evaluation function in the region of the solution to eliminate the solution found once an optimum is found. The GA continues the search for new solutions without restarting the population.

Demes

The second technique is to split the population into subpopulations or demes. The demes evolve independently except for an occasional migration of individuals between demes. In the cellular GA [256], local mating explores the peak in each deme, and finds and maintains multiple solutions. The forking GA [1104] is such a technique for multimodal optimization. Depending on the convergence status and the solution obtained so far, the forking GA divides the whole search space into subspaces. It is a multipopulation scheme that includes one parent population and one or more child populations, each exploiting a subspace [1104].

Local Selection

In the local selection scheme [794], fitness is the result of an individual's interaction with the environment and its finite shared resources. Individual fitnesses are compared to a fixed threshold, rather than to one another, to decide as to who gets the opportunity to reproduce. Local selection is an implicitly niched scheme. It maintains genetic diversity in a way similar to, yet generally more efficient than, fitness sharing.

Local selection minimizes interactions among individuals, and is thus suitable for parallel implementations. Local selection can effectively avoid premature convergence and it applies minimal selection pressure upon the population.

Constraint-satisfaction Optimization

When dealing with optimization problems with constraints, two kinds of constraints, namely, equality and inequality constraints, may arise. Any vector $\mathbf{x} \in R^n$ that satisfies all the constraints is called a *feasible solution*. Only feasible solutions are meaningful. For pure equality constraints, one can use the Lagrange multiplier method. Inequality constraints can be converted into equality constraints by introducing extra slack variables, and the Lagrange multiplier method then applied. This is the KKT method.

The penalty function method is usually used to convert equality and/or inequality constraints into a new objective function, so that beyond the constraints the objective function is abruptly reduced. The penalty function method transforms constraint optimization problems into unconstrained optimization problems by defining a new objective function in the form such as

$$f'(\mathbf{x}) = f(\mathbf{x}) + f_\mathrm{p}(\mathbf{x}) \qquad (9.32)$$

where $f(\mathbf{x})$ is the objective function, and $f_\mathrm{p}(\mathbf{x}) < 0$ is a penalty for infeasible solutions and zero for feasible solutions.

In EA implementations of constrained optimization, one needs to handle infeasible solutions in a population. One can simply reject the infeasible individuals or penalize the infeasible individuals. The latter is more common. A survey on handling constraints with the GA is given in [801].

9.9 Evolutionary Algorithms vs. Simulated Annealing

9.9.1 Comparison Between Evolutionary Algorithms and Simulated Annealing

EAs and the SA are similar in some ways. They are both stochastic global optimization methods. They can escape from local minima and converge to a solution to an arbitrary accuracy. They are both guided search methods directed toward an increasing or decreasing cost.

The process of an EA is also similar to that of the SA. In the SA, at each search there are two possibilities of selecting the activation of a neuron, which is controlled by a random function. In an EA, this is achieved by the crossover and mutation operations. The capability of an EA to converge to a premature local minimum or a global optimum is usually controlled by suitably selecting the probabilities of crossover and mutation. This is comparable to the controlled lowering of the temperature in the SA. Thus, the SA can be viewed as a subset of EAs with a population of one individual and a changing mutation rate.

The differences between the two paradigms are obvious. The SA is a naturally serial algorithm, while EAs involve a selection process that requires global coordination. The SA is too slow for practical use. EAs are much more effective in finding the global minimum due to their simplicity and parallel nature. The simplicity and generality of EAs arise from the coding and simple genetic operations. There is no need to design complex search operators. The binary representation is also suitable for hardware implementation. EAs explore the domain of the target function at many points and thus can escape from local minima. The inherent parallel property also offsets their high computational cost. However, for some well-defined numerical optimization problems, simple hill-climbing algorithms used in the SA usually outperform EAs [292].

9.9.2 Synergy of Evolutionary Algorithms and Simulated Annealing

Combination of the SA and EAs inherits the parallelization of EAs and avoids the computational bottleneck of EAs by incorporating elements of the SA. The hybrid retains the best properties of both paradigms. Many efforts in the synergy of the two approaches have been made in the past decade [1236, 197, 295, 296, 232].

The guided evolutionary SA (GESA) [1236] incorporates the idea of SA into the selection process of evolutionary computation, in place of arbitrary heuristics. The hybrid method is practically a number of parallel SA processes. The concept of family is introduced, where a family is defined as a parent together with its children. Competitions within a family and between families exist. The GESA is a practicable method that yields consistent and good near-optimal solutions, superior to the results of the ES.

The genetic SA [197] provides a completely parallel, easily scalable hybrid GA/SA method. The hybrid method combines the recombinative power of the GA and annealing schedule of the SA. This method does not require parallelization of any problem-specific portions of a serial implementation, *i.e.* an existing serial implementation can be incorporated as is. The performance of the algorithm scales up linearly with increasing number of processing elements, which enables the algorithm to utilize massive parallel architecture with maximum effectiveness. In addition, careful choice of control parameters is not required, and this is a significant advantage over the SA and the GA.

In the hybrid SA/EA system [295, 296], at each temperature, a separate SA operator is used to create an offspring for each individual until a predefined condition is reached. After the offspring are created, parents for the next generation are selected. The iteration continues by following a temperature schedule until the temperature is nearly zero. In [232], each individual in the population can intelligently plan its own annealing schedule in an adaptive fashion to the given problem at hand.

9.10 Constructing Neural Networks Using Evolutionary Algorithms

Neural-network learning is a search process for the minimization of a criterion or error function. In order to make use of existing learning algorithms, one needs to select a lot of parameters, such as the number of hidden layers, the number of units in each hidden layer, the type of learning rule, the transfer function, as well as learning parameters. In general, one usually selects an effective architecture by hand, and thus the procedure is time consuming. Moreover, gradient-based algorithms usually run multiple times to avoid local minima and also gradient information must be available. There are also adaptive methods for automatically constructing neural networks such as the upstart algorithm [373] and the growing neural tree method [996].

Evolution can be introduced into neural networks at many different levels. The evolution of connection weights provides a global approach to connection weight training. When EAs are used to construct neural networks, a drastic reduction in development time and simpler designs can be achieved. The optimization capability of EAs can lead to a minimal configuration that reduces the total training time as well as the performing time for new patterns.

In most cases, an individual in an EA is selected as a whole network. Competition occurs among those individual networks, based on the performance of each network. Within this framework, an EA can be used to evolve the structure, the parameters, and/or the nonlinear activation function of a neural network. A recent survey on evolving neural networks is given in [1225].

There are also occasions when EAs are used to evolve one hidden unit of the network at a time. The competing units are individual units. During each successive run, candidate hidden units compete so that the optimal single unit is added in that run. The search space of EAs at each run is much smaller. However, the entire set of hidden units may not be the global optimal placement.

9.10.1 Permutation Problem

Network parameters can be permuted without affecting their function. Permutation results in a topological symmetry, and consequently in a high number of symmetries in the error function. Thus, the number of local minima

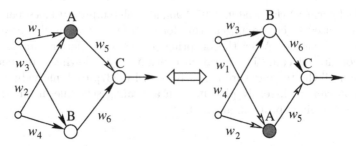

Fig. 9.6. An illustration of the permutation problem. Two networks with permuted weights and neurons are equivalent, but their chromosomes are quite different.

is high. This is the so-called *permutation problem* [448]. Fig. 9.6 shows two functionally equivalent networks that order their hidden units differently in their chromosomes. This, however, results in a coded string that looks quite different.

The difficulties for EAs using crossover in problems with explicit symmetries are well known. The permutation problem makes the crossover operator very inefficient and ineffective in producing good offspring. For two networks with permuted parameters, crossover almost certainly leads nowhere or converges very slowly. The permutation problem can be resolved by sorting the strings appropriately before crossover [448]. When evolving the architecture of the network, crossover is usually avoided and only mutations are adopted [1226, 1225].

9.10.2 Hybrid Training

As far as the computation speed is concerned, it is hard to say whether EAs can compete with the gradient-descent method or not. There is no clear winner in terms of the best training algorithm, since the best method is always problem dependent.

For large networks, EAs may be inefficient. When gradient information is readily available, it can be used to speed up the evolutionary search. The process of learning facilitates the process of evolution. The hybrid of evolution and gradient search is an effective alternative to the gradient-descent method in learning tasks, when the global optimum is at a premium. The hybrid method is more efficient than either an EA or the gradient-descent method used alone.

As is well known, EAs are inefficient in fine tuning local search although they are good at global search. This is especially true for the GA. By incorporating a local-search procedure such as the gradient descent into the evolution, the efficiency of evolutionary training can be improved significantly. Neural networks can be trained by alternating two steps, where an EA step is first used to locate a near-optimal region in the weight space and a local-search step such as the gradient-descent step is then used to find a local-optimal solution in that region [816, 672].

Hybridization of EAs and local search can be based either on the Lamarckian strategy or on the Baldwin effect. Local search corresponds to the phenotypic plasticity in biological evolution. Since Hinton and Nowlan constructed the first computational model of the Baldwin effect in 1987 [486], many authors have reported excellent results using hybrid methods (see [816, 437, 1225, 638], and the references given in [1225]).

The hybrid methods based on the Lamarckian strategy and the Baldwin effect are very successful with numerous implementations. Although the Lamarckian evolution is biologically implausible, it has proved effective within computer applications. Nevertheless, the Lamarckian strategy has been pointed out to distort the population so that the schema theorem no longer applies [1181]. The Baldwin effect only alters the fitness landscape and the basic evolutionary mechanism remains purely Darwinian. Thus, the schema theorem still applies to the Baldwin effect [1106].

9.10.3 Evolving Network Parameters

EAs are robust search and optimization techniques, and can locate the near-global optimum in a multimodal landscape. They can be used to optimize neural-network structure and parameters, or to optimize specific network performance and algorithmic parameters. EAs are suitable for learning networks with nondifferentiable activation function. Considerable research has been conducted on the evolution of connection weights, see [1225] and the references therein.

EAs evolve network parameters such as weights based on a fitness measure for the whole network. The fitness function can usually be defined as $\frac{1}{1+E}$, where E is the error or criterion function for network training. The complete set of network parameters is coded as a chromosome **s** with a fitness function $f(\mathbf{s}) = \frac{1}{1+E(D(\mathbf{s}))}$, where $D(\mathbf{s})$ is a decoding transformation.

EAs can be used to train all kinds of neural networks, irrespective of the network topology. The general applicability saves a lot of human effort in developing different training algorithms for different types of neural networks. To date, EAs are mainly used for training FNNs [1225, 457].

Coding

Coding of network parameters is most important from the point of view of the convergence speed of search. When using the binary GA, the fixed-point coding is shown to be usually superior to the floating-point coding of the parameters [816]. For crossover, it is usually better to only exchange the parameters between two chromosomes, but not to change the bits of each parameter [816]. The modifications of network parameters can be conducted by mutation.

Due to the limitation of the binary coding, real numbers are usually used to represent network parameters directly [796]. Each individual is a real vector, and crossover and mutation are specially defined for real-coded EAs. The ES and the EP are particularly well suited, and are also widely applied for continuous function optimization. These mutation-based approaches reduce the negative impact of the permutation problem.

Each instance of the neural network is encoded by the concatenation of all the network parameters in one chromosome. A heuristic concerning the order of the concatenation of the network parameters is to put connection weights terminating at the same unit together. Hidden units are in essence feature extractors and detectors. Separating inputs to the same hidden unit far apart might increase the difficulty of constructing useful feature detectors because they might be destroyed by crossover operations.

9.10.4 Evolving Network Architecture

The architecture of a neural network is referred to as its topological structure, *i.e. connectivity*. The network architecture is usually predefined and fixed. Design of the optimal architecture can be treated as a search problem in the architecture space, where each point represents an architecture. Given certain performance criteria, such as minimal training error and lowest network complexity, the performance levels of all architectures form a discrete surface in the space. The performance surface is nondifferentiable due to a discrete number of nodes, and multimodal since different architectures have similar performance.

Direct and indirect encodings are two methods for encoding architecture. For the direct encoding, every connection of the architecture is encoded into the chromosome. For the indirect encoding, only the most important parameters of an architecture, such as the number of hidden layers and the number of hidden units in each hidden layer, are encoded. Only the architecture of a network is evolved, whereas other parameters of the architecture such as the connection weights have to be learned after a near-optimal architecture is found.

Direct Encoding

In direct encoding, each parameter c_{ij}, the connectivity from nodes i to j of the architecture, can be represented by a bit denoting the presence or absence of a connection. An N_n-node architecture is represented by an $N_n \times N_n$ matrix, $\mathbf{C} = [c_{ij}]$. If c_{ij} is represented by real-valued connection weights, both the architecture and connection weights of the network are evolved simultaneously. The binary string representing the architecture is the concatenation of all the rows of the matrix. For an FNN, only the upper triangle of the matrix will have nonzero entries, and thus only this part of the connectivity matrix needs to be encoded into the chromosome. As an example, a 2-2-1 FNN is shown

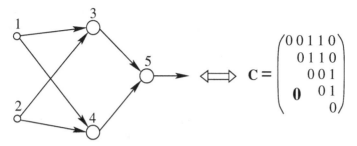

Fig. 9.7. Direct encoding of a 2-2-1 FNN architecture. The number above each node denotes the cardinal of the node.

in Fig. 9.7. Only the upper triangle of the connectivity matrix needs to be encoded in the chromosome, and we get "0110 110 01 1".

A chromosome is required to be converted back to a neural network in order to evaluate the fitness of each chromosome. The neural network is then trained after being initialized with random weights. The training error is used to measure the fitness. In this way, EAs explore all possible connectivities.

The direct encoding scheme has the problem of scalability. A large network would require a very large matrix and thus, the computation time of the evolution is increased. Prior knowledge can be used to reduce the size of the matrix. For example, for the MLP, two adjacent layers are in complete connection, and therefore its architecture can be encoded by the number of hidden layers and the number of hidden units in each layer. This leads to indirect encoding.

Indirect Encoding

Indirect encoding can effectively reduce the chromosome length of the architecture by encoding only some characteristics of the architecture. The details of each connection are either predefined or specified by some rules.

Indirect encoding may not be very good at finding a compact network with good generalization ability. Each network architecture may be encoded by a chromosome consisting of a set of parameters such as the number of hidden layers, the number of hidden nodes in each layer, and the number of connections between two layers. In this case, EAs can only search a limited subset of the whole feasible architecture space. This parametric representation method is most suitable when the type of architecture is known.

Developmental rule representation is another method for constructing architectures, and results in a more compact genotypical representation [608, 437, 1225]. Instead of direct optimization of architectures, a development rule is encoded and optimized, and is then used to construct architectures. This approach is capable of preserving promising building blocks found so far, and can thus lessen the damage of crossover [608]. The connectivity pattern of the architecture in the form of a matrix is constructed from a basis, *i.e.* a single-element matrix, by repeatedly applying suitable developmental rules to nonterminal elements in the current matrix until the matrix contains only terminal elements that indicate the presence or absence of a connection, that is, until a connectivity pattern is fully specified. The developmental rule representation method normally separates the evolution of architectures from that of connection weights.

As a consequence, the direct encoding scheme of network architectures is very good at fine tuning and generating a compact architecture, while the indirect encoding scheme is suitable for finding a particular type of network architecture quickly.

9.10.5 Simultaneously Evolving Architecture and Parameters

One major problem with the evolution of architectures without connection weights is noisy fitness evaluation [1226]. The noise is dependent on the random initialization of the weights and the training algorithm used. The noise identified is caused by the one-to-many mapping from genotypes to phenotypes. Thus, the evolution of architectures without any weight information is inaccurate for evaluating fitness, and the evolution would be very inefficient. This drawback can be alleviated by simultaneously evolving network architectures and connection weights. A considerable amount of work has been conducted in this field [1225].

In [690], an improved GA is used for training a three-layer FNN with switches at its links. Both the nonlinear mapping and the network architecture can be learned. The weights of the links govern the input-output mapping, while the switches of the links govern the network architecture. A given fully connected FNN may become a partially connected network after learning.

The GA equipped with the BP can be used to train both the architecture and the weights of the MLP [816, 171, 172]. In [816], the Quickprop algorithm [346] is used to tune a solution and to reach the nearest local minimum from the solution found by the GA. The population is initialized with chromosomes of different hidden-layer sizes, which are modified by specifically designed operators. Mutation, multipoint crossover, addition of a hidden neuron, elimination of a hidden neuron, and Quickprop training are used as genetic operators. The GA searches and optimizes the architecture, the initial weight settings for that architecture, as well as the learning rate. The operators of Quickprop training and elimination of hidden neurons can be treated as the Lamarckian strategy on evolution.

In the genetic backpropagation (G-Prop) method [171], the GA selects the initial weights and changes the number of neurons in the hidden layer through the application of five specific genetic operators, namely, mutation, multipoint crossover, addition, elimination and substitution of hidden units. The BP, on the other hand, is used to train from these weights. This makes a clean division between global and local search. This strategy attempts to avoid Lamarckism.

The G-Prop method [171] is modified in [172] by integrating an MLP learning algorithm as a training operator. As in [816], this training operator is used to train the individual network for a certain number of generations using the Quickprop. This method thus implements the Lamarckian evolution. It is pointed out in [638] that the Lamarkian approach is more effective than the two-phase and the Baldwinian approaches in difficult problems.

The EPNet [1226] is a hybrid evolutionary/learning method for fully connected FNNs using the EP and the Lamarkian strategy. Five mutation operators are used, namely, hybrid training, node deletion, connection deletion, connection addition, and node addition. Behaviors between parents and their offspring are linked by various mutations, such as partial training and node splitting. The evolved neural network is parsimonious by preferring the node/connection deletion operations to the node/connection addition operations. The hybrid training operator that consists of a modified BP with adaptive learning rates and the SA is used for modifying the connection weights. After the evolution, the best evolved neural network is further trained using the modified BP on the combined training and validation set.

9.10.6 Evolving Activation Functions and Learning Rules

There are also studies on evolving node activation functions or learning rules, since different activation functions have different properties and different learning rules have different performance [1225]. For example, the learning rate and the momentum factor of the BP algorithm can be evolved [603], and learning rules evolved to generate new learning rules [185, 87]. EAs are also used to select proper input variables for neural networks from a raw data space of a large dimension, that is, to evolve input features [441].

The activation functions can be evolved by selecting among some popular nonlinear functions such as the Heaviside, sigmoidal, and Gaussian functions. A neural network with evolutionary neurons has been proposed in [27]. The activation function $\phi(\cdot)$ for each neuron is an unknown general function, whose adequate functional form is achieved by symbolic regression used in the GP [625] during the learning period. The parallel processing of the neural network considered is defined by its architecture and the evolutionary process involved in the computation of the adequate functions. Some mathematical operators are used as the operators of the nodes of the tree.

9.11 Constructing Fuzzy Systems Using Evolutionary Algorithms

FISs are highly nonlinear systems with many input and output variables. A crucial issue in FISs is the generation of fuzzy rules. EAs can be employed for generating fuzzy rules and adjusting MFs of fuzzy sets. Sufficient system information must be encoded and the representation must be easy for evaluation and reproduction. A fuzzy rule base can be evolved by encoding the number of rules and the MFs comprising those rules into one chromosome. All the input and output variables and their corresponding MFs in the fuzzy rules are encoded. The genetic coding of each rule is the concatenation of the shape and location parameters of the MFs of all the variables. For example, for the Gaussian MF, one needs to encode the center and width of its base. If the fuzzy rule is given by

$$\text{IF } x_1 \text{ is } \mathcal{A}_1 \text{ and } x_2 \text{ is } \mathcal{A}_2, \text{ THEN } y \text{ is } \mathcal{B}$$

where x_1 and x_2 are the input variables, y is the output variable, and $\mathcal{A}_1$, $\mathcal{A}_2$ and $\mathcal{B}$ are linguistic variables, described by MFs, then the rule base can be encoded as shown in Fig. 9.8.

As in neural networks, FISs also have the *permutation problem*. If some rules in two individuals are ordered in different manners, the rules should be aligned before reproduction. This leads to much less time for evolution [261]. In addition to existing genetic operators, specific genetic operators such as rule insertion or rule deletion, where a whole rule is added or deleted at a specified point of the string, are also applied [261]. In [858], automatic optimal design of fuzzy systems is conducted by using the GESA [1236].

The ES approach is more suitable for the design of fuzzy systems due to their direct coding for real parameter optimization. Consequently, the length of the objective vector increases linearly with the number of variables. Conversely, when the GA is employed, all the parameters need to be converted into fixed-length binary strings.

9.12 Constructing Neurofuzzy Systems Using Evolutionary Algorithms

In neurofuzzy systems, parameter learning usually employs the gradient-descent method, which may converge to a local minimum. Global optimiza-

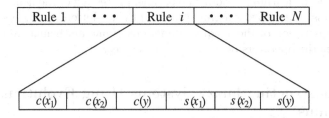

Fig. 9.8. Genetic encoding of a fuzzy rule base. $c(\cdot)$ and $s(\cdot)$, respectively, represent the center and shape (width) parameters of the Gaussian MF of a fuzzy variable.

tion algorithm such as EAs can replace the gradient-descent technique for training neurofuzzy systems. EAs can optimize both the architecture and the parameters of neurofuzzy systems. In other words, EAs are used to evolve both the fuzzy rules and their respective MFs and connection weights [902, 982, 983, 25, 863, 244].

The fuzzy neural network [902] has the architecture of a standard two-level OR/AND representation of Boolean functions of symbols. The fuzzy neural network can be trained using a three-phase procedure [901]. The architectural optimization is performed by using the GP, whereas the ensuing parameters are optimized by gradient-based learning. A collection of fuzzy sets are first selected, and kept unchanged during the successive phases of the model development. The GP is applied to search an optimal architecture of the fuzzy model. After the architecture is selected, the network is subject to some parametric refinement by optimizing the weights using gradient-based learning.

The fuzzy genetic neural system (FuGeNeSys) [982] and the genetic fuzzy rule extractor (GEFREX) [983] are synergetic models of fuzzy logic, neural networks, and the GA. They both are general methods for fuzzy supervised learning of MIMO systems. The Mamdani fuzzy model is used, and the weighted mean defuzzification method applied. Each individual in the population is made up of a set of N_r rules, with each rule comprising n inputs (antecedents) and m outputs (consequents). The fine-grained GA [256] is used, in which the individuals are typically placed on a planar grid. Both global and local selections are used. Global selection is fitness-proportional and identifies the center of the subpopulation (deme), while local selection uses the same selection method but only in the subpopulation (deme). A hill-climbing operator starts whenever an individual is generated with a fitness value higher than the best one obtained so far. The two methods use different genetic coding techniques and fitness functions.

In the GEFREX [983], the genetic coding involves only the premises of the fuzzy rules. This coding is a mix of floating-point and binary parts, where the floating-point part codes the significant values of the MFs and the binary part is devoted to feature selection. The number of inputs n depends on the application fields, whereas the number of rules N_r is specified by the user of the GEFREX. At most $N_r \times n$ antecedents are required. The binary part requires only bits, which are used when the feature-selection option is enabled during the learning phase. The GEFREX is also able to detect significant features when requested during the learning phase. The optimal consequences can be obtained using the SVD. Different crossover and mutation operators are used for the binary genes and real-coded genes.

In [25], the ES and the SA are combined to simultaneously optimize the number of fuzzy rules of a given network while training its parameters. In [863], the training of hybrid fuzzy polynomial neural networks (HFPNN) is comprised of both a structural phase using the GA and the ensuing parametric phase using the LS-based learning. The fuzzy adaptive learning control network (FALCON) [244] is a five-layer neurofuzzy system. The FALCON-GA is a three-phase hybrid learning algorithm for structure/parameter learning, namely, the fuzzy ART for clustering supervised training data, the GA for finding proper fuzzy rules, and the BP for tuning input/output MFs.

9.13 Constructing Evolutionary Algorithms Using Fuzzy Logic

To complete the discussion of the synergy of the three softcomputing paradigms, we describe the application of fuzzy logic to the construction of EAs in this section.

9.13.1 Fuzzy Encoding for Genetic Algorithms

Conventional GA parameter coding is static for the entire search. Real parameter values are commonly encoded as the chromosomes that are bit strings of the binary, Gray or floating-point coding. The chromosomes are then modified by genetic operators. Special care is taken to keep each parameter in a chromosome within the desired range. This results in a slow convergence. Greater accuracy in the final solution is obtained and convergence speed increased by dynamically controlling the coding of the search space. The addition of fuzzy rules to control coding changes provides a more uniform performance in the GA search. Examples of fuzzy encoding techniques for the GA are the fuzzy GA parameter coding [1060] and the fuzzy coding for chromosomes [1023]. These methods use an intermediate mapping between the genetic strings and the search space parameters.

The fuzzy coding [1023] provides the value of a parameter on the basis of the optimum number of selected fuzzy sets and their effectiveness in terms of the degree of membership. It represents the knowledge associated with each parameter and is an indirect method of encoding. By partitioning each parameter into fuzzy sets, at different parts of fuzzy partitioning, MFs of different shapes are selected. The GA optimizes MFs and the number of fuzzy sets, while the actual parameter value is obtained through defuzzification. Fuzzy encoding with a suitable combination of MFs is able to find better optimized parameters than both the GA using binary encoding methods and the gradient-descent technique for parameter learning of neural networks [1023].

Compared with other coding methods, each parameter in the fuzzy coding always falls within the desired range, thus removing the additional overheads on the genetic operators. Prior knowledge from the problem domain can be integrated easily. The GA only optimizes the fuzzy sets and the MFs associated with a parameter, the actual value being obtained through a selected distribution within a given range. The genetic operators are thus independent of the actual value and can be applied to a wide range of problems. The only knowledge required is the type of parameters and their ranges. Each parameter has similar encoding and this gives uniformity in representation, even though they are different in terms of contributions and ranges.

In the fuzzy-coding approach, each parameter is encoded in two sections. In the first section, the fuzzy sets associated with each parameter are encoded in bits, with one representing the corresponding fuzzy set selected. For example, in order to optimize a system with two parameters x_1 and x_2, we can give a complete fuzzy partitioning of the interval of either parameter as three fuzzy sets $\mathcal{A}_i$, $i = 1, 2, 3$. The second section lists the corresponding degrees of membership $\mu_{\mathcal{A}_i}(x_j)$, evaluated at x_j. The corresponding chromosome using the fuzzy coding is shown in Fig. 9.9.

Fuzzy encoding has two advantages over binary encoding [900]. First, the resulting codebooks are made highly nonuniform so as to capture all necessary domain knowledge to orient toward promising search areas and reduce futile search effort. Secondly, fuzzy encoding supports a so-called *weak encoding* of optimized structures that could be helpful in the representation of neural networks. The weak encoding, in contrast to its strong counterpart, does not imply a one-to-one correspondence between the genotype and the phenotype. Rather, a single genotype induces a fuzzy family of phenotypes.

9.13.2 Adaptive Parameter Setting Using Fuzzy Logic

Adaptive GAs dynamically adjust selected control parameters or genetic operators during the evolution. Their objective is to offer the most appropriate exploration and exploitation behavior to avoid the premature convergence problem and improve the final results.

The dynamic parametric GA [670] and fuzzy adaptive GA (FAGA) [477] are GAs whose control parameters are adjusted using fuzzy logic. The main idea is to use a fuzzy controller with inputs as any combination of current performance measures and current control parameters of the GA, and outputs as new control parameters of the GA. In [670], the FIS used for adjusting the control parameters of the GA is by itself adapted by another GA. Some aspects of the FAGA, such as the steps for their design and a taxonomy for the FAGA, are reviewed and analyzed in [477].

Binary coded fuzzy sets		Degrees of membership						
$\mathcal{A}_1$ $\mathcal{A}_2$ $\mathcal{A}_3$	$\mathcal{A}_1$ $\mathcal{A}_2$ $\mathcal{A}_3$	$\mu_{\mathcal{A}_1}(x_1)$	$\mu_{\mathcal{A}_2}(x_1)$	$\mu_{\mathcal{A}_3}(x_1)$	$\mu_{\mathcal{A}_1}(x_2)$	$\mu_{\mathcal{A}_2}(x_2)$	$\mu_{\mathcal{A}_3}(x_2)$
x_1	x_2	x_1			x_2		

Fig. 9.9. The fuzzy coding representation of a chromosome for a two-parameter optimization problem. The first section contains the binary-coded fuzzy sets, and the second section contains the corresponding degrees of membership. $\mathcal{A}_i$, $i = 1, 2, 3$, are the fuzzy partitions of the parameters.

Fuzzy encoding, fuzzy crossover operations, and fuzzy fitness can be integrated into the EA paradigm [900]. Some adaptive fuzzy crossover operators, which are based on fuzzy connectives for real-coded GAs, are defined in [478].

9.14 Computer Experiments

This chapter concludes with two examples. Example 9.1 is the well-known Rosenbrock's valley function, which is usually used as a test function for optimization algorithms. We compare the performance of the simple GA and the real-coded GA. In Example 9.2, the real-coded GA is used to train an MLP. Simulations are conducted on a PC with a Pentium III 600MHz CPU and 128M RAM by using Matlab©.

9.14.1 Example 9.1: Optimization of Rosenbrock's Function

In this example, we illustrate the application of the simple GA to solve an optimization problem. Rosenbrock's valley function

$$f(\mathbf{x}) = \sum_{i=1}^{n-1} 100 \left(x_{i+1} - x_i^2\right)^2 + (1 - x_i)^2$$

is selected for optimization. The global minimum is taken at $x_i = 1$, $i = 1, \cdots, n$, with $f(\mathbf{x}) = 0$. Our simulation is limited to the two-dimensional case ($n = 2$), with $x_1, x_2 \in [-2048, 2048]$. The landscape of this function for one-hundredth of the region, namely, $x_1, x_2 \in [-204.8, 204.8]$, is shown in Fig. 9.10a, and a close-up of the regions having local minima is shown in Fig. 9.10b.

For the simple GA without the elite strategy, the size of population is 100, and the representation for each variable is 30-bit Gray coding. Single-point crossover with $P_c = 0.7$ is applied, and $P_m = 0.01$. The selection scheme is the roulette-wheel selection. In each generation, 80% of the population is newly generated. The evolution for 300 generations for a typical random run is shown in Fig. 9.11. At the 300th generation, the best solution is $x_1 = 9.6709$, $x_2 = 93.5334$, and $f^* = 75.1893$. The computation time is 16.75 s. By increasing the population size to 300 or the representation precision to 90 bits per variable, the performance cannot be improved significantly. Although an elitism strategy can considerably improve the result, the simple GA cannot avoid premature convergence and always converges to a local minimum in this example.

For most numerical optimization problems, real coding can usually generate a performance better than that of the binary coding. Here we include the elitism strategy in the real-coded GA realization. Our numerical testing shows that a big mutation rate can usually yield good results. When the population size is selected as 40, $P_c = 0.3$, $P_m = 0.9$. The selection scheme

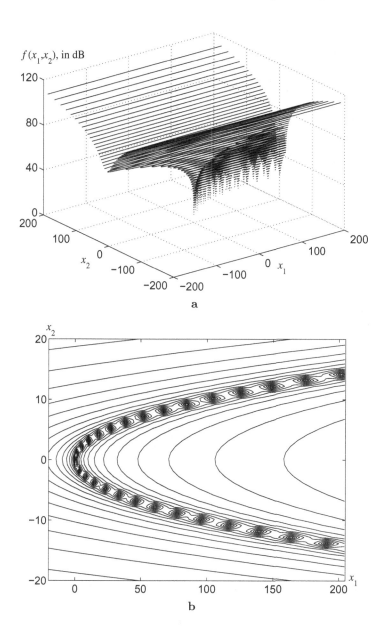

Fig. 9.10. The landscape of Rosenbrock's function $f(\mathbf{x})$ with two variables, and x_1, $x_2 \in [-204.8, 204.8]$. The spacing of the grid is set as 1. There are many local minima, and the global minimum is situated at $(1, 1)$ with $f(1, 1) = 0$. (**a**) The contour in the three-dimensional space. (**b**) A close-up contour in the two-dimensional space.

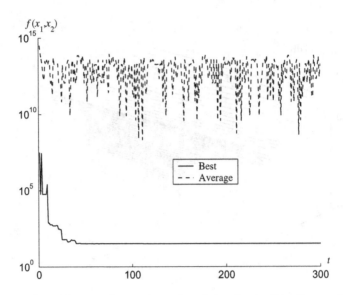

Fig. 9.11. The evolution of a random run of the simple GA: the maximum and average objectives. t corresponds to the number of generations.

is the roulette-wheel selection. The evolution for $T = 1000$ generation for a typical random run is shown in Fig. 9.12. The crossover operator generates, by averaging two parents, only one offspring. The mutation is the one-point mutation, only one gene being modified by each mutation operation. The mutation operator rejects infeasible chromosomes that are beyond the domain. An annealing variance $\sigma = \sigma_0(1 - \frac{t}{T}) + \sigma_1$ is selected for mutation, where $\sigma_0 = 10$, and $\sigma_1 = 0.5$. All chromosomes in the old generations are replaced by the new offspring except that the largest chromosome of the old generation remains in the new generation. At the 1000th generation, the solution of a typical run is $x_1 = 0.9830$, $x_2 = 0.9663$, and $f^* = 2.9116 \times 10^{-4}$. The time for 1000 generations is 33.84 s. The adaptation is shown in Fig. 9.12. The real-coded GA typically leads to a performance better than that of the simple GA.

9.14.2 Example 9.2: Iris Classification

We now revisit the Iris classification problem given in Example 3.2. Here, we use the GA to train the 4-4-3 MLP and hope to find a global optimum solution for the weights.

There are a total of 28 weights in the network, which are encoded as a string of 28 numbers. The fitness function is defined as $f = \frac{1}{1+E}$, where E is the training error, that is, the MSE function. In the case of binary encoding, crossover through the middle of the string is not permitted since the permu-

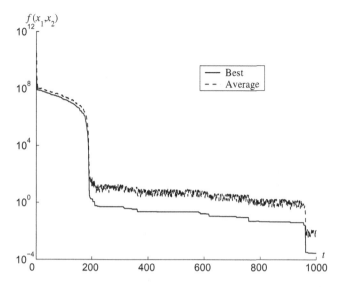

Fig. 9.12. The evolution of a random run of the real-code GA with the elitism strategy: the maximum and average objectives. t corresponds to the number of generations.

tation problem may occur. For real encoding, crossover is defined as a linear combination of the same parameters in two parents. As a result, the permutation problem does not arise. Mutation is realized by adding a stochastic deviation.

In the experiment, we employ real encoding. A fixed population of 100 is applied. The selection scheme is the roulette-wheel selection. Only mutation is employed. Only one random gene of a chromosome is mutated by adding Gaussian noise with variance $\sigma = \sigma_0 \left(1 - \frac{t}{T}\right) + \sigma_1$. The initial population is randomly generated with components of all chromosomes as random numbers in $(0, 1)$. σ_0 and σ_1 are, respectively, selected as 10 and 0.5. The elitism strategy is adopted to guarantee that the best individual always stays in the populations. The results for a typical random run are shown in Figs. 9.13 and 9.14. The computation time is 2427.2 s for 500 generations. Although the training error is relatively large, $E = 2.8557$, it is interesting to note that the rate of correct classification is 99.33%.

Naturally, the learning results are not better than those with descent algorithms simulated in Example 3.2, and descent algorithms are selected whenever mathematical models are available. EAs are useful when no such models are available. EAs are also preferred if the global optimum is at a premium. In the above implementation, the selection of variance σ is of vital importance. In ESs, σ itself is evolved, and some other measures beneficial to numerical optimization are also used. ESs typically generate a performance better than that of the GAs for numerical optimizations [53].

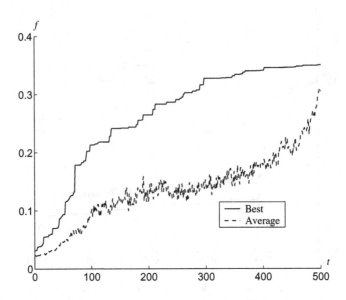

Fig. 9.13. The evolution of the real-coded GA for training a 4-4-3 MLP: the fitness and average fitness. t corresponds to the number of generations.

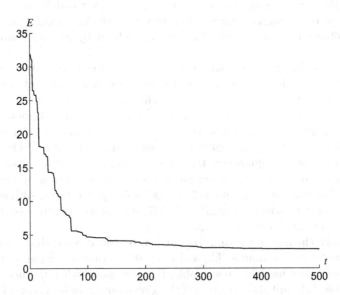

Fig. 9.14. The evolution of the real-coded GA for training a 4-4-3 MLP: the training error. t corresponds to the number of generations.

10

Discussion and Outlook

A brief summary of the book is first presented in this chapter. The SVM, a very popular and recent method, is then described. This is followed by a brief introduction to some other neural-network models such as the CNN and softcomputing methods including the tabu search. Some recent research topics are also presented.

10.1 A Brief Summary

So far, we have introduced some popular neural-network models. These models are adequate for the majority of neural-network applications. Due to their general-purpose nature and proven excellent properties such as adaptivity, robustness and easy hardware implementation, the neural-network approach provides a powerful means to solve information processing and optimization problems. Fuzzy logic and evolutionary computation are, respectively, introduced as a powerful approach to information processing and an optimization tool for the learning of neural networks.

We now give a brief summary of the neural-network models described in this book. The Hopfield model is a dynamic model that is suitable for hardware implementation and can converge to the result in the same order of time as the circuit time constant. It is especially useful for associative memory and optimization. The Boltzmann machine is a generalization of the Hopfield model obtained by integrating the global optimization capability of the SA. The MLP and the RBFN are FNN models with universal approximation capability. The MLP is a global model, while the RBFN is a local one. The MLP is the most studied and used model. The RBFN is orders of magnitude faster than the MLP in learning, and is a good alternative method to the MLP. Clustering neural networks such as the SOM and the ART and clustering algorithms such as the C-means and the FCM are based on competitive learning. Clustering is an important tool for data analysis, and finds wide applications in VQ, classification and pattern recognition. Clustering is also a classical method for determining the prototypes of the RBFN. The family of neural networks and algorithms for the PCA, ICA, MCA and SVD deserve particular attention. These approaches provide simple neural algorithms for data analysis and signal processing, which are conventionally treated by the computationally intensive SVD or EVD.

Fuzzy sets serve as information granules quantifying a given input or output variable, and fuzzy logic is a means of knowledge representation. The integration of fuzzy logic into neural networks yields neurofuzzy systems, which capture the merits of both paradigms. The learning capability of neural networks is exploited to adapt the knowledge base from the given data, and this work is traditionally conducted by human experts. The application of fuzzy logic endows neural networks with the capability of explaining their actions. Neurofuzzy models usually achieve a faster convergence speed with a smaller network size, when compared with their neural-network counterparts.

EAs are a family of general-purpose global optimization algorithms based on the principles of natural evolution and genetics. The application of EAs to neural networks or neurofuzzy systems helps in finding optimal neural or neurofuzzy systems. Both the architecture and parameters of a network can be evolved, and EAs are sometimes used to evolve learning algorithms as well. Furthermore, evolution itself can be accelerated by integrating learning, either in the form of the Lamarckian strategy or based on the Baldwin effect. Other population-based techniques such as the PSO, the immune algorithm, and the ACO have also been introduced as global optimization methods.

10.2 Support Vector Machines

The SVM is a three-layer FNN. It implements the structural risk-minimization principle, which has its foundation in statistical learning theory [1118, 1120]. Instead of minimizing the training error, the SVM purports to minimize an upper bound of the generalization error and maximizes the margin between a separating hyperplane and the training data.

Nonlinear kernel functions are used to overcome the curse of dimensionality. The space of the input examples, $\{\mathbf{x}\} \subset R^n$, is mapped onto a high-dimensional feature space so that the optimal separating hyperplane built on this space allows a good generalization capacity. By choosing an adequate mapping, the input examples become linearly or almost linearly separable in the high-dimensional space. This transforms the SVM learning into a quadratic optimization problem, which has one global solution.

The SVM has been shown to be a universal approximator for various kernels [446]. It has aroused a surge of interest recently, and is being more and more frequently used for classification, regression, and clustering. One of the main features of the SVM is the absence of local minima. The SVM is defined in terms of a subset of the learning data, called *support vector*. The SVM model is a sparse representation of the training data, and allows

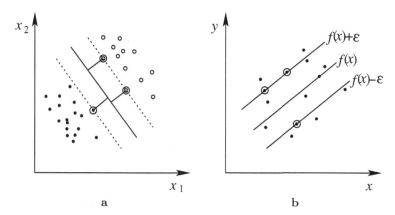

Fig. 10.1. An illustration of the hyperplane in the two-dimensional feature space of the SVM. (**a**) SVM for classification: The margin is defined by the distance between the hyperplane and the nearest of the examples in the two classes. (**b**) SVM for regression: The objective function penalizes examples whose y values are not within $(f(x) - \varepsilon, f(x) + \varepsilon)$. Those examples in circles are support vectors, against which the margin pushes.

the extraction of a condensed data set based on the support vectors. Recent reviews on the SVM are given in [148, 832, 1007][1].

10.2.1 Support Vector Machines for Classification

The SVM was originally proposed for binary classification. It aims at finding the optimal hyperplane that maximizes the margin between the examples of two different classes, as illustrated in Fig. 10.1a.

The optimal hyperplane can be constructed by solving the following QP problem [1119]:
Minimize

$$E_0(\mathbf{w}, \boldsymbol{\xi}) = \frac{1}{2}\|\mathbf{w}\|^2 + C\sum_{i=1}^{N}\xi_p \tag{10.1}$$

subject to

$$y_p\left(\mathbf{w}^{\mathrm{T}}\mathbf{x}_p + \theta\right) \geq 1 - \xi_p \tag{10.2}$$

$$\xi_p \geq 0 \tag{10.3}$$

for $p = 1, \cdots, N$, where $\boldsymbol{\xi} = (\xi_1, \cdots, \xi_N)^{\mathrm{T}}$, ξ_p, $p = 1, \cdots, N$, being slack variables, $\mathbf{w}$ and θ are the weight and bias parameters for determining the hyperplane, $y_p \in \{-1, +1\}$ is the described output of the classifier, and C is a parameter that trades off wide margin with a small number of margin failures.

---

[1] A selected collection of tutorials, publications, computer codes for SVM and kernel methods can be found at *http://www.kernel-machines.org*.

By applying the Lagrange multiplier method and replacing $\mathbf{x}_p^T \mathbf{x}$ by the kernel function $\kappa(\mathbf{x}_p, \mathbf{x})$, the ultimate objective of SVM learning is to find α_p, $p = 1, \cdots, N$, so as to minimize the dual quadratic form [1119]

$$E_{\text{SVM}} = \frac{1}{2} \sum_{p=1}^{N} \sum_{i=1}^{N} y_p y_i \kappa(\mathbf{x}_p, \mathbf{x}_i) \alpha_p \alpha_i - \sum_{p=1}^{N} \alpha_p \tag{10.4}$$

subject to

$$\sum_{p=1}^{N} y_p \alpha_p = 0 \tag{10.5}$$

$$0 \leq \alpha_p \leq C, \quad p = 1, \cdots, N \tag{10.6}$$

where α_p is the weight for the kernel corresponding to the pth example.

The output of the SVM gives the decision on the classification, which is defined by

$$u(\mathbf{x}) = \text{sgn} \left(\sum_{p=1}^{N} \alpha_p y_p \kappa(\mathbf{x}_p, \mathbf{x}) + \theta \right) \tag{10.7}$$

The QP problem given by (10.4) through (10.6) will terminate when all of the KKT conditions [362] are fulfilled

$$\begin{cases} y_p u_p \geq 1, & \alpha_p = 0 \\ y_p u_p = 1, & 0 < \alpha_p < C \\ y_p u_p \leq 1, & \alpha_p = C \end{cases} \tag{10.8}$$

where u_p is the SVM output for the pth example. Those patterns with non-zero α_p are the support vectors, which lie on the margin. The kernel function $\kappa(\mathbf{x}_p, \mathbf{x}) = \varphi^T(\mathbf{x}_p) \varphi(\mathbf{x})$, where the form of $\varphi(\cdot)$ is implicitly defined by the choice of the kernel function and does not need to be given. When $\kappa(\cdot)$ is a linear function, that is, $\kappa(\mathbf{x}_p, \mathbf{x}) = \mathbf{x}_p^T \mathbf{x}$, the SVM reduces to be the linear SVM [1118]. The popular polynomial, Gaussian, and sigmoidal kernels, are, respectively, given by (7.91) through (7.93).

It is shown in [1024] that the separating hyperplane for two-class classification obtained by the SVM is equivalent to the solution obtained by the LDA on the set of support vectors.

Solving the Quadratic Programming Problem

Due to its immense size, the QP problem cannot be easily solved via the standard QP techniques. The chunking technique [1118] was designed to solve this QP problem. Chunking breaks down the large QP problem into a series of smaller QP subproblems, whose ultimate goal is to identify all nonzero α_p, since training examples with $\alpha_p = 0$ do not change the solution. There are some methods that decompose the large QP problem into smaller QP subproblems. Each QP subproblem is initialized with the results of the previous subproblem. However, decomposition requires the use of a numerical QP algorithm such as the standard projected conjugate gradient (PCG) algorithm, which is notoriously slow since the QP problem is polynomial-time solvable with the time growing rapidly with the size of the data set. Furthermore, chunking still cannot handle large-scale training problems due to restriction on computer memory.

The sequential minimal optimization (SMO) [920] breaks this large QP problem into a series of smallest possible QP problems. Each small QP problem involves only two α_p and is solved analytically. This avoids using a time-consuming numerical QP optimization as an inner loop. The amount of memory required for the SMO is linear in the size of the training set, and this allows the SMO to handle very large training sets. The SMO has a computational complexity of somewhere between linear and quadratic in the size of the training set for various test problems, while that of a chunking algorithm based on standard QP is somewhere between linear and cubic in the size of the training set. The computation complexity of the SMO is dominated by SVM evaluation, thus the SMO is the fastest for linear SVMs and sparse data sets [920].

Multiclass Classification

The SVM can be used as a single classifier as well as combined with different types of classifiers. To solve a classification problem with m classes with $m > 2$, many strategies can be adopted [1170, 268]. The most simple and intuitive one decomposes the problem into m binary classification problems. For the ith two-class problem, the original m-class training data are labeled as belonging to or not belonging to class i and are used for training. When an example is classified into only one of these binary classifiers, the example is considered correctly classified. Otherwise it is unclassified.

10.2.2 Support Vector Regression

The SVM has been extended to regression. An illustration of SVM for regression is given in Fig. 10.1b. The objective is to find a linear regression

$$f(\mathbf{x}) = \mathbf{w}^{\mathrm{T}}\mathbf{x} + \theta \tag{10.9}$$

such that

$$E_1(\mathbf{w}) = \frac{1}{2}\|\mathbf{w}\|^2 + C\sum_{p=1}^{N}\|y_p - f(\mathbf{x}_p)\|_{\varepsilon} \tag{10.10}$$

is minimized, where $\|\cdot\|_{\varepsilon}$ is the ε-insensitive loss function, defined by

$$\|\mathbf{x}\|_{\varepsilon} = \max\{0, \|\mathbf{x}\| - \varepsilon\} \tag{10.11}$$

and $\varepsilon > 0$ and $C > 0$ are prespecified. Other robust statistics-based loss functions such as Huber's function can also be used [514].

This problem can be transformed into a QP problem [1119]:
Minimize

$$E_2(\mathbf{w}, \boldsymbol{\xi}, \boldsymbol{\zeta}) = \frac{1}{2}\|\mathbf{w}\|^2 + C\sum_{p=1}^{N}(\xi_p + \zeta_p) \tag{10.12}$$

subject to

$$\left(\mathbf{w}^{\mathrm{T}}\mathbf{x}_p + \theta\right) - y_p \leq \varepsilon + \xi_p \tag{10.13}$$

$$y_p - \left(\mathbf{w}^{\mathrm{T}}\mathbf{x}_p + \theta\right) \leq \varepsilon + \zeta_p \tag{10.14}$$

$$\xi_p \geq 0, \quad \zeta_p \geq 0 \tag{10.15}$$

for $p = 1, \cdots, N$. When the error is smaller than ε, the slack variables ξ_p and ζ_p take the value zero.

By generalizing the linear regression to kernel-based regression estimation and applying the Lagrange multiplier method, we get the following optimization problem [1119, 1120]:
Minimize

$$E_{\mathrm{SVR}}(\boldsymbol{\alpha}, \boldsymbol{\beta}) = \frac{1}{2}\sum_{p,i=1}^{N}(\alpha_p - \beta_p)(\alpha_i - \beta_i)\kappa(\mathbf{x}_p, \mathbf{x}_i)$$

$$+ \sum_{p=1}^{N}(\alpha_p - \beta_p)y_p + \sum_{p=1}^{N}(\alpha_p + \beta_p)\varepsilon \tag{10.16}$$

subject to

$$\sum_{p=1}^{N}(\alpha_p - \beta_p) = 0 \tag{10.17}$$

$$0 \leq \alpha_p, \beta_p \leq C \tag{10.18}$$

for $p = 1, \cdots, N$, where the Lagrange multipliers α_p and β_p, respectively, correspond to (10.13) and (10.14).

The output of the SVM generates the regresssion

$$u(\mathbf{x}) = f(\mathbf{x}) = \sum_{p=1}^{N}(\beta_p - \alpha_p)\kappa(\mathbf{x}_p, \mathbf{x}) + \theta \tag{10.19}$$

where θ can be solved using the boundary conditions.

The SVR is a very robust method due to the introduction of the ε-insensitive loss function. Varying ε influences the number of support vectors and thus controls the complexity of the model. Kernel selection is application specific. The RBFN has a structure similar to that of the SVM, but with a Gaussian kernel, and the sigmoidal function in the MLP can also be treated as a kind of kernel function.

There are a number of efficient batch and incremental learning algorithms for SVR. The ν-SVR [188] is a batch learning algorithm. Examples of adaptive learning algorithms include various improved versions of the SMO for SVR [1028, 359], and the accurate online SVR (AOSVR) [743]. The AOSVR can efficiently update a trained SVR function whenever a sample is added to or removed from the training set. The updated SVR function is identical to that produced by a batch algorithm, but the update is faster [743]. In [652], the SVR has been extended to simultaneously approximate a function and its derivatives by introducing additional constraints on the derivatives.

10.2.3 Support Vector Clustering

The SVC [97] is an SVM-based clustering method that uses a Gaussian kernel to transform the data points into a high-dimensional feature space. Clustering is conducted in the feature space and is then mapped back to the data space.

The SVC attempts to find in the feature space the smallest sphere of radius R that encloses all the data points in a set $\{\mathbf{x}_p\}$ of size N. The SVC can be described by the minimization of

$$E_3\left(R, \mathbf{c}, \boldsymbol{\xi}\right) = R^2 + C \sum_{p=1}^{N} \xi_p \tag{10.20}$$

subject to

$$\left\|\varphi\left(\mathbf{x}_p\right) - \mathbf{c}\right\|^2 \leq R^2 + \xi_p \tag{10.21}$$

$$\xi_p \geq 0 \tag{10.22}$$

for $p = 1, \cdots, N$, where $\varphi(\cdot)$ maps a pattern onto the feature space, ξ_p is a slack variable for the pth data point, $\mathbf{c}$ is the center of the enclosing sphere, and C is a penalty constant controlling the noise.

Based on the Lagrange multiplier method, the problem is transformed into the minimization of

$$E_{\text{SVC}} = \sum_{p=1}^{N}\sum_{i=1}^{N} \alpha_p \alpha_i \kappa\left(\mathbf{x}_p, \mathbf{x}_i\right) - \sum_{p=1}^{N} \alpha_p \kappa\left(\mathbf{x}_p, \mathbf{x}_i\right) \tag{10.23}$$

subject to

$$\sum_{p=1}^{N} \alpha_p = 1 \tag{10.24}$$

$$0 \leq \alpha_p \leq C \tag{10.25}$$

where $\kappa(\cdot)$ is selected as the Gaussian kernel and α_p is the Lagrange multiplier corresponding to the pth data point. The width of the Gaussian kernel, σ, controls the cluster scale while the soft margin ξ_p helps in coping with the outliers and overlapping clusters. By varying the two parameters, the SVC maintains a minimal number of support vectors so as to generate smooth cluster boundaries of arbitrary shape.

The distance between the mapping of an input pattern and the spherical center can be computed as

$$d^2(\mathbf{x}, \mathbf{c}) = \|\varphi(\mathbf{x}) - \mathbf{c}\|^2$$

$$= \kappa(\mathbf{x}, \mathbf{x}) - 2 \sum_{p=1}^{N} \alpha_p \kappa(\mathbf{x}_p, \mathbf{x}) + \sum_{p=1}^{N} \sum_{i=1}^{N} \alpha_p \alpha_i \kappa(\mathbf{x}_p, \mathbf{x}_i) \quad (10.26)$$

Those data points that are on the boundary of the contours are support vectors.

The multisphere SVC [224] is an extension to the SVC [97] obtained by creating multiple spheres to adaptively represent individual clusters. It is an adaptive cell-growing method, which essentially identifies dense regions in the data space by finding their corresponding spheres with minimal radius in the feature space. The multisphere SVC can obtain cluster prototypes as well as cluster memberships. In [153], the clusters obtained by the batch C-means algorithm are further iteratively refined using a two-class SVM.

10.3 Other Neural-network Models and Softcomputing Approaches

In addition to the popular models we have treated so far, there are also some neural-network models and softcomputing paradigms that are attractive for some specific application areas.

10.3.1 Generalized Single-layer Networks

The generalized single-layer network (GSLN) [495], also known as the *generalized linear discriminant*, has a three-layer architecture similar to the RBFN. Each node in the hidden layer contains a nonlinear activation function $\phi_i(\cdot)$, and the output nodes implement linear combinations of the nonlinear kernel functions of the inputs.

For a J_1-J_2-J_3 network, given input $\mathbf{x}_p$, the output of the network is $y_{p,j} = \sum_{i=1}^{J_2} w_{ij} \phi_i(\mathbf{x}_p)$, $j = 1, \cdots, J_3$. Depending on the set of kernel functions used, the GSLN may have broad approximation capabilities. The RBFN and the Volterra network [940] are two types of GSLNs that have this property. Like the OLS method for RBFN learning [208], orthogonal methods in conjunction with some information criteria are usually used for self-structuring the GSLN to generate a parsimonious, yet accurate, network [11].

The generalization ability of the GSLN is analyzed in [495] by using the theory of PAC learning [1115] and the concept of VC dimension [1121]. Necessary and sufficient conditions on the number of training examples are derived to guarantee a particular generalization performance of the GSLN [495]. Some types of GSLNs, such as the RBFN, have the drawback of using an excessive number of kernel functions, thus requiring a high number of training data and often leading to a poor ability to generalize.

10.3.2 Cellular Neural Networks

The CNN is a two- or higher-dimensional array of cells, with only local interactions that can be programmed by a template matrix [239]. Each cell is connected only to the cells in its nearest neighborhood. Each cell comprises a linear resistive summing input unit, an RC linear dynamical unit, and a three-region, symmetrical, piecewise-linear resistive output unit. Thus, each cell has its own dynamics whose evolution is dependent on its circuit time constant $\tau = RC$.

The CNN is a generalization of the Hopfield network [502], and can be used to solve a more generalized optimization problem. An important part of designing a neural optimization approach is the incorporation of constraints. The common method is to add a penalized energy item of constraints, and then to use the Lagrange multiplier method. The CNN overcomes the massive interconnection problem of parallel distributed processing. The key features are asynchronous parallel processing, continuous time dynamics, and local interactions among the network elements.

Due to its local interconnectivity property, CNN chips can have high-density cells, and some physical implementations such as analog CMOS, emulated digital CMOS and optical implementations, are available. The CNN universal machine [965] is the analog cellular computer for processing analog array signals. It has a computational power of tera (10^{12}) or peta (10^{15}) analog operations per second on a single CMOS chip [240].

The CNN with a two-dimensional array architecture is a natural candidate for image processing or simulation of partial differential equations. Any input-output function to be realized by the CNN can also be visualized as an image-processing task, where the external input, the initial condition and the output, arranged as two-dimensional arrays, are, respectively, the external input, initial and output images. The external input image together with the initial image constitutes the input images of the CNN. Using different cloning templates, namely, the representation of the local interconnection patterns, different operations can be conducted on an image. The CNN has now become an important method for image processing.

10.3.3 Wavelet Neural Networks

The wavelet neural network (WNN) [1258, 1257, 1256] has the same structure as the RBFN, and can be treated as a generalization of the RBFN. The only difference between the two models lies in the activation function for the hidden units: The RBFN uses RBFs, while the WNN uses wavelet functions.

Due to the localized properties in both the time and frequency domains of wavelet functions, wavelets are locally receptive field functions that approximate discontinuous or rapidly changing functions. The WNN has become a popular tool for universal function approximation.

Since wavelet decomposition provides a method to represent a signal in multiple scales, wavelets have the multiresolution property. This property is very useful for function approximation. Wavelets with coarse resolution can capture the global or low-frequency feature easily, while wavelets with fine resolution can capture the local or high-frequency feature of the function accurately. This distinguished characteristic leads the WNN to fast convergence, easy training and high accuracy. A fuzzy wavelet network has also been introduced for approximating arbitrary nonlinear functions in [492].

10.3.4 Tabu Search

Tabu search is a metaheuristic method. It is a general stochastic global optimization method, which was originally developed for very large combinatorial optimization tasks [415, 416] and was later extended to continuous function optimization [271]. Tabu search uses a set of strategies and learned information to mimic human insights for problem solving.

Tabu search exploits the search space by introducing flexible memory structures and strategic restrictions as well as aspiration levels. It uses the adaptive memory of its past history and leads to the best solution. A tabu list is used to keep the information about the past steps of the search, and to create and exploit new solutions in the search space. Tabu search starts searching with a present solution and constructs a set of feasible solutions from the present one based on neighborhood by using the tabu list. The solutions constructed are evaluated and the one with the highest evaluation value is selected as the next solution. The tabu list is then updated. To prevent cycling on previously visited solutions, the algorithm sets as tabu every step that has led to an improvement of the objective function during a certain number of iterations, depending on the size of the tabu list. The adaptive memory is implemented with reference to short-term and long-term records, and an aspiration level specifies as to what is a good enough solution. The integration of STMs and LTMs enables a suitable balance between a coarse exploration of the design space and a localized search for the optimum.

Tabu search is conceptually much simpler than EAs or the SA and is easy to implement. Tabu search, for many optimization problems, is superior to the SA and EAs both in terms of the computation time for a solution and in terms of the solution quality. Tabu search pioneered the systematic exploration of memory in search processes, while EAs pioneered the idea of combining solutions.

The converging speed of tabu search to the global optimum is dependent on the initial solution, since it is a form of iterative search. By introducing parallelism, tabu search can find the promising regions of the search space very quickly. A parallel tabu search model, which is based on the crossover operator of the GA, has been described in [568].

10.3.5 Rough Set

The theory of rough sets [894, 895] was proposed by Pawlak in 1982 as a mathematical tool for managing ambiguity, vagueness, and general uncertainty that arise from granularity in the universe of discourse. The theory of rough sets can be approached as an extension to the classical theory of sets.

The objects of the universe of discourse $\mathcal{U}$, called *rough sets*, can be identified only within the limits determined by the knowledge represented by a given indiscernibility relation $\mathcal{R}$. The indiscernibility relation defines a partition in $\mathcal{U}$. A rough set is an imprecise representation of a concept (set) in terms of a pair of subsets, a lower approximation and an upper approximation. The approximations themselves can be crisp, imprecise or fuzzy. The lower approximation is the set of objects definitely belonging to the vague concept, whereas the upper approximation is the set of objects possibly belonging to the vague concept. These approximations are used to define the notions of discernibility matrices, discernibility functions, reducts, and dependency factors, all of which are necessary for the reduction of knowledge [810].

Rough sets have been successfully applied for such areas as knowledge discovery, knowledge acquisition, data mining, machine learning, inductive reasoning, and pattern recognition. Hybridizations for rule generation and exploiting the characteristics of rough sets with neural, fuzzy, and evolutionary approaches are available in the literature (see [1228, 63, 885, 883] and some of the references in [810]).

10.3.6 Emerging Computing Paradigms

Recently, more and more nature-inspired new computing paradigms are being proposed for learning and optimization. Among them, DNA computing [893], membrane computing [892], and quantum computing [491] deserve mentioning. They are all computation techniques at the molecular level.

DNA computing is based on the information processing capabilities of organic molecules. Membrane computing abstracts from the structure and functioning of living cells[2]. Quantum computing is based on the theory of quantum physics, which describes the behavior of particles of atomic size.

Other nature-inspired computing paradigms include those inspired by chemical reaction and diffusion at the molecular level, such as the autocatalytic network [369].

---

[2] Interested readers are referred to *http://psystems.disco.unimib.it for more information.*

10.4 Some Research Topics

Neural networks and softcomputing attract considerable global research efforts due to their effectiveness and general-purpose nature in computing. These research activities are fruitful and numerous techniques as well as their applications are available.

The MLP and the RBFN are the most widely studied and applied neural-network models. Due to their popularity, there are still many efforts aimed at improving their learning algorithms and many new models are emerging by exploiting their structure or algorithms. Thanks to their universal approximation capability, they are widely used for regression and prediction, control and adaptation, classification, and signal processing. They will still dominate in the applications of neural networks.

RNN models such as the Hopfield model and the Boltzmann machine are popular associative memory models and are useful for pattern recognition or pattern completion. They are also useful for combinatorial optimization. The SA and mean-field annealing provide these models with global optimization capability. Future research on these models may be focused on the introduction of new storing algorithms for associative memory. Hardware implementation of these models is also an important topic, since it can lead to convergence within a period of the order of the circuit time constant.

Clustering is a fundamental method in data analysis, knowledge extraction and pattern recognition. Currently, an important area is clustering a data set that exhibits arbitrary shapes. Curve detection is particularly useful in segmenting images for image analysis, understanding and coding. Accordingly, cluster validity measures are required to be defined for clusters with different shapes.

PCA, ICA, and APCA neural networks and algorithms are extremely important for data analysis and information processing, which are conventionally based on linear-algebra techniques, such as the computationally demanding EVD and SVD. The TLS are directly related to the PCA. These neural networks provide simple neural algorithms for solving EVD and SVD. These simple neural-network models significantly reduce the cost for adaptive signal, speech, image, and video processing, and pattern recognition, and will continue to be of great research interest. The ICA technique is attracting more and more research attention. This is due to the fact that it can be used for BSS. BSS for MIMO systems is recently a very active and commercially driven topic in signal processing, communications and medical systems.

The theory of fuzzy sets and logic, as a mathematical extension to the classical theory of sets and binary logic, has become a general mathematical tool for data analysis. Fuzzy logic is a paradigm for modeling human reasoning, and is a basic tool for machine learning and expert systems. In addition to basic research on the theory of fuzzy sets and logic, the hybridization of fuzzy logic and neural networks is a major research area. Existing neurofuzzy systems are mainly customized for clustering, classification, and regression, and more research efforts are needed along this line so as to bring forth more efficient techniques for data analysis and machine learning.

EAs and similar population-based methods are simple, parallel, general-purpose, global optimization methods. They are useful for any optimization problem, particularly when conventional optimization techniques are invalid. They will remain as active and efficient global optimization methods. The high computational cost of EAs can be reduced by introducing learning into EAs, depending on the prior knowledge of a given optimization problem. More importance has recently been attached to the investigation of dynamics of evolutionary computation and the solution to such hard optimization problems as the multimodal, multiobjective and constrained optimizations. In addition, more and more nature-inspired computing paradigms are emerging. EAs have become a fundamental approach to the structure and parameter adaptation of neural networks and fuzzy systems.

10.4.1 Face Recognition

In the area of pattern recognition, recently there are more research activities in the recognition of human faces, fingerprints and car license tags, in view of the importance of global security. An image is first segmented to find the region of interest. This can be conducted by using a clustering technique. The region of interest is then preprocessed so as to extract prominent features for further recognition. Although any feature-extraction technique such as the DCT or the FFT can be used, neural techniques such as the PCA and the LDA are local methods and thus, are more attractive and widely used. The Hopfield network, the MLP, the RBFN and various clustering techniques, can all be used for classification, depending on the specific classification problem. The MLP and the RBFN are supervised classification techniques, while the Hopfield network and clustering are used for unsupervised classification.

10.4.2 Data Mining

The wealth of information in huge databases or the Web has aroused tremendous interest in the area of data mining, also known as *knowledge discovery in databases (KDD)* [539]. Data mining emerges as a rapidly growing area, and is gaining popularity as a business information-management tool. Data mining is to automatically search large stores of data for consistent patterns and/or relationships between variables so as to predict future behavior. For example, data mining can help a retail company find the common preferences of its customers, or help people to navigate, search, and visualize the contents of the Web. The process of data mining consists of three phases, namely, data preprocessing and exploration, model selection and validation, as well as final deployment.

Clustering, neurofuzzy systems and rough sets, and evolution-based global optimization methods are usually used for data mining. Neurofuzzy systems and rough sets are ideal tools for knowledge representation. Data mining needs first to discover the structural features in a database, and exploratory techniques through self-organization such as clustering are particularly promising.

Some of the data-mining approaches that use clustering are database segmentation, predictive modeling, and visualization of large databases [539]. Structured databases have well-defined features and data mining can easily succeed with good results. Web mining is more difficult since the world wide web (WWW) is a less structured database [343]. The topology-preserving property for the SOM makes it particularly suitable for web information processing.

10.4.3 Functional Data Analysis

Functional data analysis (FDA) is an extension of the traditional data analysis to functional data [934]. The FDA characterizes a series of data points as a single piece of data. Examples of functional data are spectra, temporal series, and spatiotemporal images. Functional data are usually represented by regular or irregular sampling as lists of input-output pairs.

The FDA is, more often than not, closely related with the multivariate statistics and regularization. Many statistical methods, such as the PCA, multivariate linear modeling and canonical correlation analysis, can be applied within the FDA framework. Regularization in the FDA can be implemented by penalizing roughness as part of the criterion functions or by using particular representations that possess inherent smoothness.

Conventional neural-network models such as the RBFN and the MLP have been extended to functional data inputs by incorporating functional processing [966, 967].

A

Appendix: Mathematical Preliminaries

In this Appendix, we provide some mathematical preliminaries that are used in the preceding chapters.

A.1 Linear Algebra

This section is based on [424, 247, 1185].

Pseudoinverse

The pseudoinverse $\mathbf{A}^{\dagger}$, also called the *Moore–Penrose generalized inverse*, of a matrix $\mathbf{A} \in R^{m \times n}$ is unique, which satisfies

$$\mathbf{A}\mathbf{A}^{\dagger}\mathbf{A} = \mathbf{A} \tag{A.1}$$

$$\mathbf{A}^{\dagger}\mathbf{A}\mathbf{A}^{\dagger} = \mathbf{A}^{\dagger} \tag{A.2}$$

$$\left(\mathbf{A}\mathbf{A}^{\dagger}\right)^{\mathrm{T}} = \mathbf{A}\mathbf{A}^{\dagger} \tag{A.3}$$

$$\left(\mathbf{A}^{\dagger}\mathbf{A}\right)^{\mathrm{T}} = \mathbf{A}^{\dagger}\mathbf{A} \tag{A.4}$$

$\mathbf{A}^{\dagger}$ can be calculated by

$$\mathbf{A}^{\dagger} = \left(\mathbf{A}^{\mathrm{T}}\mathbf{A}\right)^{-1}\mathbf{A}^{\mathrm{T}} \tag{A.5}$$

if $\mathbf{A}^{\mathrm{T}}\mathbf{A}$ is nonsingular, and

$$\mathbf{A}^{\dagger} = \mathbf{A}^{\mathrm{T}}\left(\mathbf{A}\mathbf{A}^{\mathrm{T}}\right)^{-1} \tag{A.6}$$

if $\mathbf{A}\mathbf{A}^{\mathrm{T}}$ is nonsingular. The pseudoinverse is directly associated with linear LS problems.

When $\mathbf{A}$ is a square nonsingular matrix, the pseudoinverse $\mathbf{A}^{\dagger}$ is its inverse $\mathbf{A}^{-1}$. For a scalar α, if $\alpha \neq 0$, $\alpha^{\dagger} = \alpha^{-1}$; if $\alpha = 0$, $\alpha^{\dagger} = 0$.

Linear Least-squares Problems

The Linear LS or L_2-norm problem is basic to many signal-processing techniques. It tries to solve an SLE, written in matrix form

$$\mathbf{Ax} = \mathbf{b} \tag{A.7}$$

where $\mathbf{A} \in R^{m \times n}$, $\mathbf{x} \in R^n$, and $\mathbf{b} \in R^m$.

This problem can be converted into the minimization of the squared error function

$$E(\mathbf{x}) = \frac{1}{2}\|\mathbf{Ax} - \mathbf{b}\|_2^2 = \frac{1}{2}(\mathbf{Ax} - \mathbf{b})^T(\mathbf{Ax} - \mathbf{b}) \tag{A.8}$$

The solution corresponds to one of the following three situations [424]

- $\text{rank}(\mathbf{A}) = n = m$. We get a unique exact solution

$$\mathbf{x}^* = \mathbf{A}^{-1}\mathbf{b} \tag{A.9}$$

  and $E(\mathbf{x}^*) = 0$.
- $\text{rank}(\mathbf{A}) = n < m$. The system is overdetermined, and has no exact solution. There is a unique solution in the LSE sense

$$\mathbf{x}^* = \mathbf{A}^{\dagger}\mathbf{b} \tag{A.10}$$

  where $\mathbf{A}^{\dagger}$ is the pseudoinverse, $\mathbf{A}^{\dagger} = (\mathbf{A}^T\mathbf{A})^{-1}\mathbf{A}^T$. In this case,

$$E(\mathbf{x}^*) = \mathbf{b}^T\left(\mathbf{I} - \mathbf{AA}^{\dagger}\right)\mathbf{b} \geq 0 \tag{A.11}$$

- $\text{rank}(\mathbf{A}) = m < n$. The system is underdetermined, and the solution is not unique. But the solution with the minimum L_2-norm $\|\mathbf{x}\|_2^2$ is unique

$$\mathbf{x}^* = \mathbf{A}^{\dagger}\mathbf{b} \tag{A.12}$$

  Here the pseudoinverse $\mathbf{A}^{\dagger}$ is given by $\mathbf{A}^{\dagger} = \mathbf{A}^T\left(\mathbf{AA}^T\right)^{-1}$. In this case, $E(\mathbf{x}^*) = 0$ and $\|\mathbf{x}^*\|_2^2 = \mathbf{b}^T\left(\mathbf{AA}^T\right)^{-1}\mathbf{b}$.

Vector Norms

A norm acts as a measure of distance. A vector norm on R^n is a mapping $f : R^n \rightarrow R$ that satisfies such properties: For any $\mathbf{x}, \mathbf{y} \in R^n$, $a \in R$,

- $f(\mathbf{x}) \geq 0$, and $f(\mathbf{x}) = 0$ iff $\mathbf{x} = 0$.
- $f(\mathbf{x} + \mathbf{y}) \leq f(\mathbf{x}) + f(\mathbf{y})$.
- $f(a\mathbf{x}) = |a|f(\mathbf{x})$.

The mapping is denoted as $f(\mathbf{x}) = \|\mathbf{x}\|$.

The p-norm or L_p-norm is a popular class of vector norms

$$\|\mathbf{x}\|_p = \left(\sum_{i=1}^{n} |x_i|^p \right)^{\frac{1}{p}} \tag{A.13}$$

with $p \geq 1$. Usually, the L_1, L_2, and L_∞ norms are more useful

$$\|\mathbf{x}\|_1 = \sum_{i=1}^{n} |x_i| \tag{A.14}$$

$$\|\mathbf{x}\|_2 = \sum_{i=1}^{n} \left(x_i^2 \right)^{\frac{1}{2}} = \left(\mathbf{x}^{\mathrm{T}} \mathbf{x} \right)^{\frac{1}{2}} \tag{A.15}$$

$$\|\mathbf{x}\|_\infty = \max_{1 \leq i \leq n} |x_i| \tag{A.16}$$

The L_2-norm is the popular Euclidean norm.

A matrix $\mathbf{Q} \in R^{m \times m}$ is said to be *orthogonal* if $\mathbf{Q}^{\mathrm{T}} \mathbf{Q} = \mathbf{I}$. The matrix $\mathbf{Q}$ is called an *orthogonal matrix* or *unitary matrix*. The 2-norm is invariant under orthogonal transforms, that is, for all orthogonal $\mathbf{Q}$ of appropriate dimensions

$$\|\mathbf{Q}\mathbf{x}\|_2 = \|\mathbf{x}\|_2 \tag{A.17}$$

Matrix Norms

A matrix norm is a generalization of the vector norm by extending from R^n to $R^{m \times n}$.

For a matrix $\mathbf{A} = [a_{ij}]_{m \times n}$, the most frequently used matrix norms are the Frobenius norm

$$\|\mathbf{A}\|_{\mathrm{F}} = \left(\sum_{i=1}^{m} \sum_{j=1}^{n} a_{ij}^2 \right)^{\frac{1}{2}} \tag{A.18}$$

and the matrix p-norm

$$\|\mathbf{A}\|_p = \sup_{\mathbf{x} \neq 0} \frac{\|\mathbf{A}\mathbf{x}\|_p}{\|\mathbf{x}\|_p} = \max_{\|\mathbf{x}\|_p = 1} \|\mathbf{A}\mathbf{x}\|_p \tag{A.19}$$

where sup is the supreme operation.

The matrix 2-norm and the Frobenius norm are invariant with respect to orthogonal transforms, that is, for all orthogonal $\mathbf{Q}_1$ and $\mathbf{Q}_2$ of appropriate dimensions

$$\|\mathbf{Q}_1 \mathbf{A} \mathbf{Q}_2\|_{\mathrm{F}} = \|\mathbf{A}\|_{\mathrm{F}} \tag{A.20}$$

$$\|\mathbf{Q}_1 \mathbf{A} \mathbf{Q}_2\|_2 = \|\mathbf{A}\|_2 \tag{A.21}$$

Eigenvalue Decomposition

Given a square matrix $\mathbf{A} \in R^{n \times n}$, if there exists a scalar λ and a nonzero vector $\mathbf{v}$ such that

$$\mathbf{Av} = \lambda\mathbf{v} \tag{A.22}$$

then λ and $\mathbf{v}$ are, respectively, called an *eigenvalue* of $\mathbf{A}$ and its corresponding *eigenvector*. All the eigenvalues $\lambda_i, i = 1, \cdots, n$, can be obtained by solving the characteristic equation

$$\det(\mathbf{A} - \lambda\mathbf{I}) = 0 \tag{A.23}$$

where $\mathbf{I}$ is an $n \times n$ identity matrix. The set of all the eigenvalues is called the *spectrum* of $\mathbf{A}$.

If $\mathbf{A}$ is nonsingular, $\lambda_i \neq 0$. If $\mathbf{A}$ is symmetric, then all the λ_is are real. The maximum and minimum eigenvalues satisfy the Rayleigh quotient

$$\lambda_{\max}(\mathbf{A}) = \max_{\mathbf{v} \neq 0} \frac{\mathbf{v}^{\mathrm{T}}\mathbf{Av}}{\mathbf{v}^{\mathrm{T}}\mathbf{v}} \tag{A.24}$$

$$\lambda_{\min}(\mathbf{A}) = \min_{\mathbf{v} \neq 0} \frac{\mathbf{v}^{\mathrm{T}}\mathbf{Av}}{\mathbf{v}^{\mathrm{T}}\mathbf{v}} \tag{A.25}$$

The trace of a matrix is equal to the sum of all its eigenvalues and the determinant of a matrix is equal to the product of its eigenvalues

$$\mathrm{tr}(\mathbf{A}) = \sum_{i=1}^{n} \lambda_i \tag{A.26}$$

$$|\mathbf{A}| = \prod_{i=1}^{n} \lambda_i \tag{A.27}$$

Singular Value Decomposition

For a matrix $\mathbf{A} \in R^{m \times n}$, there exist real unitary matrices $\mathbf{U} = [\mathbf{u}_1 \ \mathbf{u}_2 \ \cdots \ \mathbf{u}_m] \in R^{m \times m}$ and $\mathbf{V} = [\mathbf{v}_1 \ \mathbf{v}_2 \ \cdots \ \mathbf{v}_n] \in R^{n \times n}$ such that

$$\mathbf{U}^{\mathrm{T}}\mathbf{AV} = \boldsymbol{\Sigma} \tag{A.28}$$

where $\boldsymbol{\Sigma} \in R^{m \times n}$ is a real pseudodiagonal $m \times n$ matrix with $\sigma_i, i = 1, \cdots, p$, $p = \min(m, n)$, $\sigma_1 \geq \sigma_2 \geq \cdots \geq \sigma_p \geq 0$, on the diagonal and zeros off the diagonal. σ_is are called the *singular values* of $\mathbf{A}$, $\mathbf{u}_i$ and $\mathbf{v}_i$ are, respectively, called the *left singular vector* and *right singular vector* for σ_i. They satisfy the relations

$$\mathbf{Av}_i = \sigma_i\mathbf{u}_i \quad \text{and} \quad \mathbf{A}^{\mathrm{T}}\mathbf{u}_i = \sigma_i\mathbf{v}_i \tag{A.29}$$

Accordingly, $\mathbf{A}$ can be written as

$$\mathbf{A} = \mathbf{U}\boldsymbol{\Sigma}\mathbf{V}^{\mathrm{T}} = \sum_{i=1}^{r} \lambda_i \mathbf{u}_i \mathbf{v}_i^{\mathrm{T}} \tag{A.30}$$

where r is the cardinality of the smallest nonzero singular value. In the special case when $\mathbf{A}$ is a symmetric non-negative definite matrix, $\boldsymbol{\Sigma} = \mathrm{diag}\left(\lambda_1^{\frac{1}{2}}, \cdots, \lambda_p^{\frac{1}{2}}\right)$, where $\lambda_1 \geq \lambda_2 \geq \cdots \lambda_p \geq 0$ are the real eigenvalues of $\mathbf{A}$, $\mathbf{v}_i$ being the corresponding eigenvectors.

The SVD is useful in many situations. The rank of $\mathbf{A}$ can be determined by the number of nonzero singular values. The power of $\mathbf{A}$ can be easily calculated by

$$\mathbf{A}^k = \mathbf{U}\boldsymbol{\Sigma}^k\mathbf{V}^{\mathrm{T}} \tag{A.31}$$

where k is a positive integer. The SVD is extensively applied in linear inverse problems. The pseudoinverse of $\mathbf{A}$ can then be described by

$$\mathbf{A}^{\dagger} = \mathbf{V}_r\boldsymbol{\Sigma}_r^{-1}\mathbf{U}_r^{\mathrm{T}} \tag{A.32}$$

where $\mathbf{V}_r$, $\boldsymbol{\Sigma}_r$, and $\mathbf{U}_r$ are the matrix partitions corresponding to the r nonzero singular values.

The Frobenius norm can thus be calculated as

$$\|\mathbf{A}\|_{\mathrm{F}} = \left(\sum_{i=1}^{p} \sigma_i^2\right)^{\frac{1}{2}} \tag{A.33}$$

and the matrix 2-norm is calculated by

$$\|\mathbf{A}\|_2 = \sigma_1 \tag{A.34}$$

QR Decomposition

For the full-rank or overdetermined linear LS case, $m \geq n$, (A.7) can also be solved by using QR decomposition procedure.

$\mathbf{A}$ is first factorized as

$$\mathbf{A} = \mathbf{QR} \tag{A.35}$$

where $\mathbf{Q}$ is an $m \times m$ orthogonal matrix, that is, $\mathbf{Q}^{\mathrm{T}}\mathbf{Q} = \mathbf{I}$, and $\mathbf{R} = \begin{bmatrix} \overline{\mathbf{R}} \\ \mathbf{0} \end{bmatrix}$ is an $m \times n$ upper triangular matrix with $\overline{\mathbf{R}} \in R^{n \times n}$.

Inserting (A.35) into (A.7) and premultiplying by $\mathbf{Q}^{\mathrm{T}}$, we have

$$\mathbf{Rx} = \mathbf{Q}^{\mathrm{T}}\mathbf{b} \tag{A.36}$$

Denoting $\mathbf{Q}^{\mathrm{T}}\mathbf{b} = \begin{bmatrix} \overline{\mathbf{b}} \\ \widetilde{\mathbf{b}} \end{bmatrix}$, where $\overline{\mathbf{b}} \in R^n$ and $\widetilde{\mathbf{b}} \in R^{m-n}$, we have

$$\overline{\mathbf{R}}\mathbf{x} = \overline{\mathbf{b}} \tag{A.37}$$

Since $\overline{\mathbf{R}}$ is a triangular matrix, $\mathbf{x}$ can be easily solved using backward substitution. This is the procedure used in the GSO procedure.

When rank$(\mathbf{A}) < n$, the rank-deficient LS problem has an infinite number of solutions, the QR decomposition does not necessarily produce an othonormal basis for range$(\mathbf{A}) = \{\mathbf{y} \in R^m : \mathbf{y} = \mathbf{A}\mathbf{x} \text{ for some } \mathbf{x} \in R^n\}$. The QR-cp can be applied to produce an orthonormal basis for range$(\mathbf{A})$.

The QR decomposition is a basic method for computing the SVD. The QR decomposition itself can be computed by means of the Givens rotation, the Householder transform, or the GSO.

Condition Numbers

The condition number of a matrix $\mathbf{A} \in R^{m \times n}$ is defined by

$$\text{cond}_p(\mathbf{A}) = \|\mathbf{A}\|_p \|\mathbf{A}^\dagger\|_p \tag{A.38}$$

where p can be selected as 1, 2, ∞, Frobenius, or any other norm. The relation, cond$(\mathbf{A}) \geq 1$, always holds. Matrices with small condition numbers are well conditioned, while matrices with large condition number are poorly conditioned or ill-conditioned. The condition number is especially useful in numerical computation, where ill-conditioned matrices are sensitive to rounding errors.

For the L_2-norm,

$$\text{cond}_2(\mathbf{A}) = \frac{\sigma_1}{\sigma_p} \tag{A.39}$$

where $p = \min(m, n)$.

Householder Reflections and Givens Rotations

Orthogonal transforms play an important role in the matrix computation such as the EVD, the SVD, and the QR decomposition. The Householder reflection, also termed the *Householder transform*, and Givens rotations, also called the *Givens transform*, are two basic operations in the orthogonalization process. These operations are easily constructed, and they introduce zeros in a vector so as to simplify matrix computations. The Householder reflection is exceedingly efficient for annihilating all but the first entry of a vector, while the Givens rotation is more effective to transform a specified entry of a vector into zero.

Let $\mathbf{v} \in R^n$ be nonzero. The Householder reflection is defined as a rank-one modification to the identity matrix

$$\mathbf{P} = \mathbf{I} - 2\frac{\mathbf{v}\mathbf{v}^T}{\mathbf{v}^T\mathbf{v}} \tag{A.40}$$

The Householder matrix $\mathbf{P} \in R^{n \times n}$ is symmetric and orthogonal. $\mathbf{v}$ is called a *Householder vector*. The Householder tranform of a matrix $\mathbf{A}$ is given by

PA. By specifying the form of the transformed matrix, one can find a suitable Householder vector $\mathbf{v}$. For example, one can define a Householder vector as $\mathbf{v} = \mathbf{u} - \alpha \mathbf{e}_1$, where $\mathbf{u} \in R^m$ is an arbitrary vector of length $|\alpha|$ and $\mathbf{e}_1 \in R^m$, wherein only the first entry is unity, all the other entries being zero. In this case, $\mathbf{Px}$ becomes a vector with only the first entry nonzero, where $\mathbf{x} \in R^n$ is a nonzero vector.

The Givens rotation $\mathbf{G}(i, k, \theta)$ is a rank-two correction to the identity matrix $\mathbf{I}$. It modifies $\mathbf{I}$ by setting the (i, i)th entry as $\cos\theta$, the (i, k)th entry as $\sin\theta$, the (k, i)th entry as $-\sin\theta$, and the (k, k)th entry as $\cos\theta$. The Givens transform $\mathbf{G}(i, k, \theta)\mathbf{x}$ applies a counterwise rotation of θ radians in the (i, k) coordinate plane. One can specify an entry in a vector to zero by applying the Givens rotation, and then calculate the rotation angle θ.

Matrix Inversion Lemma

The matrix inversion lemma is also called the *Sherman–Morrison–Woodbury formula*. It is useful in deriving many iterative algorithms. Assume that the relationship between the matrix at iterations t and $t + 1$ is given as

$$\mathbf{A}(t + 1) = \mathbf{A}(t) + \Delta\mathbf{A}(t) \qquad (A.41)$$

where $\mathbf{A} \in R^{n \times n}$. If $\Delta\mathbf{A}(t)$ can be expressed as $\mathbf{UV}^{\mathrm{T}}$, where $\mathbf{U} \in R^{n \times m}$ and $\mathbf{V} \in R^{m \times n}$, it is referred to as a *rank-m update*. The matrix inversion lemma gives [424]

$$\mathbf{A}^{-1}(t + 1) = \mathbf{A}^{-1}(t) - \Delta\mathbf{A}^{-1}(t)$$
$$= \mathbf{A}^{-1}(t) - \mathbf{A}^{-1}(t)\mathbf{U}\left(\mathbf{I} + \mathbf{V}^{\mathrm{T}}\mathbf{A}^{-1}(t)\mathbf{U}\right)^{-1}\mathbf{V}^{\mathrm{T}}\mathbf{A}^{-1}(t) \quad (A.42)$$

where both $\mathbf{A}(t)$ and $\left(\mathbf{I} + \mathbf{V}^{\mathrm{T}}\mathbf{A}^{-1}(t)\mathbf{U}\right)$ are assumed to be nonsingular. Thus, a rank-m correction to a matrix results in a rank-m correction to its inverse.

Some modifications to the formula are available, and one popular update is given here. If $\mathbf{A}$ and $\mathbf{B}$ are two positive-definite matrices, which have the relation

$$\mathbf{A} = \mathbf{B}^{-1} + \mathbf{CDC}^{\mathrm{T}} \qquad (A.43)$$

where $\mathbf{C}$ and $\mathbf{D}$ are also matrices. The matrix inversion lemma gives the inverse of $\mathbf{A}$ as

$$\mathbf{A}^{-1} = \mathbf{B} - \mathbf{BC}(\mathbf{D} + \mathbf{C}^{\mathrm{T}}\mathbf{BC})^{-1}\mathbf{C}^{\mathrm{T}}\mathbf{B} \qquad (A.44)$$

A.2 Stability of Dynamic Systems

For a dynamic system described by a set of ordinary differential equations, the stability of the system can be examined by Lyapunov's second theorem or the Lipschitz condition.

Theorem A.1 (Lyapunov's Second Theorem). *For a dynamic system described by a set of differential equations*

$$\frac{\mathrm{d}\mathbf{x}}{\mathrm{d}t} = \mathbf{f}(\mathbf{x}) \tag{A.45}$$

where $\mathbf{x} = (x_1(t), x_2(t), \cdots, x_n(t))^{\mathrm{T}}$ *and* $\mathbf{f} = (f_1, f_2, \cdots, f_n)^{\mathrm{T}}$, *if there exists a positive definite function* $E = E(\mathbf{x})$, *called a* Lyapunov function *or* energy function, *so that*

$$\frac{\mathrm{d}E}{\mathrm{d}t} = \sum_{j=1}^{n} \frac{\partial E}{\partial x_j} \frac{\mathrm{d}x_j}{\mathrm{d}t} \leq 0 \tag{A.46}$$

with $\frac{\mathrm{d}E}{\mathrm{d}t} = 0$ *only for* $\frac{\mathrm{d}\mathbf{x}}{\mathrm{d}t} = \mathbf{0}$, *then the system is stable, and the trajectories* $\mathbf{x}$ *will asymptotically converge to stationary points as* $t \to \infty$.

The stationary points are also known as *equilibrium points* and *attractors*. The crucial step in applying the Lyapunov's second theorem is to find a suitable energy function.

Theorem A.2 (Lipschitz Condition). *For a dynamic system described by (A.45), a sufficient condition that guarantees the existence and uniqueness of the solution is given by the Lipschitz condition*

$$\|\mathbf{f}(\mathbf{x}_1) - \mathbf{f}(\mathbf{x}_2)\| \leq \gamma \|\mathbf{x}_1 - \mathbf{x}_2\| \tag{A.47}$$

where γ *is any positive constant, called the* Lipschitz constant, *and* $\mathbf{x}_1, \mathbf{x}_2$ *are any two variables in the domain of the function vector* $\mathbf{f}$. $\mathbf{f}(\mathbf{x})$ *is said to be* Lipschitz continuous.

If $\mathbf{x}_1$ and $\mathbf{x}_2$ are in some neighborhood of $\mathbf{x}$, then they are said to satisfy the Lipschitz condition locally and will reach a unique solution in the neighborhood of $\mathbf{x}$. The unique solution is a trajectory that will converge to an attractor asymptotically and reach it only at $t \to \infty$.

A.3 Probability Theory and Stochastic Processes

This section is based on [1185].

Gaussian Distribution

The Gaussian distribution, known as the *normal distribution*, is the most common assumption for error distribution. The PDF of the normal distribution is defined as

$$p(x) = \frac{1}{\sigma\sqrt{2\pi}} \mathrm{e}^{-\frac{(x-\mu)^2}{2\sigma^2}} \tag{A.48}$$

for $x \in R$, where μ is the mean and $\sigma > 0$ is the standard deviation. For the Gaussian distribution, 99.73% of the data are within the range of $[\mu - 3\sigma, \mu + 3\sigma]$. The Gaussian distribution has its first-order moment as μ, second-order moment as σ^2, and higher-order moments as zero. If $\mu = 0$ and $\sigma = 1$, the distribution is called the *standard normal distribution*. The PDF is also known as the *likelihood function*. An ML estimator is a set of values (μ, σ) that maximizes the likelihood function for a fixed value of x.

The cumulative distribution function (CDF) is defined as the probability that a random variable is less than or equal to a value x, that is

$$F(x) = \int_{-\infty}^{x} p(t)dt \qquad (A.49)$$

The standard normal CDF, conventionally denoted Φ, is given by setting $\mu = 0$ and $\sigma = 1$. The standard normal CDF is usually expressed by

$$\Phi(x) = \frac{1}{2}\left[1 + \mathrm{erf}\left(\frac{x}{\sqrt{2}}\right)\right] \qquad (A.50)$$

where the error function $\mathrm{erf}(x)$ is a nonelementary function, which is defined by

$$\mathrm{erf}(x) = \frac{2}{\sqrt{\pi}} \int_{0}^{x} e^{-t^2} dt \qquad (A.51)$$

When vector $\mathbf{x} \in R^n$, the PDF of the normal distribution is then defined by

$$p(\mathbf{x}) = \frac{1}{(2\pi)^{\frac{n}{2}}|\mathbf{\Sigma}|}e^{-\frac{1}{2}(\mathbf{x}-\boldsymbol{\mu})^{\mathrm{T}}\mathbf{\Sigma}^{-1}(\mathbf{x}-\boldsymbol{\mu})} \qquad (A.52)$$

where $\boldsymbol{\mu}$ and $\mathbf{\Sigma}$ are the mean vector and the covariance matrix, respectively.

Cauchy Distribution

The Cauchy distribution, also known as the *Cauchy–Lorentzian distribution*, is another popular data-distribution model. The density of the Cauchy distribution is defined as

$$p(x) = \frac{1}{\pi\sigma\left[1 + \left(\frac{x-\mu}{\sigma}\right)^2\right]} \qquad (A.53)$$

for $x \in R$, where μ specifies the location of the peak and σ is the scale parameter that specifies the half-width at the half-maximum. When $\mu = 0$ and $\sigma = 1$, the distribution is called the *standard Cauchy distribution*.

Accordingly, the CDF of the Cauchy distribution is calculated by

$$F(x) = \frac{1}{\pi}\arctan\left(\frac{x-\mu}{\sigma}\right) + \frac{1}{2} \qquad (A.54)$$

None of the moments is defined for the Cauchy distribution. The median of the distribution is equal to μ. The Cauchy distribution has a longer tail than the Gaussian distribution, and this makes it more valuable in stochastic search algorithms by searching larger subspaces in the data space.

Markov Processes, Markov Chains, and Markov-chain Analysis

Given a stochastic process $\{X(t) : t \in \mathcal{T}\}$, where t is time, $X(t)$ is a state in the state space $\mathcal{S}$. A Markov process is defined as a stochastic process that satisfies the relation characterized by the conditional distribution

$$P\left[X\left(t_0 + t_1\right) \le x \middle| X\left(t_0\right) = x_0, X(\tau) = x_\tau, -\infty < \tau < t_0\right]$$
$$= P\left[X\left(t_0 + t_1\right) \le x \middle| X\left(t_0\right) = x_0\right] \quad \text{(A.55)}$$

for any value of t_0 and for $t_1 > 0$. The future distribution of the process is determined by the present value of $X(t_0)$ only.

When $\mathcal{T}$ and $\mathcal{S}$ are discrete, a Markov process is called a *Markov chain*. Conventionally, time is indexed using integers, and a Markov chain is a set of random variables that satisfy

$$P\left[X_n = x_n \middle| X_{n-1} = x_{n-1}, X_{n-2} = x_{n-2}, \cdots\right]$$
$$= P\left[X_n = x_n \middle| X_{n-1} = x_{n-1}\right] \quad \text{(A.56)}$$

This definition can be extended for multistep Markov chains, where a chain state has conditional dependency on only a finite number of its previous states.

For a Markov chain, $P\left[X_n = j \middle| X_{n-1} = i\right]$ is the transition probability of state i to j at time $n - 1$. If

$$P\left[X_n = j \middle| X_{n-1} = i\right] = P\left[X_{n+m} = j \middle| X_{n+m-1} = i\right] \quad \text{(A.57)}$$

for $m \ge 0$ and $i, j \in \mathcal{S}$, the chain is said to be *time homogeneous*. In this case, one can denote

$$P_{i,j} = P\left[X_n = j \middle| X_{n-1} = i\right] \quad \text{(A.58)}$$

and the transition probabilities can be represented by a matrix, called the *transition matrix*, $\mathbf{P} = [P_{i,j}]$, where $i, j = 0, 1, \cdots$. For finite $\mathcal{S}$, $\mathbf{P}$ has a finite dimension.

In the Markov-chain analysis, the transition probability after k step transitions is $\mathbf{P}^k$. The *stationary distribution* or *steady-state distribution* is a vector that satisfies

$$\mathbf{P}^{\mathrm{T}} \boldsymbol{\pi}^* = \boldsymbol{\pi}^* \quad \text{(A.59)}$$

That is, $\boldsymbol{\pi}^*$ is the left eigenvector of $\mathbf{P}$ corresponding to the eigenvalue 1.

If $\mathbf{P}$ is irreducible and aperiodic, that is, every state is accessible from every other state and in the process none of the states repeats itself periodically, then $\mathbf{P}^k$ converges elementwise to a matrix each row of which is the unique stationary distribution $\boldsymbol{\pi}^*$, with

$$\lim_{k \to \infty} \left(\mathbf{P}^k\right)^{\mathrm{T}} \boldsymbol{\pi} = \boldsymbol{\pi}^* \quad \text{(A.60)}$$

Many modeling applications are Markovian, and the Markov-chain analysis is widely used for convergence analysis for algorithms.

A.4 Numerical Optimization Techniques

Although optimization problems can be solved analytically in some cases, numerical optimization techniques are usually more powerful and are also indispensible for all disciplines in science and engineering [362].

Optimization problems discussed in this book are mainly unconstrained continuous optimization problems, COPs, and QP problems. To deal with constraints, the KKT theorem, as a generalization to the Lagrange multiplier method, introduces a slack variable into each inequality constraint before applying the Lagrange multiplier method. The conditions derived from the procedure are known as the *KKT conditions* [362].

A Simple Taxonomy

Optimization techniques can generally be divided into derivative methods and nonderivative methods [842], depending on whether or not derivatives of the objective function are required for the calculation of the optimum.

Derivative methods can be either gradient-search methods or second-order methods. Gradient-search methods include the gradient descent[1], CG methods, and the natural-gradient method. Examples of second-order methods are Newton's method, the Gauss–Newton method, quasi-Newton methods, the trust-region method, the limited-memory quasi-Newton methods, the LM method, and the truncated Newton method. CG methods can also be viewed as a reduced form of the quasi-Newton method, with systematic reinitializations of $\mathbf{H}_t$ to the identity matrix.

Derivative methods can also be classified into *model-based* and *metric-based* methods. Model-based methods improve the current point by a local approximating model. Newton and quasi-Newton methods are model-based methods. Metric-based methods perform a transformation of the variables and then apply a gradient search method to improve the point. The steepest-descent method, quasi-Newton methods, and CG methods belong to this latter category.

Typical nonderivative methods for multivariable functions are random-restart hill-climbing[2], the SA, EAs, random search, tabu search, the PSO, the ACO, and their hybrids. Other nonderivative search methods include univariant search parallel to an axis, sequential simplex method, and acceleration methods in direct search such as the Hooke–Jeeves method, Powell's method and Rosenbrock's method. The Hooke–Jeeves method accelerates in distance,

---

[1] The gradient descent is also known as *steepest descent*. It searches for a local minimum by taking steps along the negative direction of the gradient of the function. If the steps are along the positive direction of the gradient, the method is known as *gradient ascent* or *steepest ascent*. The gradient-descent method is credited to Cauchy.

[2] Hill-climbing attempts to optimize a discrete or continuous function for a local optimum. When operating on continuous space, it is called *gradient ascent*.

the Powell's method accelerates in direction, and the Rosenbrock method accelerates in both direction and distance.

Lagrange Multiplier Method

The Lagrange multiplier method can be used to analytically solve continuous function optimization subject to equality constraints [362]. Let $f(\mathbf{x})$ be the objective function and $h_i(\mathbf{x}) = 0$, $i = 1, \cdots, m$ be the constraints. The Lagrange function can be constructed as

$$L\left(\mathbf{x}; \lambda_1, \cdots, \lambda_m\right) = f(\mathbf{x}) + \sum_{i=1}^{m} \lambda_i h_i(\mathbf{x}) \tag{A.61}$$

where λ_i, $i = 1, \cdots, m$, are called the *Lagrange multipliers*.

The constraned optimization problem is converted into an unconstrained optimization problem: Optimize $L\left(\mathbf{x}; \lambda_1, \cdots, \lambda_m\right)$. By equating $\frac{\partial}{\partial \mathbf{x}} L\left(\mathbf{x}; \lambda_1, \cdots, \lambda_m\right)$ and $\frac{\partial}{\partial \lambda_i} L\left(\mathbf{x}; \lambda_1, \cdots, \lambda_m\right)$, $i = 1, \cdots, m$ to zero and solving the resulting set of equations, we can obtain the $\mathbf{x}$ position at the extremum of $f(\mathbf{x})$ under the constraints.

Line Search

The popular quasi-Newton and CG methods implement a line search at each iteration. The efficiency of the line-search method significantly affects the performance of these methods.

Bracketing and sectioning are two elementary operations for any line search method. A bracket is an interval (α_1, α_2) that contains an optimal value of α. Any three values of α that satisfy $\alpha_1 < \alpha_2 < \alpha_3$ form a bracket when the values of the function $f(\alpha)$ satisfies $f(\alpha_2) \leq \min\left(f(\alpha_1), f(\alpha_3)\right)$. Sectioning is applied to reduce the size of the bracket at a uniform rate. Once a bracket is identified, it can be contracted by using sectioning or interpolation techniques or their combinations. Popular sectioning techniques are the gloden-section search, the Fibonacci search, the secant method, Brent's quadratic approximation, and Powell's quadratic-convergence search without derivatives. The Newton–Raphson search is an analytical line-search technique based on the gradient of the objective function. Wolfe's conditions are two inequality conditions for performing inexact line search. Wolfe conditions enable an efficient selection of the step size without minimizing $f(\alpha)$.

References

1. Aarts E, Korst J (1989) Simulated annealing and Boltzmann machines. John Wiley, Chichester
2. Abe S, Kawakami J, Hirasawa K (1992) Solving inequality constrained combinatorial optimization problems by the Hopfield neural networks. Neural Netw 5:663–670
3. Abe S, Lan MS (1995) Fuzzy rules extraction directly from numerical data for function approximation. IEEE Trans Syst Man Cybern **25**(1):119–129
4. Abe Y, Iiguni Y (2003) Fast computation of RBF coefficients for regularly sampled inputs. Electron Lett **39**(6):543–544
5. Abed-Meraim K, Attallah S, Chkeif A, Hua Y (2000) Orthogonal Oja algorithm. IEEE Signal Process Lett **7**(5):116–119
6. Aberbour M, Mehrez H (1998) Architecture and design methodology of the RBF-DDA neural network. In: Proc IEEE Int Symp Circuits Syst (ISCAS'98), **3**:199–202
7. Abid S, Fnaiech F, Najim M (2001) A fast feedforward training algorithm using a modified form of the standard backpropagation algorithm. IEEE Trans Neural Netw **12**(2):424–430
8. Abu-Mostafa Y, St Jacques J (1985) Information capability of the Hopfield network. IEEE Trans Info Theory **31**(4):461–464
9. Ackley DH, Hinton GE, Sejnowski TJ (1985) A learning algorithm for Boltzmann machines. Cognitive Sci **9**:147–169
10. Ada GL, Nossal GJV (1987) The clonal selection theory. Sci Amer **257**(2):50–57
11. Adeney KM, Korenberg MJ (2000) Iterative fast orthogonal search algorithm for MDL-based training of generalized single-layer networks. Neural Netw **13**:787–799
12. Ahalt SC, Krishnamurty AK, Chen P, Melton DE (1990) Competitive learning algorithms for vector quantization. Neural Netw **3**(3):277–290
13. Aiguo S, Jiren L (1998) Evolving Gaussian RBF network for nonlinear time series modelling and prediction. Electron Lett **34**(12):1241–1243
14. Aihara K, Takabe T, Toyoda M (1990) Chaotic neural networks. Phys Lett A **144**(6,7):333–340
15. Aires F, Schmitt M, Chedin A, Scott N (1999) The "weight smoothing" regularization of MLP for Jacobian stabilization. IEEE Trans Neural Netw **10**(6):1502–1510

16. Aiyer SVB, Niranjan N, Fallside F (1990) A theoretical investigation into the performance of the Hopfield model. IEEE Trans Neural Netw **1**(2):204–215

17. Akaike H (1969) Fitting autoregressive models for prediction. Ann Institute Statist Math **21**:425–439

18. Akaike H (1974) A new look at the statistical model identification. IEEE Trans Automat Contr **19**:716–723

19. Akhmetov DF, Dote Y, Ovaska SJ (2001) Fuzzy neural network with general parameter adaptation for modeling of nonlinear time-series. IEEE Trans Neural Netw **12**(1):148–152

20. Akiyama Y, Yamashita A, Kajiura M, Aiso H (1989) Combinatorial optimization with Gaussian machines. In: Proc IEEE Int Joint Conf Neural Netw, Washington DC, USA, 533–540

21. Albus JS (1975) A new approach to manipulator control: Cerebellar model articulation control (CMAC). Trans on ASME J Dyna Syst Measurement Contr **97**:220–227

22. Alexandridis A, Sarimveis H, Bafas G (2003) A new algorithm for online structure and parameter adaptation of RBF networks. Neural Netw **16**(7):1003–1017

23. Al-kazemi B, Mohan CK (2002) Training feedforward neural networks using multi-phase particle swarm optimization. In: Wan L, Rajapakse JC, Fukusbima K, Lee SY, Yao X (eds) Proc 9th Int Conf on Neural Info Processing (ICONIP'O2), **5**:2615–2619

24. Almeida LB (1987) A learning rule for asynchronous perceptrons with feedback in combinatorial environment. In: Proc IEEE 1st Int Conf Neural Netw, San Diego, 609–618

25. Alpaydin G, Dundar G, Balkir S (2002) Evolution-based design of neural fuzzy networks using self-adapting genetic parameters. IEEE Trans Fuzzy Syst **10**(2):211–221

26. Alspector J, Jayakumar A, Ngo B (1992) An electronic parallel neural CAM for decoding. In: IEEE Workshop Neural Netw for Signal Processing II, Amsterdam, Denmark, 581–587

27. Alvarez A (2002) A neural network with evolutionary neurons. Neural Process Lett **16**:43–52

28. Amari SI (1972) Learning patterns and pattern sequences by self-organizing nets of threshold elements. IEEE Trans Computers **21**:1197–1206

29. Amari SI (1998) Natural gradient works efficiently in learning. Neural Computat **10**:251–276

30. Amari S, Murata N, Muller KR, Finke M, Yang H (1996) Statistical theory of overtraining–Is cross-validation asymptotically effective? In: Touretzky DS, Mozer MC, Hasselmo ME (eds) Advances in neural information processing systems **8**, 176–182. MIT Press, Cambridge, MA

31. Amari SI, Cichocki A, Yang H (1996) A new learning algorithm for blind signal separation. In: Touretzky DS, Mozer MC, Hasselmo ME, (eds) Advances in neural information processing systems **8**, 757–763. MIT Press, Cambridge, MA

32. Ampazis N, Perantonis SJ (2002) Two highly efficient second-order algorithms for training feedforward networks. IEEE Trans Neural Netw **13**(5):1064–1074

33. Anastasiadis AD, Magoulas GD, Vrahatis MN (2005) New globally convergent training scheme based on the resilient propagation algorithm. Neurocomput **64**:253–270

34. Anderson JA (1972) A simple neural network generating interactive memory. Math Biosci **14**:197–220

35. Anderson IA, Bezdek JC, Dave R (1982) Polygonal shape description of plane boundaries. In: Troncale L (ed) Systems science and science, **1**:295–301. SGSR, Louisville, KY

36. Anderson JA, Silverstein JW, Ritz SA, Jones RS (1977) Distinctive features, categorical perception, and probability learning: Some applications of a neural model. Psychological Rev **84**:413–451

37. Andras P (1999) Orthogonal RBF neural network approximation. Neural Process Lett **9**:141–151

38. Andrew L (1996) Implementing the robustness of winner-take-all cellular neural network. IEEE Trans Circuits Syst–II **43**(4):329–334

39. Angelov PP, Filev DP (2004) An approach to online identification of Takagi-Sugeno fuzzy models. IEEE Trans Syst Man Cybern–B **34**(1):484–498

40. Anouar F, Badran F, Thiria S (1998) Probabilistic self-organizing maps and radial basis function networks. Neurocomput **20**:83–96

41. Anthony M, Biggs N (1992) Computational learning theory. Cambridge University Press, Cambridge, UK

42. Arabas J, Michalewicz Z, Mulawka J (1994) GAVaPS–A genetic algorithm with varying population size. In: Proc 1st IEEE Int Conf Evol Computat, Orlando, 73–78

43. Aras N, Oommen BJ, Altinel IK (1999) The Kohonen network incorporating explicit statistics and its application to the travelling salesman problem. Neural Netw **12**:1273–1284

44. Asanovic K, Morgan N (1991) Experimental determination of precision requirements for back-propagation training of artificial neural networks. TR-91–036, Int Computer Sci Institute, Berkeley, CA

45. Astrom KJ, Wittenmark B (1995) Adaptive control. Addison-Wesley, Reading, MA

46. Atlan H, Cohen IR (1989) Theories of immune networks. Spriner-Verlag, Berlin

47. Attallah S, Abed-Meraim K (2001) Fast algorithms for subspace tracking. IEEE Signal Process Lett **8**(7):203–206

48. Auer P, Herbster M, Warmuth MK (1996) Exponentially many local minima for single neurons. In: Touretzky DS, Mozer MC, Hasselmo ME (eds) Advances in neural information processing systems **8**, 316–322. MIT Press, Cambridge, MA

49. Azeem MF, Hanmandlu M, Ahmad N (2000) Generalization of adaptive neuro-fuzzy inference systems. IEEE Trans Neural Netw **11**(6):1332–1346

50. Azencott R, Doutriaux A, Younes L (1993) Synchronous Boltzmann machines and curve identification tasks. Network **4**:461–480

51. Azimi-Sadjadi R, Liou RJ (1992) Fast learning process of multilayer neural networks using recursive least squares method. IEEE Trans Signal Process **40**(2):446–450

52. Back AD, Trappenberg TP (2001) Selecting inputs for modeling using normalized higher order statistics and independent component analysis. IEEE Trans Neural Netw **12**(3):612–617

53. Back T, Schwefel H (1993) An overview of evolutionary algorithms for parameter optimization. Evol Computat **1**(1):1–23

54. Back T, Rudolph G, Schwefel HP (1993) Evolutionary programming and evolutionary strategies: similarities and differences. In: Fogel DB, Atmar W (eds) Proc 2nd Annual Conf Evol Programming, La Jolla, CA, 11–22

55. Badeau R, Richard G, David B (2004) Sliding window adaptive SVD algorithms. IEEE Trans Signal Process **52**(1):1–10

56. Baermann F, Biegler-Koenig F (1992) On a class of efficient learning algorithms for neural networks. Neural Netw **5**(1):139–144

57. Baird B (1990) Associative memory in a simple model of oscillating cortex. In: Touretzky DS (ed) Advances in neural information processing systems **2**, 68–75. Morgan Kaufmann, San Mateo, CA

58. Baldi P, Hornik K (1989) Neural networks for principal component analysis: Learning from examples without local minima. Neural Netw **2**:53–58

59. Baldwin JM (1896) A new factor in evolution. Amer Naturalist **30**:441–451

60. Ball GH, Hall DJ (1967) A clustering technique for summarizing multivariate data. Behav Sci **12**:153–155

61. Bandyopadhyay S, Maulik U, Pakhira MK (2001) Clustering using simulated annealing with probabilistic redistribution. Int J Pattern Recogn Artif Intell **15**(2):269–285

62. Bandyopadhyay S, Maulik U (2002) An evolutionary technique based on k-means algorithm for optimal clustering. Info Sci **146**:221–237

63. Banerjee M, Mitra S, Pal SK (1998) Rough fuzzy MLP: Knowledge encoding and classification. IEEE Trans Neural Netw **9**:1203–1216

64. Bannour S, Azimi-Sadjadi MR (1995) Principal component extraction using recursive least squares learning. IEEE Trans Neural Netw **6**(2):457–469

65. Baraldi A, Parmiggiani F (1997) Novel neural network model combining radial basis function, competitive Hebbian learning rule, and fuzzy simplified adaptive resonance theory. In: Proc SPIE **3165**, Applications of fuzzy logic technology IV, San Diego, CA, 98–112.

66. Baraldi A, Blonda P (1999) A survey of fuzzy clustering algorithms for pattern recognition–Part II. IEEE Trans Syst Man Cybern–B **29**(6):786–801

67. Baraldi A, Alpaydin E (2002) Constructive feedforward ART clustering networks–Part I; Part II. IEEE Trans Neural Netw **13**(3):645–677

68. Barhen J, Protopopescu V, Reister D (1997) TRUST: A deterministic algorithm for global optimization. Science **276**:1094–1097

69. Barnard E (1992). Optimization for training neural nets. IEEE Trans Neural Netw **3**(2):232–240

70. Barron AR (1991) Complexity regularization with applications to artificial neural networks. In: Roussas G (ed) Nonparametric functional estimation and related topics, 561–576. Kluwer, Dordrecht, Netherland

71. Barron AR (1993) Universal approximation bounds for superpositions of a sigmoidal function. IEEE Trans Info Theory **39**(3):930–945

72. Bartlett PL (1993) Lower bounds on the Vapnik-Chervonenkis dimension of multi-layer threshold networks. In: Proc 6th Annual ACM Conf Computat Learning Theory, ACM Press, New York, 144–150

73. Bartlett PL, Maass W (2003) Vapnik-Chervonenkis Dimension of Neural Nets. In: Arbib MA (ed) The handbook of brain theory and neural networks, 1188–1192. 2nd edn, MIT Press, Cambridge

74. Barto AG (1992) Reinforcement learning and adaptive critic methods. In: White DA, Sofge DA (eds) Handbook of intelligent control: neural, fuzzy, and adaptive approaches, 469–471. Van Nostrand Reinhold, New York

75. Barto AG, Sutton RS, Anderson CW (1983) Neuronlike adaptive elements that can solve difficult learning control problems. IEEE Trans Syst Man Cybern **13**:834–846

76. Basak J, De RK, Pal SK (1998) Unsupervised feature selection using neuro-fuzzy approach. Pattern Recogn Lett **19**:997–1006

77. Basak J, Mitra S (1999) Feature selection using radial basis function networks. Neural Comput Applic **8**:297–302

78. Battiti R (1989) Accelerated backpropagation learning: Two optimization methods. Complex Syst **3**:331–342

79. Battiti R (1992) First- and second-order methods for learning: Between steepest descent and Newton methods. Neural Computat **4**(2):141–166

80. Battiti R (1994) Using mutual information for selecting features in supervised neural net learning. IEEE Trans Neural Netw **5**(4):537–550

81. Battiti R, Masulli F (1990) BFGS optimization for faster automated supervised learning. In: Proc Int Neural Network Conf, Paris, France, **2**:757–760. Kluwer, Dordrecht, Netherland

82. Battiti R, Masulli G, Tecchiolli G (1994) Learning with first, second, and no derivatives: A case study in high energy physics. Neurocomput **6**(2):181–206

83. Battiti R, Tecchiolli G (1995) Training neural nets with the reactive Tabu search. IEEE Trans Neural Netw **6**(5):1185–1200

84. Baturone I, Sanchez-Solano S, Barriga A, Huertas JL (1997) Implementation of CMOS fuzzy controllers as mixed-signal integrated circuits. IEEE Trans Fuzzy Syst **5**(1):1–19

85. Baum EB, Haussler D (1989) What size net gives valid generalization? Neural Computat **1**:151–160

86. Baum EB, Wilczek F (1988) Supervised learning of probability distributions by neural networks. In: Anderson DZ (ed) Neural information processing systems, 52–61. American Institute Physics, New York

87. Baxter J (1992) The evolution of learning algorithms for artificial neural networks. In: Green D, Bosso-maier T (eds) Complex systems, 313–326. IOS Press, Amsterdam, Netherlands

88. Baykal N, Erkmen AM (2000) Resilient backpropagation for RBF networks. In: Proc 4th Int Conf Knowledge-Based Intell Engineering Syst & Allied Technologies, Brighton, UK, 624–627

89. Bean J (1994) Genetic algorithms and random keys for sequence and optimization. ORSA J Comput **6**(2):154–160

90. Beasley D, Bull DR, Martin RR (1993) A sequential niche technique for multimodal function optimization. Evol Computat **1**(2):101–125

91. Beigi HSM (1993) Neural network learning through optimally conditioned quadratically convergent methods requiring no line search. In: Proc IEEE 36th Midwest Symp Circuits Syst, Detroit, MI, **1**:109–112

92. Beiu V, Taylor JG (1996) On the circuit complexity of sigmoid feedforward neural networks. Neural Netw **9**(7):1155–1171

93. Beiu V (2003) A survey of perceptron circuit complexity results. In: Proc Int Joint Conf Neural Netw, Portland, Oregon, **2**:989–994

94. Bell AJ, Sejnowski TJ (1995) An information-maximization approach to blind separation and blind deconvolution. Neural Computat **7**:1129–1159

95. Belue LM, Bauer KWJ (1995) Determining input features for multilayered perceptrons. Neurocomput **7**:111–121

96. Benaim M (1994) On functional approximation with normalized Gaussian units. Neural Computat **6**(2):319–333
97. Ben-Hur A, Horn D, Siegelmann H, Vapnik V (2001) Support vector clustering. J Mach Learn Res **2**:125–137
98. Beni G, Liu X (1994) A least biased fuzzy clustering method. IEEE Trans Pattern Anal Mach Intell **16**(9):954–960
99. Benitez JM, Castro JL, Requena I (1997) Are artificial neural networks black boxes? IEEE Trans Neural Netw **8**(5):1156–1164
100. Benvenuto N, Piazza F (1992) On the complex backpropagation algorithm. IEEE Trans Signal Process **40**(4):967–969
101. Berthold MR (1994) A time delay radial basis function network for phoneme recognition. In: Proc IEEE Int Conf Neural Netw, Orlando, **7**:4470–4473
102. Berthold MR, Diamond J (1995) Boosting the performance of RBF networks with dynamic decay adjustment. In: Tesauro G, Touretzky DS, Leen T (eds) Advances in neural information processing systems **7**, 521–528. MIT Press, New York
103. Bezdek J (1974) Cluster validity with fuzzy sets. J Cybern **3**(3):58–71
104. Bezdek J (1981) Pattern recognition with fuzzy objective function algorithms. Plenum Press, New York
105. Bezdek JC, Pal NR (1995) Two soft relatives of learning vector quantization. Neural Netw **8**(5):729–743
106. Bezdek JC, Pal NR (1998) Some new indexes or cluster validity. IEEE Trans Syst Man Cybern **28**(3):301–303
107. Bezdek JC, Tsao EC, Pal NR (1992) Fuzzy Kohonen clustering networks. In: Proc 1st IEEE Int Conf Fuzzy Syst, San Diego, CA, 1035–1043
108. Bhandari D, Pal NR, Pal SK (1994) Directed mutation in genetic algorithms. Info Sci **79**(3/4):251–270
109. Bhaya A, Kaszkurewicz E (2004) Steepest descent with momentum for quadratic functions is a version of the conjugate gradient method. Neural Netw **17**:65–71
110. Bianchini M, Fanelli S, Gori M, Maggini M (1997) Terminal attractor algorithms: A critical analysis. Neurocomput **15**:3–13
111. Bilchev G, Parmee IC (1995) The ant colony metaphor for searching continuous design spaces. In: Fogarty TC (ed) Proc AISB Workshop Evol Computing, Sheffield, UK, LNCS **993**, 25–39. Springer-Verlag, London, UK
112. Billings SA, Zheng GL (1995) Radial basis network configuration using genetic algorithm. Neural Netw **8**(6):877–890
113. Billings SA, Hong X (1998) Dual-orthogonal radial basis function networks for nonlinear time series prediction. Neural Netw **11**:479–493
114. Bilski J, Rutkowski L (1998) A fast training algorithm for neural networks. IEEE Trans Circuits Syst–II **45**(6):749–753
115. Bingham E, Hyvarinen A (2000) ICA of complex valued signals: A fast and robust deflationary algorithm. In: Proc Int Joint Conf Neural Netw (IJCNN), Como, Italy, **3**:357–362
116. Bishop CM (1992) Exact calculation of the Hessian matrix for the multilayer perceptron. Neural Computat **4**(4):494–501
117. Bishop CM (1995) Neural networks for pattern recogonition. Oxford Press, New York
118. Bishop CM (1995) Training with noise is equivalent to Tikhonov regularization. Neural Computat **7**(1):108–116

119. Blanco A, Delgado M, Pegalajar MC (2000) Extracting rules from a (fuzzy/crisp) recurrent neural network using a self-organizing map. Int J Intell Syst **15**(7):595–621

120. Blum AL, Rivest RL (1992) Training a 3-node neural network is NP-complete. Neural Netw **5**(1):117–127

121. Bobrow JE, Murray W (1993) An algorithm for RLS identification of parameters that vary quickly with time. IEEE Trans Automatic Contr **38**(2):351–354

122. Boley D (1998) Principal direction divisive partitioning. Data Mining & Knowledge Discov **2**(4):325–344

123. Bolle D, Shim GM (1995) Nonlinear Hebbian training of the perceptron. Network **6**:619–633

124. Bonabeau E, Dorigo M, Theraulaz G (1999) Swarm intelligence: From natural to artificial systems. Oxford Press, New York

125. Borghese NA, Ferrari S (1998) Hierarchical RBF networks and local parameter estimate. Neurocomput **19**:259–283

126. Bortoletti A, Di Fiore C, Fanelli S, Zellini P (2003) A new class of quasi-Newtonian methods for optimal learning in MLP-networks. IEEE Trans Neural Netw **14**(2):263–273

127. Bors GA, Pitas I (1996) Median radial basis function neural network. IEEE Trans Neural Netw **7**(6):1351–1364

128. Bouras S, Kotronakis M, Suyama K, Tsividis Y (1998) Mixed analog-digital fuzzy logic controller with continuous-amplitude fuzzy inferences and defuzzification. IEEE Trans Fuzzy Syst **6**(2):205–215

129. Bourlard H, Kamp Y (1988) Auto-association by multilayer perceptrons and singular value decomposition. Biol Cybern **59**:291–294

130. Boyd GD (1987) Optically excited synapse for neural networks. Appl Optics **26**:2712–2719

131. Bradley PS, Fayyad UM, Reina CA (1998) Scaling EM (expectation-maximization) clustering to large databases. MSR-TR-98-35, Microsoft Research

132. Bradley PS, Mangasarian OL, Steet WN (1996) Clustering via Concave minimization. In: Touretzky DS, Mozer MC, Hasselmo ME (eds) Advances in neural information processing systems **8**, 368–374. MIT Press, Cambridge, MA

133. Broomhead DS, Lowe D (1988) Multivariable functional interpolation and adaptive networks. Complex Syst **2**:321–355

134. Brouwer RK (1997) Training a feed-forward network by feeding gradients forward rather than by back-propagation of errors. Neurocomput **16**:117–126

135. Brown M, Harris CJ, Parks P (1993) The interpolation capabilities of the binary CMAC. Neural Netw **6**(3):429–440

136. Bruck J (1990) On the convergence properties of the Hopfield model. Proc IEEE **78**(10):1579–1585

137. Bruck J, Roychowdhury WP (1990) On the number of spurious memories in the Hopfield model. IEEE Trans Info Theory **36**(2):393–397

138. Brunelli R (1994) Training neural nets through stochastic minimization. Neural Netw **7**:1405–1412, 1579–1585

139. Bruske J, Sommer G (1995) Dynamic Cell Structure. In: Tesauro G, Touretzky DS, Leen TK (eds) Advances in neural information processing systems **7**, 497–504. MIT Press, Cambridge, MA

140. Bruzzone L, Prieto DF (1998) Supervised training technique for radial basis function neural networks. Electron Lett **34**(11):1115–1116

141. Buckley JJ (1989) Fuzzy complex numbers. Fuzzy Sets Syst **33**:333–345

142. Buckley JJ (1993) Sugeno type controllers are universal controllers. Fuzzy Sets Syst **53**:299–304

143. Buckley JJ, Hayashi Y, Czogala E (1993) On the equivalence of neural nets and fuzzy expert systems. Fuzzy Sets Syst **53**:129–134

144. Buckley JJ, Eslami E (2002) An introduction to fuzzy logic and fuzzy sets. Physica-Verlag, Heidelberg

145. Bugmann G (1998) Normalized Gaussian radial basis function networks. Neurocomput **20**:97–110

146. Buhmann J, Kuhnel H (1993) Vector quantization with complexity costs. IEEE Trans Info Theory **39**(4):1133–1145

147. Burel G (1992) Blind separation of sources: A nonlinear neural algorithm. Neural Netw **5**:937–947

148. Burges CJC (1998) A tutorial on support vector machines for pattern recognition. Data Mining & Knowledge Discov **2**(2):121–167

149. Burke LI (1991) Clustering characterization of adaptive resonance. Neural Netw **4**(4):485–491

150. Burnet FM (1959) The clonal selection theory of acquired immunity. Cambridge University Press, Cambridge, UK

151. Burton RM, Mpitsos GJ (1992) Event dependent control of noise enhances learning in neural networks. Neural Netw **5**:627–637

152. Calvert BD, Marinov CA (2000) Another K-winners-take-all analog neural network. IEEE Trans Neural Netw **11**(4):829–838

153. Camastra F, Verri A (2005) A novel kernel method for clustering. IEEE Trans Pattern Anal Mach Intell **27**(5):801–805

154. Cancelo G, Mayosky M (1998) A parallel analog signal processing unit based on radial basis function networks. IEEE Trans Nuclear Sci **45**(3):792–797

155. Cao Y, Wu J (2002) Projective ART for clustering data sets in high dimensional spaces. Neural Netw **15**:105–120

156. Cardoso JF, Laheld BH (1996) Equivariant adaptive source separation. IEEE Trans Signal Process **44**(12):3017–3030

157. Carpenter GA (1997) Distributed learning, recognition, and prediction by ART and ARTMAP neural networks. Neural Netw **10**:1473–1494

158. Carpenter GA (2003) Default ARTMAP. In: Proc Int Joint Conf Neural Netw, Portland, Oregon, **2**:1396–1401

159. Carpenter GA, Grossberg S (1987) A massively parallel architecture for a self-organizing neural pattern recognition machine. Computer Vision, Graphics, Image Process **37**:54–115

160. Carpenter GA, Grossberg S (1987) ART 2: Self-organization of stable category recognition codes for analog input patterns. Appl Optics **26**:4919–4930

161. Carpenter GA, Grossberg S (1988) The ART of adaptive pattern recognition by a self-organizing neural network. Computer **21**:77–88

162. Carpenter GA, Grossberg S (1990) ART 3: Hierarchical search using chemical transmitters in self-organizing pattern recognition architectures. Neural Netw **3**:129–152

163. Carpenter G, Grossberg S, Rosen DB (1991) Fuzzy ART: Fast stable learning and categorization of analog patterns by an adaptive resonance system. Neural Netw **4**:759–771

164. Carpenter G, Grossberg S, Rosen DB (1991) ART 2-A: An adaptive resonance algorithm for rapid category learning and recognition. In: Proc Int Joint Conf Neural Netw, Seattle, WA, **2**:151–156. Also: Neural Netw **4**:493–504

165. Carpenter GA, Grossberg S, Reynolds JH (1991) ARTMAP: Supervised real-time learning and classification of nonstationary data by a self-organizing neural network. Neural Netw **4**(5):565–588

166. Carpenter GA, Grossberg S, Markuzon N, Reynolds JH, Rosen DB (1992) Fuzzy ARTMAP: A neural network architecture for incremental supervised learning of analog multidimensional maps. IEEE Trans Neural Netw **3**:698–713

167. Carpenter GA, Markuzon N (1998) ARTMAP-IC and medical diagnosis: Instance counting and inconsistent cases. Neural Netw **11**:323–336

168. Carpenter GA, Ross WD (1995) ART-EMAP: A neural network architecture for object recognition by evidence accumulation. IEEE Trans Neural Netw **6**(4):805–818

169. Cartwright HM, Harris SP (1993) The application of the genetic algorithm to two-dimensional strings: The source apportionment problem. In: Proc Int Conf Genetic Algorithms, Urbana-Champaign, IL, 631

170. Castellano G, Fanelli AM, Pelillo M (1997) An iterative pruning algorithm for feedforward neural networks. IEEE Trans Neural Netw **8**(3):519–531

171. Castillo PA, Merelo JJ, Rivas V, Romero G, Prieto A (2000) G-Prop: Global optimization of multilayer perceptrons using GAs. Neurocomput **35**:149–163

172. Castillo PA, Carpio J, Merelo JJ, Prieto A (2000) Evolving multilayer perceptrons. Neural Process Lett **12**:115–127

173. Castro JL (1995) Fuzzy logic controllers are universal approximators. IEEE Trans Syst Man Cybern **25**(4):629–635

174. Castro JL, Delgado M, Mantas CJ (2001) A fuzzy rule-based algorithm to train perceptrons. Fuzzy Sets Syst **118**:359–367

175. Castro JL, Mantas CJ, Benitez J (2002) Interpretation of artificial neural networks by means of fuzzy rules. IEEE Trans Neural Netw **13**(1):101–116

176. Catalan JA, Jin JS, Gedeon T (1999) Reducing the dimensions of texture features for image retrieval using multi-layer neural networks. Pattern Anal Appl **2**:196–203

177. Cechin A, Epperlein U, Koppenhoefer B, Rosenstiel W (1996) The extraction of Sugeno fuzzy rules from neural networks. In: Verleysen M (ed) Proc European Symp Artif Neural Netw, Bruges, Belgium, 49–54

178. Celeux G, Govaert G (1992) A classification EM algorithm for clustering and two stochastic versions. Computat Statist Data Anal **14**:315–332

179. Cetin BC, Burdick JW, Barhen J (1993) Global descent replaces gradient descent to avoid local minima problem in learning with artificial neural networks. In: Proc IEEE Int Conf on Neural Netw, San Francisco, 836–842

180. Cetin BC, Barhen J, Burdick JW (1993) Terminal repeller unconstrained subenergy tunneling (TRUST) for fast global optimization. J Optim Theory Appl **77**:97–126

181. Cevikbas IC, Ogrenci AS, Dundar G, Balkir S (2000) VLSI implementation of GRBF (Gaussian radial basis function) networks. In: Proc IEEE Int Symp Circuits Syst (ISCAS), Geneva, Switzerland, 646–649

182. Cha I, Kassam SA (1995) Channel equalization using adaptive complex radial basis function networks. IEEE J Selected Areas Commun **13**(1):122–131

183. Chakraborty K, Mehrotra KG, Mohan CK, Ranka S (1992) An optimization network for solving a set of simultaneous linear equations. In: Proc Int Joint Conf Neural Netw, Baltimore, MD, **2**:516–521

184. Chakraborty UK, Janikow CZ (2000) An analysis of Gray versus binary encoding in genetic search. Info Sci **156**:253–269

185. Chalmers DJ (1990) The evolution of learning: An experiment in genetic connectionism. In: Touretzky DS, Elman JL, Hinton GE (eds) Proc 1990 Connectionist Models Summer School, 81–90. Morgan Kaufmann, San Mateo, CA

186. Chandra P, Singh Y (2004) An activation function adapting training algorithm for sigmoidal feedforward networks. Neurocomput **61**:429–437

187. Chandrasekaran H, Chen HH, Manry MT (2000) Pruning of basis functions in nonlinear approximators. Neurocomput **34**:29–53

188. Chang CC, Lin CJ (2002) Training ν-support vector regression: Theory and algorithms. Neural Computat **14**:1959–1977

189. Charalambous C (1992) Conjugate gradient algorithm for efficient training of artificial neural networks. IEE Proc–G **139**(3):301–310

190. Chatterjee C, Roychowdhury VP, Ramos J, Zoltowski MD (1997) Self-organizing algorithms for generalized eigen-decomposition. IEEE Trans Neural Netw **8**(6):1518–1530.

191. Chatterjee C, Roychowdhury VP, Chong EKP (1998) On relative convergence properties of principal component analysis algorithms. IEEE Trans Neural Netw **9**(2):319–329

192. Chatterjee C, Kang Z, Roychowdhury VP (2000) Algorithms for accelerated convergence of adaptive PCA. IEEE Trans Neural Netw **11**(2):338–355

193. Chauvin Y (1989) Principal component analysis by gradient descent on a constrained linear Hebbian cell. In: Proc Int Joint Conf Neural Netw, Washington DC, 373–380

194. Chen CL, Chen WC, Chang FY (1993) Hybrid learning algorithm for Gaussian potential function networks. Proc IEE–D **140**(6):442–448

195. Chen CL, Nutter RS (1991) Improving the training speed of three-layer feedforward neural nets by optimal estimation of the initial weights. In: Proc Int Joint Conf Neural Netw, Seattle, WA, **3**:2063–2068

196. Chen DS, Jain RC (1994) A robust backpropagation learning algorithm for function approximation. IEEE Trans Neural Netw **5**(3):467–479

197. Chen H, Flann NS, Watson DW (1998) Parallel genetic simulated annealing: A massively parallel SIMD algorithm. IEEE Trans Parallel Distributed Syst **9**(2):126–136

198. Chen H, Liu RW (1994) An on-line unsupervised learning machine for adaptive feature extraction. IEEE Trans Circuits Syst–II **41**(2):87–98

199. Chen HH, Manry MT, Chandrasekaran H (1999) A neural network training algorithm utilizing multiple sets of linear equations. Neurocomput **25**:55–72

200. Chen JL, Chang JY (2000) Fuzzy perceptron neural networks for classifiers with numerical data and linguistic rules as inputs. IEEE Trans Fuzzy Syst **8**(6):730–745

201. Chen L, Aihara K (1995) Chaotic simulated annealing by a neural-network model with transient chaos. Neural Netw **8**(6):915–930

202. Chen LH, Chang S (1995) An adaptive learning algorithm for principal component analysis. IEEE Trans Neural Netw **6**(5):1255–1263

203. Chen MS, Liou RJ (1999) An efficient learning method of fuzzy inference system. In: Proc IEEE Int Fuzzy Syst Conf, Seoul, Korea, 634–638

204. Chen S (1995) Nonlinear time series modelling and prediction using Gaussian RBF networks with enhanced clustering and RLS learning. Electron Lett **31**(2):117–118

205. Chen S, Billings SA, Luo W (1989) Orthorgonal least squares methods and their applications to non-linear system identification. Int J Contr **50**(5):1873–1896

206. Chen S, Billings SA, Cowan CFN, Grant PM (1990) Practical identification of NARMAX models using radial basis functions. Int J Contr **52**(6):1327–1350

207. Chen S, Chng ES, Alkadhimi K (1996) Regularized orthogonal least squares algorithm for constructing radial basis function networks. Int J Contr **64**(5):829–837

208. Chen S, Cowan C, Grant P (1991) Orthogonal least squares learning algorithm for radial basis function networks. IEEE Trans Neural Netw **2**(2):302–309

209. Chen W, Er MJ, Wu S (2005) PCA and LDA in DCT domain. Pattern Recogn Lett **26**:2474–2482

210. Chen S, Grant PM, Cowan CFN (1992) Orthogonal least squares learning algorithm for training multioutput radial basis function networks. IEE Proc–F **139**(6):378–384

211. Chen S, Grant PM, McLaughlin S, Mulgrew B (1993) Complex-valued radial basis function networks. In: Proc IEE 3rd Int Conf Artif Neural Netw, Brighton, UK, 148–152

212. Chen T (1997) Modified Oja's algorithms for principal subspace and minor subspace extraction. Neural Process Lett **5**:105–110

213. Chen T, Amari SI, Lin Q (1998) A unified algorithm for principal and minor components extraction. Neural Netw **11**:385–390

214. Chen T, Amari SI, Murata N (2001) Sequential extraction of minor components. Neural Process Lett **13**:195–201

215. Chen Y, Hou C (1992) High resolution adaptive bearing estimation using a complex-weighted neural network. In: Proc IEEE Int Conf Acoustics Speech Signal Processing (ICASSP), San Francisco, CA, **2**:317–320

216. Chen S, Sun T (2005) Class-information-incorporated principal component analysis. Neurocomput **69**:216–223

217. Chen S, Wigger J (1995) Fast orthogonal least squares algorithm for efficient subset model selection. IEEE Trans Signal Process **43**(7):1713–1715

218. Chen YX, Wilamowski BM (2002) TREAT: A trust-region-based error-aggregated training algorithm for neural networks. In: Proc Int Joint Conf Neural Netw, **2**:1463–1468

219. Cheng TW, Goldgof DB, Hall LO (1998) Fast fuzzy clustering. Fuzzy Sets Syst **93**:49–56

220. Cheng CB, Lee ES (2001) Fuzzy regression with radial basis function network. Fuzzy Sets Syst **119**:291–301

221. Cherkauer KJ (1992) Genetic search for nearest-neighbor exemplars. In: Proc4th Midwest Artif Intell Cognitive Science Society Conf, Utica, IL, 87–91

222. Chester DL (1990) Why two hidden layers are better than one. In: Proc Int Joint Conf Neural Netw, Washington DC, 265–268

223. Cheung YM (2003) k*-Means: A new generalized k-means clustering algorithm. Pattern Recogn Lett **24**:2883–2893

224. Chiang J, Hao P (2003) A new kernel-based fuzzy clustering approach: Support vector clustering with cell growing. IEEE Trans Fuzzy Syst 11(4):518–527

225. Chinrunrueng C, Sequin CH (1995). Optimal adaptive k-means algorithm with dynamic adjustment of learning rate. IEEE Trans Neural Netw 6(1):157–169

226. Chiu S (1994) Fuzzy model identification based on cluster estimation. J Intell Fuzzy Syst 2(3):267–278

227. Chiu SL (1994) A cluster estimation method with extension to fuzzy model identification. In: Proc IEEE Int Conf Fuzzy Syst, Orlando, FL, 2:1240–1245

228. Chiueh TD, Goodman RM (1991) Recurrent correlation associative memories. IEEE Trans Neural Netw 2(2):275–284

229. Chiueh TD, Tsai HK (1993) Multivalued associative memories based on recurrent networks. IEEE Trans Neural Netw 4(2):364–366

230. Choi DI, Park SH (1994) Self-creating and organizing neural network. IEEE Trans Neural Netw 5(4):561–575

231. Cho KB, Wang BH (1996) Radial basis function based adaptive fuzzy systems and their applications to system identification and prediction. Fuzzy Sets Syst 83:325–339

232. Cho HJ, Oh SY, Choi DH (1998) Population-oriented simulated annealing technique based on local temperature concept. Electron Lett 34(3):312–313

233. Choi J, Sheu BJ, Chang JCF (1994) A Gaussian synapse circuit for analog VLSI neural networks. IEEE Trans VLSI Syst 2(1):129–133

234. Choi JJ, Arabshahi P, Marks II RJ, Caudell TP (1992) Fuzzy parameter adaptation in neural systems. In: Proc Int Joint Conf Neural Netw, Baltimore, MD, 1:232–238

235. Choi S, Cichocki A, Amari S (2002) Equivariant nonstationary source separation. Neural Netw 15:121–130

236. Chow MY, Altrug S, Trussell HJ (1999) Heuristic constraints enforcement for training of and knowledge extraction from a fuzzy/neural architecture—Part I: Foundations. IEEE Trans Fuzzy Syst 7(2):143–150

237. Choy CST, Siu WC (1998) A class of competitive learning models which avoids neuron underutilization problem. IEEE Trans Neural Netw 9(6):1258–1269

238. Choy CST, Siu WC (1998) Fast sequential implementation of "neural-gas" network for vector quantization. IEEE Trans Commun 46(3):301–304

239. Chua LO, Yang L (1988) Cellular neural network–Part I: Theory; Part II: Applications. IEEE Trans Circuits Syst 35:1257–1290

240. Chua LO, Roska T (2002) Cellular neural network and visual computing–Foundation and applications. Cambridge University Press, Cambridge, UK

241. Chuang CC, Su SF, Hsiao CC (2000) The annealing robust backpropagation (ARBP) learning algorithm. IEEE Trans Neural Netw 11(5):1067–1077

242. Chuang CC, Jeng JT, Lin PT (2004) Annealing robust radial basis function networks for function approximation with outliers. Neurocomput 56:123–139

243. Chung FL, Lee T (1994) Fuzzy competitive learning. Neural Netw 7(3):539–551

244. Chung IF, Lin CJ, Lin CT (2000) A GA-based fuzzy adaptive learning control network. Fuzzy Sets Syst 112:65–84

245. Churcher S, Murray AF, Reekie HM (1993) Programmable analogue VLSI for radial basis function networks. Electron Lett 29(18):1603–1605

246. Cibas T, Soulie FF, Gallinari P, Raudys S (1996) Variable selection with neural networks. Neurocomput **12**:223–248.

247. Cichocki A, Unbehauen R (1992) Neural networks for optimization and signal processing. Wiley, New York

248. Cichocki A, Swiniarski RW, Bogner RE (1996) Hierarchical neural network for robust PCA computation of complex valued signals. In: Proc 1996 World Congress Neural Netw, San Diego, 818–821

249. Ciocoiu IB (2002) RBF networks training using a dual extended Kalman filter. Neurocomput **48**:609–622

250. Clerc M, Kennedy J (2002) The particle swarm–Explosion, stability, and convergence in a multidimensional complex space. IEEE Trans Evol Computat **6**(1):58–73

251. Cohen MA, Grossberg S (1983) Absolute stability of global pattern formation and parallel memory storage by competitive neural networks. IEEE Trans Syst Man Cybern **13**:815–826

252. Corchado J, Fyfe C (2000) A comparison of kernel methods for instantiating case based reasoning systems. Comput Info Syst **7**:29–42

253. Corne DW, Knowles JD, Oates MJ (2000) The pareto envelope-based selection algorithm for multiobjective optimisation. In: Schoenauer M, Deb K, Rudolph G, Yao X, Lutton E, Guervos JJM, Schwefel HP (eds) Parallel problem solving from nature (PPSN VI), 839–848. Springer, Berlin

254. Cilingiroglu U, Pamir B, Gunay ZS, Dulger F (1997) Sampled-analog implementation of application-specific fuzzy controllers. IEEE Trans Fuzzy Syst **5**(3):431–442

255. Coello CAC, Pulido GT, Lechuga MS (2004) Handling multiple objectives with particle swarm optimization. IEEE Trans Evol Computat **8**(3):256–279

256. Collins RJ, Jefferson DR (1991) Selection in massively parallel genetic algorithms. In: Belew RK, Booker LB (eds) Proc 4th Int Conf Genetic Algorithms, 249–256. Morgan Kaufmann, San Diego, CA

257. Collins M, Dasgupta S, Schapire RE (2002) A generalization of principal component analysis to the exponential family. In: Dietterich TD, Becker S, Ghahramani Z (eds) Advances in neural information processing systems **14**, 617–624. MIT Press, Cambridge, MA

258. Comon P (1994) Independent component analysis—A new concept? Signal Process **36**(3):287–314

259. Coombes S, Taylor JG (1994) Using generalized principal component analysis to achieve associative memory in a Hopfield net. Network **5**:75–88

260. Coombes S, Campbell C (1996) Efficient learning beyond saturation by single-layered neural networks. Technical report 96.6, Bristol Center for Applied Nonlinear Mathematics, University of Bristol, UK

261. Cooper MG, Vidal JJ (1996) Genetic design of fuzzy controllers. In: Pal SK, Wang PP (eds) Genetic algorithms for pattern recognition, 283–298. CRC Press, Boca Raton, FL

262. Costa A, De Gloria A, Farabosch P, Pagni A, Rizzotto G (1995) Hardware solutions of fuzzy control. Proc IEEE **83**(3):422–434

263. Costa P, Larzabal P (1999) Initialization of supervised training for parametric estimation. Neural Process Lett **9**:53–61

264. Cotter NE, Guillerm TJ (1992) The CMAC and a theorem of Kolmogorov. Neural Netw **5**(2):221–228

265. Cottrell GW, Munro P, Zipser D (1987) Learning internal representations from gray-scale images: An example of extensional programming. In: Proc 9th Annual Conf of Cognitive Sci Society, Seattle, 462–473
266. Cottrell M, Ibbou S, Letremy P (2004) SOM-based algorithms for qualitative variables. Neural Netw **17**:1149–1167
267. Cover TM (1965) Geometrical and statistical properties of systems of linear inequalities with applications in pattern recognition. IEEE Trans Electronics Computers **14**:326–334
268. Crammer K, Singer Y (2001) On the algorithmic implementation of multiclass kernel-based vector machines. J Mach Learning Res **2**:265–292
269. Cruz CS, Dorronsoro JR (1998) A nonlinear discriminant algorithm for feature extraction and data classification. IEEE Trans Neural Netw **9**(6):1370–1376
270. Culhane AD, Peckerar MC, Marrian CRK (1989) A neural net approach to discrete Hartley and Fourier transforms. IEEE Trans Circuits Syst **36**(5):695–702
271. Cvijovic D, Klinowski J (1995) Taboo search: An approach to the multiple minima problem. Science **267**(3):664–666
272. Cybenko G (1988) Continuous valued neural networks with two hidden layers are sufficient. Technical report, Dept of Computer Science, Tufts University, Medford, MA, USA
273. Cybenko G (1989) Approximation by superposition of a sigmoid function. Math of Contr, Signals & Syst **2**:303–314
274. Czech ZJ (2001) Three parallel algorithms for simulated annealing. In: Wyrzykowski R, Dongarra J, Paprzycki M, Waniewski J (eds) Proc 4th Int Conf Parallel Processing & Applied Math, Naczow, Poland, LNCS **2328**, 210–217. Springer-Verlag, London
275. Dai YH, Yuan Y (1999) A nonlinear conjugate gradient method with a strong global convergence property. SIAM J Optimization **10**:177–182
276. Darken C, Moody JE (1992) Towards faster stochastic gradient search. In: Moody JE, Hanson SJ, Lippmann RP (eds) Advances in neural information processing systems **4**, 1009–1016. Morgan Kaufmann, San Mateo, CA, USA
277. Dave RN (1990) Fuzzy-shell clustering and applications to circle detection in digital images. Int J General Syst **16**:343–355
278. Dave RN (1991) Characterization and detection of noise in clustering. Pattern Recogn Lett **12**:657–664
279. Dave RN, Krishnapuram R (1997) Robust clustering methods: A unified view. IEEE Trans Fuzzy Syst **5**(2):270–293
280. Dave RN, Sen S (2002) Robust fuzzy clustering of relational data. IEEE Trans Fuzzy Syst **10**(6):713–727
281. Davidor Y (1991) A naturally occurring niche and species phenomenon: the model and first results. In: Belew RK, Booker LB (eds) Proc 4th Int Conf Genetic Algorithms, 257–262. Morgan Kaufmann, San Mateo, CA
282. Davies DL, Bouldin DW (1979) A cluster separation measure. IEEE Trans Pattern Anal Mach Intell **1**(4):224–227
283. Davis L (1991) Hybridization and numerical representation. In: Davis L (ed) Handbook of genetic algorithms, 61–71. Van Nostrand Reinhold, New York
284. Davis L (1991) Bit-climbing, representational bias, and test suite design. In: Belew R, Booker L (eds) Proc 4th Int Conf Genetic Algorithms, 18–23. Morgan Kaufmann, San Mateo, CA

285. Davis L, Grefenstette JJ (1991) Concerning GENESIS and OOGA. In: Davis L (ed) Handbook of genetic algorithms, 374–377. Van Nostrand Reinhold, New York

286. Dawkins R (1976) The selfish gene. Oxford University Press, Oxford, UK

287. De Castro LN, Von Zuben FJ (2002) Learning and optimization using the clonal selection principle. IEEE Trans Evol Computat **6**(3):239–251

288. De Castro MCF, De Castro FCC, Amaral JN, Franco PRG (1998) A complex valued Hebbian learning algorithm. In: Proc Int Joint Conf Neural Netw (IJCNN), Anchorage, AK, **2**:1235–1238

289. Deb K, Agrawal S, Pratap A, Meyarivan T (2000) A fast elitist non-dominated sorting genetic algorithm for multi-objective optimization: NSGA-II. In: Schoenauer M, Deb K, Rudolph G, Yao X, Lutton E, Guervos JJM, Schwefel HP (eds) Parallel problem solving from nature (PPSN VI), 849–858. Springer, Berlin

290. Deb K, Pratap A, Agarwal S, Meyarivan T (2002) A fast and elitist multiobjective genetic algorithm: NSGA-II. IEEE Trans Evol Computat **6**:182–197

291. De Jong K (1975) An analysis of the behavior of a class of genetic adaptive systems. PhD Thesis, University of Michigan, Ann Arbor

292. De Jong K (1993) Genetic algorithms are NOT function optimizers. In: Whitley LD (ed) Foundations of genetic algorithms **2**, 5–17. Morgan Kaufmann, San Mateo, CA

293. Delbruck T (1991) 'Bump' circuits for computing similarity and dissimilarity of analog voltage. In: Proc IEEE Int Joint Conf Neural Netw, Seattle, WA, **1**:475–479

294. Delgado M, Mantas CJ, Moraga C (1999) A fuzzy rule based backpropagation method for training binary multilayer perceptron. Info Sci **113**:1–17

295. Delport V (1996) Codebook design in vector quantisation using a hybrid system of parallel simulated annealing and evolutionary selection. Electron Lett **32**(13):1158–1160

296. Delport V (1998) Parallel simulated annealing and evolutionary selection for combinatorial optimization. Electron Lett **34**(8):758–759

297. Demir GK, Ozmehmet K (2005) Online local learning algorithms for linear discriminant analysis. Pattern Recogn Lett **26**:421–431

298. Dempsey GL, McVey ES (1993) Circuit implementation of a peak detector neural network. IEEE Trans Circuits Syst–II **40**:585–591

299. Dempster AP, Laird NM, Rubin DB (1977) Maximum likelihood from incomplete data via the EM algorithm. J Roy Stat Soc B **39**(1):1–38.

300. Denker JS, Schwartz D, Wittner B, Solla SA, Howard R, Jackel L, Hopfield J (1987) Large automatic learning, rule extraction, and generalization. Complex Syst **1**:877–922

301. Denoeux T, Lengelle R (1993) Initializing backpropagation networks with prototypes. Neural Netw **6**(3):351–363

302. Denoeux T, Masson MH (2004) Principal component analysis of fuzzy data using autoassociative neural networks. IEEE Trans Fuzzy Syst **12**(3):336–349

303. Desieno D(1988) Adding a conscience to competitive learning. In: Proc IEEE Int Conf Neural Netw, **1**:117 –124

304. Devijver PA, Kittler J (1982) Pattern recognition: A statistics approach. Prentice-Hall, Englewood Cliffs, NJ

305. Diamantaras KI, Kung SY (1994) Cross-correlation neural network models. IEEE Trans Signal Process **42**(11):3218–3323

306. Dick S (2005) Toward complex fuzzy logic. IEEE Trans Fuzzy Syst **13**(3):405–414

307. Dickerson JA, Kosko B (1993) Fuzzy function learning with covariance ellipsoids. In: Proc IEEE Int Conf Neural Netw, San Francisco, **3**:1162–1167

308. Diederich S, Opper M (1987) Learning of correlated patterns in spin-glass networks by local learning rules. Physical Rev Lett **58**:949–952

309. Ding C, He X (2004) Cluster structure of k-means clustering via principal component analysis. In: Dai H, Srikant R, Zhang C (eds) Proc 8th Pacific-Asia Conf on Advances in Knowledge Discov Data Mining (PAKDD 2004), Sydney, Australia, 414–418

310. Dixon LCW (1975) Conjugate gradient algorithms: Quadratic termination properties without linear searches. IMA J Appl Math **15**: 9–18

311. Dorffner G (1994) Unified framework for MLPs and RBFNs: Introducing conic section function networks. Cybern & Syst **25**:511–554

312. Dorigo M, Maniezzo V, Colorni A (1991) Positive feedback as a search strategy. Dipartimento di Elettronica, Politecnico di Milano, Milan, Italy, Tech Rep 91-016

313. Dorigo M, Gambardella LM (1997) Ant colony system: A cooperative learning approach to the traveling salesman problem. IEEE Trans Evol Computat **1**(1):53–66

314. Dorigo M, Di Caro G, Gambardella LM (1999) Ant algorithms for discrete optimization. Artif Life **5**(2):137–172

315. Dorigo M, Stutzle T (2004) Ant colony optimization. MIT Press, Cambridge, MA

316. Douglas SC, Kung S, Amari S (1998) A self-stabilized minor subspace rule. IEEE Signal Process Lett **5**(12):328–330

317. Draghici S (2002) On the capabilities of neural networks using limited precision weights. Neural Netw **15**:395–414

318. Drago G, Ridella S (1992) Statistically controlled activation weight initialization (SCAWI). IEEE Trans Neural Netw **3**(4):627–631

319. Drucker H, Le Cun Y (1992) Improving generalization performance using double backpropagation. IEEE Trans Neural Netw **3**(6):991–997

320. Du KL, Cheng KKM, Swamy MNS (2002) A fast neural beamformer for antenna arrays. In: Proc IEEE Int Conf Commun, New York, **1**:139–144

321. Du KL, Cheng KKM, Lai AKY, Swamy MNS (2002) Neural methods for antenna array sinal processing: A review. Signal Process **82**(4):547–561

322. Du KL, Huang X, Wang M, Zhang B, Hu J (2000) Robot impedance learning of the peg-in-hole dynamic assembly process. Int J Robotics Automation **15**(3):107–118

323. Du KL, Swamy MNS (2004) Simple and practical cyclostationary beamforming algorithms. IEE Proc–VIS **151**(3):175–179

324. Dualibe C, Verleysen M, Jespers PGA (2003) Design of analog fuzzy logic controllers in CMOS technology. Kluwer, Netherlands

325. Duch W (2005) Uncertainty of data, fuzzy membership functions, and multilayer perceptrons. IEEE Trans Neural Netw **16**(1):10–23

326. Duch W, Jankowski N (1999) Survey of neural transfer functions. Neural Comput Surveys **2**:163–212

327. Duda RO, Hart PE (1973) Pattern classification and scene analysis. Wiley, New York

328. Dunn JC (1974) Some recent investigations of a new fuzzy partitioning algorithm and its application to pattern classification problems. J Cybern **4**:1–15

329. Edelman GM (1978) Group selection and phasic reentrant signaling: a theory of higher brain function. In: Edelman GM, Mountcastle VB (eds) The mindful brain: Cortical organization and the group-selective theory of higher brain function, 51–100. MIT Press, Cambridge, MA, USA

330. Eitzinger C, Plach H (2003) A new approach to perceptron training. IEEE Trans Neural Netw **14**(1):216–221

331. El-Sonbaty Y, Ismail M (1998) Fuzzy clustering for symbolic data. IEEE Trans Fuzzy Syst **6**(2):195–204

332. El Zooghby AH, Christodoulou CG, Georgiopoulos M (1998) Neural network-based adaptive beamforming for one- and two-dimensional antenna arrays. IEEE Trans Anten Propagat **46**(12):1891–1893

333. Engel J (1988) Teaching feed-forward neural networks by simulated annealing. Complex Syst **2**(6):641–648

334. Engelbrecht AP (2001) A new pruning heuristic based on variance analysis of sensitivity information. IEEE Trans Neural Netw **12**(6):1386–1399

335. Eom K, Jung K, Sirisena H (2003) Performance improvement of backpropagation algorithm by automatic activation function gain tuning using fuzzy logic. Neurocomput **50**:439–460

336. Er MJ, Wu S (2002) A fast learning algorithm for parsimonious fuzzy neural systems. Fuzzy Sets Syst **126**:337–351

337. Erdem MH, Ozturk Y (1996) A new family of multivalued networks. Neural Netw **9**(6):979–989

338. Ergezinger S, Thomsen E (1995) An accelerated learning algorithm for multilayer perceptrons: Optimization layer by layer. IEEE Trans Neural Netw **6**(1):31–42

339. Eshelman LJ (1991) The CHC adaptive search algorithm: How to have safe search when engaging in nontraditional genetic recombination. In: Rawlins GJE (ed) Foundations of genetic algorithms, 265–283. Morgan Kaufmann, San Mateo, CA

340. Eshelman LJ, Schaffer JD (1993) Real-coded genetic algorithms and interval-schemata. In: Whitley LD (ed) Foundations of genetic algorithms **2**, 187–202. Morgan Kaufmann, San Mateo, CA

341. Esposito A, Marinaro M, Oricchio D, Scarpetta S (2000) Approximation of continuous and discontinuous mappings by a growing neural RBF-based algorithm. Neural Netw **13**:651–665

342. Ester M, Kriegel HP, Sander J, Xu X (1996) A density-based algorithm for discovering clusters in large spatial databases with noise. In: Proc 2nd Int Conf Knowledge Discovery & Data Mining (KDD-96), Portland, Oregon, 226–231

343. Etzioni O (1996) The World-Wide Web: Quagmire or gold mine? Commun of ACM **39**(11):65–68

344. Fabri S, Kadirkamanathan V (1996) Dynamic structure neural networks for stable adaptive control of nonlinear systems. IEEE Trans Neural Netw **7**(5):1151–1167

345. Farhat NH, Psaltis D, Prata A, Paek E (1985) Optical implementation of the Hopfield model. Appl Optics **24**:1469–1475

346. Fahlman SE (1988) Fast learning variations on back-propation: an empirical study. In: Touretzky DS, Hinton GE, Sejnowski T (eds) Proc 1988 Connec-

tionist Models Summer School, Pittsburgh, 38–51. Morgan Kaufmann, San Mateo, CA

347. Fahlman SE, Lebiere C (1990) The cascade-correlation learning architecture. In: Touretzky DS (ed) Advances in neural information processing systems **2**, 524–532. Morgan Kaufmann, San Mateo, CA

348. Farooq O, Datta S (2003) Phoneme recognition using wavelet based features. Info Sci **150**:5–15

349. Feng DZ, Bao Z, Shi WX (1998) Cross-correlation neural network model for the smallest singular component of general matrix. Signal Process **64**:333–346

350. Feng DZ, Bao Z, Jiao LC (1998) Total least mean squares algorithm. IEEE Trans Signal Process **46**(8):2122–2130

351. Ferrari S, Maggioni M, Borghese NA (2004) Multiscale approximation with hierarchical radial basis functions networks. IEEE Trans Neural Netw **15**(1):178–188

352. Ferrari-Trecate G, Rovatti R (2002) Fuzzy systems with overlapping Gaussian concepts: Approximation properties in Sobolev norms. Fuzzy Sets Syst **130**:137–145

353. Figueiredo M, Gomides F, Rocha A, Yager R (1993) Comparison of Yager's level set method for fuzzy logic control with Mamdani and Larsen methods. IEEE Trans Fuzzy Syst **2**:156–159

354. Finnoff W (1994) Diffusion approximations for the constant learning rate backpropagation algorithm and resistance to local minima. Neural Computat **6**(2):285–295

355. Fiori S (2000) An experimental comparison of three PCA neural networks. Neural Process Lett **11**:209–218

356. Fiori S (2000) Blind separation of circularly-distributed sources by neural extended APEX algorithm. Neurocomput **34**:239–252

357. Fiori S (2003) Extended Hebbian learning for blind separation of complex-valued sources sources. IEEE Trans Circuits Syst–II **50**(4):195–202

358. Fiori S, Piazza F (1998) A general class of ψ-APEX PCA neural algorithms. IEEE Trans Circuits Syst–I **47**(9):1394–1397

359. Flake GW, Lawrence S (2002) Efficient SVM regression training with SMO. Mach Learning **46**:271–290

360. Flanagan JA (1996) Self-organization in Kohonen's SOM. Neural Netw **9**(7):1185–1197

361. Fleisher M (1988) The Hopfield model with multi-level neurons. In: Anderson DZ (ed) Neural information processing systems, 278–289. American Institute Physics, New York

362. Fletcher R (1991) Practical methods of optimization. Wiley, New York

363. Fletcher R, Reeves CW (1964) Function minimization by conjugate gradients. Computer J **7**:148–154

364. Fogarty TC (1989) Varying the probability of mutation in the genetic algorithm. In: Proc 3rd Int Conf Genetic Algorithms, Fairfax, VA, 104–109.

365. Fogel DB (1992) An analysis of evolutionary programming. In: Fogel DB, Atmar JW (eds) Proc 1st Annual Conf Evol Programming, 43–51. Evol Programming Society, La Jolla, CA

366. Fogel DB (1995) Evolutionary computation. IEEE Press, New Jersy

367. Fogel L, Owens J, Walsh M (1966) Artificial intelligence through simulated evolution. Wiley, New York

368. Foldiak P (1989) Adaptive network for optimal linear feature extraction. In: Proc Int Joint Conf Neural Netw (IJCNN), Washington DC, **1**:401–405

369. Forrest S (ed) (1991) Emergent computation. MIT Press, Cambridge

370. Fox BR, McMahon MB (1991) Genetic operators for sequencing problems. In: Rawlins GJE (ed) Foundations of genetic algorithms, 284–300. Morgan Kaufmann, San Mateo, CA

371. Frantz DR (1972) Non-linearities in genetic adaptive search. PhD thesis, University of Michigan, Ann Arbor

372. Frascon P, Cori M, Maggini M, Soda G (1996) Representation of finite state automata in recurrent radial basis function networks. Mach Learning **23**:5–32

373. Frean M (1990) The upstart algorithm: A method for constructing and training feedforward neural networks. Neural Computat **2**(2):198–209

374. Frean M (1992) A thermal perceptron learning rule. Neural Computat **4**(6):946–957

375. Friedrichs F, Schmitt M (2005) On the power of Boolean computations in generalized RBF neural networks. Neurocomput **63**:483–498

376. Frigui H, Krishnapuram R (1999) A robust competitive clustering algorithm with applications in computer vision. IEEE Trans Pattern Anal Mach Intell **21**(5):450–465

377. Fritzke B (1994) Growing cell structures—A self-organizing neural networks for unsupervised and supervised learning. Neural Netw **7**(9):1441–1460

378. Fritzke B (1994) Supervised learning with growing cell structures. In: Cowan JD, Tesauro G, Alspector J (eds). Advances in neural information processing systems **6**, 255–262. Morgan Kaufmann, San Mateo, CA

379. Fritzke B (1994) Fast learning with incremental RBF Networks. Neural Process Lett **1**(1):2–5

380. Fritzke B (1995) A growing neural gas network learns topologies. In: Tesauro G, Touretzky DS, Leen TK (eds) Advances in neural information processing systems **7**, 625–632. MIT Press, Cambridge, MA

381. Fritzke B (1995) Growing grid–A self-organizing network with constant neighborhood range and adaptation strength. Neural Process Lett **2**(5):9–13

382. Fritzke B (1997) A self-organizing network that can follow nonstationary distributions. In: Gerstner W, Germond A, Hasler M, Nicoud JD (eds) Proc Int Conf Artif Neural Netw, Lausanne, Switzerland, LNCS **1327**, 613–618. Springer, Berlin

383. Fritzke B (1997) The LBG-U method for vector quantization–An improvement over LBG inspired from neural networks. Neural Process Lett **5**(1):35–45

384. Fu Z, Dowling EM (1995) Conjugate gradient eigenstructure tracking for adaptive spectral estimation. IEEE Trans Signal Process **43**(5):1151–1160

385. Furao S, Hasegawa O (2006) An incremental network for on-line unsupervised classification and topology learning. Neural Netw **19**:90–106

386. Fukuoka Y, Matsuki H, Minamitani H, Ishida A (1998) A modified backpropagation method to avoid false local minima. Neural Netw **11**:1059–1072

387. Fukushima K (1975) Cognition: A self-organizing multulayered neural network. Biol Cybern **20**:121–136

388. Fukushima K (1980) Neocognitron: A self-organizing neural network model for a mechanism of pattern recognition unaffected by shift in position. Biol Cybern **36**:193–202

389. Funahashi K (1989) On the approximate realization of continuous mappings by neural networks. Neural Netw 2(3):183–192

390. Funahashi KI, Nakamura Y (1993) Approximation of dynamical systems by continuous time recurrent neural networks. Neural Netw 6(6):801–806

391. Gabrays B, Bargiela A (2000) General fuzzy min-max neural networks for clustering and classification. IEEE Trans Neural Netw 11(3):769–783

392. Gadea R, Cerda J, Ballester F, Mocholi A (2000) Artificial neural network implementation on a single FPGA of a pinelined on-line backprogation. Proc 13th Int Symp Syst Synthesis, Madrid, Spain, 225–230

393. Galland CC (1993) The limitations of deterministic Boltzmann machine learning. Network 4:355–380

394. Gallant SI (1988) Connectionist expert systems. Commun of ACM 31(2):152–169

395. Gallant SI (1990) Perceptron-based learning algorithms. IEEE Trans Neural Netw 1(2):179–191

396. Gallinari P, Cibas T (1999) Practical complexity control in multilayer perceptrons. Signal Process 74:29–46

397. Gao K, Ahmad MO, Swamy MNS (1990) A neural network least-square estimator. In: Proc Int Joint Conf Neural Netw, Washington DC, 3:805–810

398. Gao K, Ahmad MO, Swamy MNS (1991) Nonlinear signal processing with self-organizing neural networks. In: Proc IEEE Int Symp Circuits Syst, Singapore, 3:1404–1407

399. Gao K, Ahmad MO, Swamy MNS (1992) A modified Hebbian rule for total least-squares estimation with complex valued arguments. In: Proc IEEE Int Symp Circuits Syst, San Diego, 1231–1234

400. Gao K, Ahmad MO, Swamy MNS (1994) A constrained anti-Hebbian learning algorithm for total least-square estimation with applications to adaptive FIR and IIR filtering. IEEE Trans Circuits Syst–II 41(11):718–729

401. Gardner E (1988) The space of the interactions in neural network models. J Physics A 21:257–270

402. Gath I, Geva AB (1989) Unsupervised optimal fuzzy clustering. IEEE Trans Pattern Anal Mach Intell 11(7):773–781.

403. Gath I, Hoory D (1995) Fuzzy clustering of elliptic ring-shaped clusters. Pattern Recogn Lett 16:727–741

404. Geman S, Geman D (1984) Stochastic relaxation, Gibbs distributions, and the Bayesian restoration of images. IEEE Trans Pattern Anal Mach Intell 6:721–741

405. Geman S, Bienenstock E, Doursat R (1992) Neural networks and the bias/variance dilemma. Neural Computat 4(1):1–58

406. Georgiou G, Koutsougeras C (1992) Complex domain backpropagation. IEEE Trans Circuits Syst–II 39(5):330–334

407. Gersho A (1979) Asymptotically optimal block quantization. IEEE Trans Info Theory 25(4):373–380

408. Geva AB (1999) Hierarchical unsupervised fuzzy clustering. IEEE Trans Fuzzy Syst 7(6):723–733

409. Ghahramani Z, Jordan M (1994) Supervised learning from incomplete data via an EM approach. In: Cowan J, Tesauro G, Alspector J (eds) Advances in neural information processing systems 6, 120–127. Morgan Kaufmann, San Mateo, CA

410. Ghodsi A, Schuurmans D (2003) Automatic basis selection techniques for RBF networks. Neural Netw **16**:809–816

411. Giannakopoulos X, Karhunen J, Oja E. An experimental comparison of neural algorithms for independent component analysis and blind separation. Int J Neural Syst **9**(2):99–114 (1999)

412. Girolami M (2002) Mercer kernel based clustering in feature space. IEEE Trans Neural Netw **13**(3):780–784

413. Girosi F, Poggio T (1990) Networks and the best approximation property. Biol Cybern **63**:169–176

414. Glauber RJ (1963) Time-dependent statistics of the Ising model. J Math Phys **4**:294–307

415. Glover F (1989) Tabu search–Part I. ORSA J Comput **1**(3):190–206

416. Glover F (1990) Tabu search–Part II. ORSA J Comput **2**(1):4–32

417. Godara LC (1997) Application of antenna arrays to mobile communications, Part II: Beam-forming and direction-of-arrival considerations. Proc IEEE **85**(8):1195–1245

418. Goh YS, Tan EC (1994) Pruning neural networks during training by backpropagation. In: Proc IEEE Region 10's Ninth Annual Int Conf (TENCON'94), Singapore, 805–808

419. Goldberg DE (1985) Optimal initial population size for binary-coded genetic algorithms. TCGA Report No. 850001. The Clearinghouse for Genetic Algorithms, University of Alabama, Tuscalossa

420. Goldberg DE (1989) Genetic algorithms in search, optimization, and machine learning. Addison-Wesley, Reading, MA, USA

421. Goldberg DE, Deb K (1990) A comparative analysis of selection schemes used in genetic algorithms. In: Rawlins GJE (ed) Foundations of genetic algorithms, 69–93. Morgan Kaufmann, San Mateo, CA

422. Goldberg DE, Deb K, Kargupta H, Harik G (1993) Rapid, accurate optimization of difficult problems using fast messy genetic algorithms. In: Proc 5th Int Conf Genetic Algorithms, Urbana-Champaign, IL, 56–64

423. Goldberg DE, Deb K, Korb B (1989) Messy genetic algorithms: motivation, analysis, and first results. Complex Syst **3**:493–530

424. Golub GH, van Loan CF (1989) Matrix computation. 2nd edn, Johns Hopkins University Press, Baltimore, MD

425. Gomm JB, Yu DL (2000) Selecting radial basis function network centers with recursive orthogonal least squares training. IEEE Trans Neural Netw **11**(2):306–314

426. Gonzalez J, Rojas I, Pomares H, Ortega J, Prieto A (2002) A new clustering technique for function approximation. IEEE Trans Neural Netw **13**(1):132–142

427. Gonzalez J, Rojas I, Ortega J, Pomares H, Fernandez FJ, Diaz AF. Multiobjective evolutionary optimization of the size, shape, and position parameters of radial basis function networks for function approximation. IEEE Trans Neural Netw **14**(6) 1478–1495 (2003)

428. Gori M, Maggini M (1996) Optimal convergence of on-line backpropagation. IEEE Trans Neural Netw **7**(1):251–254

429. Gorinevsky D (1997) An approach to parametric nonlinear least square optimization and application to task-level learning control. IEEE Trans Automat Contr **42**(7):912–927

430. Goryn D, Kaveh M (1989) Conjugate gradient learning algorithms for multi-layer perceptrons. In: Proc IEEE 32nd Midwest Symp Circuits Syst, Champaign, IL, 736–739

431. Grefenstette J, Gopal R, Rosmaita B, Gucht D (1985) Genetic algorithms for the travelling saleman problem. In: Grefenstette J (ed) Proc 2rd Int Conf Genetic Algorithms & Their Appl, 160–168. Lawrence Erlbaum Associates, Mahwah, NJ

432. Greville T (1960) Some applications of pseudo-inverse of matrix. SIAM Rev **2**:15–22

433. Grefenstette JJ (1993) Deception considered harmful. In: Whitley LD (ed) Foundations of genetic algorithms **2**, 75–91. Morgan Kaufmann, San Mateo, CA

434. Grossberg S (1972) Neural expectation: Cerebellar and retinal analogues of cells fired by unlearnable and learnable pattern classes. Kybernetik **10**:49–57

435. Grossberg S (1976) Adaptive pattern classification and universal recording: I. Parallel development and coding of neural feature detectors; II. Feedback, expectation, olfaction, and illusions. Biol Cybern **23**:121–134 & 187–202

436. Grossberg S (1987) Competitive learning: From iterative activation to adaptive resonance. Cognitive Sci **11**:23–63

437. Gruau F, Whitley D (1993) Adding learning to the cellular development of neural networks: evolution and the Baldwin effect. Evol Computat **1**(3):213–233

438. Guedalia ID, London M, Werman M (1999) An on-line agglomerative clustering method for nonstationary data. Neural Computat **11**:521–540

439. Guha S, Rastogi R, Shim K (2001) CURE: An efficient clustering algorithm for large databases. Info Syst **26**(1):35–58

440. Guillaume S (2001) Designing fuzzy inference systems from data: An interpretability-oriented review. IEEE Trans Fuzzy Syst **9**(3):426–443

441. Guo Z, Uhrig RE (1992) Using genetic algorithms to select inputs for neural networks. In: Whitley D, Schaffer JD (eds) Proc IEEE Int Workshop Combinations of Genetic Algorithms & Neural Netw (COGANN-92), Baltimore, MD, 223–234

442. Gupta A, Lam SM (1998) Weight decay backpropagation for noisy data. Neural Netw **11**:1127–1137

443. Gustafson DE and Kessel W (1979) Fuzzy clustering with a fuzzy covariance matrix. In: Proc IEEE Conf Decision Contr, San Diego, CA, 761–766

444. Hagan MT, Menhaj MB (1994) Training feedforward networks with the Marquardt algorithm. IEEE Trans Neural Netw **5**(6):989–993

445. Hamker FH (2001) Life-long learning cell structures—Continuously learning without catastrophic interference. Neural Netw **14**:551–573

446. Hammer B, Gersmann K (2003) A note on the universal approximation capability of support vector machines. Neural Process Lett **17**:43–53

447. Hammer B, Micheli A, Sperduti A, Strickert M (2004) Recursive self-organizing network models. Neural Netw **17**:1061–1085

448. Hancock PJB (1992) Genetic algorithms and permutation problems: A comparison of recombination operators for neural net structure specification. In: Whitley D, Schaffer JD (eds) Proc IEEE Int Workshop Combinations of Genetic Algorithms & Neural Netw (COGANN-92), Baltimore, MD, 108–122

449. Hanna MT (2000) On the stability of a Tank and Hopfield type neural network in the general case of complex eigenvalues. IEEE Trans Signal Process **48**(1):289–293

450. Hanna AI, Mandic DP (2002) A normalised complex backpropagation algorithm. In: Proc IEEE Conf Acoustics, Speech, Signal Processing (ICASSP), Orlando, FL, 977–980

451. Hannan JM, Bishop JM (1996) A class of fast artificial NN training algorithms. Tech report JMH-JMB 01/96, Dept of Cybernetics, University of Reading, UK

452. Hannan JM, Bishop JM (1997) A comparison of fast training algorithms over two real problems. In: Proc IEE Conf Artif Neural Netw, Cambridge, UK, 1–6

453. Hansen N, Ostermeier A (1996) Adapting arbitrary normal mutation distributions in evolution strategies: the covariance matrix adaptation. In: Proc IEEE Int Conf Evol Computat, Nagoya, Japan, 312–317

454. Hansen N, Ostermeier A (2001) Completely derandomized self-adaptation in evolutionary strategies. Evol Computat **9**(2):159–195

455. Hanson SJ, Burr DJ (1988) Minkowski back-propagation: Learning in connectionist models with non-Euclidean error signals. In: Anderson DZ (ed) Neural information processing systems, 348–357. American Institute Physics, New York

456. Haritopoulos M, Yin H, Allinson NM (2002) Image denoising using self-organizing map-based nonlinear independent component analysis. Neural Netw **15**:1085–1098

457. Harpham C, Dawson CW, Brown MR (2004) A review of genetic algorithms applied to training radial basis function networks. Neural Comput Applic **13**(3):193–201

458. Hartman E (1991) A high storage capacity neural network content-addressable memory. Network **2**:315–334

459. Harth E (1976) Visual perceptron: A dynamic theorey. Biol Cybern **22**:169–180

460. Hassibi B, Stork DG, Wolff GJ (1992) Optimal brain surgeon and general network pruning. In: Proc IEEE Int Conf Neural Netw, San Francisco, 293–299

461. Hassoun MH (1995) Fundamentals of artificial neural networks. MIT Press, Cambridge, MA, USA

462. Hassoun MH, Song J (1992) Adaptive Ho-Kashyap rules for perceptron training. IEEE Trans Neural Netw **3**(1):51–61

463. Hassoun MH, Watta PB (1996) The Hamming associative memory and its relation to the exponential capacity DAM. In: Proc IEEE Int Conf Neural Netw, Washington DC, **1**:583–587

464. Hathaway RJ, Bezdek JC (1994) NERF c-means: Non-Euclidean relational fuzzy clustering. Pattern Recogn **27**:429–437

465. Hathaway RJ, Bezdek JC (2000) Generalized fuzzy c-means clustering strategies using L_p norm distances. IEEE Trans Fuzzy Syst **8**(5):576–582

466. Hathaway RJ, Bezdek JC (2001) Fuzzy c-means clustering of incomplete data. IEEE Trans Syst Man Cybern–B **31**(5):735–744

467. Haussler D. Probably approximately correct learning. In: Proc 8th National Conf Artif Intell, Boston, MA, **2**:1101–1108

468. Hawkins DM (1980) Identification of outliers. Chapman & Hall, London, UK
469. Hayashi Y, Buckley JJ, Czogala E (1993) Fuzzy neural network with fuzzy signals and weights. Int J Intell Syst **8**(4):527–537
470. Haykin S (1999) Neural networks: A comprehensive foundation. 2nd edn, Prentice Hall, Upper Saddle River, New Jersey
471. He J, Tan AH, Tan CL (2004) Modified ART 2A growing network capable of generating a fixed number of nodes. IEEE Trans Neural Netw **15**(3):728–737
472. He Y (2002) Chaotic simulated annealing with decaying chaotic noise. IEEE Trans Neural Netw **13**(6):1526–1531
473. Hebb DO (1949) The organization of behavior. Wiley, New York
474. Hecht-Nielsen R (1987) Kolmogorov's mapping neural network existence theorem. In: Proc IEEE 1st Int Conf Neural Netw, San Diego, CA, **3**:11–14
475. Heiss M, Kampl S (1996) Multiplication-free radial basis function network. IEEE Trans Neural Netw **7**(6):1461–1464
476. Herdy M (1991) Application of the evolution strategy to discrete optimization problems. In: Schwefel HP, Manner R (eds) Parallel problem solving from nature, LNCS **496**, 188–192. Springer-Verlag, Berlin
477. Herrera F, Lozano M (2003) Fuzzy adaptive genetic algorithms: design, taxonomy, and future directions. Soft Comput **7**:545–562
478. Herrera F, Lozano M, Verdegay JL (1996) Dynamic and heuristic fuzzy connectives-based crossover operators for controlling the diversity and convergence of real-coded genetic algorithms. Int J Intell Syst **11**:1013–1041
479. Herrmann M, Yang HH (1996) Perspectives and limitations of self-organizing maps in blind separation of source signals. In: Progress in neural information processing: Proc ICONIP'96, Hong Kong, 1211–1216
480. Hesser J, Manner R (1991) Towards an optimal mutation probability for genetic algorithms. In: Schwefel HP, Manner R (eds) Parallel problem solving from nature, LNCS **496**, 23–32. Spinger, Berlin
481. Hestenes MR, Stiefel E (1952) Methods of conjugate gradients for solving linear systems. J of Res of National Bureau of Standards–B **49**:409–436
482. Hikawa H (2003) A digital hardware pulse-mode neuron with piecewise linear activation function. IEEE Trans Neural Netw **14**(5):1028–1037
483. Hinton GE (1987) Connectionist learning procedures. Technical Report CMU-CS-87-115, Carnegie-Mellon University, Computer Sci Dept, Pittsburgh, PA
484. Hinton GE (1989) Connectionist learning procedure. Artif Intell **40**:185–234
485. Hinton GE (1989) Deterministic Boltzmann learning performs steepest descent in weight-space. Neural Computat **1**:143-150
486. Hinton GE, Nowlan SJ (1987) How learning can guide evolution. Complex Syst **1**:495–502
487. Hinton GE, Sejnowski TJ (1986) Learning and relearning in Boltzmann machines. In: Rumelhart DE, McClelland JL (eds) Parallel distributed processing: Explorations in microstructure of cognition, **1**:282–317. MIT Press, Cambridge, MA
488. Hinton GE, van Camp D (1993) Keeping neural networks simple by minimizing the description length of the weights. In: Proc 6th annual ACM Conf Computat Learning Theory, Santa Cruz, CA, 5–13
489. Hirose A (1992) Dynamics of fully complex-valued neural networks. Electron Lett **28**(16):1492–1494
490. Hirose Y, Yamashita K, Hijiya S (1991) Back-propagation algorithm which varies the number of hidden units. Neural Netw **4**:61–66

491. Hirvensalo M (2001) Quantum computing. Springer-Verlag, Berlin
492. Ho DWC, Zhang PA, Xu J (2001) Fuzzy wavelet networks for function learning. IEEE Trans Fuzzy Syst **9**(1):200–211
493. Ho YC, Kashyap RL (1965) An algorithm for linear inequalities and its applications. IEEE Trans Electronic Computers **14**:683–688
494. Hoeppner F (1997) Fuzzy shell clustering algorithms in image processing: fuzzy C-rectangular and 2-rectangular shells. IEEE Trans Fuzzy Syst **5**(4):599–613
495. Holden SB, Rayner PJW (1995) Generalization and PAC learning: some new results for the class of generalized single-layer networks. IEEE Trans Neural Netw **6**(2):368–380
496. Holland J (1975) Adaptation in natural and artificial systems. University of Michigan Press, Ann Arbor, Michigan
497. Holm JEW, Botha EC (1999) Leap-frog is a robust algorithm for training neural networks. Network **10**:1–13
498. Holmblad P, Ostergaard J (1982) Control of a cement kiln by fuzzy logic. In: Gupta MM, Sanchez E (eds) Fuzzy information and decision processes, 389–399. North-Holland, Amsterdam
499. Hong SG, Oh SK, Kim MS, Lee JJ (2001) Nonlinear time series modelling and prediction using Gaussian RBF network with evolutionary structure optimisation. Electron Lett **37**(10):639–640
500. Hong T, Lee C (1996) Induction of fuzzy rules and membership functions from training examples. Fuzzy Sets Syst **84**:33–37
501. Hong X, Billings SA (1997) Givens rotation based fast backward elimination algorithm for RBF neural network pruning. Proc IEE–Contr Theory Appl **144**(5):381–384
502. Hopfield JJ (1982) Neural networks and physical systems with emergent collective computational abilities. Proc Nat Acad Sci **79**:2554–2558
503. Hopfield JJ (1984) Neurons with graded response have collective computational properties like those of two-state neurons. Proc Nat Acad Sci **81**:3088–3092
504. Hopfield JJ, Tank DW (1985) Neural computation of decisions in optimization problems. Biol Cybern **52**:141–152
505. Hopfield JJ, Tank DW (1986) Computing with neural circuits: A model. Science **233**:625–633
506. Horel JD (1984) Complex principal component analysis: Theory and examples. J Climate & Appl Meteorology **23**:1660–1673
507. Horn J, Nafpliotis N, Goldberg DE (1994) A niched pareto genetic algorithm for multiobjective optimization. In: Proc of 1st IEEE Conf Evol Comput, IEEE World Congress Comput Intell, Orlando, FL, **1**:82–87
508. Hornik KM, Stinchcombe M, White H (1989) Multilayer feedforward networks are universal approximators. Neural Netw **2**:359–166
509. Howland P, Park H (2004) Generalizing discriminant analysis using the generalized singular value decomposition. IEEE Trans Pattern Anal Mach Intell **26**(8):995–1006
510. Huang GB (2003) Learning capability and storage capacity of two-hidden-layer feedforward networks. IEEE Trans Neural Netw **14**(2):274–281
511. Huang GB, Saratchandran P, Sundararajan N (2004) An efficient sequential learning algorithm for growing and pruning RBF (GAP-RBF) Networks. IEEE Trans Syst Man Cybern–B **34**(6):2284–2292

512. Huang GB, Saratchandran P, Sundararajan N (2005) A generalized growing and pruning RBF (GGAP-RBF) neural network for function approximation. IEEE Trans Neural Netw 16(1):57–67

513. Huang GB, Zhu QY, Siew CK (2004) Extreme learning machine: A new learning scheme of feedforward neural networks. In: Proc Int Joint Conf Neural Netw, Budapest, Hungary, 2:985–990

514. Huber PJ (1981) Robust statistics. Wiley, New York

515. Hueter GJ (1988) Solution of the traveling salesman problem with an adaptive ring. In: Proc IEEE Int Conf Neural Netw, San Diego, 85–92

516. Hung C, Lin S (1995) Adaptive Hamming net: a fast-learning ART 1 model without searching. Neural Netw 8(4):605–618

517. Hunt SD, Deller JR Jr (1995) Selective training of feedforward artificial neural networks using matrix perturbation theory. Neural Netw 8(6):931–944

518. Hunt KJ, Haas R, Murray-Smith R (1996) Extending the functional equivalence of radial basis function networks and fuzzy inference systems. IEEE Trans Neural Netw 7(3):776–781

519. Huntsberger TL, Ajjimarangsee P (1990) Parallel self-organizing feature maps for unsupervised pattern recognition. Int J General Syst 16:357–372

520. Hurdle JF (1997) The synthesis of compact fuzzy neural circuits. IEEE Trans Fuzzy Syst 5(1):44–55

521. Hush DR, Salas JM (1988) Improving the learning rate of back-propagation with the gradient reuse algorithm. In: Proc IEEE Int Conf Neural Netw, San Diego, CA, 1:441–447

522. Hush DR, Horne B, Salas JM (1992) Error surfaces for multilayer perceptrons. IEEE Trans Syst Man Cybern 22(5):1152–1161

523. Hyvarinen A (1999) Fast and robust fixed-point algorithms for independent component analysis. IEEE Trans Neural Netw 10(3):626–634

524. Hyvarinen A (2001) Blind source separation by nonstationarity of variance: A cumulant-based approach. IEEE Trans Neural Netw 12(6):1471–1474

525. Hyvarinen A, Bingham E (2003) Connection between multilayer perceptrons and regression using independent component analysis. Neurocomput 50:211–222

526. Hyvarinen A, Oja E (1997) A fast fixed-point algorithm for independent component analysis. Neural Computat 9(7):1483–1492

527. Hyvarinen A, Pajunen P (1999) Nonlinear independent component analysis: existence and uniqueness results. Neural Netw 12:429–439

528. Ibnkahla M (2000) Applications of neural networks to digital communications — A survey. Sgnal Process 80:1185–1215

529. Igel C, Husken M (2003) Empirical evaluation of the improved Rprop learning algorithms. Neurocomput 50, 105–123

530. IIguni Y, Sakai H, Tokumaru H (1992) A real-time learning algorithm for a multilayered neural network based on the extended Kalman filter. IEEE Trans Signal Process 40(4):959–967

531. Ikeda N, Watta P, Artiklar M, Hassoun MH (2001) A two-level Hamming network for high performance associative memory. Neural Netw 14:1189–1200

532. Ingber L (1989) Very fast simulated re-annealing. Math & Computer Modelling 12(8):967–973

533. Ishikawa M (1995) Learning of modular structured networks. Artif Intell 75:51–62

534. Ishikawa M (2000) Rule extraction by successive regularization. Neural Netw **13**(10):1171–1183

535. Ishibuchi H, Fujioka R, Tanaka H (1993) Neural networks that learn from fuzzy IF-THEN rules. IEEE Trans Fuzzy Syst **1**:85–97

536. Jacobs RA (1988) Increased rates of convergence through learning rate adaptation. Neural Netw **1**:295–307

537. Jacobsson H (2005) Rule extraction from recurrent neural networks: A taxonomy and review. Neural Computat **17**(6): 1223–1263

538. Jagota A, Mandziuk J (1998) Experimental study of Perceptron-type local learning rule for Hopfield associative memory. Info Sci **111**:65–81

539. Jain A, Murty M, Flynn P (1999) Data clustering: A review. ACM Comput Surveys **31**(3):264–323

540. Jang JS, Lee SY, Shin SY (1988) An optimization network for matrix inversion. In: Anderson DZ (ed) Neural information processing systems, 397–401. American Institute Physics, New York

541. Jang JSR (1993) ANFIS: Adaptive-network-based fuzzy inference systems. IEEE Trans Syst Man Cybern **23**(3):665–685

542. Jang JSR, Mizutani E (1996) Levenberg-Marquardt method for ANFIS learning. In: Proc 1996 Biennial Conf North American Fuzzy Info Processing Society (NAFIPS), Berkeley, CA, 87–91

543. Jang JSR, Sun CI (1993) Functional equivalence between radial basis function Networks and fuzzy inference systems. IEEE Trans Neural Netw **4**(1):156–159

544. Jang JSR, Sun CI (1995) Neuro-fuzzy modeling and control. Proc IEEE **83**(3):378–406

545. Jankovic M, Ogawa H (2004) Time-oriented hierarchical method for computation of principal components using subspace learning algorithm. Int J Neural Syst **14**(5):313–323

546. Jankowski S, Lozowski A, Zurada JM (1996) Complex-valued multi-state neural associative memory. IEEE Trans Neural Netw **7**(6):1491–1496

547. Jankowski N, Kadirkamanathan V (1997) Statistical control of growing and pruning in RBF-like neural networks. In: Proc 3rd Conf Neural Netw & Their Appl, Kule, Poland, 663–670

548. Janssen P, Stoica P, Soderstrom T, Eykhoff P (1988) Model structure selection for multivariable systems by cross-validation. Int J Contr **47**:1737–1758

549. Jerne NK (1974) Towards a network theory of the immune system. Annales d'Immunologie (Paris) **125C**:373–389

550. Jha S, Durrani T (1991) Direction of arrival estimation using artificial neural networks. IEEE Trans Syst Man Cybern **21**(5):1192–1201

551. Jiang X, Chen M, Manry MT, Dawson MS, Fung AK (1994) Analysis and optimization of neural networks for remote sensing. Remote Sensing Rev **9**:97–144

552. Jianping D, Sundararajan N, Saratchandran P (2002) Communication channel equalization using complex-valued minimal radial basis function neural networks. IEEE Trans Neural Netw **13**(3):687–696

553. Jiang M, Yu X (2001) Terminal attractor based back propagation learning for feedforward neural networks. In: Proc IEEE Int Symp Circuits Syst (ISCAS 2001), Sydney, Australia, **2**:711–714

554. Jiao L, Wang L (2000) A novel genetic algorithm based on immunity. IEEE Trans Syst Man Cybern–A **30**(5):552–561

555. Jin Y (2003) Advanced fuzzy systems design and applications. Physica-Verlag, Heidelberg

556. Johansson EM, Dowla FU, Goodman DM (1991) Backpropagation learning for multilayer feedforward neural networks using the conjugate gradient method. Int J Neural Syst 2(4):291–301

557. Juang CF, Lin CT (1998) An on-line self-constructing neural fuzzy inference network and its application. IEEE Trans Fuzzy Syst 6(1):12–32

558. Jutten C, Herault J (1991) Blind Separation of Sources, Part I: An adaptive algorithm based on a neuromimetic architecture. Signal Process 24(1):1–10

559. Kappen HJ, Rodriguez FB (1997) Efficient learning in Boltzmann machines using linear response theory. Neural Computat 10:1137–1156

560. Knjazew D, Goldberg DE (2000) OMEGA - Ordering messy GA: Solving permutation problems with the fast messy genetic algorithm and random keys. In: Whitley LD, Goldberg DE, Cantu-Paz E, Spector L, Parmee IC, Beyer HG (eds) Proc Genetic & Evol Computat Conf, Las Vegas, Nevada, 181–188. Morgan Kaufmann, San Francisco, CA, USA

561. Knowles JD, Corne DW (2000) Approximating the nondominated front using the Pareto archived evolution strategy. Evol Computat 8(2): 149–172

562. Kobayashi K (1991) On the capacity of a neuron with a non-monotone output function. Network 2:237–243

563. Kolen J, Hutcheson T (2002) Reducing the time complexity of the fuzzy C-means algorithm. IEEE Trans Fuzzy Syst 10(2):263–267

564. Kong H, Wang L, Teoh EK, Wang JG, Venkateswarlu R (2004) A Framework of 2D Fisher discriminant analysis: Application to face recognition with small number of training samples. In: Proc IEEE Computer Society Conf Computer Vision & Pattern Recogn (CVPR), San Diego, CA, 2:1083–1088

565. Kadirkamanathan V (1994) A statistical inference based growth criterion for the RBF network. In: Proc IEEE Workshop Neural Netw for Signal Processing, Ermioni, 12–21

566. Kadirkamanathan V, Niranjan M (1993) A function estimation approach to sequential learning with neural network. Neural Computat 5(6):954–975

567. Kaelbling LP, Littman MH, Moore AW (1996) Reinforcement learning: A survey. J Artif Intell Res 4:237–285

568. Kalinli A, Karaboga D (2004) Training recurrent neural networks by using parallel tabu search algorithm based on crossover operation. Engineering Appl of Artif Intell 17:529–542

569. Kam M, Cheng R (1989) Convergence and pattern stabilization in the Boltzmann machine. In: Touretzky DS (ed) Advances in neural information processing systems 1, 511–518. Morgan Kaufmann, San Mateo, CA

570. Kamarthi SV, Pittner S (1999) Accelerating neural network training using weight extrapolations. Neural Netw 12:1285–1299

571. Kambhatla N, Leen TK (1993) Fast non-linear dimension reduction. In: Proc IEEE Int Conf Neural Netw, San Francisco, CA, 3, 1213–1218

572. Kaminski W, Strumillo P (1997) Kernel orthonormalization in radial basis function neural networks. IEEE Trans Neural Netw 8(5):1177–1183

573. Kamp Y, Hasler M (1990) Recursive neural networks for associative memory. Wiley, New York

574. Kang Z, Chatterjee C, Roychowdhury VP (2000) An adaptive quasi-Newton algorithm for eigensubspace estimation. IEEE Trans Signal Process 48(12):3328–3333

575. Kanjilal PP, Banerjee DN (1995) On the application of orthogonal transformation for the design and analysis of feedforward networks. IEEE Trans Neural Netw **6**(5):1061–1070

576. Kanter I, Sompolinsky H (1987) Associative recall of memory without errors. Phys Rev A **35**(1):380–392

577. Kantsila A, Lehtokangas M, Saarinen J (2004) Complex RPROP-algorithm for neural network equalization of GSM data bursts. Neurocomput **61**:339–360

578. Kanungo T, Mount DM, Netanyahu NS, Piatko CD, Silverman R, Wu AY (2002) An efficient k-means clustering algorithm: Analysis and implementation. IEEE Trans Pattern Anal Mach Intell **24**(7):881–892

579. Karayiannis NB (1997) A methodology for constructing fuzzy algorithms for learning vector quantization. IEEE Trans Neural Netw **8**(3):505–518

580. Karayiannis NB (1999) Reformulated radial basis neural networks trained by gradient descent. IEEE Trans Neural Netw **10**(3):657–671

581. Karayiannis NB (1999) An axiomatic approach to soft learning vector quantization and clustering. IEEE Trans Neural Netw **10**(5):1153–1165

582. Karayiannis NB, Bezdek JC (1997) An integrated approach to fuzzy learning vector quantization and fuzzy c-means clustering. IEEE Trans Fuzzy Syst **5**(4):622–628

583. Karayiannis NB, Bezdek JC, Pal NR, Hathaway RJ, Pai PI (1996) Repair to GLVQ: A new family of competitive learning schemes. IEEE Trans Neural Netw **7**(5):1062–1071

584. Karayiannis NB, Mi GW (1997) Growing radial basis neural networks: Merging supervised and unsupervised learning with network growth techniques. IEEE Trans Neural Netw **8**(6):1492–1506

585. Karayiannis NB, Pai PI (1996) Fuzzy algorithms for learning vector quantization. IEEE Trans Neural Netw **7**:1196–1211

586. Karayiannis NB, Randolph-Gips MM (2003) Soft learning vector quantization and clustering algorithms based on non-Euclidean norms: multinorm algorithms. IEEE Trans Neural Netw **14**(1):89–102

587. Karhunen J, Joutsensalo J (1995) Generalizations of principal component analysis, optimization problems, and neural networks. Neural Netw **8**(4):549–562

588. Karhunen J, Malaroiu S (1999) Locally linear independent component analysis. In: Proc Int Joint Conf Neural Netw, Washington DC, 882–887

589. Karhunen J, Oja E, Wang L, Vigario R, Joutsensalo J (1997) A class of neural networks for independent component analysis. IEEE Trans Neural Netw **8**(3):486–504

590. Karhunen J, Pajunen P (1997) Blind source separation and tracking using nonlinear PCA criterion: A least-squares approach. In: Proc IEEE Int Conf Neural Netw (ICNN'97), Houston, Texas, **4**:2147–2152

591. Karhunen J, Pajunen P, Oja E (1998) The nonlinear PCA criterion in blind source separation: Relations with other approaches. Neurocomput **22**(1):5–20

592. Karnin ED (1990) A simple procedure for pruning back-propagation trained neural networks. IEEE Trans Neural Netw **1**(2):239–242

593. Karypis G, Han EH, Kumar V (1999) Chameleon: Hierarchical clustering using dynamic modeling cover feature. Computer **12**:68–75

594. Kasuba T (1993) Simplified fuzzy ARTMAP. AI Expert **8**(11):18–25

595. Kaymak U, Setnes M (2002) Fuzzy clustering with volume prototypes and adaptive cluster merging. IEEE Trans Fuzzy Syst 10(6):705–712
596. Kechriotis G, Manolakos E (1994) Training fully recurrent neural networks with complex weights. IEEE Trans Circuits Syst–II 41(3):235–238
597. Keller JM, Hunt DJ (1985) Incorporating fuzzy membership functions into the perceptron algorithm. IEEE Trans Pattern Anal Mach Intell 7(6):693–699
598. Kennedy J, Eberhart R (1995) Particle swarm optimization. In: Proc IEEE Int Conf Neural Netw, Perth, WA, 4:1942–1948
599. Kennedy J, Eberhart R (2001) Swarm intelligence. Morgan Kaufmann, San Francisco, CA
600. Ker-Chau L (1985) From Stein's unbiased risk estimates to the method of generalized cross validation. Ann Statistics 13(4):1352–1377
601. Kersten PR (1999). Fuzzy order statistics and their application to fuzzy clustering. IEEE Trans Fuzzy Syst 7(6):708–712
602. Kim DW, Lee KY, Lee D, Lee KH (2005) A kernel-based subtractive clustering method. Pattern Recogn Lett 26:879–891
603. Kim HB, Jung SH, Kim TG, Park KH (1996) Fast learning method for backpropagation neural network by evolutionary adaptation of learning rates. Neurocomput 11(1):101–106
604. Kim J, Kasabov N (1999) HyFIS: Adaptive neuro-fuzzy inference systems and their application to nonlinear dynamical systems. Neural Netw 12:1301–1319
605. Kim T, Adali T (2002) Fully complex multi-layer perceptron network for nonlinear signal processing. J VLSI Signal Process 32(1):29–43
606. Kim T, Adali T (2002) Universal approximation of fully complex feedforward neural networks. In: Proc IEEE ICCASP'02, Orlando, FL, 1:973–976
607. Kirkpatrick S, Gelatt Jr CD, Vecchi MP (1983) Optimization by simulated annealing. Science 220:671–680
608. Kitano H (1990) Designing neural networks using genetic algorithms with graph generation system. Complex Syst 4(4):461–476
609. Kohonen T (1972) Correlation matrix memories. IEEE Trans Computers 21:353–359
610. Kohonen T (1982) Self-organized formation of topologically correct feature maps. Biol Cybern 43:59–69
611. Kohonen T (1989) Self-organization and associative memory. Springer, Berlin
612. Kohonen T (1990) The self-organizing map. Proc IEEE 78:1464–1480
613. Kohonen T (1996) Emergence of invariant-feature detectors in the adaptive-subspace self-organizing map. Biol Cybern 75:281–291
614. Kohonen T (1997) Self-organizing maps. Springer, Berlin
615. Kohonen T, Kangas J, Laaksonen J, Torkkola K (1992) LVQPAK: A program package for the correct application of learning vector quantization algorithms. In: Proc Int Joint Conf Neural Netw, Baltimore, MD, 1:725–730
616. Kohonen T, Oja E, Simula O, Visa A, Kangas J (1996) Engineering applications of the self-organizing map. Proc IEEE 84(10):1358–1384
617. Koiran P, Sontag ED (1996) Neural networks with quadratic VC dimension. In: Touretzky DS, Mozer MC, Hasselmo ME (eds) Advances in neural information processing systems 8,197–203. MIT Press, Cambridge, MA, USA
618. Koldovsky Z, Tichavsky P, Oja E (2005) Cramer-Rao lower bound for linear independent component analysis. In: Proc ICASSP'05, Philadelphia, 3:581–584

619. Kolen JF, Pollack JB (1990) Backpropagation is sensitive to initial conditions. Complex Syst **4**(3):269–280

620. Kolmogorov AN (1957) On the representation of continuous functions of several variables by superposition of continuous functions of one variable and addition. Doklady Akademii Nauk USSR **114**(5):953–956

621. Kosko B (1987) Adaptive bidirectional associative memories. Appl Optics **26**:4947–4960

622. Kosko B (1992) Fuzzy system as universal approximators. In: Proc IEEE Int Conf Fuzzy Syst, San Diego, CA, 1153–1162

623. Kosko B (ed) (1992) Neural networks for signal processing. Prentice Hall, Englewood Cliffs, NJ

624. Kosko B (1997) Fuzzy engineering. Prentice Hall, Englewood Cliffs, NJ

625. Koza JR (1993) Genetic programming. MIT Press, Cambridge, MA

626. Kozma R, Sakuma M, Yokoyama Y, Kitamura M (1996) On the accuracy of mapping by neural networks trained by backpropagation with forgetting. Neurocomput **13**:295–311

627. Kraaijveld MA, Duin RPW (1991) Generalization capabilities of minimal kernel-based networks. In: Proc Int Joint Conf Neural Netw, Seattle, WA, **1**:843–848

628. Kramer MA (1991) Nonlinear principal component analysis using autoassociative neural networks. AIChE J **37**(2):233–243

629. Kreinovich V, Nguyen HT, Yam Y (2000) Fuzzy systems are universal approximators for a smooth function and its derivatives. Int J Intell Syst **15**:565–574

630. Krishna K, Murty MN (1999) Genetic k-means algorithm. IEEE Trans Syst Man and Cybern–B **29**(3):433–439

631. Krishnamurthy AK, Ahalt SC, Melton DE, Chen P (1990) Neural networks for vector quantization of speech and images. IEEE J Select Areas Commun **8**(8):1449–1457

632. Krishnapuram R, Nasraoui O, Frigui H (1992) The fuzzy c spherical shells algorithm: A new approach. IEEE Trans Neural Netw **3**(5):663–671

633. Krishnapuram R, Kim J (2000) Clustering algorithms based on volume criteria. IEEE Trans Fuzzy Syst **8**(2):228–236

634. Krishnapuram R, Keller JM (1993) A possibilistic approach to clustering. IEEE Trans Fuzzy Syst **1**(2):98–110

635. Krogh A, Hertz JA (1990) Hebbian learning of principal components. In: Eckmiller R, Hartmann G, Hauske G (eds) Parallel processing in neural systems and computers, 183–186. North-Holland, Amsterdam

636. Kruschke JK, Movellan JR (1991) Benefits of gain: Speeded learning and minimal layers in back-propagation networks. IEEE Trans Syst Man Cybern **21**(1):273–280

637. Krzyzak A, Linder T (1998) Radial basis function networks and complexity regularization in function learning. IEEE Trans Neural Netw **9**(2):247–256

638. Ku KWC, Mak MW, Siu WC (2003) Approaches to combining local and evolutionary search for training neural networks: A review and some new results. In: Ghosh A, Tsutsui S (eds) Advances in evolutionary computing: Theory and applications, 615–641. Springer-Verlag, Berlin

639. Kuncheva LI (1997) Initializing of an RBF network by a genetic algorithm. Neurocomput **14**:273–288

640. Kung SY (1990) Constrained principal component analysis via an orthogonal learning network. In: Proc IEEE Int Symp Circuits Syst, New Orleans, LA, 1:719–722

641. Kung SY, Diamantaras KI (1990) A neural network learning algorithm for adaptive principal components extraction (APEX). In: Proc IEEE ICCASP, Albuquerque, NM, 861–864

642. Kung SY, Diamantaras KI, Taur JS (1994) Adaptive principal components extraction (APEX) and applications. IEEE Trans Signal Process 42(5):1202–1217

643. Kuo YH, Chen CL (1998) Generic LR fuzzy cells for fuzzy hardware synthesis. IEEE Trans Fuzzy syst 6(2):266–285

644. Kurita N, Funahashi KI (1996) On the Hopfield neural networks and mean field theory. Neural Netw 9:1531–1540

645. Kwan HK (1992) Simple sigmoid-like activation function suitable for digital hardware implementation. Electron Lett 28:1379–1380

646. Kwok T, Smith KA (1999) A unified framework for chaotic neural-network approaches to combinatorial optimization. IEEE Trans Neural Netw 10(4):978–981

647. Kwok TY, Yeung DY (1997) Objective functions for training new hidden units in constructive neural networks. IEEE Trans Neural Netw 8(5):1131–1148

648. Kwok TY, Yeung DY (1997) Constructive algorithms for structure learning in feedforward neural networks for regression problems. IEEE Trans Neural Netw 8(3):630–645

649. Langari R, Wang L, Yen J (1997) Radial basis function networks, regression weights, and the expectation-maximization algorithm. IEEE Trans Syst Man Cybern–A 27(5):613–623

650. la Tendresse I, Gottlieb J, Kao O (2001) The effects of partial restarts in evolutionary search. In: Proc 5th Int Conf Artif Evolution, Le Creusot, France, LNCS 2310, 117–127. Springer

651. Lazaro M, Santamaria I, Pantaleon C (2003) A new EM-based training algorithm for RBF networks. Neural Netw 16:69–77

652. Lazaro M, Santamaria I, Perez-Cruz F, Artes-Rodriguez A (2005) Support vector regression for the simultaneous learning of a multivariate function and its derivatives. Neurocomput 69:42–61

653. Lazzaro J, Lyckebusch S, Mahowald MA, Mead CA (1989) Winner-take-all networks of O(n) complexity. In: Touretzky DS (ed) Advances in neural information processing systems 1, 703–711. Morgan Kaufmann, San Mateo, CA

654. Le Cun Y, Boser B, Denker JS, Henderson D, Howard RE, Hubbard W, Jackel LD (1989) Back-propagation applied to handwritten zipcode recognition. Neural Computat 1(4):541–551

655. Le Cun Y, Boser B, Denker JS, Henderson D, Howard RE, Hubbard W, Jackel LD (1990) Handwritten digit recognition with a back-propagation network. In: Touretzky DS (ed) Advances in neural information processing systems 2, 396–404. Morgan Kaufmann, San Mateo, CA

656. Le Cun Y, Denker JS, Solla SA (1990) Optimal brain damage. In: Touretzky DS (ed) Advances in neural information processing systems 2, 598–605. Morgan Kaufmann, San Mateo, CA

657. Le Cun Y, Kanter I, Solla SA (1991) Second order properties of error surfaces: learning time and generalization. In: Lippmann RP, Moody JE, Touretzky DS

(eds) Advances in neural information Processing systems **3**, 918–924. Morgan Kaufmann, San Mateo, CA

658. LeCun Y, Simard PY, Pearlmutter B (1993) Automatic learning rate maximization by on-line estimation of the Hessian's eigenvectors. In: Hanson SJ, Cowan JD, Giles CL (eds) Advances in neural information processing systems **5**, 156–163. Morgan Kaufmann, San Mateo, CA

659. Lee BW, Shen BJ (1992) Design and analysis of analog VLSI neural networks. In: Kosko B (ed) Neural networks for signal processing, 229–284. Prentice-Hall, Englewood Cliffs, NJ

660. Lee BW, Shen BJ (1993) Parallel hardware annealing for optimal solutions on electronic neural networks. IEEE Trans Neural Netw **4**(4):588–599

661. Lee CC, Chung PC, Tsai JR, Chang CI (1999) Robust radial basis function neural networks. IEEE Trans Syst Man Cybern–B **29**(6):674–685

662. Lee CW, Shin YC (2003) Construction of fuzzy systems using least-squares method and genetic algorithm. Fuzzy Sets Syst **137**:297–323

663. Lee CY (2003) Entropy-Boltzmann selection in the genetic algorithms. IEEE Trans Syst Man Cybern–B **33**(1):138–142

664. Lee DL (2001) Improving the capacity of complex-valued neural networks with a modified gradient descent learning rule. IEEE Trans Neural Netw **12**(2):439–443

665. Lee DY, Kim BM, Cho HS (1999) A self-organized RBF network combined with ART II. In: Proc IEEE Int Joint Conf Neural Netw, Washington DC, **3**:1963–1968

666. Lee HM, Chen CM, Huang TC (2001) Learning efficiency improvement of back-propagation algorithm by error saturation prevention method. Neurocomput **41**:125–143

667. Lee J (2003) Attractor-based trust-region algorithm for efficient training of multilayer perceptrons. Electron Lett **39**(9):727–728

668. Lee J, Chiang HD (2002) Theory of stability regions for a class of nonhyperbolic dynamical systems and its application to constraint satisfaction problems. IEEE Trans Circuits Syst–I **49**(2):196–209

669. Lee KY, Jung S (1999) Extended complex RBF and its application to M-QAM in presence of co-channel interference. Electron Lett **35**(1):17–19

670. Lee MA, Takagi H (1993) Dynamic control of genetic algorithms using fuzzy logic techniques. In: Proc 5th Int Conf Genetic Algorithms (ICGA'93), Urbana-Champaign, IL, 76–83

671. Lee SJ, Hou CL (2002) An ART-based construction of RBF networks. IEEE Trans Neural Netw **13**(6):1308–1321

672. Lee SW (1996) Off-line recognition of totally unconstrained handwritten numerals using multilayer cluster neural network. IEEE Trans Pattern Anal Mach Intell **18**(6):648–652

673. Lee Y, Oh SH, Kim MW (1991) The effect of initial weights on premature saturation in back-propagation training. In: Proc IEEE Int Joint Conf Neural Netw, Seattle, WA, **1**:765–770

674. Le Gall A, Zissimopoulos V (1999) Extended Hopfield models for combinatorial optimization. IEEE Trans Neural Netw **10**(1):72–80

675. Lehtokangas M (1999) Modelling with constructive backpropagation. Neural Netw **12**:707–716

676. Lehtokangas M, Korpisaari P, Kaski K (1996) Maximum covariance method for weight initialization of multilayer perceptron networks. In: Proc European symp Artif Neural Netw (ESANN'96), Bruges, Belgium, 243–248

677. Lehtokangas M, Saarinen J (1998) Centroid based multilayer perceptron networks. Neural Process Lett 7:101–106

678. Lehtokangas M, Saarinen J, Huuhtanen P, Kaski K (1995) Initializing weights of a multilayer perceptron network by using the orthogonal least squares algorithm. Neural Computat 7:982–999

679. Lehtokangas M, Saarinen J, Kaski K (1995) Accelerating training of radial basis function networks with cascade-correlation algorithm. Neurocomput 9:207–213

680. Lehtokangas M, Salmela P, Saarinen J, Kaski K (1998) Weights initialization techniques. In: Leondes CT (ed) Algorithms and architecture, 87–121. Academic Press, London, UK

681. Lemaitre L, Patyra M, Mlynek D (1994) Analysis and design of CMOS fuzzy logic controller in current mode. IEEE J Solid-State Circuits 29(3):317–322

682. Lendaris GG, Mathia K, Saeks R (1999) Linear Hopfield networks and constrained optimization. IEEE Trans Syst Man Cybern-B 29(1):114–118

683. Leng G, Prasad G, McGinnity TM (2004) An on-line algorithm for creating self-organizing fuzzy neural networks. Neural Netw 17:1477–1493

684. Leonardis A, Bischof H (1998) An efficient MDL-based construction of RBF networks. Neural Netw 11:963–973

685. Leski J (2003) Towards a robust fuzzy clustering. Fuzzy Sets Syst 137:215–233

686. Leski JM (2003) Generalized weighted conditional fuzzy clustering. IEEE Trans Fuzzy Syst 11(6):709–715

687. Leung H, Haykin S (1991) The complex backpropagation algorithm. IEEE Trans Signal Process 3(9):2101–2104

688. Leung ACS, Wong KW, Tsoi AC (1997) Recursive algorithms for principal component extraction. Network 8:323–334

689. Leung CS, Wong KW, Sum PF, Chan LW (2001) A pruning method for the recursive least squared algorithm. Neural Netw 14:147–174

690. Leung FHF, Lam HK, Ling SH, Tam PKS (2003) Tuning of the structure and parameters of a neural network using an improved genetic algorithm. IEEE Trans Neural Netw 14(1):79–88

691. Levin AU, Leen TK, Moody JE (1994) Fast pruning using principal components. In: Cowan JD, Tesauro G, Alspector J (eds) Advances in neural information processing systems 6, 35-42, Morgan Kaufman, San Francisco, CA

692. Levy BC, Adams MB (1987) Global optimization with stochastic neural networks. In: Proc 1st IEEE Conf Neural Netw, San Diego, CA, 3:681–689

693. Li HX, Chen CLP (2000) The equivalence between fuzzy logic systems and feedforward neural networks. IEEE Trans Neural Netw 11(2):356–365

694. Li JH, Michel AN, Parod W (1989) Analysis and synthesis of a class of neural networks: Linear systems operating on a closed hypercube. IEEE Trans Circuits Syst 36(11):1405–1422

695. Li LK (1992) Approximation theory and recurrent networks. In: Proc Int Joint Conf Neural Netw, Baltimore, MD, 266–271

696. Li M, Yuan B (2005) 2D-LDA:A statistical linear discriminant analysis for image matrix. Pattern Recogn Lett 26:527–532

697. Li W, Swetits JJ (1998) The linear L_1 estimator and the Huber M-estimator. SIAM J Optim **8**(2):457–475
698. Li X, Yu W (2002) Dynamic system identification via recurrent multilayer perceptrons. Info Sci **147**:45–63
699. Li XD, Ho JKL, Chow TWS (2005) Approximation of dynamical time-variant systems by continuous-time recurrent neural networks. IEEE Trans Circuits Syst **52**(10):656–660
700. Li Y, Deng JM (1998) WAV—A weight adaptation algorithm for normalized radial basis function networks. In: Proc IEEE Int Conf Electronics Circuits Syst, Lisboa, Portugal, **2**:117–120
701. Liang Y (1996) Combinatorial optimization by Hopfield networks using adjusting neurons. Info Sci **94**:261–276
702. Liang YC, Feng DP, Lee HP, Lim SP, Lee KH (2002) Successive approximation training algorithm for feedforward neural networks. Neurocomput **42**:311–322
703. Liano K (1996) Robust error measure for supervised neural network learning with outliers. IEEE Trans Neural Netw **7**(1):246–250
704. Liao Y, Fang SC, Nuttle HLW (2003) Relaxed conditions for radial-basis function networks to be universal approximators. Neural Netw **16**:1019–1028
705. Lim YC, Liu B, Evans JB (1990) VLSI circuits for decomposing binary integers into power-of-two terms. In: Proc IEEE Int Symp Circuits Syst, New Orleans, 2304–2307
706. Lin CT, Lee CSG (1995) A multi-valued Boltzmann machine. IEEE Trans Syst Man Cybern **25**(4):660–669
707. Lin CT, Lu YC (1996) A neural fuzzy system with fuzzy supervised learning. IEEE Trans Syst Man Cybern-B **26**(5):744–763
708. Lin FJ, Wai RJ (2001) Hybrid control using recurrent fuzzy neural network for linear-induction motor servo drive. IEEE Trans Fuzzy Syst **9**(1):102–115
709. Lin JK, Grier DG, Cowan JD (1997) Source separation and density estimation by faithful equivariant SOM. In: Mozer MC, Jordan MI, Petsche T (eds) Advances in neural information processing systems **9**, 536–542. MIT Press, Cambridge, MA
710. Lin JS (1999) Fuzzy clustering using a compensated fuzzy Hopfield network. Neural Process Lett **10**:35–48
711. Lin JS, Cheng KS, Mao CW (1996) A fuzzy Hopfield neural network for medical image segmentation. IEEE Trans Nucl Sci **43**(4):2389–2398
712. Lin S, Kernighan BW (1973) An effective heuristic algorithm for the traveling salesman problem. Oper Res **21**:498–516
713. Lin SY, Huang RJ, Chiueh TD (1998) A tunable Gaussian/square function computation circuit for analog neural networks. IEEE Trans Circuits Syst–II **45**(3):441–446
714. Linde Y, Buzo A, Gray RM (1980) An algorithm for vector quantizer design. IEEE Trans Commun **28**:84–95
715. Linsker R (1986) From basic network principles to neural architecture. Proc Nat Acad Sci USA **83**:7508–7512, 8390–8394, 9779–8783
716. Linsker R (1988) Self-organization in a perceptual network. IEEE Computer **21**(3):105–117
717. Lippman RP (1987) An introduction to computing with neural nets. IEEE ASSP Magazine **4**(2):4–22
718. Lippman RP (1989) Review of neural networks for speech recognition. Neural Computat **1**(1):1–38

719. Little WA (1974) The existence of persistent states in the brain. Math Biosci
     19:101–120
720. Liu B, Si J (1994) The best approximation to C^2 functions and its error
     bounds using regular-center Gaussian networks. IEEE Trans Neural Netw
     5(5):845–847
721. Liu CS, Tseng CH (1999) Quadratic optimization method for multilayer neu-
     ral networks with local error-backpropagation. Int J Syst Sci **30**(8):889–898
722. Liu D, Chang TS, Zhang Y (2002) A constructive algorithm for feedfor-
     ward neural networks with incremental training. IEEE Trans Circuits Syst–I
     49(12):1876–1879
723. Liu P (2000) Max-min fuzzy Hopfield neural networks and an efficient learning
     algorithm. Fuzzy Sets Syst **112**:41–49
724. Liu P, Li H (2004) Efficient learning algorithms for three-layer regular feed-
     forward fuzzy neural networks. IEEE Trans Neural Netw **15**(3):545–558
725. Liu P, Li H (2005) Hierarchical TS fuzzy system and its universal approxi-
     mation. Info Sci **169**:279–303
726. Liu SH, Lin JS (2000) A compensated fuzzy Hopfield neural network for
     codebook design in vector quantization. Int J Pattern Recogn & Artif Intell
     14(8):1067–1079
727. Liu Y, You Z, Cao L (2005) A simple functional neural network for computing
     the largest and smallest eigenvalues and corresponding eigenvectors of a real
     symmetric matrix. Neurocomput **67**:369–383
728. Liu Y, Zheng Q, Shi Z, Chen J (2004) Training radial basis function networks
     with particle swarms. In: Yin F, Wang J, Guo C (eds) Proc Int Symp Neural
     Netw, Dalian, China, LNCS **3173**, 317–322. Springer, Berlin
729. Liu ZQ, Glickman M, Zhang YJ (2000) Soft-competitive learning paradigms.
     In: Liu ZQ, Miyamoto S (eds) Soft computing and human-centered machines,
     131–161. Springer-Verlag, New York
730. Ljung L (1977) Analysis of recursive stochastic algorithm. IEEE Trans Au-
     tomat Contr **22**:551–575
731. Lo ZP, Bavarian B (1991) On the rate of convergence in topology preserving
     neural networks. Biol Cybern **65**:55–63
732. Locatelli M (2001) Convergence and first hitting time of simulated annealing
     algorithms for continuous global optimization. Math Methods of Oper Res
     54:171–199
733. Loeve M (1963) Probability theory. 3rd edn, Van Nostrand, New York
734. Lowe D (1995) On the use of nonlocal and non-positive definite basis func-
     tions in radial basis function networks. Proc IEE Int Conf Artif Neural Netw,
     Cambridge, UK, 206–211
735. Lu W, Rajapakse JC (2005) Approach and applications of constrained ICA.
     IEEE Trans Neural Netw **16**(1):203–212
736. Luk A, Lien S (1998) Learning with lotto-type competition. In: Proc Int Joint
     Conf Neural Netw, Anchorage, AK, **2**:1143–1146
737. Luk A, Lien S (1999) Lotto-type competitive learning and its stability. In:
     Proc Int Joint Conf Neural Netw, Washington DC, **2**:1425–1428
738. Luo FL, Unbehauen R (1998) A minor subspace analysis algorithm. IEEE
     Trans Neural Netw **8**(5):1149–1155
739. Luo FL, Unbehauen R, Li YD (1995) A principal component analysis algo-
     rithm with invariant norm. Neurocomput **8**:213–221

740. Luo FL, Unbehauen R, Cichocki A (1997) A minor component analysis algorithm. Neural Netw **10**(2):291–297
741. Ma J (1997) The stability of the generalized Hopfield networks in randomly asynchronous mode. Neural Netw **10**:1109–1116
742. Ma J (1999) The asymptotic memory capacity of the generalized Hopfield network. Neural Netw **12**:1207–1212
743. Ma J, Theiler J, Perkins S (2003) Accurate online support vector regression. Neural Computat**15**(11):2683–2703
744. Ma S, Ji C (1999) Performance and efficiency: Recent advances in supervised learning. Proc IEEE **87**(9):1519–1535
745. Ma S, Ji C, Farmer J (1997) An efficient EM-based training algorithm for feedforward neural networks. Neural Netw **10**:243–256
746. MacQueen JB (1967) Some methods for classification and analysis of multivariate observations. In: Proc 5th Berkeley Symp on Math Statistics and Probability, University of California Press, Berkeley, 281–297
747. Madrenas J, Verleysen M, Thissen P, Voz JL (1996) A CMOS analog circuit for Gaussian functions. IEEE Trans Circuits Syst–II **43**(1):70–74
748. Maffezzoni P, Gubian P (1994) VLSI design of radial functions hardware generator for neural computations. Proc IEEE 4th Int Conf Microelectronics for Neural Netw & Fuzzy Syst, Turin, Italy, 252–259
749. Magoulas GD, Plagianakos VP, Vrahatis MN (2002) Globally convergent algorithms with local learning rates. IEEE Trans Neural Netw **13**(3):774–779
750. Magoulas GD, Vrahatis MN, Androulakis GS (1997) Effective backpropagation training with variable stepsize. Neural Netw **10**(1):69–82
751. Maiorov V, Pinkus A (1999) Lower bounds for approximation by MLP neural networks. Neurocomput **25**:81–91
752. Majani E, Erlanson R, Abu-Mostafa Y (1989) On the k-winners-take-all network. In: Touretzky DS (ed) Advances in neural information processing systems **1**, 634–642. Morgan Kaufmann, San Mateo, CA
753. Mamdani EH (1974) Application of fuzzy algorithms for control of a simple dynamic plant. Proc IEEE **12**(1):1585–1588
754. Man Y, Gath I (1994) Detection and separation of ring-shaped clusters using fuzzy clustering. IEEE Trans Pattern Anal Mach Intell **16**(8):855–861
755. Manderick B, Spiessens P (1989) Fine-grained parallel genetic algorithms. In: Schaffer JD (ed) Proc 3rd Int Conf Genetic Algorithms, 428–433. Morgan Kaufmann, San Mateo, CA
756. Mandic DP, Chambers JA (2000) A normalised real time recurrent learning algorithm. Signal Process **80**:1909–1916
757. Mann JR, Gilbert S (1989) An analog self-organizing neural network chip. In: Touretzky DS (ed) Advances in neural information processing systems **1**, 739–747. Morgan Kaufmann, San Mateo, CA
758. Manry MT, Apollo SJ, Allen LS, Lyle WD, Gong W, Dawson MS, Fung AK (1994) Fast training of neural networks for remote sensing. Remote Sensing Rev **9**:77–96
759. Mansfield AJ (1991) Training perceptrons by linear programming. National Physical Laboratory, NPL Report DITC 181/91, Teddington, Middlesex, UK
760. Mao J, Jain AK (1995) Artificial neural networks for feature extraction and multivariate data projection. IEEE Trans Neural Netw **6**(2):296–317
761. Mao J, Jain AK (1996) A self-organizing network for hyperellipsoidal clustering (HEC). IEEE Trans Neural Netw **7**(1):16–29

762. Mao KZ (2002) RBF neural network center selection based on Fisher ratio class separability measure. IEEE Trans Neural Netw **13**(5):1211–1217
763. Marchesi ML, Piazza F, Uncini A (1996) Backpropagation without multiplier for multilayer neural networks. IEE Proc–Circuits Devices Syst **143**(4):229–232
764. Marquardt D (1963) An algorithm for least-squares estimation of nonlinear parameters. SIAM J Appl Math **11**:431–441
765. Martens JP, Weymaere N (2002) An equalized error backpropagation algorithm for the on-line training of multilayer perceptrons. IEEE Trans Neural Netw **13**(3):532–541
766. Martin GL, Pittman JA (1991) Recognizing hand-printed letters and digits using backpropagation learning. Neural Computat **3**(2):258–267
767. Martinetz TM (1993) Competitive Hebbian learning rule forms perfectly topology preserving maps. In: Proc Int Conf Artif Neural Netw (ICANN), Amsterdam, 427–434
768. Martinetz TM, Berkovich SG, Schulten KJ (1993) Neural-gas network for vector quantization and its application to time-series predictions. IEEE Trans Neural Netw **4**(4):558–569
769. Martinetz TM, Schulten KJ (1994) Topology representing networks. Neural Netw **7**:507–522
770. Martinez D, Bray A (2003) Nonlinear blind source separation using kernels. IEEE Trans Neural Netw **14**(1):228–235
771. Mashor MY (2001) Adaptive fuzzy c-means clustering algorithm for a radial basis function network. Int J Syst Sci **32**(1):53–63
772. Massey L (2003) On the quality of ART1 text clustering. Neural Netw **16**:771–778
773. Mastorocostas PA (2004) Resilient back propagation learning algorithm for recurrent fuzzy neural networks. Electron Lett **40**(1):57–58
774. Mathew G, Reddy VU (1996) A quasi-Newton adaptive algorithm for generalized symmetric eigenvalue problem. IEEE Trans Signal Process **44**(10):2413–2422
775. Mathew G, Reddy VU, Dasgupta S (1995) Adaptive estimation of eigensubspace. IEEE Trans Signal Process **43**(2):401–411
776. Mathias K, Whitley LD (1995) Changing representations during search: A comparative study of delta coding. Evol Computat **2**(3):249–278
777. Matsuda S (1998) "Optimal" Hopfield network for combinatorial optimization with linear cost function. IEEE Trans Neural Netw **9**(6):1319–1330
778. Matsuoka K, Ohya M, Kawamoto M (1995) A neural net for blind separation of nonstationary signals. Neural Netw **8**(3):411–419
779. Matsuoka K, Yi J (1991) Backpropagation based on the logarithmic error function and elimination of local minima. In: Proc IEEE Int Joint Conf Neural Netw, Seattle, WA, 1117–1122
780. Mayes DJ, Murray AF, Reekie HM (1989) Non-Gaussian kernel circuits in analog VLSI: Implication for RBF network performance. IEE Proc–Circuits Devices Syst **146**(4):169–175
781. Mayes DJ, Murray AF, Reekie HM (1996) Pulsed VLSI for RBF neural networks. In: Proc 5th IEEE Int Conf Microelectronics for Neural Netw, Evol & Fuzzy Syst, Lausanne, Switzerland, 177–184
782. Mays CH (1963) Adaptive threshold logic. PhD thesis, Stanford University

783. McCulloch WS, Pitts W (1943) A logical calculus of the ideas immanent in nervous activity. Bull of Math Biophysics **5**:115–133
784. McEliece RJ, Posner EC, Rodemich ER, Venkatesh SS (1987) The capacity of the Hopfield associative memory. IEEE Trans Info Theory **33**(4):461–482
785. McGarry KJ, MacIntyre J (1999) Knowledge extraction and insertion from radial basis function networks. IEE Colloquium Applied Statist Pattern Recogn, Birmingham, UK, 15/1–15/6
786. McLachlan G, Basford K (1988) Mixture models: Inference and application to clustering. Marcel Dekker, New York
787. McLoone S, Brown MD, Irwin G, Lightbody G (1998) A hybrid linear/nonlinear training algorithm for feedforward neural networks. IEEE Trans Neural Netw **9**(4):669–684
788. McLoone SF, Irwin GW (1997) Fast parallel off-line training of multilayer perceptrons. IEEE Trans Neural Netw **8**(3):646–653
789. McLoone S, Irwin G (1999) A variable memory quasi-Newton training algorithm. Neural Process Lett **9**:77–89
790. McLoone S, Irwin G (2001) Improving neural network training solutions using regularisation. Neurocomput **37**:71–90
791. McLoone SF, Asirvadam VS, Irwin GW (2002) A memory optimal BFGS neural network training algorithm. In: Proc Int Joint Conf Neural Netw, Honolulu, HI, **1**:513–518
792. Meinguet J (1979) Multivariate interpolation at arbitrary points made simple. J Appl Math Phys **30**:292–304
793. Meireles MRG, Almeida PEM, Simoes MG (2003) A comprehensive review for industrial applicability of artificial neural networks. IEEE Trans Indu Electron **50**(3):585–601
794. Menczer F, Belew RK (1998) Local selection. In: Proc 7th Int Conf Evol Programming, San Diego, CA, LNCS **1447**, 703–712. Springer, Berlin
795. Menczer F, Degeratu M, Steet WN (2000) Efficient and scalable Pareto optimization by evolutionary local selection algorithms. Evol Computat **8**(2):223–247
796. Menczer F, Parisi D (1992) Evidence of hyperplanes in the genetic learning of neural networks. Biol Cybern **66**:283–289
797. Metropolis N, Rosenbluth A, Rosenbluth M, Teller A, Teller E (1953) Equations of state calculations by fast computing machines. J Chemical Physics **21**(6):1087–1092
798. Mezard M, Nadal JP (1989) Learning in feedforward layered networks: The tiling algorithm. J Physics, **A22**: 2191–2203
799. Miao Y, Hua Y (1998) Fast subspace tracking and neural network learning by a novel information criterion. IEEE Trans Signal Process **46**(7):1967–1979
800. Micchelli CA (1986) Interpolation of scattered data: distance matrices and conditionally positive definite functions. Constr Approx **2**:11–22
801. Michalewicz Z (1995) A survey of constraint handling techniques in evolutionary computation methods. In: McDonnell J, Reynolds JR, Fogel D (eds) Evolutionary Programming 4. MIT Press, Cambridge MA: 135–155
802. Michalewicz Z (1996) Genetic algorithms + data structure = evolution programs. 3rd edn, Springer, Berlin
803. Michel AN, Si J, Yen G (1991) Analysis and synthesis of a class of discrete-time neural networks described on hypercubes. IEEE Trans Neural Netw **2**(1):32–46

804. Mika S, Ratsch G, Weston J, Scholkopf B, Muller KR (1999) Fisher discriminant analysis with kernels. In: Hu YH, Larsen J, Wilson E, Douglas S (eds) Neural Networks for Signal Processing IX. IEEE, Piscataway, NJ: 41–48

805. Milano M, Koumoutsakos P, Schmidhuber J (2004) Self-organizing nets for optimization. IEEE Trans Neural Netw 15(3):758–765

806. Miller WT, Glanz FH, Kraft LG (1990) CMAC: An associative neural network alternative to backpropagation. Proc IEEE 78(10):1561–1567

807. Minai AA, Williams RD (1990) Backpropagation heuristics: A study of the extended delta-bar-delta algorithm. In: Proc IEEE Int Conf on Neural Netw, San Diego, CA, 1:595–600

808. Minsky ML, Papert S (1969) Perceptrons. MIT Press, Cambridge, MA

809. Mitra S, Basak J (2001) FRBF: A fuzzy radial basis function network. Neural Comput Appl 10:244–252

810. Mitra S, Hayashi Y (2000) Neuro-fuzzy rule generation: Survey in soft computing framework. IEEE Trans Neural Netw 11(3):748–768

811. Mizutani E, Jang JS (1995) Coactive neural fuzzy modeling. In: Proc IEEE Int Conf Neural Netw, Perth, Australia, 2:760–765

812. Moller MF (1993) A scaled conjugate gradient algorithm for fast supervised learning. Neural Netw 6(4):525–533

813. Moller R, Hoffmann H (2004) An extension of neural gas to local PCA. Neurocomput 62:305–326

814. Moller R, Konies A (2004) Coupled principal component analysis. IEEE Trans Neural Netw 15(1):214–222

815. Moller R (2006) First-order approximation of Gram-Schmidt orthonormalization beats deflation in coupled PCA learning rules. Neurocomput (in print)

816. Montana DJ, Davis L (1989) Training feedforward networks using genetic algorithms. In: Sridhara N (ed) Proc 11th Int Joint Conf Artif Intell, Detroit, 762–767. Morgan Kaufmann, San Mateo, CA

817. Moody J, Darken CJ (1989) Fast learning in networks of locally-tuned processing units. Neural Computat 1(2):281–294

818. Moody JO, Antsaklis PJ (1996) The dependence identification neural network construction algorithm. IEEE Trans Neural Netw 7(1):3–13

819. Moody JE, Rognvaldsson T (1997) Smoothness regularizers for projective basis function networks. In: Mozer MC, Jordan MI, Petsche T (eds) Advances in neural information processing systems 9, 585–591. Morgan Kaufmann, San Mateo, CA

820. Moore B (1988) ART and pattern clustering. In: Touretzky D, Hinton G, Sejnowski T (eds) Proc 1988 Connectionist Model Summer School, 174–183. Morgan Kaufmann, San Mateo, CA

821. More JJ (1977) The Levenberg-Marquardt algorithm: Implementation and theory. In: Watson GA (ed) Numerical Analysis, Lecture Notes in Mathematics 630, 105–116. Springer-Verlag, Berlin

822. Morita M (1993) Associative memory with nonmonotonicity dynamics. Neural Netw 6:115–126

823. Moscato P (1989) On evolution, search, optimization, genetic algorithms and martial arts: Towards memetic algorithms. Tech Report. 826, Caltech Concurrent Computation Program, Calif Institute Technology, Pasadena, California

824. Moscato P (1999) Memetic algorithms: A short introduction. In: Corne D, Glover F, Dorigo M (eds) New ideas in optimization, 219–234. McGraw-Hill, London

825. Moses D, Degani O, Teodorescu HN, Friedman M, Kandel A (1999) Linguistic coordinate transformations for complex fuzzy sets. Proc IEEE Int Conf Fuzzy Syst, Seoul, Korea, **3**:1340–1345

826. Mozer MC, Smolensky P (1989) Using relevance to reduce network size automatically. Connection Sci **1**(1):3–16

827. Muezzinoglu MK, Guzelis C, Zurada JM (2003) A new design method for the complex-valued multistate Hopfield associative memory. IEEE Trans Neural Netw **14**(4):891–899

828. Muhlenbein H (1989) Parallel genetic algorithms, population genetics and combinatorial optimization. In: Schaffer JD (ed) Proc 3rd Int Conf Genetic Algorithms, 416–421. Morgan Kaufman, San Mateo, CA

829. Mulenbein H, Schlierkamp-Voose D (1995) Analysis of selection, mutation and recombination in genetic algorithms. In: Banzhaf W, Eechman FH (eds) Evolution and biocomputation, LNCS **899**, 142–168. Springer, Berlin

830. Mulder SA, Wunsch II DC (2003) Million city traveling salesman problem solution by divide and conquer clustering with adaptive resonance neural networks. Neural Netw **16**:827–832

831. Muller B, Reinhardt J, Strickland M (1995) Neural networks: An introduction. 2nd edn, Springer-Verlag, Berlin

832. Muller KR, Mika S, Ratsch G, Tsuda K, Scholkopf B (2001) An introduction to kernel-based learning algorithms. IEEE Trans Neural Netw **12**(2):181–201

833. Musavi MT, Ahmed W, Chan KH, Faris KB, Hummels DM (1992) On the training of radial basis function classifiers. Neural Netw **5**(4):595–603

834. Muselli M (1997) On convergence properties of pocket algorithm. IEEE Trans Neural Netw **8**(3):623–629

835. Nagaraja G, Bose RPJC (2006) Adaptive conjugate gradient algorithm for perceptron training. Neurocomput **69**:368–386

836. Nagaraja G, Krishna G (1974) An algorithm for the solution of linear inequalities. IEEE Trans Computers **23**(4):421–427

837. Narayan S (1997) The generalized sigmoid activation function: Competitive supervised learning. Info Sci **99**:69–82

838. Narendra KS, Parthasarathy K (1990) Identification and control of dynamic systems using neural networks. IEEE Trans Neural netw **1**(1):4–27

839. Natarajan BK (1989) On learning sets and functions. Mach Learning **4**(1):67–97.

840. Nauck D, Klawonn F, Kruse R (1997) Foundations of neuro-fuzzy systems. Wiley, New York

841. Nauck D, Kruse R (1992) A neural fuzzy controller learning by fuzzy error propagation. In: Proc Workshop North American Fuzzy Info Processing Society (NAFIPS92), Puerto Vallarta, Mexico, 388–397

842. Nazareth JL (2003) Differentiable optimization and equation solving. Springer, New York

843. Nemoto I, Kubono M (1996) Complex associative memory. Neural Netw **9**(2):253–261

844. Ng SC, Leung SH, Luk A (1995) Fast and global convergent weight evolution algorithm based on modified back-propagation. In: Proc IEEE Int Conf Neural Netw, Perth, Australia, 3004–3008

845. Ng SC, Leung SH, Luk A (1999) Fast convergent generalized back-propagation algorithm with constant learning rate. Neural Process Lett **9**:13–23

846. Ngia LSH, Sjoberg J (2000) Efficient training of neural nets for nonlinear adaptive filtering using a recursive Levenberg-Marquardt algorithm. IEEE Trans Signal Process **48**(7):1915–1927

847. Nguyen D, Widrow B (1990) Improving the learning speed of 2-layer neural networks by choosing initial values of the adaptive weights. In: Proc Int Joint Conf Neural Netw, San Diego, CA, **3**:21–26

848. Nikov A, Stoeva S (2001) Quick fuzzy backpropagation algorithm. Neural Netw **14**:231–244

849. Nishiyama K, Suzuki K (2001) H_∞-learning of layered neural networks. IEEE Trans Neural Netw **12**(6):1265–1277

850. Nitta T (1997) An extension to the back-propagation algorithm to complex numbers. Neural Netw **10**(8):1391–1415

851. Nix A, Vose MD (1992) Modeling genetic algorithms with Markov chains. Ann Math & Artif Intell **5**:79–88

852. Niyogi P, Girosi F (1996) On the relationship between generalization error, hypothesis complexity, and sample complexity for radial basis functions. Neural Computat **8**:819–842

853. Niyogi P, Girosi F (1999) Generalization bounds for function approximation from scattered noisy data. Advances in Computat Math **10**:51–80

854. Nomura H, Hayashi I, Wakami N (1992) A learning method of fuzzy inference rules by descent method. In: Proc IEEE Int Conf Fuzzy Syst, San Diego, CA, 203–210

855. Nowlan SJ (1990) Maximum likelihood competitive learning. In: Touretzky DS (ed) Advances in neural information processing systems **2**, 574–582. Morgan Kaufmann, San Mateo, CA

856. Nowlan SJ, Hinton GE (1992) Simplifying neural networks by soft weight-sharing. Neural Computat **4**(4):473–493

857. Nurnberger A, Nauck D, Kruse R (1999) Neuro-fuzzy control based on the NEFCON-model: Recent developments. Soft Comput **2**:168–182

858. Nyberg M, Pao YH (1995) Automatic optimal design of fuzzy systems based on universal approximation and evolutionary programming. In: Li H, Gupta MM (eds) Fuzzy logic and intelligent systems, 311–366. Kluwer, Norwell, MA

859. Obermayer K, Ritter H, Schulten K (1991) Development and spatial structure of cortical feature maps: a model study. In: Lippmann RP, Moody JE, Touretzky DS (eds) Advances in neural information processing systems **3**, 11–17. Morgan Kaufmann, San Mateo, CA

860. Ochoa G, Harvey I, Buxton H (1999) On recombination and optimal mutation rates. In: Banzhaf W, Daida J, Eiben AE (eds) Proc of Genetic & Evol Computat Conf (GECCO'99), 488-495. Morgan Kaufmann, CA

861. Odorico R (1997) Learning vector quantization with training count (LVQTC). Neural Netw **10**(6):1083–1088

862. Oh SH (1997) Improving the error back-propagation algorithm with a modified error function. IEEE Trans Neural Netw **8**(3):799–803

863. Oh SK, Pedrycz W, Park HS (2005) Multi-layer hybrid fuzzy polynomial neural networks: A design in the framework of computational intelligence. Neurocomput **64**:397–431

864. Oja E (1982) A simplified neuron model as a principal component analyzer. J Math & Biology **15**:267–273

865. Oja E (1992) Principal components, minor components, and linear neural networks. Neural Netw **5**:929–935

866. Oja E, Karhunen J (1985) On stochastic approximation of the eigenvectors and eigenvalues of the expectation of a random matrix. J Math Anal & Appl **104**:69–84

867. Oja E, Ogawa H, Wangviwattana J (1991) Learning in non-linear constrained Hebbian networks. In: Kohonen T, Makisara K, Simula O, Kangas J (eds) Proc Int Conf Artif Neural Netw (ICANN'91), **1**:385–390. North-Holland, Amsterdam

868. Oja E, Ogawa H, Wangviwattana J (1992) Principal component analysis by homogeneous neural networks. IEICE Trans Info Syst **E75-D**:366–382

869. Oja E, Valkealahti K (1997) Local independent component analysis by the self-organizing map. In: Proc Int Conf Artif Neural Netw, Lausanne, Switzerland, 553–558

870. Oliveira ALI, Melo BJM, Meira SRL (2005) Improving constructive training of RBF networks through selective pruning and model selection. Neurocomput **64**:537–541

871. Omlin CW, Giles CL (1996) Extraction of rules from discrete-time recurrent neural networks. Neural Netw **9**:41–52

872. Omlin CW, Thornber KK, Giles CL (1998) Fuzzy finite-state automata can be deterministically encoded into recurrent neural networks. IEEE Trans Fuzzy Syst **6**:76–89

873. Orr MJL (1995) Regularization in the selection of radial basis function centers. Neural Computat **7**(3):606–623

874. Osowski S (1993) New approach to selection of initial values of weights in neural function approximation. Electron Lett **29**:313–315

875. Ouyang S, Bao Z, Liao G (1999) Adaptive step-size minor component extraction algorithm. Electron Lett **35**(6):443–444

876. Ouyang S, Bao Z, Liao G (2000) Robust recursive least squares learning algorithm for principal component analysis. IEEE Trans Neural Netw **11**(1):215–221

877. Ouyang S, Bao Z (2002) Fast principal component extraction by a weighted information criterion. IEEE Trans Signal Process **50**(8): 1994–2002

878. Ouyang S, Ching PC, Lee T (2003) Robust adaptive quasi-Newton algorithms for eigensubspace estimation. IEE Proc–VIS **150**(5):321–330

879. Paetz J (2004) Reducing the number of neurons in radial basis function networks with dynamic decay adjustment. Neurocomput **62**:79–91

880. Pal NR, Bezdek JC, Tsao ECK (1993) Generalized clustering networks and Kohonen's self-organizing scheme. IEEE Trans Neural Netw **4**(2):549–557

881. Pal NR, Chakraborty D (2000) Mountain and subtractive clustering method: Improvements and generalizations. Int J Intell Syst **15**:329–341

882. Pal SK, Basak J, De RK (1998) Fuzzy feature evaluation index and connectionist realisation. Info Sci **105**:173–188

883. Pal SK, Dasgupta B, Mitra P (2004) Rough self organizing map. Appl Intell **21**:289–299

884. Pal SK, Mitra S (1992) Multilayer perceptron, fuzzy sets, and classification. IEEE Trans Neural Netw **3**(5):683–697

885. Pal SK, Mitra S, Mitra P (2003) Rough-fuzzy MLP: Modular evolution, rule generation, and evaluation. IEEE Trans Knowledge Data Eng **15**(1):14–25

886. Parisi R, Di Claudio ED, Orlandim G, Rao BD (1996) A generalized learning paradigm exploiting the structure of feedforward neural networks. IEEE Trans Neural Netw **7**(6):1450–1460

887. Park J, Sanberg IW (1991) Universal approximation using radial-basis-function networks. Neural Computat **3**:246–257

888. Parlos AG, Femandez B, Atiya AF, Muthusami J, Tsai WK (1994) An accelerated learning algorithm for multilayer perceptron networks. IEEE Trans Neural Netw **5**(3):493–497

889. Patane G, Russo M (2001) The enhanced-LBG algorithm. Neural Netw **14**:1219–1237

890. Patnaik LM, Mandavilli S (1996) Adaptation in genetic algorithms. In: Pal SK, Wang PP (eds) Genetic algorithms for pattern recognition, 45–64. CRC Press, Boca Raton, Florida

891. Patyra MJ, Grantner JL, Koster K (1996) Digital fuzzy logic controller: Design and implementation. IEEE Trans Fuzzy Syst **4**(4):439–459

892. Paun G (2002) Membrane computing: An introduction. Springer-Verlag, Berlin

893. Paun G, Rozenberg G, Salomaa A (1998) DNA computing. Springer-Verlag, Berlin

894. Pawlak Z (1982) Rough sets. Int J Computer & Info Sci **11**:341–356

895. Pawlak Z (1991) Rough sets—Theoretical aspects of reasoning about data. Kluwer, Dordrecht, Netherlands

896. Pearlmutter BA (1989) Learning state space trajectories in recurrent neural networks. In: Proc IEEE Int Joint Conf Neural Netw, Washington DC, 365–372

897. Pearlmutter BA (1995) Gradient calculations for dynamic recurrent neural networks: A surrey. IEEE Trans Neural Netw **6**(5):1212–1228

898. Pearlmutter BA, Hinton GE (1986) G-maximization: An unsupervised learning procedure for discovering regularities. In: Denker JS (ed) Conf Proc Neural Netw for Computing, No. **151**, 333–338. American Institute Physics, Snowbird, Utah, USA

899. Pedrycz W (1998) Conditional fuzzy clustering in the design of radial basis function neural networks. IEEE Trans Neural Netw **9**(4):601–612

900. Pedrycz W (1998) Fuzzy evolutionary computing. Soft Comput **2**:61–72

901. Pedrycz W, Reformat M (2003) Evolutionary fuzzy modeling. IEEE Trans Fuzzy Syst **11**(5):652–665

902. Pedrycz W, Rocha AF (1993) Fuzzy-set based models of neurons and knowledge-based networks. IEEE Trans Fuzzy Syst **1**(4):254–266

903. Pedrycz W, Waletzky J (1997) Fuzzy clustering with partial supervision. IEEE Trans Syst Man Cynern-B **27**(5):787–795

904. Peng H, Ozaki T, Haggan-Ozaki V, Toyoda Y (2003) A parameter optimization method for radial basis function type models. IEEE Trans Neural Netw **14**(2):432–438

905. Peper F, Noda H (1996) A symmetric linear neural network that learns principal components and their variances. IEEE Trans Neural Netw **7**(4):1042–1047

906. Perantonis SJ, Virvilis V (1999) Input feature extraction for multilayered perceptrons using supervised principal component analysis. Neural Process Lett **10**:243–252

907. Perantonis SJ, Virvilis V (2000) Efficient perceptron learning using constrained steepest descent. Neural Netw **13**(3):351–364

908. Perantonis SJ, Ampazis N, Spirou S (2000) Training feedforward neural networks with the dogleg method and BFGS Hessian updates. In: Proc Int Joint Conf Neural Netw, Como, Italy, 138–143

909. Personnaz L, Guyon I, Dreyfus G (1986) Collective computational properties of neural networks: New learning mechanism. Phys Rev A **34**(5):4217–4228

910. Pernia-Espinoza AV, Ordieres-Mere JB, Martinez-de-Pison FJ, Gonzalez-Marcos A (2005) TAO-robust backpropagation learning algorithm. Neural Netw **18**: 191–204

911. Peterson C (1990) Applications of mean field theory neural networks. In: Lima R, Streit R, Mendes RV (eds) Dynamic and stochastic processes: Theory and applications, Springer Notes in Physics **355**, 141–173. Springer, Berlin

912. Peterson C, Anderson JR (1987) A mean field learning algorithm for neural networks. Complex Syst **1**(5):995–1019

913. Pfister M, Rojas R (1993) Speeding-up backpropagation—A comparison of orthogonal techniques. In: Proc Int Joint Conf Neural Netw, Nagoya, Japan, **1**:517–523

914. Pfister M, Rojas R (1994) Qrprop–a hybrid learning algorithm which adaptively includes second order information. In: Proc 4th Dortmund Fuzzy Days, 55–62

915. Phatak DS, Koren I (1994) Connectivity and performance tradeoffs in the cascade correlation learning architecture. IEEE Trans on Neural Netw **5**(6):930–935

916. Phua PKH, Ming D (2003) Parallel nonlinear optimization techniques for training neural networks. IEEE Trans Neural Netw **14**(6):1460–1468

917. Picone J (1993) Signal modeling techniques in speech recognition. Proc IEEE **81**(9):1215–1247

918. Pineda FJ (1987) Generalization of back-propagation to recurrent neural networks. Physical Rev Lett **59**:2229–2232

919. Platt J (1991) A resource allocating network for function interpolation. Neural Computat **3**(2);213–225

920. Platt J (1998) Sequential minimal optimization: A fast algorithm for training support vector machines. MSR-TR-98-14, Microsoft Research

921. Plumbley M (1995) Lyapunov function for convergence of principal component algorithms. Neural Netw **8**:11–23,

922. Plumbley MD (2003) Algorithms for nonnegative independent component analysis. IEEE Trans Neural Netw **14**(3):534–543

923. Poggio T, Girosi F (1990) Networks for approximation and learning. Proc IEEE **78**(9):1481–1497

924. Polak E (1971) Computational methods in optimization: A unified approach. Academic Press, New York

925. Ponnapalli PVS, Ho KC, Thomson M (1999) A formal selection and pruning algorithm for feedforward artificial neural network optimization. IEEE Trans Neural Netw **10**(4):964–968

926. Powell MJD (1987) Radial basis functions for multivariable interpolation: A review. In: Mason JC, Cox MG (eds) Algorithms for Approximation, 143–167. Clarendon Press, Oxford

927. Prechelt L (1998) Automatic early stopping using cross validation: Quantifying the criteria. Neural Netw **11**:761–767

928. Prugel-Bennett A (2004) Symmetry breaking in population-based optimization. IEEE Trans Evol Computat **8**(1):63–79

929. Puskorius GV, Feldkamp LA (1991) Decoupled extended Kalman filter training of feedforward layered networks. In: Proc Int Joint Conf Neural Netw, Seattle, WA, **1**:771–777

930. Quinlan JR (1993) C4.5: Programs for machine learning. Morgan Kaufmann, San Francisco, CA, USA
931. Raju GVS, Zhou J, Kisner RA (1991) Hierarchical fuzzy control. Int J Contr 54(5):1201–1216
932. Ramot D, Friedman M, Langholz G, Kandel A (2003) Complex fuzzy logic. IEEE Trans Fuzzy Syst 11(4):450–461
933. Ramot D, Milo R, Friedman M, Kandel A (2002) Complex fuzzy sets. IEEE Trans Fuzzy Syst 10(2):171–186
934. Ramsay J, Silverman B (1997) Functional data analysis. Springer, New York
935. Rao KD, Swamy MNS, Plotkin EI (2000) Complex EKF neural network for adaptive equalization. In: Proc IEEE Int Symp Circuits Syst, Geneva, Switzerland, 349–352
936. Rao YN, Principe JC, Wong TF (2004) Fast RLS-like algorithm for generalized eigendecomposition and its applications. J VLSI Signal Process 37:333–344
937. Rastogi R (1987) Array signal processing with interconnected neuron-like elements. Proc IEEE ICASSP'87, 2328–2331
938. Rathbun TF, Rogers SK, DeSimio MP, Oxley ME (1997) MLP iterative construction algorithm. Neurocomput 17:195–216
939. Rattan SSP, Hsieh WW (2005) Complex-valued neural networks for nonlinear complex principal component analysis. Neural Netw 18:61–69
940. Rayner P, Lynch MR (1989) A new connectionist model based on a non-linear adaptive filter. Proc ICASSP89, Glasgow, UK, 1191–1194
941. Rechenberg I (1973) Evolutionsstrategie–optimierung technischer systeme nach prinzipien der biologischen information. Formman Verlag, Freiburg, Germany
942. Reed R (1993) Pruning algorithms—A survey. IEEE Trans Neural Netw 4(5):740–747
943. Reed R, Marks II RJ, Oh S (1995) Similarities of error regularization, sigmoid gain scaling, target smoothing, and training with jitter. IEEE Trans Neural Netw 6(3):529–538
944. Reeke GN, Sporns O, Edelman GM (1990) Synthetic neural modelling: The 'Darwin' series of recognition automata. Proc IEEE 18(9):1498–1530
945. Reilly DL, Cooper LN, Elbaum C (1982) A neural model for category learning. Biol Cybern 45:35–41
946. Reyneri LM (2003) Implementation issues of neuro-fuzzy hardware: Going toward HW/SW codesign. IEEE Trans Neural Netw 14(1):176–194
947. Reynolds RG (1994) An introduction to cultural algorithms. In: Sebald AV, Fogel LJ (eds) Proc 3rd Annual Conf Evol Programming, 131–139. World Scientific, River Edge, NJ
948. Richardt J, Karl F, Muller C (1998) Connections between fuzzy theory, simulated annealing, and convex duality. Fuzzy Sets Syst 96:307–334
949. Riedmiller M, Braun H (1993) A direct adaptive method for faster backpropagation learning: The RPROP algorithm. In: Proc IEEE Int Conf Neural Netw, San Francisco, CA, 586–591
950. Rigler AK, Irvine JM, Vogl TP (1991) Rescaling of variables in back propagation learning. Neural Netw 4(2):225–229
951. Rissanen J (1978) Modeling by shortest data description. Automatica 14(5):465–477

952. Rissanen J (1999) Hypothesis selection and testing by the MDL principle. Computer J **42**(4):260–269

953. Ritter H (1995) Self-organizing feature maps: Kohonen maps. In: Arbib MA (ed) The handbook of brain theory and neural networks, 846–851. MIT Press, Cambridge, MA

954. Ritter H (1999) Self-organizing maps in non-Euclidean spaces. In: Oja E, Kaski S (eds) Kohonen Maps, 97–108. Springer, Berlin

955. Rivals I, Personnaz L (1998) A recursive algorithm based on the extended Kalman filter for the training of feedforward neural models. Neurocomput **20**:279–294

956. Rizzi A, Panella M, Mascioli FMF (2002) Adaptive resolution min-max classifiers. IEEE Trans Neural Netw **13**(2):402–414

957. Rojas I, Pomares H, Bernier JL, Ortega J, Pino B, Pelayo FJ, Prieto A (2002) Time series analysis using normalized PG-RBF network with regression weights. Neurocomput **42**:267–285

958. Rojas I, Pomares H, Ortega J, Prieto A (2000) Self-organized fuzzy system generation from training examples. IEEE Trans Fuzzy Syst **8**(1):23–36

959. Rojas R (1996) Neural networks: A systematic introduction. Springer, Berlin

960. Rose K (1998) Deterministic annealing for clustering, compression, classification, regression, and related optimization problems. Proc IEEE **86**(11):2210–2239

961. Rose K, Gurewitz E, Fox GC (1990) A deterministic annealing approach to clustering. Pattern Recogn Lett **11**(9): 589–594

962. Rosenblatt R (1958) The Perceptron: A probabilistic model for information storage and organization in the brain. Psychological Rev **65**: 386–408

963. Rosenblatt R (1962) Principles of neurodynamics. Spartan Books, New York

964. Rosipal R, Koska M, Farkas I (1998) Prediction of chaotic time-series with a resource-allocating RBF network. Neural Process Lett **7**:185–197

965. Roska T, Chua LO (1993) The CNN universal machine: An analogic array computer. IEEE Trans Circuits Syst–II **40**(3):163–173

966. Rossi F, Conan-Guez B (2005) Functional multi-layer perceptron: A nonlinear tool for functional data analysis. Neural Netw **18**:45–60

967. Rossi F, Delannay N, Conan-Guez B, Verleysen M (2005) Representation of functional data in neural networks. Neurocomput **64**:183–210

968. Rovetta S, Zunino R (1999) Efficient training of neural gas vector quantizers with analog circuit implementation. IEEE Trans Circuits Syst–II **46**(6):688–698

969. Roy A, Govil S, Miranda R (1995) An algorithm to generate radial basis functions (RBF)-like nets for classification problems. Neural Netw **8**(2):179–201

970. RoyChowdhury P, Singh YP, Chansarkar RA (1999) Dynamic tunneling technique for efficient training of multilayer perceptrons. IEEE Trans Neural Netw **10**(1):48–55

971. Royden HL (1968) Real analysis. 2nd edn, Macmillan, New York

972. Rubanov NS (2000) The layer-wise method and the backpropagation hybrid approach to learning a feedforward neural network. IEEE Trans Neural Netw **11**(2):295–305

973. Rubner J, Schulten K (1990) Development of feature detectors by self-organization. Biol Cybern **62**:193–199

974. Rubner J, Tavan P (1989) A self-organizing network for principal-component analysis. Europhysics Lett **10**:693–698

975. Ruck DW, Rogers SK, Kabrisky M (1990) Feature selection using a multilayer perceptron. Neural Netw Comput **2**(2):40–48

976. Ruck DW, Rogers SK, Kabrisky M, Maybeck PS, Oxley ME (1992) Comparative analysis of backpropagation and the extended Kalman filter for training multilayer perceptrons. IEEE Trans Pattern Anal Mach Intell **14**(6):686–691

977. Rudolph G (1994) Convergence analysis of canonical genetic algorithm. IEEE Trans Neural Netw **5**(1):96–101

978. Rumelhart DE, Durbin R, Golden R, Chauvin Y (1995) Backpropagation: the basic theory. In: Chauvin Y, Rumelhart DE (eds) Backpropagation: Theory, architecture, and applications. Lawrence Erlbaum, Hillsdale, NJ: 1–34

979. Rumelhart DE, Hinton GE, Williams RJ (1986) Learning internal representations by error propagation. In: Rumelhart DE, McClelland JL (eds) Parallel distributed processing: Explorations in the microstructure of cognition, **1**: Foundation, 318–362. MIT Press, Cambridge

980. Rumelhart DE, Zipser D (1985) Feature discovery by competititve learning. Cognitive Sci **9**:75–112

981. Runkler TA, Bezdek JC (1999) Alternating cluster estimation: A new tool for clustering and function approximation. IEEE Trans Fuzzy Syst **7**(4):377–393

982. Russo M (1998) Fugenesys—A fuzzy genetic neural system for fuzzy modeling. IEEE Trans Fuzzy Syst **6**(3):373–388

983. Russo M (2000) Genetic fuzzy learning. IEEE Trans Evol Computat **4**(3):259–273

984. Rutenbar RA (1989) Simulated annealling algorithms: An overview. IEEE Circuits Devices Magazine **5**(1):19–26

985. Ryad Z, Daniel R, Noureddine Z (2001) The RRBF: dynamic representation of time in radial basis function network. In: Proc 8th IEEE Int Conf Emerging Technol & Factory Automation, Antibes-Juan les Pins, **2**:737–740

986. Saarinen S, Bramley R, Cybenko G (1993) Ill conditioning in neural network training problems. SIAM J Sci Computing **14**(3):693–714

987. Saegusa R, Sakano H, Hashimoto S (2004) Nonlinear principal component analysis to preserve the order of principal components. Neurocomput **61**:57–70

988. Saito K, Nakano R (1990) Rule extraction from facts and neural networks. In: Proc Int Neural Network Conf, Paris, France, 379–382. Kluwer, Dordrecht, the Netherlands

989. Salapura V (2000) A fuzzy RISC processor. IEEE Trans Fuzzy Syst **8**(6):781–790

990. Salmeron M, Ortega J, Puntonet CG, Prieto A (2001) Improved RAN sequential prediction using orthogonal techniques. Neurocomput **41**:153–172

991. Sanchez A VD (1995) Second derivative dependent placement of RBF centers. Neurocomput **7**:311–317

992. Sanchez A VD (1995) Robustization of a learning method for RBF networks. Neurocomput **9**:85–94

993. Sanchez A VD (2002) Frontiers of research in BSS/ICA. Neurocomput **49**:7–23

994. Sanger TD (1989) Optimal unsupervised learning in a single-layer linear feedforward neural network. Neural Netw **2**:459–473

995. Sanger TD (1989) An optimality principle for unsupervised learning. In: Touretzky DS (ed) Advances in neural information processing systems **1**, 11–19. Morgan Kaufmann, San Mateo, CA

996. Sanger TD (1991) A tree-structured adaptive network for function approximation in high dimensional space. IEEE Trans Neural Netw **2**(2):285–293

997. Sarimveis H, Alexandridis A, Bafas G (2003) A fast training algorithm for RBF networks based on subtractive clustering. Neurocomput **51**:501–505

998. Sato Y, Ochiai T (1995) 2-D genetic algorithms for determining neural network structure and weights. In: McDonnel JR, Reynolds RG, Fogel DB (eds) Proc 4th Annual Conf Evol Programming, San Diego, CA, 789–804. MIT Press, Cambridge, MA

999. Sato A, Yamada K (1995) Generalized learning vector quantization. In: Tesauro G, Touretzky D, Leen T (eds) Advances in neural information processing systems **7**, 423–429. MIT Press, Cambridge, MA

1000. Scalero RS, Tepedelenlioglu N (1992) A fast new algorithm for training feedforward neural networks. IEEE Trans Signal Process **40**(1):202–210

1001. Schaffer JD (1984) Some experiments in machine learning using vector evaluated genetic algorithms. PhD Thesis, Vanderbilt University, Nashville, TN

1002. Schaffer JD, Caruana RA, Eshelman LJ, Das R (1989) A study of control parameters affecting online performance of genetic algorithms for function optimisation. In: Schaffer JD (ed) Proc 3rd Int Conf Genetic Algorithms, Arlington, VA, 70–79. Morgan Kaufmann, San Mateo, CA

1003. Schapire RE (1990) The strength of weak learnability. Mach Learning **5**:197–227

1004. Schilling RJ, Carroll JJ Jr, Al-Ajlouni AF (2001) Approximation of nonlinear systems with radial basis function neural networks. IEEE Trans Neural Netw **12**(1):1–15

1005. Schneider RS, Card HC (1998) Analog hardware implementation issues in deterministic Boltzmann machines. IEEE Trans Circuits Syst–II **45**(3):352–360

1006. Schmidt RO (1986) Multiple emitter location and signal parameter estimation. IEEE Trans Anten Propagat **34**:276–280

1007. Scholkopf B (1997) Support vector learning. R Oldenbourg Verlag, Munich, Germany

1008. Scholkopf B, Smola A, Muller KR (1998) Nonlinear component analysis as a kernel eigenvalue problem. Neural Computat **10**:1299–1319

1009. Schraudolph NN, Belew RK (1992) Dynamic parameter encoding for genetic algorithms. Mach Learning **9**(1):9–21

1010. Schwarz G (1978) Estimating the dimension of a model. Ann Statistics **6**:461–464

1011. Schwefel HP (1981) Numerical optimization of computer models. Wiley, Chichester

1012. Schwenker F, Kestler HA, Palm G (2001) Three learning phases for radial-basis-function networks. Neural Netw **14**:439–458

1013. Seiler G, Nossek J (1993) Winner-take-all cellular neural networks. IEEE Trans Circuits Syst–II **40**(3):184–190

1014. Sejnowski T, Rosenberg C (1986). NETtalk: A parallel network that learns to read aloud. Technical Report JHU/EECS-86/01, Johns Hopkins University

1015. Serrano-Gotarredona T, Linares-Barranco B (1996) A modified ART 1 algorithm more suitable for VLSI implementations. Neural Netw **9**(6):1025–1043

1016. Setiono R (2000) Extracting M of N rules from trained neural networks. IEEE Trans Neural Netw **11**(2):512–519

1017. Setiono R, Hui LCK (1995) Use of quasi-Newton method in a feed-forward neural network construction algorithm. IEEE Trans Neural Netw **6**(1):273–277

1018. Setnes M, Babuska R, Kaymak U, van Nauta Remke HR (1998). Similarity measures in fuzzy rule base simplification. IEEE Trans Syst Man Cybern–B **28**(3):376–386

1019. Shah S, Palmieri F (1990) MEKA—A fast, local algorithm for training feed-forward neural networks. In: Proc Int Joint Conf Neural Netw (IJCNN), San Diego, CA, **3**:41–46

1020. Shamma SA (1989) Spatial and temporal processing in cellular auditory network. In: Koch C, Segev I (eds) Methods in neural modeling. MIT Press, Cambridge, MA:247–289

1021. Shang Y, Wah BW (1996) Global optimization for neural network training. IEEE Computer **29**:45–54

1022. Shanno D (1978) Conjugate gradient methods with inexact searches. Math of Oper Res **3**:244–256

1023. Sharma SK, Irwin GW (2003) Fuzzy coding of genetic algorithms. IEEE Trans Evol Computat **7**(4):344–355

1024. Shashua AA (1999) On the equivalence between the support vector machine for classification and sparsified Fisher's linear discriminant. Neural Process Lett **9**(2):129–139

1025. Shawe-Taylor J (1995) Sample sizes for sigmoidal neural networks. In: Proc 8th Annual Conf Computat Learning Theory, Santa Cruz, CA, 258–264

1026. Shawe-Taylor JS, Cohen DA (1990) Linear programming algorithm for neural networks. Neural Netw **3**(5):575–582

1027. Sheta AF, De Jong K (2001) Time series forecasting using GA-tuned radial basis functions. Info Sci **133**:221–228

1028. Shevade SK, Keerthi SS, Bhattacharyya C, Murthy KRK (1999) Improvements to SMO algorithm for SVM regression. Technical Report CD-99-16, National University of Singapore

1029. Shi D, Gao J, Yeung DS, Chen F (2004) Radial basis function network pruning by sensitivity analysis. In: Tawfik AY, Goodwin SD (eds) Proc 17th Conf Canadian Society for Computat Studies of Intell: Advances in Artif Intell, London, Ontario, Canada. Lecture notes in artificial intelligence **3060**, 380–390. Springer, Berlin

1030. Shi Y, Eberhart RC (1998) A modified particle swarm optimizer. In: Proc IEEE Conf Evol Computat, Anchorage, AK, 69–73

1031. Shih FY, Moh J, Chang FC (1992) A new ART-based neural architecture for pattern classification and image enhancement without prior knowledge. Pattern Recogn **25**(5):533–542

1032. Si J, Michel AN (1995) Analysis and synthesis of a class of discrete-time neural networks with multilevel threshold neurons. IEEE Trans Neural Netw **6**(1):105–116

1033. Siegelmann HT, Sontag ED (1995) On the computational power of neural nets. J Computer & Syst Sci **50**(1):132–150

1034. Sietsma J, Dow RJF (1991) Creating artificial neural networks that generalize. Neural Netw **4**:67–79

1035. Silva FM, Almeida LB (1990) Speeding-up backpropagation. In: Eckmiller R (ed) Advanced neural computers, 151–156. North-Holland, Amsterdam

1036. Sima J (1996) Back-propagation is not efficient. Neural Netw **9**(6):1017–1023

1037. Simon D (2002) Training radial basis neural networks with the extended Kalman filter. Neurocomput **48**:455–475

1038. Simpson PK (1992) Fuzzy min-max neural networks–Part I. classification. IEEE Trans Neural Netw **3**:776–786

1039. Simpson PK (1993) Fuzzy min-max neural networks–Part II: clustering. IEEE Trans Fuzzy Syst **1**(1):32–45

1040. Singhal S, Wu L (1989) Training feedforward networks with the extended Kalman algorithm. In: Proc IEEE ICASSP-89, Glasgow, UK, **2**:1187–1190

1041. Sisman-Yilmaz NA, Alpaslan FN, Jain L (2004) ANFIS-unfolded-in-time for multivariate time series forecasting. Neurocomput **61**:139–168

1042. Sjoberg J, Zhang Q, Ljung L, Benveniste A, Delyon B, Glorennec PY, Hjalmarsson H, Juditsky A (1995) Nonlinear black-box modeling in system identification: A unified overview. Automatica **31**(12):1691–1724

1043. Smith J, Vavak F (1999) Replacement strategies in steady state genetic algorithms: Static environments. In: Banzhaf W, Reeves C (eds) Foundations of genetic algorithms **5**, 219–233. Morgan Kaufmann, CA

1044. Smyth SG (1992) Designing multilayer perceptrons from nearest neighbor systems. IEEE Trans Neural Netw **3**(2):329–333

1045. Snyman JA (1983) An improved version of the original leap-frog dynamic method for unconstrained minimization: LFOP1(b). Appl Math Modelling **7**:216–218

1046. Sohn I, Ansari N (1998) Configuring RBF neural networks. Electron Lett **34**(7):684–685

1047. Solis FJ, Wets JB (1981) Minimization by random search techniques. Math of Oper Res **6**:19–30

1048. Solla SA, Levin E, Fleisher M (1988) Accelerated learning in layered neural networks. Complex Syst **2**:625–640

1049. Soria-Olivas E, Martin-Guerrero JD, Camps-Valls G, Serrano-Lopez AJ, Calpe-Maravilla J, Gomez-Chova L (2003) A low-complexity fuzzy activation function for artificial neural networks. IEEE Trans Neural Netw **14**(6):1576–1579

1050. Specht DF (1990) Probabilistic neural networks. Neural Netw **3**:109–118

1051. Sperduti A, Starita A (1993) Speed up learning and networks optimization with extended back propagation. Neural Netw **6**(3):365–383

1052. Srinivas N, Deb K (1995) Multiobjective optimization using nondominated sorting in genetic algorithms. Evol Computat **2**(3):221–248

1053. Stahlberger A, Riedmiller M (1997) Fast network pruning and feature extraction using the unit-OBS algorithm. In: Mozer MC, Jordan MI, Petsche T (eds) Advances in neural information processing systems **9**, 655–661. MIT Press, Cambridge, MA

1054. Staiano A, Tagliaferri R, Pedrycz W (2006) Improving RBF networks performance in regression tasks by means of a supervised fuzzy clustering. Neurocomput (in press)

1055. Stan O, Kamen E (2000) A local linearized least squares algorithm for training feedforward neural networks. IEEE Trans Neural Netw **11**(2):487–495

1056. Stein MC (1981) Estimation of the mean of a multivariate normal distribution. Ann Statistics **9**(6):1135–1151

1057. Stoeva S, Nikov A (2000) A fuzzy backpropagation algorithm. Fuzzy Sets Syst 112:27–39

1058. Storkey AJ (1997) Increasing the capacity of the Hopfield network without sacrificing functionality. In: Gerstner W, Germond A, Hastler M, Nicoud J (eds) ICANN97, LNCS **1327**, 451–456. Springer, Berlin

1059. Storkey AJ, Valabregue R (1997) Hopfield learning rule with high capacity storage of time-correlated patterns. Electron Lett **33**(21):1803–1804

1060. Streifel RJ, Marks II RJ, Reed R, Choi JJ, Healy M (1999) Dynamic fuzzy control of Genetic Algorithm parameter coding. IEEE Trans Syst Man Cybern–B **29**(3):426–433

1061. Strickert M, Hammer B (2005) Merge SOM for temporal data. Neurocomput **64**:39–71

1062. Su MC, Chou CH (2001) A modified version of the K-means algorithm with a distance based on cluster symmetry. IEEE Trans Pattern Anal Mach Intell **23**(6):674–680

1063. Su MC, Liu YC (2005) A new approach to clustering data with arbitrary shapes. Pattern Recogn **38**:1887–1901

1064. Sugeno M (1974) The Theory of fuzzy integrals and its applications. PhD Thesis, Tokyo Institute Technology

1065. Sum JPF, Leung CS, Tam PKS, Young GH, Kan WK, Chan LW (1999) Analysis for a class of winner-take-all model. IEEE Trans Neural Netw **10**(1):64–71

1066. Sum J, Leung CS, Young GH, Kan WK (1999) On the Kalman filtering method in neural network training and pruning. IEEE Trans Neural Netw **10**:161–166

1067. Sun CT (1994) Rule-base structure identification in an adaptive-network-based inference system. IEEE Trans Fuzzy Syst **2**(1):64–79

1068. Sutanto EL, Mason JD, Warwick K (1997) Mean-tracking clustering algorithm for radial basis function centre selection. Int J Contr **67**(6):961–977

1069. Swamy MNS, Thulasiraman K (1981) Graphs, networks, and algorithms. Wiley, New York

1070. Syswerda G (1989) Uniform crossover in genetic algorithms. In: Schaffer JD (ed) Proc 3rd Int Conf Genetic Algorithms, Fairfax, VA, 2–9

1071. Szu HH, Hartley RL (1987) Nonconvex optimization by fast simulated annealing. Proc IEEE **75**:1538–1540 Also published: Szu H. Fast simulated annealing, Physics Lett A, **122**:152–162

1072. Tabatabai MA, Argyros IK (1993) Robust estimation and testing for general nonlinear regression models. Appl Math & Computat **58**:85–101

1073. Tagliaferri R, Eleuteri A, Meneganti M, Barone F (2001) Fuzzy min-max neural networks: from classification to regression. Soft Comput **5**:69–76

1074. Tagliarini GA, Christ JF, Page EW (1991) Optimization using neural networks. IEEE Trans Computers **40**(12):1347–1358

1075. Takagi T, Sugeno M (1985) Fuzzy identification of systems and its applications to modelling and control. IEEE Trans Syst Man Cybern **15**(1):116–132

1076. Taleb A, Cirrincione G (1999) Against the convergence of the minor component analysis neurons. IEEE Trans Neural Netw **10**(1):207–210

1077. Tam PKS, Sum J, Leung CS, Chan LW (1996) Network response time for a general class of WTA. In: Progress in neural information processing: Proc Int Conf Neural Info Processing, Hong Kong, **1**:492–495

1078. Tamura S, Tateishi M (1997) Capabilities of a four-layered feedforward neural network: four layers versus three. IEEE Trans Neural Netw **8**(2):251–255

1079. Tan Y, Wang J (2001) Nonlinear blind source separation using higher order statistics and a genetic algorithm. IEEE Trans Evol Computat **5**(6):600–612

1080. Tan Y, Wang J, Zurada JM (2001) Nonlinear blind source separation using a radial basis function network. IEEE Trans Neural Netw **12**(1):134–144

1081. Tanaka K, Sugeno M (1992) Stability analysis and design of fuzzy control systems. Fuzzy Sets Syst **45**:135–150

1082. Tang Z, Koehler GJ (1994) Deterministic global optimal FNN training algorithms. Neural Netw **7**(2):301–311

1083. Tank DW, Hopfield JJ (1986) Simple "neural" optimization networks: An A/D converter, signal decision circuit, and a linear programming circuit. IEEE Trans Circuits Syst **33**:533–541

1084. Tesauro G, Janssens B (1988) Scaling relationships in back-propagation learning. Complex Syst **2**:39–44

1085. Thawonmas R, Abe S (1999) Function approximation based on fuzzy rules extracted from partitioned numerical data. IEEE Trans Syst Man Cybern–B **29**(4):525–534

1086. Thimm G, Fiesler E (1997) High-order and multilayer perceptron initialization. IEEE Trans Neural Netw **8**(2):349–359

1087. Thulasiraman K, Swamy MNS (1992) Graphs: Theory and algorithms. Wiley, New York

1088. Tickle A, Andrews R, Golea M, Diederich J (1998) The truth will come to light: Direction and challenges in extracting the knowledge embedded within trained artificial neural networks. IEEE Trans Neural Netw **9**(6):1057–1068

1089. Tikhonov AN (1963) On solving incorrectly posed problems and method of regularization. Doklady Akademii Nauk USSR **151**:501–504

1090. Tipping ME, Lowe D (1998) Shadow targets: A novel algorithm for topographic projections by radial basis functions. Neurocomput **19**:211–222

1091. Todorovic B, Stankovic M (2001) Sequential growing and pruning of radial basis function network. In: Proc Int Joint Conf Neural Netw (IJCNN01), Washington DC, 1954–1959

1092. Todorovic B, Stankovic M, Moraga C (2002) Extended Kalman filter trained recurrent radial basis function network in nonlinear system identification. In: Dorronsoro JR (ed) Proc Int Conf on Artif Neural Netw (ICANN), Madrid, Spain, 819–824

1093. Tokuda I, Tokunaga R, Aihara K (2003) Back-propagation learning of infinite-dimensional dynamical systems. Neural Netw **16**:1179–1193

1094. Tollenaere T (1990) SuperSAB: fast adaptive backpropation with good scaling properties. Neural Netw **3**(5):561–573

1095. Tontini G, de Queiroz AA (1996) RBF FUZZY-ARTMAP: a new fuzzy neural network for robust on-line learning and identification of patterns. In: Proc IEEE Int Conf Syst Man Cybern **2**:1364–1369

1096. Tou JT, Gonzalez RC (1976) Pattern recognition principles. Addison Wesley, Reading, MA

1097. Towsey M, Alpsan D, Sztriha L (1995) Training a neural network with conjugate gradient methods. In: Proc IEEE Int Conf Neural Netw, Perth, Australia, 373–378

1098. Treadgold NK, Gedeon TD (1998) Simulated annealing and weight decay in adaptive learning: the SARPROP algorithm. IEEE Trans Neural Netw **9**(4):662–668

1099. Tresp V, Hollatz J, Ahmad S (1997) Representing probabilistic rules with networks of Gaussian basis functions. Mach Learning **27**:173–200

1100. Tresp V, Neuneier R, Zimmermann HG (1997) Early brain damage. In: Mozer M, Jordan MI, Petsche P (eds) Advances in neural information processing systems **9**, 669–675. MIT Press, Cambridge, MA

1101. Tsallis C, Stariolo DA (1996) Generalized simulated annealing. Physica A **233**:395–406

1102. Tsekouras G, Sarimveis H, Kavakli E, Bafas G (2004) A hierarchical fuzzy-clustering approach to fuzzy modeling. Fuzzy Sets Syst **150**(2):245–266

1103. Tsoi AC, Back AD (1994) Locally recurrent globally feedforward networks: A critical review of architectures. IEEE Trans Neural Netw **5**(2):229–239

1104. Tsutsui S, Fujimoto Y, Ghosh A (1997). Forking genetic algorithms: GAs with search space division schemes. Evol Computat **5**(1):61–80

1105. Tsypkin YZ (1973) Foundations of the theory of learning. Academic Press, New York

1106. Turney P (1996) Myths and legends of the Baldwin effect. In: Proc 13th Int Conf Mach Learning, Bari, Italy, 135–142

1107. Uehara K, Fujise M (1993) Fuzzy inference based on families of α-level sets. IEEE Trans Fuzzy Syst **1**(2):111–124

1108. Ukrainec A, Haykin S (1996) A mutual information-based learning strategy and its application to radar. In: Chen CH (ed) Fuzzy Logic and Neural Network Handbook, 12.3–12.26. McGraw-Hill, New York

1109. Ultsch A, Mantyk R, Halmans G (1993) Connectionist knowledge acquisition tool: CONKAT. In: Hand DJ (ed) Artificial intelligence frontiers in statistics: AI and statistics **III**, 256–263. Chapman & Hall, London

1110. Uncini A, Vecci L, Campolucci P, Piazza F (1999) Complex-valued neural networks with adaptive spline activation functions. IEEE Trans Signal Process **47**(2):505–514

1111. Urahama K, Nagao T (1995) K-winners-take-all circuit with O(N) complexity. IEEE Trans Neural Netw **6**:776–778

1112. Uykan Z, Guzelis C, Celebi ME, Koivo HN (2000) Analysis of input-output clustering for determining centers of RBFN. IEEE Trans Neural Netw **11**(4):851–858

1113. Vakil-Baghmisheh MT, Pavesic N (2003) A fast simplified fuzzy ARTMAP network. Neural Process Lett **17**:273–316

1114. Vakil-Baghmisheh MT, Pavesic N (2005) Training RBF networks with selective backpropagation. Neurocomput, **62**:39–64

1115. Valiant P (1984) A theory of the learnable. Commun of ACM **27**(11):1134–1142

1116. Vallet F (1989) The Hebb rule for learning linearly separable Boolean functions: learning and generalisation. Europhys Lett **8**(8):747–751

1117. van der Smagt P (1994) Minimisation methods for training feed-forward neural networks. Neural Netw **7**(1):1–11

1118. Vapnik VN (1982) Estimation of dependences based on empirical data. Springer-Verlag, New York

1119. Vapnik VN (1995) The nature of statistical learning theory. Springer, New York

1120. Vapnik VN (1998) Statistical learning theory. Wiley, New York
1121. Vapnik VN, Chervonenkis AJ (1971) On the uniform convergence of relative frequencies of events to their probabilities. Theory of Probability & its Appl **16**:264–280
1122. Varela F, Sanchez-Leighton V, Coutinho A (1989) Adaptive strategies gleaned from immune networks: Viability theory and comparison with classifier systems. In: Goodwin B, Saunders PT (eds) Theoretical biology: Epigenetic and evolutionary order (a Waddington Memorial Conf), 112–123. Edinburgh University Press, Edinburgh, UK
1123. Veitch AC, Holmes G (1991) A modified quickprop algorithm. Neural Computat **3**:310–311
1124. Venkatesh SS, Psaltis D (1989) Linear and logarithmic capacities in associative memory. IEEE Trans Info Theory **35**:558–568
1125. Ventura D, Martinez T (2000) Quantum associative memory. Info Sci **124**:273–296
1126. Venugopal KP, Pandya AS (1991) Alopex algorithm for training multilayer neural networks. In: Proc Int Joint Conf Neural Netw, Seattle, WA, **1**:196–201
1127. Verikas A, Gelzinis A (2000) Training neural networks by stochastic optimisation. Neurocomput **30**:153–172
1128. Verzi SJ, Heileman GL, Georgiopoulos M, Anagnostopoulos GC (2003) Universal approximation with fuzzy ART and fuzzy ARTMAP. In: Proc Int Joint Conf Neural Netw, Portland, Oregon, USA, **3**:1987–1992
1129. Vesanto J, Alhoniemi E (2000) Clustering of the self-organizing map. IEEE Trans Neural Netw **11**(3):586–600
1130. Vesin J (1993) An amplitude-dependent autoregressive signal model based on a radial basis function expansion. In: Proc ICASSP'93, Minneapolis, MN, **3**:129–132
1131. Vitela JE, Reifman J (1997) Premature saturation in backpropagation networks: mechanism and necessary condition. Neural Netw **10**(4):721–735
1132. Vogl TP, Mangis JK, Rigler AK, Zink WT, Alkon DL (1988) Accelerating the convergence of the backpropagation method. Biol Cybern **59**:257–263
1133. von der Malsburg C (1973) Self-organizing of orientation sensitive cells in the striata cortex. Kybernetik **14**:85–100
1134. von Lehman A, Paek EG, Liao PF, Marrakchi A, Patel JS (1988) Factors influencing learning by back-propagation. In: Proc IEEE Int Conf Neural Netw, San Diego, 765–770
1135. Vose M, Liepins G (1991) Punctuated equilibria in genetic search. Complex Syst **5**:31–44
1136. Vuorimaa P (1994) Fuzzy self-organizing map. Fuzzy Sets Syst **66**(2):223–231
1137. Waibel A, Hanazawa T, Hinton G, Shikano K, Lang KJ (1989) Phoneme recognition using time-delay neural networks. IEEE Trans Acoust Speech Signal Process **37**(3):328–339
1138. Wallace M, Tsapatsoulis N, Kollias S (2005) Intelligent initialization of resource allocating RBF networks. Neural Netw **18**(2):117–122
1139. Wan EA (1990) Temporal backpropagation for FIR neural networks. In: Proc IEEE Int Joint Conf Neural Netw, San Diego, CA, 575–580
1140. Wan EA (1994) Time series prediction by using a connectionist network with internal delay lines. In: Weigend AS, Gershenfeld NA (eds) Time series prediction: Forcasting the future and understanding the past, 195–217. Addison-Wesley, Reading, MA

1141. Wang C, Principe JC (1999) Training neural networks with additive noise in the desired signal. IEEE Trans Neural Netw **10**(6):1511–1517

1142. Wang C, Venkatesh S, Judd JD (1994) Optimal stopping and effective machine complexity in learning. In: Cowan J, Tesauro G, Alspector J (eds) Advances in neural information processing systems **6**, 303–310. Morgan Kaufmann, San Mateo, CA

1143. Wang D, Chaudhari NS (2003) Binary neural network training algorithms based on linear sequential learning. Int J Neural Syst **13**(5):333–351

1144. Wang J, Li H (1994) Solving simultaneous linear equations using recurrent neural networks. Info Sci **76**(3/4):255–277

1145. Wang JH, Rau JD (2001) VQ-agglomeration: A novel approach to clustering. IEE Proc–VIS **148**(1):36–44

1146. Wang L, Karhunen J (1996) A simplified neural bigradient algorithm for robust PCA and MCA. Int J Neural Syst **7**(1):53–67

1147. Wang L, Li S, Tian F, Fu X (2004) A noisy chaotic neural network for solving combinatorial optimization problems: stochastic chaotic simulated annealing. IEEE Trans Syst Man Cybern–B **34**(5):2119–2125

1148. Wang L, Smith K (1998) On chaotic simulated annealing. IEEE Trans Neural Netw **9**:716–718

1149. Wang LX (1992) Fuzzy systems are universal approximators. In: Proc IEEE Int Conf Fuzzy Syst, San Diego, CA, 1163–1170

1150. Wang LX (1999) Analysis and design of hierarchical fuzzy systems. IEEE Trans Fuzzy Syst **7**(5):617–624

1151. Wang LX, Mendel JM (1992) Generating fuzzy rules by learning from examples. IEEE Trans System Man Cybern **22**(6):1414–1427

1152. Wang LX, Mendel JM (1992) Fuzzy basis functions, universal approximation, and orthogonal least-squares learning. IEEE Trans Neural Netw **3**(5):807–814

1153. Wang LX, Wei C (2000) Approximation accuracy of some neuro-fuzzy approaches. IEEE Trans Fuzzy Syst **8**(4):470–478

1154. Wang RL, Tang Z, Cao QP (2002) A learning method in Hopfield neural network for combinatorial optimization problem. Neurocomput **48**:1021–1024

1155. Wang S, Lu H (2003) Fuzzy system and CMAC network with B-spline membership/basis functions are smooth approximators. Soft Comput **7**:566–573

1156. Wang SD, Hsu CH (1991) Terminal attractor learning algorithms for back propagation neural networks. In: Proc Int Joint Conf Neural Netw, Seattle, WA, 183–189

1157. Wang XG, Tang Z, Tamura H, Ishii M (2004) A modified error function for the backpropagation algorithm. Neurocomput **57**:477–484

1158. Wang XX, Chen S, Brown DJ (2004) An approach for constructing parsimonious generalized Gaussian kernel regression models. Neurocomput **62**:441–457

1159. Wang YJ, Lin CT (1998) A second-order learning algorithm for multilayer networks based on block Hessian matrix. Neural Netw **11**:1607–1622

1160. Watkins SS, Chau PM (1992) A radial basis function neuroncomputer implemented with analog VLSI circuits. In: Proc Int Joint Conf Neural Netw, Baltimore, MD, **2**:607–612

1161. Webb AR (1994) Functional approximation in feed-forward networks: a least-squares approach to generalization. IEEE Trans Neural Netw **5**:363–371

1162. Webb AR, Lowe D (1990) The optimized internal representation of multilayer classifier networks performs nonlinear discriminant analysis. Neural Netw **3**:367–375

1163. Weingessel A, Bischof H, Hornik K, Leisch F (1997) Adaptive combination of PCA and VQ networks. IEEE Trans Neural Netw **8**(5):1208–1211

1164. Weingessel A, Hornik K (2003) A robust subspace algorithm for principal component analysis. Int J Neural Syst **13**(5):307–313

1165. Weigend AS, Rumelhart DE, Huberman BA (1991) Generalization by weight-elimination with application to forecasting. In: Lippmann RP, Moody JE, Touretzky DS (eds) Advances in neural information processing systems **3**, 875–882. Morgan Kaufmann, San Mateo, CA

1166. Werbos PJ (1974) Beyond regressions: New tools for prediction and analysis in the behavioral sciences. PhD Thesis, Harvard University, Cambridge, MA

1167. Werbos PJ (1990) Backpropagation through time: What it does and how to do it. Proc IEEE **78**(10):1550–1560

1168. Werbos PJ (1990) Consistency of HDP applied to a simple reinforcement learning problem. Neural Netw **3**:179–189

1169. Wessels LFA, Barnard E (1992) Avoiding false local minima by proper initialization of connections. IEEE Trans Neural Netw **3**(6):899–905

1170. Weston J, Watkins C (1999) Support vector machines for multi-class pattern recognition. In: Proc European Symp Artif Neural Netw (ESANN), Bruges, Belgium, 219–224

1171. Wettschereck D, Dietterich T (1992) Improving the performance of radial basis function networks by learning center locations. In: Moody JE, Hanson SJ, Lippmann RP (eds) Advances in neural information processing systems **4**, 1133–1140. Morgan Kaufmann, San Mateo, CA

1172. Weymaere N, Martens JP (1994) On the initializing and optimization of multilayer perceptrons. IEEE Trans Neural Netw **5**:738–751

1173. White H (1989) Learning in artificial neural networks: A statistical perspective. Neural Computat **1**(4):425–469

1174. Whitehead BA, Choate TD (1994) Evolving space-filling curves to distribute radial basis functions over an input space. IEEE Trans Neural Netw **5**(1):15–23

1175. Whitehead BA (1996) Cooperative-competitive genetic evolution of radial basis function centers and widths for time series prediction. IEEE Trans Neural Netw **7**(4):869–880

1176. Whitley D (1989) The GENITOR algorithm and selective pressure. In: Schaffer JD (ed) Proc of 3rd Int Conf Genetic Algorithms, 116–121. Morgan Kaufmann, San Mateo, CA

1177. Whitley D, Starkweather T (1990). GENITOR II: A distributed genetic algorithm. J Expt Theor Artif Intell **2**(3):189–214

1178. Whitley D, Starkweather T, Fuquay D (1989) Scheduling problems and traveling salesmen: The genetic edge recombination operator. In: Schaffer JD (ed) Proc 3rd Int Conf Genetic Algorithms, 133–140. Morgan Kaufmann, San Mateo, CA

1179. Whitley D, Yoo NW (1995) Modeling simple genetic algorithms for permutation problems. In: Whitley D, Vose M (eds) Foundations of genetic algorithms **3**, 163–184. Morgan Kaufmann, San Mateo, CA

1180. Whitley LD (1991) Fundamental principals of deception in genetic search. In: Rawlins GJE (ed) Foundations of genetic algorithms, 221–241. Morgan Kaufmann, San Mateo, CA

1181. Whitley LD, Gordon VS, Mathias KE (1994) Lamarckian evolution, the Baldwin effect and function optimization. In: Davidor Y, Schwefel HP, Manner R (eds) Parallel problem solving from nature III, LNCS **866**, 6–15. Springer-Verlag, London, UK

1182. Widrow B, Hoff ME (1960) Adaptive switching circuits. IRE Eastern Electronic Show & Convention (WESCON1960), Convention Record, 4:96–104

1183. Widrow B, Lehr MA (1990) 30 years of adaptive neural networks: Perceptron, Madaline, and backpropagation. Proc IEEE **78**(9):1415–1442

1184. Widrow B, Stearns SD (1985) Adaptive signal processing. Prentice-Hall, Englewood Cliffs, NJ

1185. Wikipedia, The Free Encyclopedia. http://en.wikipedia.org

1186. Wilamowski BM, Iplikci S, Kaynak O, Efe MO (2001) An algorithm for fast convergence in training neural networks. In: Proc Int Joint Conf Neural Netw, Wahington DC, **3**:1778–1782

1187. Williams RJ, Zipser D (1989) A learning algorithm for continually running fully recurrent neural networks. Neural Computat **1**(2):270–280

1188. Williams RJ, Zipser D (1995) Gradient-based learning algorithms for recurrent networks and their computational complexity. In: Chauvin Y, Rumelhart DE (eds) Backpropagation: theory, architecture, and applications, 433–486. Lawrence Erlbaum, Hillsdale, NJ

1189. Williamson J (1996) Gaussian ARTMAP: A neural network for fast incremental learning of noisy multidimensional maps. Neural Netw **9**(5):881–897

1190. Wilpon JG, Rabiner LR (1985) A modified K-means clustering algorithm for use in isolated work recognition. IEEE Trans Acoust Speech Signal Process **33**(3):587–594

1191. Wilson DR, Martinez TR (2003) The general inefficiency of batch training for gradient descent learning. Neural Netw **16**:1429–1451

1192. Wolpert DH, Macready WG (1995) No free lunch theorems for search. SFI-TR-95-02-010, Santa Fe Institute

1193. Wright AH (1991) Genetic algorithms for real parameter optimization. In: Rawlins G (ed) Foundations of genetic algorithms, 205–218. Morgan Kaufmann, San Mateo, CA

1194. Wu JM (2004) Annealing by two sets of interactive dynamics. IEEE Trans Syst Man Cybern–B **34**(3):1519–1525

1195. Wu S, Er MJ (2000) Dynamic fuzzy neural networks—A novel approach to function approximation. IEEE Trans Syst Man Cybern–B **30**(2):358–364

1196. Wu Y, Batalama SN (2000) An efficient learning algorithm for associative memories. IEEE Trans Neural Netw **11**(5):1058–1066

1197. Xiang C, Ding SQ, Lee TH (2005) Geometrical interpretation and architecture selection of MLP. IEEE Trans Neural Netw **16**(1):84–96

1198. Xie XL, Beni G (1991) A validity measure for fuzzy clustering. IEEE Trans Pattern Anal Mach Intell **13**(8):841–847

1199. Xiong H, Swamy MNS, Ahmad MO, King I (2004) Branching competitive learning network: a novel self-creating model. IEEE Trans Neural Netw **15**(2):417–429

1200. Xiong H, Swamy MNS, Ahmad MO (2005) Optimizing the kernel in the empirical feature space. IEEE Trans Neural Netw **16**(2):460–474

1201. Xiong H, Swamy MNS, Ahmad MO (2005) Two-dimension FLD for face recognition. Pattern Recogn **38**(7):1121–1124

1202. Xu D, Principe JC, Wu HC (1998) Generalized eigendecomposition with an on-line local algorithm. IEEE Signal Process Lett **5**(11):298–301

1203. Xu L (1993) Least mean square error reconstruction principle for self-organizing neural-nets. Neural Netw **6**:627–648

1204. Xu L, Krzyzak A, Oja E (1993) Rival penalized competitive learning for clustering analysis, RBF net, and curve detection. IEEE Trans Neural Netw **4**(4):636–649

1205. Xu L, Oja E, Suen CY (1992) Modified Hebbian learning for curve and surface fitting. Neural Netw **5**:441–457

1206. Xu L, Yuille AL (1995) Robust principal component analysis by self-organizing rules based on statistical physics approach. IEEE Trans Neural Netw **6**(1):131–143

1207. Xu R, Wunsch II D(2005) Survey of clustering algorithms. IEEE Trans Neural Netw **16**(3):645–678

1208. Yager R, Filev D (1994) Generation of fuzzy rules by mountain clustering. J Intell Fuzzy Syst **2**(3):209–219

1209. Yager RR, Filev D (1994) Approximate clustering via the mountain method. IEEE Trans Syst Man Cybern **24**(8):1279–1284

1210. Yair E, Zeger K, Gersho A (1992) Competitive learning and soft competition for vector quantizer design. IEEE Trans Signal Process **40**(2):294–309

1211. Yam JYF, Chow TWS (2000) A weight initialization method for improving training speed in feedforward neural network. Neurocomput **30**:219–232

1212. Yam JYF, Chow TWS (2001) Feedforward networks training speed enhancement by optimal initialization of the synaptic coefficients. IEEE Trans Neural Netw **12**(2):430–434

1213. Yam YF, Chow TWS (1993) Extended backpropagation algorithm. Electron Lett **29**(19):1701–1702

1214. Yam YF, Chow TWS, Leung CT (1997) A new method in determining the initial weights of feedforward neural networks. Neurocomput **16**:23–32

1215. Yam YF, Leung CT, Tam PKS, Siu WC (2002) An independent component analysis based weight initialization method for multilayer perceptrons. Neurocomput **48**:807–818

1216. Yan H (1991) Stability and relaxation time of Tank and Hopfield's neural network for solving LSE problems. IEEE Trans Circuits Syst **38**(9):1108–1110

1217. Yang B (1995) Projection approximation subspace tracking. IEEE Trans Signal Process **43**(1):95–107

1218. Yang B (1995) An extension of the PASTd algorithm to both rank and subspace tracking. IEEE Signal Process Lett **2**(9):179–182

1219. Yang HH, Amari SI, Cichocki A (1997) Information backpropagation for blind separation of sources from nonlinear mixture. Proc IEEE Int Conf Neural Netw, Houston, TX, **4**:2141–2146

1220. Yang J, Zhang D, Frangi AF, Yang JY (2004) Two-dimensional PCA: A new approach to appearance-based face representation and recognition. IEEE Trans Pattern Anal Mach Intell **26**(1):131–137

1221. Yang L, Yu W (1993) Backpropagation with homotopy. Neural Computat **5**(3):363–366

1222. Yang MS (1993) On a class of fuzzy classification maximum likelihood procedures. Fuzzy Sets Syst **57**:365–375

542   References

1223. Yang TN, Wang SD (2004) Competitive algorithms for the clustering of noisy data. Fuzzy Sets Syst **141**:281–299
1224. Yang WH, Chan KK, Chang PR (1994) Complex-valued neural network for direction of arrival estimation. Electron Lett **30**(7):574–575
1225. Yao X (1999) Evolving artificial neural networks. Proc IEEE **87**(9):1423–1447
1226. Yao X, Liu Y (1997) A new evolutionary system for evolving artificial neural networks. IEEE Trans Neural Netw **8**(3):694–713
1227. Yao X, Liu Y, Liang KH, Lin G (2003) Fast evolutionary algorithms. In: Ghosh S, Tsutsui S (eds) Advances in evolutionary computing: Theory and applications, 45–94. Springer-Verlag, Berlin
1228. Yasdi R (1995) Combining rough sets learning and neural learning method to deal with uncertain and imprecise information. Neurocomput **7**:61–84
1229. Yen J (1999) Fuzzy logic—A modern perspective. IEEE Trans Knowledge Data Eng **11**(1):153–165
1230. Yen JC, Guo JI, Chen HC (1998) A new k-winners-take-all neural network and its array architecture. IEEE Trans Neural Netw **9**(5):901–912
1231. Yi Z, Fu Y, Tang HJ (2004) Neural networks based approach for computing eigenvectors and eigenvalues of symmetric matrix. Computer Math w/ Appl **47**:1155–1164
1232. Yi Z, Ye M, Lv JC, Tan KK (2005) Convergence analysis of a deterministic discrete time system of Oja's PCA learning algorithm. IEEE Trans Neural Netw **16**(6):1318–1328
1233. Yildirim T, Marsland JS (1996) A conic section function network synapse and neuron implementation in VLSI hardware. Proc IEEE Int Conf Neural Netw, Washingto DC, **2**:974–979
1234. Yingwei L, Sundararajan N, Saratchandran P (1997) A sequential learning scheme for function approximation by using minimal radial basis function neural networks. Neural Computat **9**(2):461–478
1235. Yingwei L, Sundararajan N, Saratchandran P (1998) Performance evaluation of a sequential minimal radial basis function (RBF) neural network learning algorithm. IEEE Trans Neural Netw **9**(2):308–318
1236. Yip PPC, Pao YH (1995) Combinatorial optimization with use of guided evolutionary simulated annealing. IEEE Trans Neural Netw **6**(2):290–295
1237. Yoshizawa S, Morita M, Amari SI (1993) Capacity of associative memory using a nonmonotonic neuron model. Neural Netw **6**:167–176
1238. You C, Hong D (1998) Nonlinear blind equalization schemes using complex-valued multilayer feedforward neural networks. IEEE Trans Neural Netw **9**(6):1442–1455
1239. Younes L (1996) Synchronous Boltzmann machines can be universal approximators. Appl Math Lett **9**(3):109–113
1240. Yu DL (2004) A localized forgetting method for Gaussian RBFN model adaptation. Neural Process Lett **20**:125–135
1241. Yu DL, Gomm JB, Williams D (1997) A recursive orthogonal least squares algorithm for training RBF networks. Neural Process Lett **5**:167–176
1242. Yu J, Yang MS (2005) Optimality test for generalized FCM and its application to parameter selection. IEEE Trans Fuzzy Syst **13**(1):164–176
1243. Yu X, Efe MO, Kaynak O (2002) A general backpropagation algorithm for feedforward neural networks learning. IEEE Trans Neural Netw **13**(1): 251–254

1244. Yu XH, Chen GA (1997) Efficient backpropagation learning using optimal learning rate and momentum. Neural Netw **10**(3):517–527

1245. Yu XH, Chen GA, Cheng SX (1995) Dynamic learning rate optimization of the backpropagation algorithm. IEEE Trans Neural Netw **6**(3):669–677

1246. Yuh JD, Newcomb RW (1993) A multilevel neural network for A/D conversion. IEEE Trans Neural Netw **4**(3):470–483

1247. Yuille AL, Kammen DM, Cohen DS (1989) Quadrature and development of orientation selective cortical cells by Hebb rules. Biol Cybern **61**:183–194

1248. Zadeh LA (1965) Fuzzy sets. Info & Contr **8**:338–353

1249. Zadeh LA (1975) The concept of a linguistic variable and its application to approximate reasoning–I, II, III. Info Sci **8**:199–249, 301–357; **9**:43–80

1250. Zahn CT (1971) Graph-theoretical methods for detecting and describing gestalt clusters. IEEE Trans Computers **20**:68–86

1251. Zak M (1989) Terminal attractors in neural networks. Neural Netw **2**:259–274

1252. Zhang GP (2000) Neural networks for classification: A survey. IEEE Trans Syst Man Cybern–C **30**(4):451–462

1253. Zhang D, Bai XL, Cai KY (2004) Extended neuro-fuzzy models of multilayer perceptrons. Fuzzy Sets Syst **142**:221–242

1254. Zhang D, Zhou ZH, Chen S (2006) Diagonal principal component analysis for face recognition. Pattern Recogn **39**(1):140-142

1255. Zheng GL, Billings SA (1999) An enhanced sequential fuzzy clustering algorithm. Int J Syst Sci **30**(3):295–307

1256. Zhang J, Walter GG, Miao Y, Lee WNW (1995) Wavelet neural networks for function learning. IEEE Trans Signal Process **43**(6):1485–1497

1257. Zhang Q (1997) Using wavelet networks in nonparametric estimation. IEEE Trans Neural Netw **8**(2):227–236

1258. Zhang Q, Benveniste A (1992) Wavelet networks. IEEE Trans Neural Netw **3**(6): 899–905

1259. Zhu Q, Cai Y, Liu L (1999) A global learning algorithm for a RBF network. Neural Netw **12**:527–540

1260. Zhang Q, Leung YW (2000) A class of learning algorithms for principal component analysis and minor component analysis. IEEE Trans Neural Netw **11**(1):200–204

1261. Zhang T, Ramakrishnan R, Livny M (1996) BIRCH: An efficient data clustering method for very large databases. In: Proc ACM SIGMOD Conf on Management of Data, Montreal, Canada, 103–114

1262. Zhang Y, Li X (1999) A fast U-D factorization-based learning algorithm with applications to nonlinear system modeling and identification. IEEE Trans Neural Netw **10**:930–938

1263. Zhang YJ, Liu ZQ (2002) Self-splitting competitive learning: A new on-line clustering paradigm. IEEE Trans Neural Netw **13**(2):369–380

1264. Zhang XM, Chen YQ, Ansari N, Shi YQ (2004) Mini-max initialization for function approximation. Neurocomput **57**:389–409

1265. Zimmermann HJ, Sebastian HJ (1995) Intelligent system design support by fuzzy-multi-criteria decision making and/or evolutionary algorithms. In: Proc IEEE Int Conf Fuzzy Syst, Yokohama, Japan, 367–374

1266. Zitzler E, Thiele L (1999) Multiobjective evolutionary algorithms: A comparative case study and the strength Pareto approach. IEEE Trans Evol Computat **3**(4):257–271

1267. Zitzler E, Laumanns M, Thiele L (2001) SPEA2: Improving the strength Pareto evolutionary algorithm. TIK-Report 103, Dept of Electrical Engineering, Swiss Federal Institute Technology
1268. Zufiria PJ (2002) On the discrete-time dynamics of the basic Hebbian neural-network node. IEEE Trans Neural Netw **13**(6):1342–1352
1269. Zurada JM, Cloete I, van der Poel E (1996) Generalized Hopfield networks for associative memories with multi-valued stable states. Neurocomput **13**:135–149
1270. Zurada JM, Malinowski A, Usui S (1997) Perturbation method for deleting redundant inputs of perceptron networks. Neurocomput **14**:177–193
1271. Zweiri YH, Whidborne JF, Seneviratne LD (2000) Optimization and stability of a three-term backpropagation algorithm. Technical Report EM-2000-01, Dept of Mechanical Engineering, King's College London, London, UK
1272. Zweiri YH, Whidborne JF, Seneviratne LD (2003) A three-term backpropagation algorithm. Neurocomput **50**:305–318

Index